Sergey Edward Lyshevski

Control Systems Theory with Engineering Applications

With 169 Figures and a CD-ROM

Birkhäuser
Boston • Basel • Berlin

Sergey Edward Lyshevski
Department of Electrical and Computer Engineering
Purdue University at Indianapolis
Indianapolis, IN 46202-5132
USA

Library of Congress Cataloging-in-Publication Data
Lyshevski, Sergey Edward.
   Control systems theory with engineering applications / Sergey Edward Lyshevski.
      p.   cm.
   Includes bibliographical references and index.
   ISBN 0-8176-4203-X (alk. paper)
   1. Automatic control.   2. Control theory.   I. Title.
TJ213.L94 2000
629.8—dc21   00-060868

Printed on acid-free paper.

© 2001 Birkhäuser Boston   *Birkhäuser* ℬ ®

ISBN 0-8176-4203-X
ISBN 3-7643-4203-X   SPIN 10772413

Production managed by Louise Farkas; manufacturing supervised by Jerome Basma.
Typeset by Archetype Publishing, Inc., Monticello, IL.
Printed and bound by Maple-Vail Book Manufacturing Group, York, PA.
Printed in the United States of America.

9 8 7 6 5 4 3 2 1

Contents

Preface

Dynamics systems (living organisms, electromechanical and industrial systems, chemical and technological processes, market and ecology, and so forth) can be considered and analyzed using information and systems theories. For example, adaptive human behavior can be studied using automatic feedback control. As an illustrative example, the driver controls a car changing the speed and steering wheels using incoming information, such as traffic and road conditions. This book focuses on the most important and manageable topics in applied multivariable control with application to a wide class of electromechanical dynamic systems. A large spectrum of systems, familiar to electrical, mechanical, and aerospace students, engineers, and scholars, are thoroughly studied to build the bridge between theory and practice as well as to illustrate the practical application of control theory through illustrative examples. It is the author's goal to write a book that can be used to teach undergraduate and graduate classes in automatic control and nonlinear control at electrical, mechanical, and aerospace engineering departments. The book is also addressed to engineers and scholars, and the examples considered allow one to implement the theory in a great variety of industrial systems. The main purpose of this book is to help the reader grasp the nature and significance of multivariable control. To fulfill the goal, I prefer being occasionally imprecise (but understandable and rigorous) than to building logically impeccable mathematical setups in which definitions, lemmas, and theorems are welded together into a formidable barrier to challenge the students and engineers. I have attempted to fill a gap in the literature on multivariable control by presenting modern concepts of control systems theory, covering the most viable approaches for continuous- and discrete-time systems.

It is known that linear dynamic systems are systems in which the input–output relationships satisfy the principle of superposition. These systems are modeled by linear differential or difference equations, and transfer function in s- and z-domains can be found. Different approaches in analysis and control of linear systems can be researched using frequency-domain and state-space (time-domain) techniques. The duality of these concepts is well known, e.g., one can find transfer functions from differential equations and vise versa. Frequency-domain and state-space concepts are covered in the junior-level classes *Signals and Systems* (at electrical- and computer-engineering departments) and *Dynamic Systems* and

Automatic Control (at mechanical- and aerospace-engineering departments). The state-space methods must be prioritized for multivariable systems. Furthermore, the state-space concept is not only useful in analysis and design of linear systems, but also is an important starting point for advanced optimal and nonlinear control.

It is difficult to find linear real-world dynamic systems; that is, the system that is modeled using linear differential or difference equations. Electric circuits and electric motors are commonly used to introduce the students to the concepts of transfer functions and differential equations. In general, simple circuits can be described by linear differential equations. However, in studying power converters or nonlinear filters, one finds that nonlinear differential equations must be used. From a great variety of electric machines, only the permanent-magnet, direct-current motor is a nonlinear dynamic system. However, nonlinear friction and the maximum allowable applied voltage to the armature winding (specified by manufacturers) lead to non-linearities. Most electrical, electromechanical, mechanical, and aerospace systems are nonlinear, and assumptions and simplifications, used to perform linearization, cannot be viewed as a constructive way in analysis and design of high-performance real-world systems. For example, it is impossible to linearize not one-to-one functions (sin or cos) in nonlinear differential equations. All systems exhibit some degree of nonlinearity, and if the control system is analyzed and designed using linear models, performance objectives and criteria may not be met. However, linear systems shall be studied first to attain readability and avoid anxiety. The focus of this book is to analyze and design linear and nonlinear multivariable systems. The keys to these developments are analysis, identification, and optimization, with a major emphasis on nonlinear phenomena.

The three major modeling paradigms are

- Linear and multilinear models (multiple linear models can be used to describe nonlinear system dynamics at the operating regions)
- Hammerstein–Wiener models (mathematical models consist of linear dynamic elements in sequence with static nonlinearities) that map the transient behavior of a wide variety of real-world dynamic systems
- Functional expansion models (block-oriented models that provide excellent approximation capabilities when compared with contestant expansion models, such as Volterra series)

Nonlinear analysis and controllers design must be performed on the basis of complete mathematical models. Many methods can be used for analyzing, designing, and optimizing control systems. Analysis and feedback control of linear systems are straightforward and well understood. Manageable analytic methods, largely based on the eigenvalue analysis and the state-transition matrix, can be straightforwardly applied to multi-input/multi-output systems. In contrast, for nonlinear systems, control and optimization problems must be researched, and further developments must be made. Even though enormous computational power (tera-scale computing) is becoming available to attain analysis and control through numerical analysis, analytical methods must be developed and thoroughly studied because in the most practical scenarios, the designer must perform multidimensional numerical–analytical studies. Pure computer-oriented approaches, as applied to nonlinear systems, frequently lead to inadequate, confusing, and conflicting results. There are increasing demands for further developments of a nonlinear control theory to facilitate thorough research in inherently difficult nonlinear problems, analysis of nonlinear phenomena and system behavior, and finally, design of control algorithms. The Hamilton–Jacobi theory and Lyapunov methods can be applied using the state-space concept in modeling of dynamic systems. The state-space concept in analysis of dynamic systems emerged more than 100 years ago, when A. Lyapunov (professor of mechanical engineering at the Kharkov University, Ukraine) developed the pioneering results in analysis of stability. In the 1950s, the state-space concept was implemented primarily for multi-input/multi-output aircraft and spacecraft. Today, owing to availability of state-of-the -art microcontrollers and digital signal processors (DSPs), complex control algorithms can be easily implemented to attain the desired system performance. This book is intended to introduce important methods and algorithms in the study, analysis, design, and optimization of multi-input/multi-output dynamic systems, and a great number of practical examples are covered, particularly those that are relevant to electrical, mechanical, and aerospace engineering.

To avoid possible obstacles, the material is presented in sufficient detail. Basic introductory theory, related to the linear state-space concept, is covered to help one to fully understand, appreciate, and apply the knowledge. A wide range of worked-out examples and qualitative illustrations, which are treated in-depth, bridge the gap between theory and practice. Step-by-step, *Control Systems Theory With Engineering Applications* guides one through the most important aspects in analysis and control: from a rigorous theoretical foundation to applications of results. Some results from linear signals and systems theory, as well as modeling in the MATLAB environment (which promotes enormous gains in productivity and creativity), are given in the introductory chapters.

Simulation is a critical and urgently important aspect of analysis and control, development and prototyping, as well as stabilization and optimization of dynamic systems. To speed analysis and design, facilitate enormous gains in productivity and creativity, integrate control laws using advanced microprocessors and DSPs, accelerate prototyping features, generate real-time C code, and visualize the results, MATLAB® can be efficiently used. The MATLAB is a computational environment that integrates a great number of toolboxes, such as SIMULINK, Real-Time Workshop^{TM}, Control System, Nonlinear Control Design, Optimization, Robust Control, Signal Processing, Symbolic Math, System Identification, and so on. A flexible, high-performance modeling and design environment, MATLAB has become a standard cost-effective tool within the engineering community. This book demonstrates MATLAB's capabilities and helps one to master this user-friendly environment to attack and solve different problems. The application of MATLAB increases designer productivity as well as shows how to use the advanced software. The MATLAB environment offers a rich set of capabilities to efficiently solve a variety of complex analyses and control, stabilization, and optimization problems, The MATLAB files and SIMULINK models, which are given in this book, can be easily modified to study application-specific problems encountered in engineering practice. The examples, modeled and simulated in this book, consist of a wide spectrum of practical electrical, electromechanical, and aerospace dynamic systems. A variety of complex nonlinear mathematical models are thoroughly studied, and SIMULINK diagrams to simulate dynamic systems and numerical results are reported. Users can easily apply these results as well as develop new MATLAB files and SIMULINK block-diagrams using the treated enterprise-wide practical examples. Through these examples, the most efficient and straightforward analysis and design methods to approach and solve motion control problems are documented. The developed scripts and models are easily assessed, and they can be straightforwardly modified.

The major objectives of this readable and user-friendly book are to give students and engineers confidence in their ability to apply advanced theoretical concepts, to enhance learning, and to provide a gradual progression from versatile theoretical to practical topics to apply the results. This book is written for engineers and students interested in nonlinear control of real-world systems. Students and engineers are not primarily interested in theoretical encyclopedic studies. They need straightforward and practical instructions on how to approach and solve the specific practical challenging problems in control of electrical, electromechanical, and aerospace systems. It is the author's goal to stress the practical aspects in order to help one to understand and apply the advanced developments in analysis and design of multivariable dynamic systems. In analysis and control of multivariable systems, modern theory and innovative methods, as well as advanced algorithms and software, are of great importance. This book presents a well-defined theoretical base with step-by-step instructions on how to apply it by thoroughly studying a great number of practical real-world problems and using numerous examples. These worked-out examples prepare one to use the analysis, identification, control, and optimization methods presented. This helps one to fully understand, appreciate, visualize, grasp, use, and finally apply the results.

Indianapolis, Indiana   *Sergey Edward Lyshevski*

Acknowledgments

First, thanks are to my family for their constant help, complete support, and love. Many of my friends, colleagues, and students have contributed to this book, and I would like to express my sincere acknowledgments and gratitude to them. It gives me great pleasure to acknowledge the help from the outstanding team of Birkhäuser Boston, especially Wayne Yuhasz (Executive Editor) and Lauren Schultz (Associate Editor) who tremendously helped me by providing valuable feedback and assistance.

Indianapolis, Indiana   *Sergey Edward Lyshevski*

1

Introduction: Modeling, Identification, Optimization, and Control

Mathematical modeling, simulation, nonlinear analysis, decision making, identification, estimation, diagnostics, and optimization have become major mainstreams in control system engineering. The designer describes physical system dynamics in the form of differential or difference equations, and the comprehensive analysis of complex dynamics systems is performed analytically or numerically solving these equations. To develop mathematical models of the system dynamics, the Newtonian mechanics, Lagrange's equations of motion, Kirchhoff's laws, and the energy conservation principles are used. It is evident that one cannot guess models of physical systems and pretend that the assumed models describe real-world systems under consideration. Chapter 2 illustrates that the designer can straight-forwardly develop mathematical models of electromechanical systems, as well as their components (actuators, transducers, power converters, electric circuits, and filters) to be simulated and controlled. The development of accurate mathematical models, in the form of differential or difference equations, with a minimum level of simplifications and assumptions is a critical first step because all subsequent steps will be mainly based on the mathematical model used. Model development efforts are driven by the final goal, which is to satisfy the desired system performance as measured against a wide spectrum of specifications and requirements imposed. That is, mathematical models must satisfy the intents and goals for which they were developed, serve the design objectives, be user-friendly and well understood, and so forth. Mathematical models should possess flexibility in terms of simulation fidelity and attain the desired degree of accuracy to meet the objectives and desired outcomes. The scope of mathematical model developments is extended to nonlinear analysis and simulations that significantly reduce resources invested into design and prototyping of novel systems. As mathematical models are derived using fundamental physical laws, the system parameters (coefficients of differential or difference equations) can be identified, and workable identification procedures are needed. System identification is linked with model validation, model reduction, analysis, and simulation. Different identification methods are reported in this book with a number of workable examples for multivariable nonlinear dynamic systems. It is shown that least-squares and time-domain error mapping–based identification algorithms allow one to identify unknown system parameters and lead to efficient solution of the identification problem for a large class of multi-input/multi-output dynamic systems.

The application of analog and digital controllers allows the designer to solve a spectrum of problems in control of dynamic systems. In general, analysis, optimization, design, verification, test, and implementation can be divided into the application-specific (requirements–specifications) phase, the software–hardware phase (virtual prototyping, testing, and validation), and the deployment (implementation, technology transfer, and commercialization) phase. The basic ingredients are

- Model development, analysis, simulation, identification, optimization, visualization, and validation of open-and closed-loop systems applying computer-aided design tools with intelligent databases and libraries developed (advanced, efficient, and user-friendly software is available to perform these tasks, and this book illustrates the application of the MATLAB environment)
- Development, testing, and implementation of advanced high-performance software and state-of-the-art hardware using emerging technologies (structural optimization, system integration through real-time interfacing, digital signal processing, data acquisition, and motion control using analog and digital controllers)
- Testing and deployment of systems

The desired (required) specifications imposed on closed-loop systems are given in the performance domain. The commonly used criteria to be achieved by closed-loop systems are

- Stability with the desired stability margins in the full operating envelope specified
- Robustness to parameter variations, structural and environmental changes
- Tracking accuracy and disturbance attenuation
- Dynamic and steady-state accuracy
- Transient response specifications, such as the settling, delay, peak times, the maximum overshoot, and so forth

These imposed performance specifications and requirements are measured as dynamic systems that are designed and tested (for example, analytical results, numerical simulation, hardware-in-the-loop modeling, and experimental results). To guarantee the desired performance, the designer faces a wide spectrum of challenging and complex problems associated with model developments and fidelity (accuracy), analysis, modeling (simulation), identification, optimization, design, and control. A great number of specifications are imposed, and in addition to stability and robustness, the output, state transient dynamics, disturbances, noises, as well as control signals must be thoroughly analyzed. As the system output dynamics is usually prioritized, the output transients and the specified evolution envelope are examined. For example, for single-output systems, the output transient response and the desired evolution envelope are illustrated in Figure 1.1 for the step-type forcing function (reference input) $u(t) = $ const. It is obvious that the

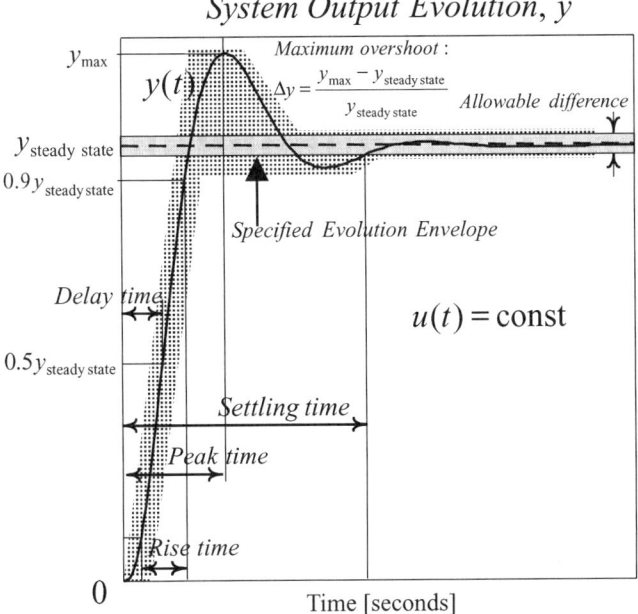

FIGURE 1.1.   Output transient response of a dynamic system, $u(t) = $ const.

system output is within the specified envelope, and the system output converges to the steady-state value $y_{\text{steady state}}$. That is, $\lim_{t \to \infty} y(t) = y_{\text{steady state}}$. One concludes that $y(t)$ converges to $u(t)$, the stability is guaranteed, and the tracking error is zero. However, the superposition principle is valid only for linear systems; stability, robustness, dynamics, accuracy, as well as other performance criteria must be thoroughly studied in the full operating envelops (assigning different reference and control inputs, studying the behavior under different initial conditions and disturbances, etc.) because the real-world systems are nonlinear.

For linear systems, and some nonlinear systems, the commonly used definitions for the settling, delay, and peak times, as well as for the maximum overshoot, are given below.

The settling time is the time needed for the system output $y(t)$ to reach and stay within the steady-state value $y_{\text{steady state}}$ (which ideally is equal to the reference input). The steady-state value $y_{\text{steady state}}$ is shown in Figure 1.1 if the reference (command) input is the step, $u(t) = $ const. The absolute allowable difference between $y(t)$ and $y_{\text{steady state}}$ is used to find the settling time, and usually this difference is specified to be 5%. That is, the settling time is the minimum time after which the system response remains within $\pm 5\%$ of the steady-state value $y_{\text{steady state}}$.

The maximum overshoot is the difference between the maximum peak value of the systems output $y(t)$ and the steady-state value $y_{\text{steady state}}$ divided by $y_{\text{steady state}}$.

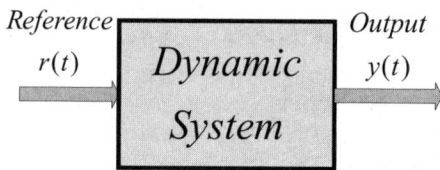

FIGURE 1.2.   Dynamic system with input $r(t)$ and output $y(t)$.

That is,

$$\Delta y = \frac{y_{\text{max}} - y_{\text{steady state}}}{y_{\text{steady state}}} \text{ or } \Delta y = \frac{y_{\text{max}} - y_{\text{steady state}}}{y_{\text{steady state}}} \times 100\%.$$

The rise time is the time needed for the system output $y(t)$ to increase from 10 to 90% of the steady-state value $y_{\text{steady state}}$ if $u(t) = \text{const}$ (step input).

The delay time is the time needed for the system output $y(t)$ to reach 50% of $y_{\text{steady state}}$ if $u(t) = \text{const}$ (step input).

The peak time is the time required for the system output $y(t)$ to reach the first peak of the overshoot.

The system performance is measured against the following criteria: stability, stability margins, robustness, sensitivity to parameter variations, transient behavior, accuracy, disturbance, noise attenuation, and so forth. The specifications are dictated by the requirements imposed on the operating envelope, and certain performance characteristics are emphasized using performance criteria. Let us assume that the designer in analyzing expected system performance, strives to optimize the system, minimizing the tracking error and settling time. Transient responses can be optimized using the integrated time and error criterion

$$J = \min_{t,e} \int_0^\infty t|e|dt$$

or the integral error performance functionals as given by

$$J = \min_e \int_0^\infty |e|dt \text{ or } J = \min_e \int_0^\infty e^2 dt.$$

Here, the tracking error vector is used, and

$$e(t) = y(t) - r(t),$$

where $r(t)$ is the reference input.

The dynamic system is documented Figure 1.2.

It is important to emphasize that owing to nonlinearities, bounds, and time-varying parameter variations, some performance criteria are stringent. In particular, the most important criteria are stability and robustness to parameter variations in the full operating envelope. A wide variety of other requirements are usually imposed.

The theory of optimal control has had difficulty in being accepted and applied by practicing engineers, primarily because real systems are nonlinear and multivariable, and practical multivariable nonlinear design has been less emphasized and studied. The major goal of this book is to educate students, engineers, and scientists on how to solve challenging problems and perform the systematic integrated research in nonlinear analysis, simulation, optimization, and control to design high-performance systems. Fundamental (analytical and numerical) and experimental studies in analysis and design of complex multivariable dynamic systems, which are modeled by linear and nonlinear differential and difference equations, are covered and illustrated using the state-space concept.

2
Mathematical Model Developments

2.1. Engineering Systems and Models

To integrate control theory in engineering practice, a bridge between real-world systems and abstract mathematical systems theory must be built. For example, applying the control theory to analyze and regulate in the desired manner the energy or information flows, the designer is confronted with the need to find adequate mathematical models of the phenomena and design controllers. Mathematical models can be found using basic physical laws. In particular, in electrical, mechanical, fluid, or thermal systems, the mechanism of storing, dissipating, transforming, and transferring energies are analyzed. We will use the Lagrange equations of motion, as well as the Kirchhoff and Newton laws to illustrate the model developments. The real-world systems integrate many components and subsystems. One can reduce interconnected systems to simple, idealized subsystems (components). However, this idealization, in most cases, is unpractical. For example, one cannot study electric motors without studying devices to be actuated, and to control electric motors, power amplifiers must be integrated as well. That is, electromechanical systems integrate mechanical systems, electromechanical motion devices (actuators and sensors), and power converters. Analyzing power converters, the designer studies switching devices (transistors or thyristors), drivers, circuits, filters, and so forth. The primary objective of this chapter is to illustrate how one can develop mathematical models of dynamic systems using basic principles and laws. Through illustrative examples, differential equations will be found to model dynamic systems. A functional block diagram of the controlled (closed-loop) dynamic systems is illustrated in Figure 2.1.1.

Multivariable dynamic systems are studied with a different level of comprehensiveness. For example, open-loop and closed-loop systems can be studied. However, in studying the mathematical models in this chapter, let us focus our efforts on the development of the differential equation to model the system transients. The designer can consider robotic manipulators, aircraft, spacecraft, submarines, as well as other space, ground, and underwater vehicles as pure mechanical systems using so-called six degree-of-freedom models in the Cartesian or other coordinate systems. However, these dynamic systems must be controlled. The following question can be asked: Why must aerospace engineers study electromechanical motion devices researching flight dynamics and motion control of flight vehicles? In advanced aircraft, as well as in other flight vehicles, control surfaces are actuated

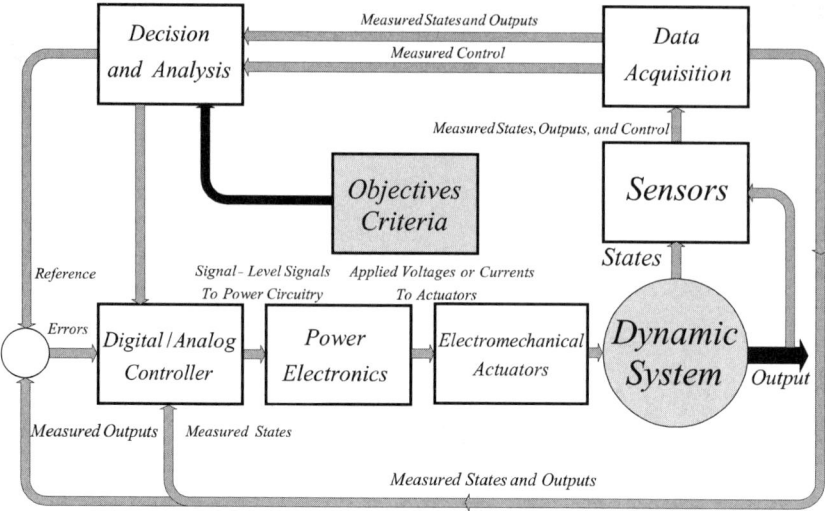

FIGURE 2.1.1.   High-level functional block diagram of the closed-loop dynamic system.

by electromechanical flight actuators. Therefore, the actuator performance must be integrated when the designer studies flight vehicles. In addition to the settling and rise time, overshoot, accuracy, mechanical limits, and deflection rates must be studied analyzing the stability. Mechanical limits on the deflection of control surfaces are imposed. The deflection rate, at which the control surface can be actuated, is a nonlinear function of the actuator's torque capabilities and hinge moment applied, deflection angle, vehicle velocity, and so on. When the control surface deflects from its free-float (zero hinge moment) position, the hinge moment opposes the electromagnetic torque developed by the actuator. The deflection rate is higher when the control surface is actuated toward its free-float position. In addition to the steady-state torque-displacement and torque-deflection rate envelopes, the actuator dynamic performance (settling and rise time, overshoot and accuracy, robustness and stability margins) is studied.

When shown the necessity of integrating electromechanical actuators in the analysis of flight vehicles, one might wonder why power electronics is a subject of our interest? Flight actuators are controlled by power amplifiers that supply the voltages to the electric motor windings to drive servo-motors. Hence, power converters and driving circuitry should be thoroughly studied. The different topologies, operating principles, mathematical models, nonlinear analysis, and simulation of power converters will be covered in this book. Mathematical models of high-frequency switching devices (usually, insulated gate bipolar and metal-oxide semiconductor field effect transistors) as well as the driving integrated circuits (transistor drivers) are complex. The simplest operational amplifier integrates more than 50 transistors, and using the so-called second-order π-model, a set of hundreds of differential equations results. However, the transient dynamics

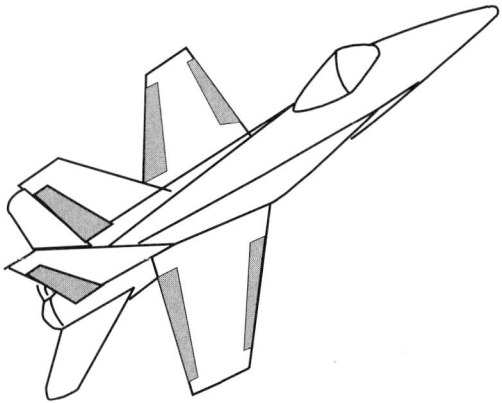

FIGURE 2.1.2.   Fighter.

of power transistors and integrated circuits are very fast (nanoseconds). Therefore, their transient behavior can be integrated in the analysis and design as the
unmodeled dynamics.

To control systems, analog and discrete controllers are used, and the information
regarding system outputs, states, reference (desired) inputs, disturbances, as well
as other data are fed to find and generate control signals. Different transducers are
applied to measure the physical variables and data of our particular interest (speed,
acceleration, temperature, pressure, etc.). Decision making and data acquisition
must be performed to attain the desired system performance and to achieve the
thorough analysis.

It was illustrated that dynamic systems integrate many components, e.g., physical systems to be controlled (aircraft, spacecraft, rocket, missile, submarine,
torpedo, car, or track), electromechanical actuators, power electronics, sensors,
controllers, data acquisition, and decision-making systems. Real-world dynamic
systems are highly coupled and nonlinear. In spite of these nonlinearities, systems
must be adequately described, modeled, identified, and simulated to design control
algorithms.

Let us study the motion control problem as applied to the advanced aircraft, as
shown in Figure 2.1.2.

The aircraft outputs are the Euler angles θ, ϕ, and Ψ. The reference inputs are
the desired (assigned by the pilot or flight computer) Euler angles that are denoted
as r_θ, r_ϕ, and r_Ψ. For rigid-body aircraft, the longitudinal and lateral dynamics are
modeled using the following state variables:

- Forward velocity v
- Angle of attack α
- Pitch rate q
- Pitch angle θ
- Sideslip angle β

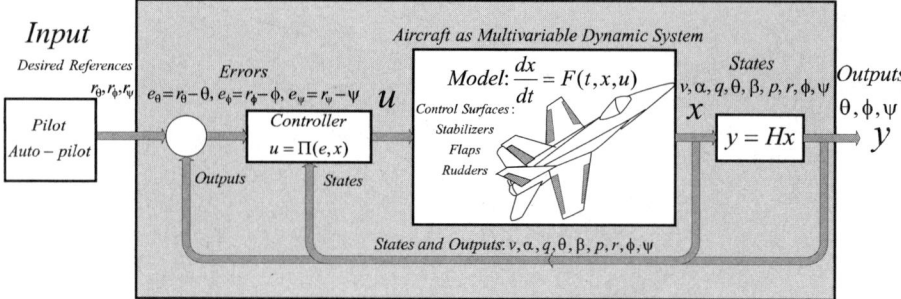

FIGURE 2.1.3.   Block diagram representation of a multi-input/multi-output closed-loop aircraft.

- Roll rate p
- Yaw rate r
- Roll angle ϕ
- Yaw angle Ψ

The aircraft is controlled by displacing the control surfaces (right and left horizontal stabilizers, right and left leading and trailing edge flaps, right and left rudders), as illustrated in Figure 2.1.2. That is, a multi-input/multi-output dynamic system (aircraft) is under our consideration. The aircraft motion within longitudinal and lateral axes is controlled by deflecting the control surfaces. The multivariable fighter dynamics, as a mechanical rigid-body system, is modeled using nine states $(v, \alpha, q, \theta, \beta, p, r, \phi, \Psi)$ $x \in \mathbb{R}^9$. The deflection of eight control surfaces (right and left horizontal stabilizers, right and left leading and trailing edge flaps, right and left rudders) are viewed as control inputs if one does not consider the flight actuators and power amplifiers. The transient and steady-state behavior of three aircraft outputs (θ, ϕ, Ψ) $y \in \mathbb{R}^3$ and reference inputs $(r_\theta, r_\phi, r_\Psi)$ $r \in \mathbb{R}^3$ are studied to perform qualitative and quantitative analysis of the aircraft performance (other state variables, obtained using comprehensive and detail mathematical models, must be also considered).

The block-diagram representation of the rigid-body aircraft with controller $u = \Pi(e, x)$ (the control input is derived by using the tracking error e and the state variables x) is illustrated in Figure 2.1.3.

It was emphasized that the aircraft is controlled by changing the angular displacement of the flight control surfaces (ailerons, elevators, canards, flaps, rudders, stabilizers), and servo-systems are used to actuate these control surfaces. To deflect ailerons, canards, fins, flaps, rudders, stabilizers, and other control surfaces, hydraulic and electric motors have been used. The minimum number of actuators is equal to the number of control surfaces needed to be actuated (in practice, at least double redundancy is needed, and two actuators are used to actuate one control surface). A light-duty control surface servo driven by a stepper motor is shown in Figure 2.1.4. The desired angular displacement of the control surface (reference

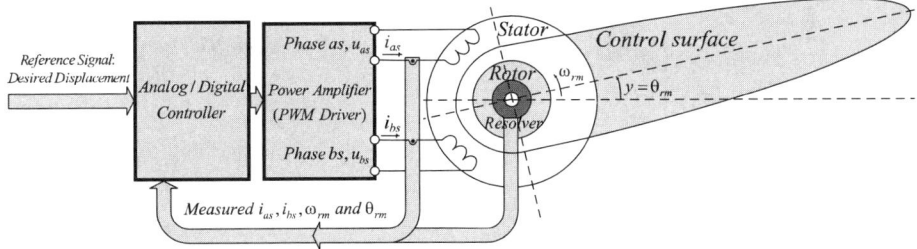

FIGURE 2.1.4.   Fly-by-wire flight servo-system actuated by stepper motor.

input) is assigned either by the pilot or the aircraft flight computer (autopilot). Using the reference signal (the specified angular displacement of the control surface), as well as the currents measured by sensors currents in the ab phase windings (i_{as} and i_{bs}), the mechanical angular velocity ω_{rm}, and the actual mechanical angular displacement θ_{rm} (measured by the resolver), the controller (analog or digital) develops signal-level signals that drive high-frequency switches. The magnitude and frequency of the applied voltages to the ab phase windings u_{as} and u_{bs} are controlled by the pulsewidth modulation (PWM) driver (power amplifier); see Figure 2.1.4.

The studied electromechanical flight servo-system integrates electromechanical motion devices (stepper motor and resolver), the power amplifier (PWM driver), transducers, and the controller. It should be emphasized that microcontrollers and digital signal processors (DSPs) are usually applied to implement control algorithms. Correspondingly, analog–digital (A/D) and digital–analog (D/A) converters are used.

Other flight servo-systems are available and used. Heavy-duty hydraulic actuators, direct-drive electric motors, and electrical cylinders can develop the torque up to thousands of Newton-meters. Figure 2.1.5 illustrates the control surface actuated by a brushless, limited-angle torque motor with permanent-magnet rotor.

The representation of the rigid-body aircraft with flight actuators is documented in Figure 2.1.6. The signal-level control signal u, developed by digital or analog controller, drives the transistor drivers integrated circuits, and power amplifiers supply the voltages to the armature windings of eight flight actuators.

The analysis performed illustrates that the designer must develop comprehensive mathematical models integrating all components of complex, multivariable, real-world dynamic systems. The state and control variables must be defined, and mathematical models must be developed with a minimum level of simplifications and assumptions. It must be emphasized that turbofan engine and thrust vectoring control, as well as other additional features, must be thoroughly studied, attacking flight dynamics and control problems. A spectrum of extremely important problems in nonlinear analysis, identification, optimization, and control of flight vehicles must be solved to guarantee the required flying and handling qualities.

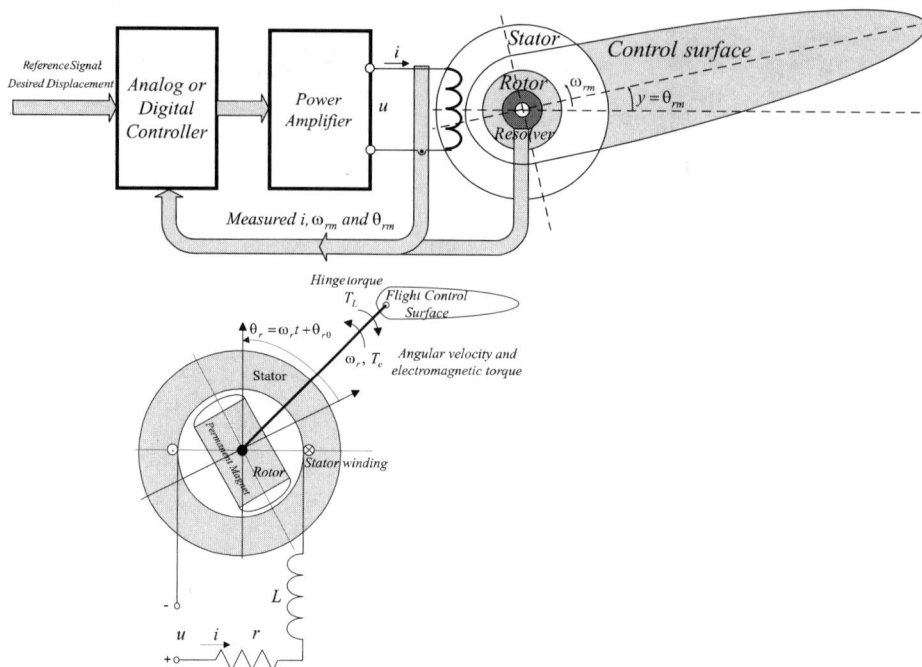

FIGURE 2.1.5.   Closed-loop system: Control surface is actuated by a limited-angle torque motor.

These requirements and specifications must be achieved for a fully functional and damaged/crippled/failed aircraft using flight management systems, and the aircraft's performance must be certified. The motion control problem integrates model developments and simulation, nonlinear analysis and real-time identification/estimation, control redesign and controller reconfiguration, as well as decision making. The high-level, closed-loop flight control system is shown in Figure 2.1.7.

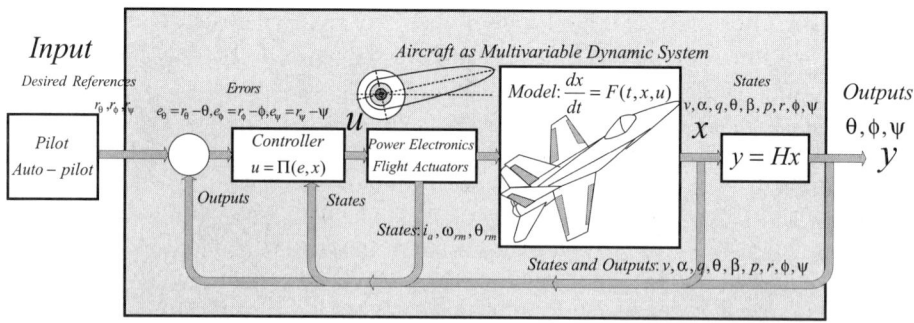

FIGURE 2.1.6.   Closed-loop system: Block diagram of a multi-input/multi-output aircraft with flight control surfaces.

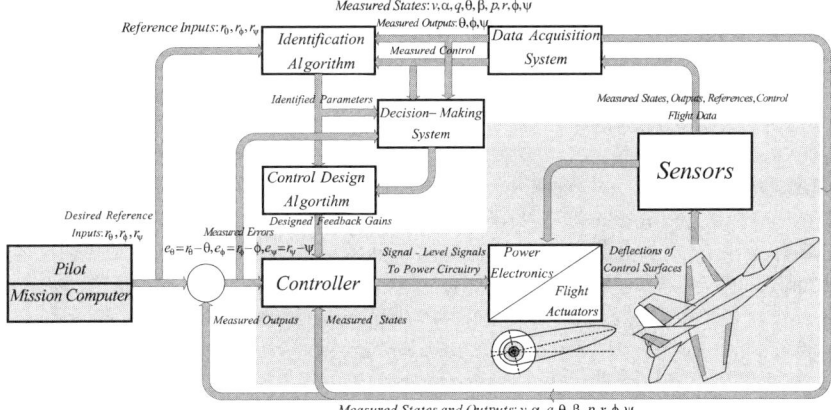

FIGURE 2.1.7.   Motion control of aircraft.

The basic generic configuration of the controlled dynamic system (robots, manipulators, electric and hybrid-electric vehicles, aircraft, helicopters, electric machines, and hydraulic actuators) was depicted in Figure 2.1.1. Modern systems rely increasingly on the digital controllers that are implemented using microprocessors and DSPs. The block diagram of the closed-loop multi-input/multi-output system is shown in Figure 2.1.8, assuming that the output and reference signals are continuous.

Dynamic systems are regulated by using the difference between the desired reference inputs $r(t)$ and the system outputs $y(t)$. The control inputs are found using the error $e(t) = r(t) - y(t)$ and the state variables.

For example, it was emphasized that for aircraft, one uses the Euler angles θ, ϕ, Ψ as the outputs; that is,

$$y(t) = \begin{cases} \theta(t) \\ \phi(t) \\ \Psi(t). \end{cases}$$

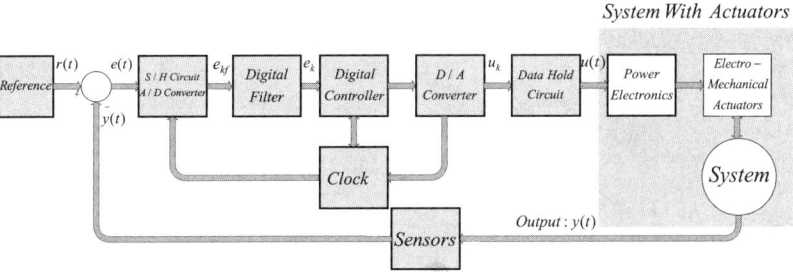

FIGURE 2.1.8.   Block diagram of a typical digitally controlled dynamic system.

The reference inputs are the desired Euler angles r_θ, r_ϕ, and r_ψ. Hence, we have

$$r(t) = \begin{cases} r_\theta(t) \\ r_\phi(t) \\ r_\psi(t). \end{cases}$$

Microprocessors and DSPs are widely used to control industrial and technological systems and processes. Specifically, microprocessors and DSPs are used to derive control signals (based on the control algorithms downloaded), perform data acquisition and filtering (based on the digital filters), attain analysis and decision making. For single-input/single-output systems, assuming that the references and outputs are continuous, the continuous-time error signal $e(t) = r(t) - y(t)$ must be converted into the digital form to perform digital filtering and regulation. The sample-and-hold circuit (S/H circuit) receives the continuous-time (analog) signal and holds this signal at the constant value for the specified period of time (the sampling period). The A/D converter converts this piecewise continuous-time signal to the digital format (binary numbers). The transformation of continuous-time signals to discrete-time signals is called sampling or discretization. That is, the input signal to the filter is the sampled version of the continuous-time error signal $e(t) = r(t) - y(t)$. The input signal of the signal-to-digital controller (microcontroller or DSP) is the digital filter output. It must be emphasized that analog filters can be used to perform the filtering. At each sampling, the discretized value of the error signal e_k in binary form is used by the digital controller to generate the control signal, which must be converted to analog form to be fed to the driving circuitry of the power converter. The D/A conversion (decoding) is performed by the D/A converter and the data-hold circuit. The A/D and D/A converters are built with finite word length. Coding and decoding are synchronized by using the clock. This brief description illustrates that the signal conversion involves multiplexing and demultiplexing, S/H, A/D (quantizing and encoding), and D/A (decoding) conversion.

2.2.   Basic Principles in Model Developments

Novel technologies and pioneering principles in system design, advanced actuators, power electronics, state-of-the-art sensors, transducers, high-performance microprocessors, DSPs have been developed and implemented to attain the required objectives and specified characteristics. The use of the newest software and hardware, state-of-the-art technologies, and concepts are motivated by the critical need to guarantee high efficiency and superior performance capabilities for industrial system, safety, compactness, simplicity, ruggedness, survivability, durability, reliability, and so on. To optimize the system performance, to attain the specified requirements, to guarantee tracking and robustness, and to expand stability margins and ensure accuracy and disturbance attenuation, there is a critical need to use complete nonlinear dynamic system models with a minimum level of simplifications and assumptions. This will allow one to rigorously approach virtual prototyping, solve the motion control and decision-making problems, and implement advanced software and hardware.

2.2.1. Newtonian Mechanics

2.2.1.1. Newtonian mechanics: Translational motion

The equations of motion for mechanical systems can be found using Newton's second law of motion, which is given as

$$\sum \vec{F}(\vec{x}, t) = m\vec{a}, \tag{2.2.1}$$

where $\vec{F}(\vec{x}, t)$ is the vector sum of all forces applied to the body ($\vec{F}$ is called the *net* force), $\vec{a}$ is the vector of acceleration of the body with respect to an inertial reference frame, and m is the mass of the body.

From (2.2.1), in the Cartesian system, we have

$$\sum \vec{F}(\mathbf{x}, t) = m\vec{a} = m\frac{d\vec{\mathbf{x}}^2}{dt^2} = m \begin{bmatrix} \dfrac{d\vec{x}^2}{dt^2} \\ \dfrac{d\vec{y}^2}{dt^2} \\ \dfrac{d\vec{z}^2}{dt^2} \end{bmatrix}.$$

Newton's second law is idealization because if the dissipated and loses energies are integrated, additional terms in (2.2.1) result. To apply Newton's law, it is convenient to use the free-body diagram.

Example 2.2.1.

Consider a body of mass m in the XY-coordinate system (xy-plane). Find the equations of motion (differential equations that model the motion dynamics). The external force $\vec{F}_a$ is applied in the x-direction. The external force is a nonlinear, time-varying function of the position and velocity. In particular, $\vec{F}_a(t, x) = \cos 2t e^{-3t} x^2 + t^2 \frac{dx}{dt}$. Assume that the viscous friction force is a linear function of velocity; that is, $F_{fr} = B_v \frac{dx}{dt}$, where B_v is the viscous friction coefficient.

The free-body diagram is shown in Figure 2.2.1.

The sum of the forces acting in the y-direction is

$$\sum \vec{F}_Y = \vec{F}_N - \vec{F}_g,$$

where $\vec{F}_g = mg$ is the gravitational force acting on the mass m; $\vec{F}_N$ is the normal force that is equal and opposite to the gravitational force, $\vec{F}_N = -\vec{F}_g$.

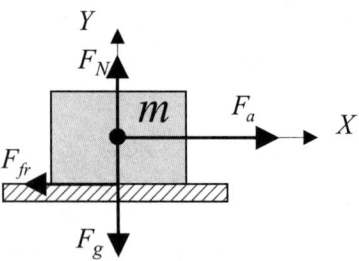

FIGURE 2.2.1.   Free-body diagram.

From (2.2.1), the equation of motion in the y-direction is

$$\vec{F}_N - \vec{F}_g = 0 = ma_y = m\frac{d^2y}{dt^2},$$

where a_y is the acceleration in the y-direction $a_y = \frac{d^2y}{dt^2}$.

The sum of the forces acting in the x-direction is found using the time-varying applied force $\vec{F}_a$ and the friction force $\vec{F}_{fr}$. We have

$$\sum \vec{F}_X = \vec{F}_a - \vec{F}_{fr},$$

Using (2.2.1), the second-order differential equation of motion in the x-direction is

$$\vec{F}_a - \vec{F}_{fr} = ma_x = m\frac{d^2x}{dt^2},$$

where a_x is the acceleration in the x-direction $a_x = \frac{d^2x}{dt^2}$.

One obtains the following second-order differential equation to model the body dynamics in the x-direction

$$\frac{d^2x}{dt^2} = \frac{1}{m}\left(F_a - B_v\frac{dx}{dt}\right).$$

Using the velocity in the x-direction $v = \frac{dx}{dt}$, a set of two first-order differential equations results, and we have

$$\frac{dx}{dt} = v, \quad \frac{dv}{dt} = \frac{1}{m}(F_a - B_v v).$$

It was assigned that the applied force is a time-varying function, and $\vec{F}_a(t, x) = \cos 2te^{-3t}x^2 + t^2\frac{dx}{dt}$. Therefore, one finds

$$\frac{dx}{dt} = v,$$

$$\frac{dv}{dt} = \frac{1}{m}(\cos 2te^{-3t}x^2 + t^2v - B_v v).$$

Example 2.2.2.

A two-mass wheel suspension system is illustrated in Figure 2.2.2. Assume that the motion is only within the y-direction. The vertical position of the wheel is denoted as y_1, and the vertical position of the frame relative to the equilibrium position is denoted as y_2. The equivalent masses of the wheel and the frame are m_1 and m_2. The constants k_{s1} and k_{s2} are the stiffness constants of the suspension spring and the tire. The shock absorber has the damping (viscous friction) coefficient B_v. Using the Newtonian mechanics, find the differential equations to model the suspension system dynamics.

For the mass m_1, the net forces, acting in the y-direction, are

$$\sum \vec{F}_{Y_1} = k_{s1}(y_2 - y_1) + B_v \left(\frac{dy_2}{dt} - \frac{dy_1}{dt} \right) - k_{s2}y_1,$$

where y_1 and y_2 are the displacement of the masses m_1 and m_2, and $\frac{dy_1}{dt}$ and $\frac{dy_2}{dt}$ are the velocities of the masses in the y-direction.

From (2.2.1), the equation of motion for the mass m_1 in the y-direction is

$$m_1 \frac{d^2 y_1}{dt^2} = k_{s1}(y_2 - y_1) + B_v \left(\frac{dy_2}{dt} - \frac{dy_t}{dt} \right) - k_{s2}y_1,$$

and hence,

$$\frac{d^2 y_1}{dt^2} = \frac{1}{m_1} \left(k_{s1}(y_2 - y_1) + B_v \left(\frac{dy_2}{dt} - \frac{dy_1}{dt} \right) - k_{s2}y_1 \right),$$

where $\frac{d^2 y_1}{dt^2}$ is the acceleration of the mass m_1.

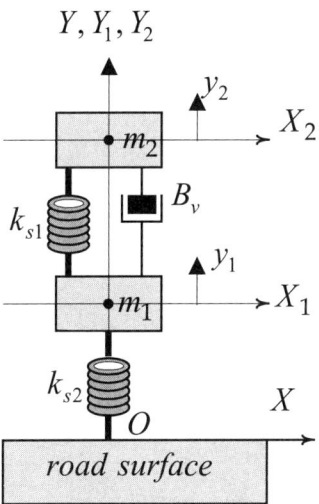

FIGURE 2.2.2.   Suspension system with two masses.

For the mass m_2, the sum of the forces acting in the y-direction is

$$\sum \vec{F}_{Y_2} = -k_{s1} (y_2 - y_1) - B_v \left(\frac{dy_2}{dt} - \frac{dy_1}{dt} \right).$$

Hence, the differential equation of motion for the mass m_2 is

$$m_2 \frac{d^2 y_2}{dt^2} = -k_{s1} (y_2 - y_1) - B_v \left(\frac{dy_2}{dt} - \frac{dy_1}{dt} \right).$$

That is,

$$\frac{d^2 y_2}{dt^2} = \frac{1}{m_2} \left(-k_{s1} (y_2 - y_1) - B_v \left(\frac{dy_2}{dt} - \frac{dy_1}{dt} \right) \right).$$

The resulting system of two second-order differential equations is

$$\frac{d^2 y_1}{dt^2} = \frac{1}{m_1} \left(k_{s1} (y_2 - y_1) + B_v \left(\frac{dy_2}{dt} - \frac{dy_1}{dt} \right) - k_{s2} y_1 \right),$$

$$\frac{d^2 y_2}{dt^2} = \frac{1}{m_2} \left(-k_{s1} (y_2 - y_1) - B_v \left(\frac{dy_2}{dt} - \frac{dy_1}{dt} \right) \right).$$

Assigning the following four state variables:

$$x_1 = y_1, x_2 = \frac{dx_1}{dt} = \frac{dy_1}{dt}, x_3 = y_2, \text{ and } x_4 = \frac{dx_3}{dt} = \frac{dy_2}{dt},$$

one finds at once a set of four first-order differential equations

$$\frac{dx_1}{dt} = x_2,$$

$$\frac{dx_2}{dt} = \frac{1}{m_1} (k_{s1} (x_3 - x_1) + B_v (x_4 - x_2) - k_{s2} x_1),$$

$$\frac{dx_3}{dt} = x_4,$$

$$\frac{dx_4}{dt} = \frac{1}{m_2} (-k_{s1} (x_3 - x_1) - B_v (x_4 - x_2)).$$

2.2.1.2. Newtonian mechanics: Rotational motion

Newton's second law of rotational motion is expressed as

$$\sum M = J\alpha = J \frac{d^2 \theta}{dt^2}, \tag{2.2.2}$$

where $\sum M$ is the sum of all moments (*net* moment) about the center of mass of a body, J is the moment of inertia about the center of mass, α is the angular acceleration of the body $\alpha = \frac{d^2 \theta}{dt^2}$, and θ is the angular displacement.

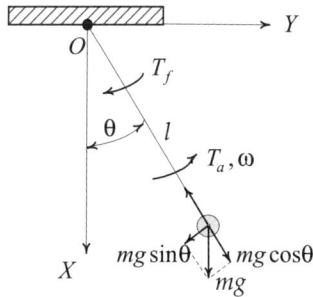

FIGURE 2.2.3.   A simple pendulum.

Example 2.2.3.

Figure 2.2.3 illustrates a point mass m (simple pendulum) suspended by a massless unstretchable string of length l. Derive the equations of motion assuming that the friction is a linear function of the angular velocity; that is, $T_f = B_m \omega$.

The restoring force, which is the tangential component, is proportional to $\sin \theta$. Using the expression for the restoring force $-mg \sin \theta$, the sum of the moments about the pivot point O is

$$\sum M = -mgl \sin \theta + T_a - B_m \omega,$$

where T_a is the applied torque (time-invariant or time-varying), l is the length of the pendulum measured from the point of rotation, and B_m is the viscous friction coeffcient.

Using (2.2.2), one obtains the differential equation of motion

$$J\alpha = J \frac{d^2\theta}{dt^2} = -mgl \sin \theta + T_a - B_m \omega,$$

where J is the moment of inertial of the mass about the point O.

That is,

$$\frac{d^2\theta}{dt^2} = \frac{1}{J}(-mgl \sin \theta + T_a - B_m \omega).$$

Using $\frac{d\theta}{dt} = \omega$, one obtains a set of two first-order differential equations

$$\frac{d\omega}{dt} = \frac{1}{J}(-mgl \sin \theta + T_a - B_m \omega),$$

$$\frac{d\theta}{dt} = \omega.$$

The moment of inertia is $J = ml^2$. Hence, we finally have

$$\frac{d\omega}{dt} = -\frac{B_m}{ml^2}\omega - \frac{g}{l}\sin \theta + \frac{1}{ml^2}T_a,$$

$$\frac{d\theta}{dt} = \omega.$$

Example 2.2.4.

Study a suspension system, as illustrated in Figure 2.2.4, and find differential equations within the y-direction. The equivalent masses of the front and rear wheels, automobile frame, and seat are denoted as $m_1, m_2, m_3,$ and m_4. The spring and damping (viscous friction) coefficients are denoted as $k_{s1}, k_{s2}, k_{s3}, k_{s4}, k_{s5}$ and $B_{v1}, B_{v2}, B_{v3}, B_{v4}$. The distances from the left and right ends to the center of mass are denoted as l_1 and l_2.

The application of Newton's translational law results in the following differential equations:

$$m_1 \frac{d^2 y_2}{dt^2} = k_{s1}(y_1 - y_2) + B_{v1}\left(\frac{dy_1}{dt} - \frac{dy_2}{dt}\right) + k_{s2}(y_3 - y_2) + B_{v2}\left(\frac{dy_3}{dt} - \frac{dy_2}{dt}\right)$$

and

$$m_2 \frac{d^2 y_6}{dt^2} = k_{s3}(y_5 - y_6) + B_{v3}\left(\frac{dy_5}{dt} - \frac{dy_6}{dt}\right) + k_{s4}(y_7 - y_6) + B_{v4}\left(\frac{dy_7}{dt} - \frac{dy_6}{dt}\right).$$

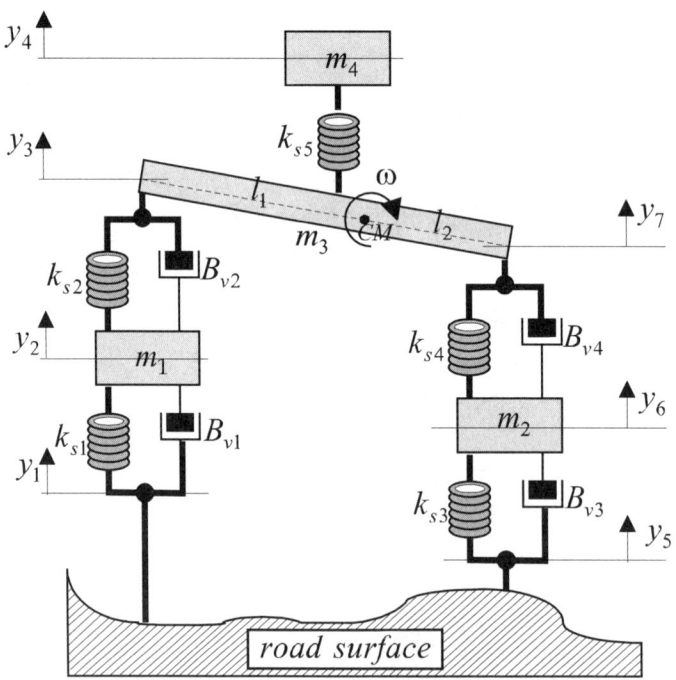

FIGURE 2.2.4.   Suspension system with two masses.

Making use of $y_{CM} = y_3 + \frac{l_1}{l_1+l_2}(y_7 - y_3)$, we have

$$m_3 \frac{d^2 y_{CM}}{dt^2} = k_{s2}(y_2 - y_3) + B_{v2}\left(\frac{dy_2}{dt} - \frac{dy_3}{dt}\right) + k_{s4}(y_6 - y_7)$$

$$+ B_{v4}\left(\frac{dy_6}{dt} - \frac{dy_7}{dt}\right) + k_{s5}\left(y_4 - y_3 - \frac{l_2}{l_1 + l_2}(y_7 - y_3)\right).$$

It is evident that

$$m_4 \frac{d^2 y_4}{dt^2} = k_{s5}\left(y_3 + \frac{l_2}{l_1 + l_2}(y_7 - y_3) - y_4\right).$$

Newton's rotational law gives

$$J\frac{d^2\theta}{dt^2} = -l_1 k_{s2}(y_2 - y_3) - l_1 B_{v2}\left(\frac{dy_2}{dt} - \frac{dy_3}{dt}\right) + l_2 k_{s4}(y_6 - y_7)$$

$$+ l_2 B_{v4}\left(\frac{dy_6}{dt} - \frac{dy_7}{dt}\right) - (l_1 - l_2)k_{s5}\left(y_4 - y_3 - \frac{l_2}{l_1 + l_2}(y_7 - y_3)\right).$$

That is, a set of differential equations was derived.

2.2.2. *Lagrange Equations of Motion*

Although Newton's laws of translational and rotational motion form the corner-stone foundation for the study of mechanical systems, they cannot be straightfor-wardly used to derive the dynamics of electromechanical systems because elec-tromagnetic and circuitry transient behavior must be considered. The designer can use the so-called electromechanical analogies. However, this avenue does not provide the desired level of viability needed. The circuitry dynamics can be found using Kirchhoff's laws, and augmenting mechanical and electrical sub-systems, mathematical models result. That is, one integrates the *torsional–mechanical* dy-namics and circuitry equations. A viable approach exists for solving the model development problems through the use of the Lagrange equations of motion. The Lagrange concept, which is based on the energy analysis, allows the designer to integrate the dynamics of mechanical and electrical sub-systems. Thus, the torsional–mechanical and circuitry dynamics are augmented. To illustrate this concept, we consider examples.

Example 2.2.5.

A body is suspended from a spring with coefficient k_s, and the damping (viscous) coefficient is B_v; see Figure 2.2.5.

Using Newton's second law, one finds the following second-order differential equation:

$$m\frac{d^2 y}{dt^2} + B_v\frac{dy}{dt} + k_s y = F_a.$$

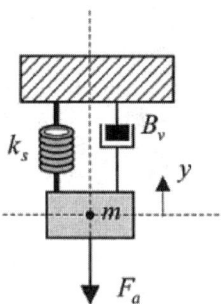

FIGURE 2.2.5.   Suspension system.

Let us find the total kinetic, dissipation, and potential energies. The total kinetic (Γ), dissipation (D), and potential (Π) energies are

$$\Gamma = \frac{1}{2}m\left(\frac{dy}{dt}\right)^2, \quad D = \frac{1}{2}B_v\left(\frac{dy}{dt}\right)^2, \quad \text{and } \Pi = \frac{1}{2}k_s y^2.$$

Using the following energy-based equation:

$$\frac{d}{dt}\left(\frac{\partial\Gamma}{\partial\dot y}\right) - \frac{\partial\Gamma}{\partial y} + \frac{\partial D}{\partial\dot y} + \frac{\partial\Pi}{\partial y} = F_a,$$

one finds $\frac{d}{dt}\left(m\frac{dy}{dt}\right) + B_v\frac{dy}{dt} + k_s y = F_a$. Assuming that the mass m is constant (time-invariant), one concludes that the derived differential equations to model the body dynamics are the same.

Example 2.2.6.

A two-mesh electric circuit is illustrated in Figure 2.2.6.
    Using Kirchhoff's law, one finds

$$L_1\frac{di_1}{dt} + \frac{1}{C_1}\int(i_1 - i_2)dt = u_a \text{ and } L_2\frac{di_2}{dt} - \frac{1}{C_1}\int(i_1 - i_2)dt + \frac{1}{C_2}\int i_2 dt = 0.$$

Using the electric charges in the first and second loops q_1 and q_2 ($q_1 = \frac{i_1}{s}$ and $q_2 = \frac{i_2}{s}$), the total kinetic and potential energies are

$$\Gamma = \frac{1}{2}L_1 i_1{}^2 + \frac{1}{2}L_2 i_2{}^2 = \frac{1}{2}L_1\dot q_1{}^2 + \frac{1}{2}L_2\dot q_2{}^2 \text{ and } \Pi = \frac{1}{2}\frac{(q_1 - q_2)^2}{C_1} + \frac{1}{2}\frac{q_2^2}{C_2}.$$

Using the following energy-based equations:

$$\frac{d}{dt}\left(\frac{\partial\Gamma}{\partial\dot q_1}\right) - \frac{\partial\Gamma}{\partial q_1} + \frac{\partial\Pi}{\partial q_1} = u_a,$$

$$\frac{d}{dt}\left(\frac{\partial\Gamma}{\partial\dot q_2}\right) - \frac{\partial\Gamma}{\partial q_2} + \frac{\partial\Pi}{\partial q_2} = 0,$$

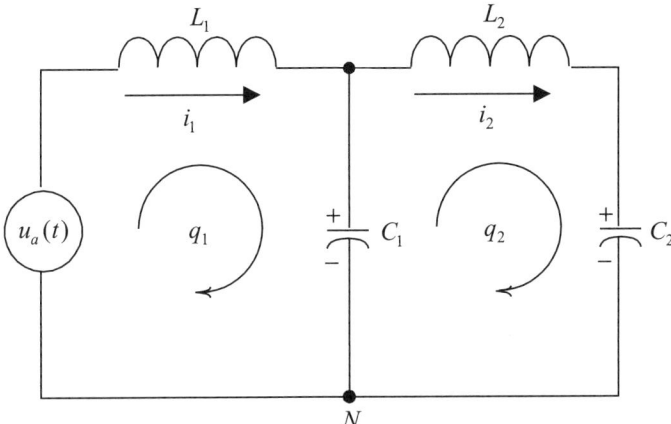

FIGURE 2.2.6.   Two-mesh circuit network.

one finds

$$\frac{d}{dt}(L_1\dot{q}_1) + \frac{q_1 - q_2}{C_1} = u_a \text{ and } \frac{d}{dt}(L_2\dot{q}_2) - \frac{q_1 - q_2}{C_2} + \frac{q_2}{C_2} = 0.$$

Taking note that $q_1 = \frac{i_1}{s}$ and $q_2 = \frac{i_2}{s}$, the equivalence of the models derived is evident.

Let us model dynamic systems in terms of the variables, which should be used to find the total kinetic, dissipation, and potential energies (Γ, D, and Π). Using the total kinetic, dissipation, and potential energies $\Gamma\left(t, q_1, \ldots, q_n, \frac{dq_1}{dt}, \ldots, \frac{dq_n}{dt}\right)$, $D\left(t, q_1, \ldots, q_n, \frac{dq_1}{dt}, \ldots, \frac{dq_n}{dt}\right)$, and $\Pi(t, q_1, \ldots, q_n)$, the Lagrange equations of motion are expressed as

$$\frac{d}{dt}\left(\frac{\partial\Gamma}{\partial\dot{q}_i}\right) - \frac{\partial\Gamma}{\partial q_i} + \frac{\partial D}{\partial\dot{q}_i} + \frac{\partial\Pi}{\partial q_i} = Q_i. \qquad (2.2.3)$$

Here, q_i and Q_i are the generalized coordinates and the generalized forces (applied forces and disturbances).

For conservative (lossless) systems, $D = 0$. Therefore, Lagrange's equations of motion are

$$\frac{d}{dt}\left(\frac{\partial\Gamma}{\partial\dot{q}_i}\right) - \frac{\partial\Gamma}{\partial q_i} + \frac{\partial\Pi}{\partial q_i} = Q_i.$$

The generalized coordinates q_i are used to derive explicit expressions for energies $\Gamma\left(t, q, \ldots, q_n, \frac{dq_1}{dt}, \ldots, \frac{dq_n}{dt}\right)$, $D\left(t, q_1, \ldots, q_n, \frac{dq_1}{dt}, \ldots, \frac{dq_n}{dt}\right)$, and $\Pi(t, q_1, \ldots, q_n)$.

Example 2.2.7.

Using the Lagrange equations of motion, let us derive the mathematical model for a simple pendulum, which was studied in Example 2.2.3. Assume that $B_m = 0$ and $B_m \neq 0$.

The generalized coordinate and force must be designated. The angular displacement is the generalized coordinate, whereas the torque applied is the generalized force.

That is, $q_1 = \theta$ and $Q_1 = T_a$.

To use the Lagrange concept, one must find explicit expressions for the kinetic and potential energies.

The kinetic energy is $\Gamma = \frac{1}{2}m(l\dot{\theta})^2$, and the potential energy is $\Pi = mgl(1 - \cos\theta)$.

Therefore, one obtains

$$\frac{\partial \Gamma}{\partial \dot{q}_1} = \frac{\partial \Gamma}{\partial \dot{\theta}} = ml^2\dot{\theta}, \quad \frac{\partial \Gamma}{\partial q_1} = \frac{\partial \Gamma}{\partial \theta} = 0$$

and

$$\frac{\partial \Pi}{\partial q_1} = \frac{\partial \Pi}{\partial \theta} = mgl\sin\theta.$$

Thus,

$$\frac{d}{dt}\left(\frac{\partial \Gamma}{\partial \dot{\theta}}\right) = ml^2\frac{d^2\theta}{dt^2} + 2ml\frac{dl}{dt}\frac{d\theta}{dt}.$$

Assume that the string is unstretchable (constant). Hence, $\frac{dl}{dt} = 0$. Therefore, from

$$ml^2\frac{d^2\theta}{dt^2} + mgl\sin\theta = T_a,$$

one finds

$$\frac{d^2\theta}{dt^2} = \frac{1}{ml^2}(-mgl\sin\theta + T_a).$$

This differential equation is in complete agreement with the mathematical model found using Newton's second law. Recall that we received

$$\frac{d^2\theta}{dt^2} = \frac{1}{J}(-mgl\sin\theta + T_a), \quad J = ml^2.$$

In real-world systems, the viscous friction coefficient should be considered, and $B_m \neq 0$. The total heat energy dissipated is $D = \frac{1}{2}B_m\dot{q}_1^2$. Hence, $\frac{\partial D}{\partial \dot{q}_3} = B_m\dot{q}_1$, and the following equation results:

$$\frac{d^2\theta}{dt^2} = \frac{1}{ml^2}\left(-mgl\sin\theta - B_m\frac{d\theta}{dt} + T_a\right).$$

Example 2.2.8.

Consider a servo-motor with two independently excited stator and rotor windings; see Figure 2.2.7. The following notations are used:

- i_s and i_r are the currents in the stator and rotor windings
- u_s and u_r are the applied voltages to the stator and rotor windings
- ω_r and θ_r are the angular velocity and position of the rotor
- T_e and T_L are the electromagnetic and load torques
- r_s, r_r, L_s, and L_r are the resistances and self-inductances of the stator and rotor windings
- N_s and N_r are the number of turns in the stator and rotor windings.

Our goal is to derive the differential equations.

Using the Lagrange concept, denote the independent generalized coordinates as q_1, q_2, and q_3, where q_1 and q_2 are the electric charges in the stator and rotor windings and q_3 is the angular displacement of the rotor.

The first derivative of the generalized coordinates $\dot{q}_1$ and $\dot{q}_2$ give the stator and rotor currents i_s and i_r, whereas q_3 is the angular velocity of the rotor ω_r.

Define the generalized forces, applied to an electromechanical system, as Q_1, Q_2, and Q_3, where Q_1 and Q_2 are the applied voltages to the stator and rotor windings and Q_3 is the load torque.

That is, one has

$$q_1 = \tfrac{i_s}{s}, \ q_2 = \tfrac{i_r}{s}, \ q_3 = \theta_r, \ \dot{q}_1 = i_s, \ \dot{q}_2 = i_r, \ \dot{q}_3 = \omega_r, \ Q_1 = u_s, \ Q_2 = u_r,$$
$$\text{and } Q_3 = -T_L.$$

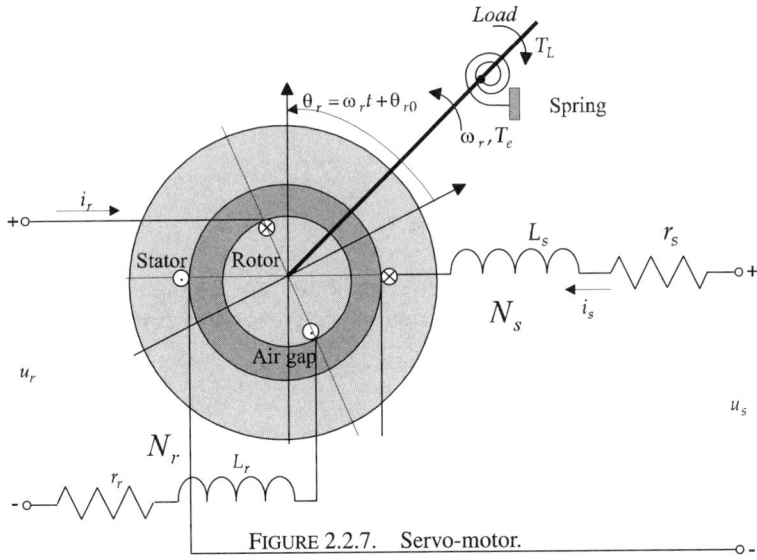

FIGURE 2.2.7.   Servo-motor.

The Lagrange equations are

$$\frac{d}{dt}\left(\frac{\partial \Gamma}{\partial \dot{q}_1}\right) - \frac{\partial \Gamma}{\partial q_1} + \frac{\partial D}{\partial \dot{q}_1} + \frac{\partial \Pi}{\partial q_1} = Q_1,$$

$$\frac{d}{dt}\left(\frac{\partial \Gamma}{\partial \dot{q}_2}\right) - \frac{\partial \Gamma}{\partial q_2} + \frac{\partial D}{\partial \dot{q}_2} + \frac{\partial \Pi}{\partial q_2} = Q_2,$$

$$\frac{d}{dt}\left(\frac{\partial \Gamma}{\partial \dot{q}_3}\right) - \frac{\partial \Gamma}{\partial q_3} + \frac{\partial D}{\partial \dot{q}_3} + \frac{\partial \Pi}{\partial q_3} = Q_3.$$

The total kinetic energy of electrical and mechanical systems is found as a sum of the total electromagnetic (electrical) Γ_E and mechanical Γ_M energies,

$$\Gamma_E = \frac{1}{2}L_s\dot{q}_1^2 + L_{sr}\dot{q}_1\dot{q}_2 + \frac{1}{2}L_r\dot{q}_2^2$$

and

$$\Gamma_M = \frac{1}{2}J\dot{q}_3^2.$$

Therefore,

$$\Gamma = \Gamma_E + \Gamma_M = \frac{1}{2}L_s\dot{q}_1^2 + L_{sr}\dot{q}_1\dot{q}_2 + \frac{1}{2}L_r\dot{q}_2^2 + \frac{1}{2}J\dot{q}_3^2.$$

The mutual inductance L_{sr} is a periodic function of the angular rotor displacement $L_{sr}(\theta) = \frac{N_s N_r}{\Re_m(\theta_r)}$.

The magnetizing reluctance $\Re_m$ is maximum if the stator and rotor windings are not displaced, and $\Re_m(\theta_r)$ is minimum if the coils are displaced by $90°$ degrees. Thus, $L_{sr\,min} \leq L_{sr}(\theta_r) \leq L_{sr\,max}$, where

$$L_{sr\,max} = \frac{N_s N_r}{\Re(90°)}$$

and

$$L_{sr\,min} = \frac{N_s N_r}{\Re_m(0°)}.$$

The mutual inductance L_{sr}, as a periodic function of the angular rotor displacement, can be approximated as a cosine function. In particular, $L_{sr\,max}(\theta_r) = L_M \cos \theta_r = L_M \cos q_3$, where the amplitude of the mutual inductance between the stator and rotor windings is

$$L_M = L_{sr\,max} = \frac{N_s N_r}{\Re_m(90°)}.$$

One has $\Gamma = \frac{1}{2}L_s\dot{q}_1^2 + L_M\dot{q}_1\dot{q}_2 \cos q_3 + \frac{1}{2}L_r\dot{q}_2^2 + \frac{1}{2}J\dot{q}_3^2.$

Therefore,

$$\frac{\partial \Gamma}{\partial q_1} = 0, \quad \frac{\partial \Gamma}{\partial \dot{q}_1} = L_s \dot{q}_1 + L_M \dot{q}_2 \cos q_3,$$

$$\frac{\partial \Gamma}{\partial q_2} = 0, \quad \frac{\partial \Gamma}{\partial \dot{q}_2} = L_M \dot{q}_1 \cos q_3 + L_r \dot{q}_2,$$

$$\frac{\partial \Gamma}{\partial q_3} = -L_M \dot{q}_1 \dot{q}_2 \sin q_3,$$

and

$$\frac{\partial \Gamma}{\partial \dot{q}_3} = J \dot{q}_3.$$

The potential energy of the spring is $\Pi = \frac{1}{2} k_s q_3^2$. Thus, $\frac{\partial \Pi}{\partial q_1} = 0$, $\frac{\partial \Pi}{\partial q_2} = 0$, and $\frac{\partial \Pi}{\partial q_3} = k_s q_3$.

The total heat energy dissipated is expressed as $D = D_E + D_M$, where the heat energy dissipated in the stator and rotor windings is $D_E = \frac{1}{2} r_s \dot{q}_1^2 + \frac{1}{2} r_r \dot{q}_2^2$, and the heat energy dissipated by a mechanical system is $D_M = \frac{1}{2} B_m \dot{q}_3^2$.

Hence, $D = \frac{1}{2} r_s \dot{q}_1^2 + \frac{1}{2} r_r \dot{q}_2^2 + \frac{1}{2} B_m \dot{q}_3^2$, and $\frac{\partial D}{\partial \dot{q}_1} = r_s \dot{q}_1$, $\frac{\partial D}{\partial \dot{q}_2} = r_r \dot{q}_2$, and $\frac{\partial D}{\partial \dot{q}_3} = B_m \dot{q}_3$.

Taking note of $q_1 = \frac{i_s}{s}$, $q_2 = \frac{i_r}{s}$, $q_3 = \theta_r$, $\dot{q}_1 = i_s$, $\dot{q}_2 = i_r$, $\dot{q}_3 = \omega_r$, $Q_1 = u_s$, $Q_2 = u_r$, and $Q_3 = -T_L$, we have differential equations for a servo-system in terms of the stator and rotor currents as well as angular displacement

$$L_s \frac{di_s}{dt} + L_M \cos \theta_r \frac{di_r}{dt} - L_M i_r \sin \theta_r \frac{d\theta_r}{dt} + r_s i_s = u_s,$$

$$L_r \frac{di_r}{dt} + L_M \cos \theta_r \frac{di_s}{dt} - L_M i_s \sin \theta_r \frac{d\theta_r}{dt} + r_r i_r = u_r,$$

$$J \frac{d^2 \theta_r}{dt^2} + L_M i_s i_r \sin \theta_r + B_m \frac{d\theta_r}{dt} + k_s \theta_r = -T_L.$$

Using the stator and rotor currents, angular velocity, and displacement as the state variables, the nonlinear differential equations in Cauchy's form are given as four nonlinear coupled equations. In particular,

$$\frac{di_s}{dt} = \frac{1}{L_s L_r - L_M^2 \cos^2 \theta_r} (-r_s L_r i_s - \frac{1}{2} L_M^2 i_s \omega_r \sin 2\theta_r + r_r L_M i_r \cos \theta_r$$

$$+ L_r L_M i_r \omega_r \sin \theta_r + L_r u_s - L_M \cos \theta_r u_r),$$

$$\frac{di_r}{dt} = \frac{1}{L_s L_r - L_M^2 \cos^2 \theta_r} (r_s L_M i_s \cos \theta_r + L_s L_M i_s \omega_r \sin \theta_r - r_r L_s i_r$$

$$- \frac{1}{2} L_M^2 i_r \omega_r \sin 2\theta_r - L_M \cos \theta_r u_s + L_s u_r),$$

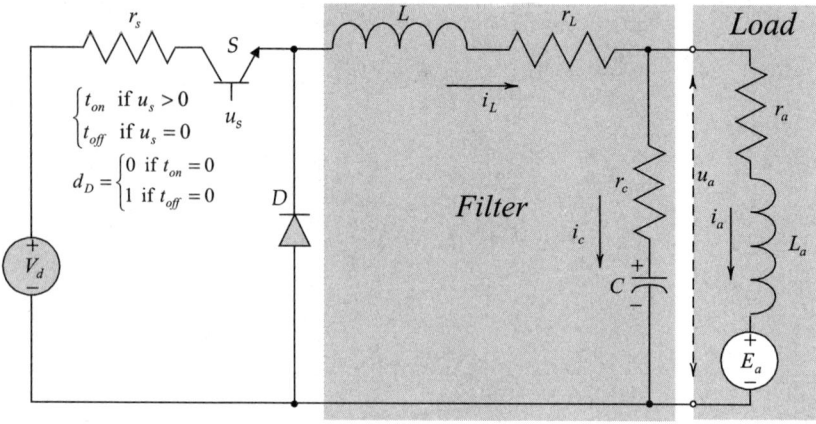

FIGURE 2.2.8.   Step-down switching converter.

$$\frac{d\omega_r}{dt} = \frac{1}{J}(-L_M i_s i_r \sin\theta_r - B_m \omega_r - k_s \theta_r - T_L),$$

$$\frac{d\theta_r}{dt} = \omega_r.$$

Some mathematical models can be linearized. However, for most of dynamic systems, the linearization cannot be performed. For example, the derived nonlinear differential equations cannot be linearized because the motor rotates, and the angular displacement θ_r varies from $-\infty$ to $+\infty$. The trigonometric functions (see terms $\sin\theta_r$ and $\cos\theta_r$) are not one-to-one functions. Hence, the linearization cannot be performed.

2.2.3.   State-Space Models in Matrix Forms: Examples of Model Developments for Simple Dynamic Systems

Let us illustrate the application of the state-space concept and express the mathematical models in the form of differential equations in the state-space form.

Example 2.2.9. Mathematical model developments for a *Buck* converter.

A buck (step-down) switching converter is illustrated in Figure 2.2.8. Switch S, inductor L, and capacitor C have resistances, which are denoted as r_s, r_L, and r_c. The load is formed by the resistor r_a and inductor L_a.

The switch S (high-frequency transistor) is opened and closed. The voltage at the load terminal is regulated by using a PWM switching. For lossless switch (if $r_s = 0$), the voltage across the diode D is equal to the supplied voltage V_d when the switch is closed, and zero when the switch is open. The voltage applied to the load u_a, is regulated by controlling the switching *on* and *off* durations (t_{on} and t_{off}), and the switching frequency is $\frac{1}{t_{on}+t_{off}}$. The *on* and *off* durations are controlled by u_s (which is the output signal of the driving circuitry), and if $u_s = 0$, the switch is closed, whereas if $u_s > 0$, the switch is open. The *averaging* concept is commonly used to model high-frequency converters. Making use of the duty ratio

$$d_D = \frac{t_{on}}{t_{on} + t_{off}},$$

$$d_D = \begin{cases} 0 \text{ if } t_{on} = 0 \\ 1 \text{ if } t_{off} = 0 \end{cases}, \qquad d_D \in [0\ 1],$$

using the averaging concept, if $r_s = 0$, one has

$$u_{dN} = \frac{t_{on}}{t_{on} + t_{off}} V_d = d_D V_d.$$

When the switch is closed, the diode is reverse biased, and using Kirchhoff's laws, we have for $t_{off} = 0$, $d_D = 1$

$$\frac{du_C}{dt} = \frac{1}{C}(i_L - i_a),$$

$$\frac{di_L}{dt} = \frac{1}{L}(-u_C - (r_L + r_c)i_L + r_c i_a - r_s i_L + V_d),$$

$$\frac{di_a}{dt} = \frac{1}{L_a}(u_C + r_c i_L - (r_a + r_c)i_a - E_a).$$

If the switch is open, the diode is forward biased. For $d_D = 0$ ($t_{on} = 0$), the following differential equations result:

$$\frac{du_C}{dt} = \frac{1}{C}(i_L - i_a),$$

$$\frac{di_L}{dt} = \frac{1}{L}(-u_C - (r_L + r_c)i_L + r_c i_a),$$

$$\frac{di_a}{dt} = \frac{1}{L_a}(u_C + r_c i_L - (r_a + r_c)i_a - E_a).$$

When the switch is closed, the duty ratio is 1, whereas if the switch is open, the duty ratio is 0. That is,

$$d_D = \begin{cases} 0 & \text{if } t_{on} = 0 \\ 1 & \text{if } t_{off} = 0. \end{cases}$$

Assuming that the switching frequency is high, one uses the duty ratio to finds the following set of differential equations to model the converter transients:

$$\frac{du_C}{dt} = \frac{1}{C}(i_L - i_a),$$

$$\frac{di_L}{dt} = \frac{1}{L}(-u_C - (r_L + r_c)i_L + r_c i_a - r_s i_L d_D + V_d d_D),$$

$$\frac{di_a}{dt} = \frac{1}{L_a}(u_C + r_c i_L - (r_a + r_c)i_a - E_a).$$

One concludes that the resulting buck converter is modeled by a set of nonlinear differential equations due to a nonlinear term $\frac{r_s}{L}i_L d_D$. The duty ratio can be considered as a control, and the converter output is the applied terminal voltage u_a, $u_a = u_C + r_c i_L - r_c i_a$. It should be emphasized that u_s should be considered as the input (control) signal because the duty ratio is adjusted by regulating u_s.

The derived mathematical model can be represented in the state-space form using the state variables and the control vector.

The vector of the state variables is

$$x = \begin{bmatrix} u_C \\ i_L \\ i_a \end{bmatrix},$$

and the control is $u = [d_D]$.

The state-space concept is widely used in analysis and modeling using the state, control (forcing function), and output variables. In addition, the state-space techniques are commonly applied in design and optimization of complex dynamic systems. Mathematical models of dynamic systems are found in the form of linear and nonlinear differential and difference equations. In general, a set of n first-order, linear ordinary differential equations with n-states $x \in \mathbb{R}^n$ and m-controls (forcing functions) $u \in \mathbb{R}^m$ is written as

$$\frac{dx_1}{dt} = a_{11}x_1 + a_{12}x_2 + \cdots + a_{1n-1}x_{n-1} + a_{1n}x_n + b_{11}u_1 + b_{12}u_2$$
$$+ \cdots + b_{1m-1}u_{m-1} + b_{1m}u_m,$$

$$\frac{dx_2}{dt} = a_{21}x_1 + a_{22}x_2 + \cdots + a_{2n-1}x_{n-1} + a_{2n}x_n + b_{21}u_1 + b_{22}u_2$$
$$+ \cdots + b_{2m-1}u_{m-1} + b_{2m}u_m,$$

$$\vdots$$

$$\frac{dx_{n-1}}{dt} = a_{n-11}x_1 + a_{n-12}x_2 + \cdots + a_{n-1n-1}x_{n-1} + a_{n-1n}x_n$$
$$+ b_{n-11}u_1 + b_{n-12}u_2 + \cdots + b_{n-1m-1}u_{m-1} + b_{n-1m}u_m,$$

$$\frac{dx_n}{dt} = a_{n1}x_1 + a_{n2}x_2 + \cdots + a_{nn-1}x_{n-1} + a_{nn}x_n + b_{n1}u_1 + b_{n2}u_2$$
$$+ \cdots + b_{nm-1}u_{m-1} + b_{nm}u_m.$$

These linear differential equations can be rewritten in matrix form as

$$\frac{dx}{dt} = \begin{bmatrix} \dfrac{dx_1}{dt} \\ \dfrac{dx_2}{dt} \\ \vdots \\ \dfrac{dx_{n-1}}{dt} \\ \dfrac{dx_n}{dt} \end{bmatrix} = \begin{bmatrix} a_{11} & a_{12} & \cdots & a_{1n-1} & a_{1n} \\ a_{21} & a_{22} & \cdots & a_{2n-1} & a_{2n} \\ \vdots & \vdots & \ddots & \vdots & \vdots \\ a_{n-11} & a_{n-12} & \cdots & a_{n-1n-1} & a_{n-1n} \\ a_{n1} & a_{n2} & \cdots & a_{nn-1} & a_{nn} \end{bmatrix} \begin{bmatrix} x_1 \\ x_2 \\ \vdots \\ x_{n-1} \\ x_n \end{bmatrix}$$

$$+ \begin{bmatrix} b_{11} & b_{12} & \cdots & b_{1m-1} & b_{1m} \\ b_{21} & b_{22} & \cdots & b_{2m-1} & b_{2m} \\ \vdots & \vdots & \ddots & \vdots & \vdots \\ b_{n-11} & b_{n-12} & \cdots & b_{n-1m-1} & b_{n-1m} \\ b_{n1} & b_{n2} & \cdots & b_{nm-1} & b_{nm} \end{bmatrix} \begin{bmatrix} u_1 \\ u_2 \\ \vdots \\ u_{m-1} \\ u_m \end{bmatrix}$$

$$= Ax + Bu$$

where $A \in \mathbb{R}^{n \times n}$ and $B \in \mathbb{R}^{n \times m}$ are the matrices of coefficients

$$A = \begin{bmatrix} a_{11} & a_{12} & \cdots & a_{1n-1} & a_{1n} \\ a_{21} & a_{22} & \cdots & a_{2n-1} & a_{2n} \\ \vdots & \vdots & \ddots & \vdots & \vdots \\ a_{n-11} & a_{n-12} & \cdots & a_{n-1n-1} & a_{n-1n} \\ a_{n1} & a_{n2} & \cdots & a_{nn-1} & a_{nn} \end{bmatrix}$$

$$\text{and } B = \begin{bmatrix} b_{11} & b_{12} & \cdots & b_{1m-1} & b_{1m} \\ b_{21} & b_{22} & \cdots & b_{2m-1} & b_{2m} \\ \vdots & \vdots & \ddots & \vdots & \vdots \\ b_{n-11} & b_{n-12} & \cdots & b_{n-1m-1} & b_{n-1m} \\ b_{n1} & b_{n2} & \cdots & b_{nm-1} & b_{nm} \end{bmatrix}.$$

For multi-input/multi-output dynamic systems (for example, aircraft, spacecraft, multi–degree-of-freedom robots, and servomechanisms), b outputs $y \in \mathbb{R}^b$ are obtained in terms of n-states $x \in \mathbb{R}^n$ and m-controls $u \in \mathbb{R}^m$. For example, in electric drives, the output is the angular velocity of the mechanism output shaft. It is evident that the output angular velocity of the shaft is proportional to the angular velocity of the electric motor. Using the gear ratio, one has $y = k_{\text{gear}}\omega_{\text{motor}}$, where ω_{motor} is the angular velocity of the motor, which is the state variable. In servo-systems, the output linear position or angular displacement are proportional to the angular rotor displacement, and the output equation is $y = k_{\text{gear}}\theta_{\text{rotor}}$.

In general, the output equation is expressed as

$$y_1 = h_{11}x_1 + h_{12}x_2 + \cdots + h_{1n-1}x_{n-1} + h_{1n}x_n,$$

$$y_2 = h_{21}x_1 + h_{22}x_2 + \cdots + h_{2n-1}x_{n-1} + h_{2n}x_n,$$

$$\vdots$$

$$y_{b-1} = h_{b-11}x_1 + h_{b-12}x_2 + \cdots + h_{b-1n-1}x_{n-1} + h_{b-1n}x_n,$$

$$y_b = h_{b1}x_1 + h_{b2}x_2 + \cdots + h_{bn-1}x_{n-1} + h_{bn}x_n$$

The following matrix form results:

$$
y = \begin{bmatrix} y_1 \\ y_2 \\ \vdots \\ y_{b-1} \\ y_b \end{bmatrix} = \begin{bmatrix} h_{11} & h_{12} & \cdots & h_{1n-1} & h_{1n} \\ h_{21} & h_{22} & \cdots & h_{2n-1} & h_{2n} \\ \vdots & \vdots & \ddots & \vdots & \vdots \\ h_{b-11} & h_{b-12} & \cdots & h_{b-1n-1} & h_{b-1n} \\ h_{b1} & h_{b2} & \cdots & h_{bn-1} & h_{bn} \end{bmatrix} \begin{bmatrix} x_1 \\ x_2 \\ \vdots \\ x_{n-1} \\ x_n \end{bmatrix} = Hx,
$$

$$
H = \begin{bmatrix} h_{11} & h_{12} & \cdots & h_{1n-1} & h_{1n} \\ h_{21} & h_{22} & \cdots & h_{2n-1} & h_{2n} \\ \vdots & \vdots & \ddots & \vdots & \vdots \\ h_{b-11} & h_{b-12} & \cdots & h_{b-1n-1} & h_{b-1n} \\ h_{b1} & h_{b2} & \cdots & h_{bn-1} & h_{bn} \end{bmatrix},
$$

where $H \in \mathbb{R}^{b \times n}$ is the matrices of coefficients.

The following state-space form:

$$
y = Hx + Du, \, D \in \mathbb{R}^{b \times m}
$$

can be used to describe the output equation.

That is, the system output is a function of states and control. However, for most dynamic systems, the output is not a function of the control inputs. Therefore,

$$
D = \begin{bmatrix} 0 & 0 & \cdots & 0 & 0 \\ 0 & 0 & \cdots & 0 & 0 \\ \vdots & \vdots & \ddots & \vdots & \vdots \\ 0 & 0 & \cdots & 0 & 0 \\ 0 & 0 & \cdots & 0 & 0 \end{bmatrix} \in \mathbb{R}^{b \times m}.
$$

Example 2.2.10.

Using the state-space concept, develop the state-space model of the series RLC circuit illustrated in Figure 2.2.9. The voltage across the capacitor is the circuit output. Find matrices A, B, H, and D of the state-space model and the output equation.

Using Kirchhoff's law, which gives

$$
C\frac{du_C}{dt} = i \text{ and } L\frac{di}{dt} = -u_C - Ri + u_a(t),
$$

a set of two first-order differential equations to model the circuitry dynamics is found as

$$
\frac{du_C}{dt} = \frac{1}{C}i,
$$

$$
\frac{di}{dt} = \frac{1}{L}(-u_C - Ri + u_a(t)).
$$

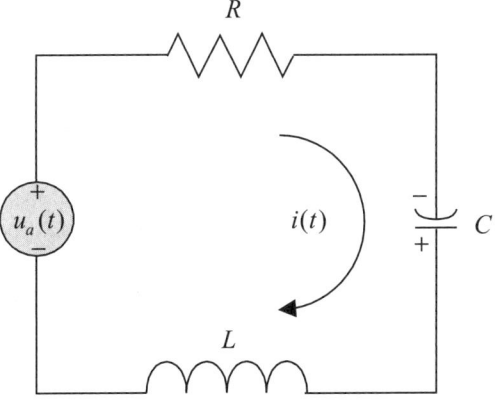

FIGURE 2.2.9.   *RLC* circuit.

We denote the state and control variables as

$$x_1(t) = u_C(t), x_2(t) = i(t) \text{ and } u(t) = u_a(t).$$

Thus, we have

$$\frac{dx}{dt} = \begin{bmatrix} \dfrac{dx_1}{dt} \\ \dfrac{dx_2}{dt} \end{bmatrix} = \begin{bmatrix} \dfrac{du_C}{dt} \\ \dfrac{di}{dt} \end{bmatrix} = \begin{bmatrix} 0 & \dfrac{1}{C} \\ -\dfrac{1}{L} & -\dfrac{R}{L} \end{bmatrix} \begin{bmatrix} u_C \\ i \end{bmatrix} + \begin{bmatrix} 0 \\ \dfrac{1}{L} \end{bmatrix} u_a = Ax + Bu.$$

The matrices of coefficients are found to be

$$A = \begin{bmatrix} 0 & \dfrac{1}{C} \\ -\dfrac{1}{L} & -\dfrac{R}{L} \end{bmatrix}$$

and

$$B = \begin{bmatrix} 0 \\ \dfrac{1}{L} \end{bmatrix}.$$

The voltage across the capacitor is the output. Hence, $y(t) = u_C(t)$.
The output equation is

$$y = \begin{bmatrix} 1 & 0 \end{bmatrix} \begin{bmatrix} u_C \\ i \end{bmatrix} + \begin{bmatrix} 0 \end{bmatrix} u_a = Hx + Du,$$

where $H = \begin{bmatrix} 1 & 0 \end{bmatrix}$ and $D = \begin{bmatrix} 0 \end{bmatrix}$.

Example 2.2.11. Mathematical model of permanent-magnet direct-current motors.

Our goal is to develop a mathematical model and build an s-domain block diagram of permanent-magnet DC motors. A schematic diagram of a permanent-magnet DC machine (motor and generator operation) is illustrated in Figure 2.2.10.

Assume that the *susceptibility* of permanent magnets is constant. Then, the flux, provided by the permanent magnets (poles), is time-invariant (the magnitude of the flux linkages is constant). By using Kirchhoff's voltage and Newton's second laws, the differential equations for permanent-magnet DC machines are derived using the motor circuitry and mechanical coupling illustrated in Figure 2.2.10. Denoting the *back emf* and *torque* constants as k_a, we have the following differential equations describing the armature winding and *torsional–mechanical* dynamics

$$\frac{di_a}{dt} = -\frac{r_a}{L_a}i_a - \frac{k_a}{L_a}\omega_r + \frac{1}{L_a}u_a \quad \text{(motor circuitry dynamics)},$$

$$\frac{d\omega_r}{dt} = \frac{k_a}{J}i_a - \frac{B_m}{J}\omega_r - \frac{1}{J}T_L \quad \text{(torsional–mechanical dynamics)}.$$

Augmenting these two first-order differential equations, in the state-space form,

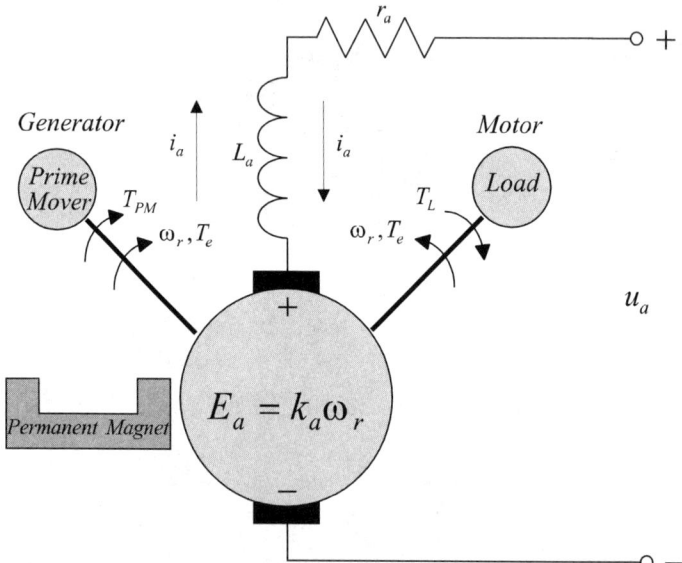

FIGURE 2.2.10.   Schematic diagram of a permanent-magnet DC motor/generator.

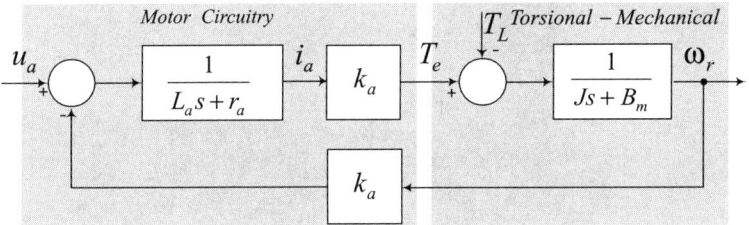

FIGURE 2.2.11.   Block diagram of permanent-magnet DC motors.

we have

$$
\begin{bmatrix} \dfrac{di}{dt} \\[2mm] \dfrac{d\omega_r}{dt} \end{bmatrix} = \begin{bmatrix} -\dfrac{r_a}{L_a} & -\dfrac{k_a}{L_a} \\[2mm] \dfrac{k_a}{J} & -\dfrac{B_m}{J} \end{bmatrix} \begin{bmatrix} i_a \\[2mm] \omega_r \end{bmatrix} + \begin{bmatrix} \dfrac{1}{L_a} \\[2mm] 0 \end{bmatrix} u_a - \begin{bmatrix} 0 \\[2mm] \dfrac{1}{J} \end{bmatrix} T_L.
$$

An s-domain block diagram of permanent-magnet DC motors is developed; see Figure 2.2.11.

3

Modeling of Dynamic Systems using MATLAB and SIMULINK

3.1. Engineering Computations Using MATLAB

The MATLAB environment, which allows one to integrate user-friendly tools with great computational capabilities, is found to be one of the most useful tools available to model complex dynamic systems, design control algorithms, optimize systems, accomplish data analysis and visualization, perform hardware-in-the-loop simulation, and deploy the developed control laws through generation of C-code using advanced microcontrollers and digital signal processors (DSPs). A family of application-specific toolboxes, with a specialized collection of m-files for solving different problems commonly encountered in engineering practice, provides comprehensiveness in linear and nonlinear analysis and design, system optimization, and synthesis. A great number of books in MATLAB and SIMULINK are available. In addition to the demonstrations and viable help available, the Math-Works educational website can be used for references; see `http://education.mathworks.com` and `http://www.mathworks.com`. This section is intended to help the designer to use MATLAB efficiently, and the author does not intend to substitute the MATLAB books available. Only the most basic features will be demonstrated to introduce MATLAB and SIMULINK as well as to show how they can be applied.

To start MATLAB, double click on the appropriate icon. After each MATLAB command, the Enter (Return) key must be pressed. Interact with MATLAB using the Command Window. The MATLAB prompt ≫ is displayed in the Command Window, and a blinking cursor appears to the right of the prompt when the Command Window is active. Type `ver`, and we have the information regarding the MATLAB version and the MATLAB toolboxes available; see Figure 3.1.1.

The use of the help and demo commands is the simplest way to find the information and help needed. Type

```
≫ help cos
```

and press the Enter key, and the following information is displayed:

```
COS    Cosine.
COS(X) is the cosine of the elements of X
```

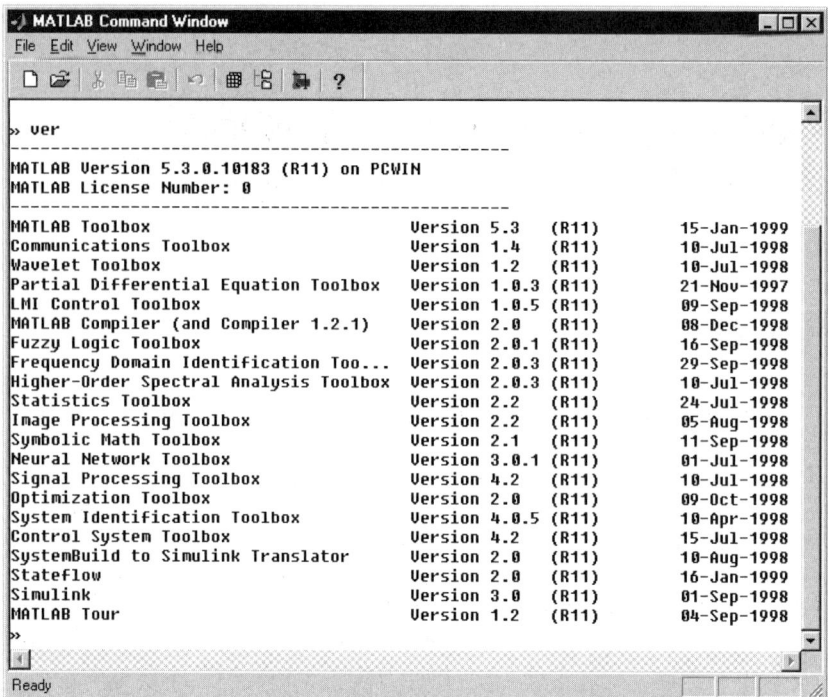

FIGURE 3.1.1.   MATLAB Command Window: MATLAB version and MATLAB toolboxes.

MATLAB offers the following help topics:

- help datafun (data analysis)
- help demo (demonstration)
- help funfun (differential equations solvers)
- help general (general purpose command)
- help graph2d and help graph3d (two- and three-dimensional graphics)
- help elmat and help matfun (matrices and linear algebra)
- help elfun and help specfun (mathematical functions)
- help lang (programming language)
- help ops (operators and special characters)
- help polyfun (polynomials)

The MATLAB Demos Window is illustrated in Figure 3.1.2.

By choosing the subtopics (Matrices, Numerics, Visualization, Language/ Graphics, Gallery, Games, Miscellaneous, and To learn more), different topics will be explained and thoroughly covered. For example, clicking the subtopic Matrices, we have the Matrices MATLAB Demos Window, as documented in Figure 3.1.3.

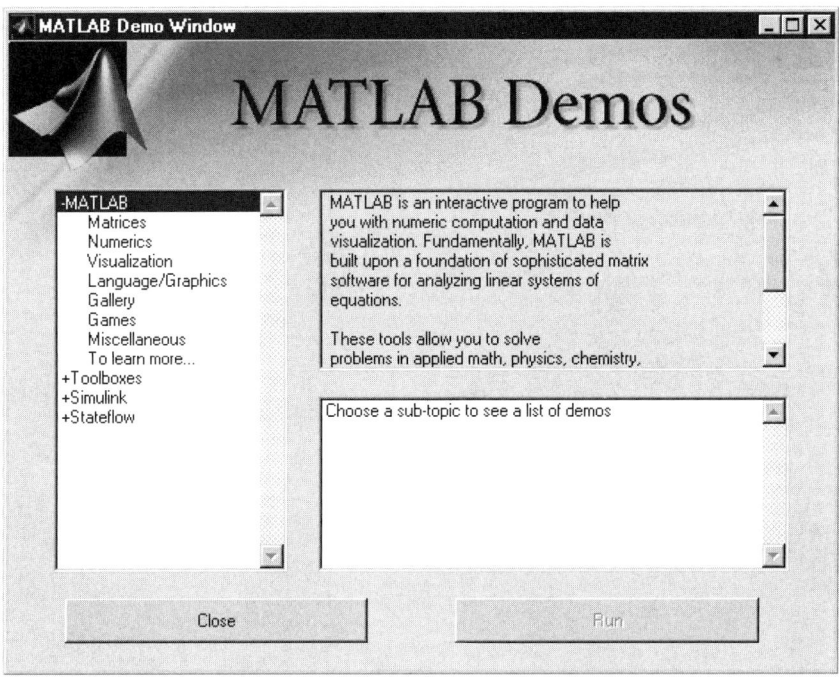

FIGURE 3.1.2.   MATLAB Demos Window.

By double clicking Basic matrix operations, Inverse of matrices, Graphs and matrices, Sparse matrices, Matrix multiplication, Eigenvalues & singular value show, and Command line demos, illustrative examples are available to explore different problems.

This book is devoted to control, and the control-oriented toolboxes are available. The commonly used toolbox is the Control System Toolbox, and the user can practice many examples that allow one to quickly learn how to efficiently use MATLAB to solve a wide variety of problems in analysis of systems in time and frequency domains, nonlinear modeling and simulation, control and optimization, filtering and data analysis, and so forth. The use of the Control System Toolbox allows the designer to quickly learn the MATLAB capabilities; see Model Analysis Example, Digital Control of a Disk Drive, and Command Line Demos in Figure 3.1.4.

To illustrate the basic arithmetic operations (addition, subtraction, multiplication, division, and exponentiation), we calculate $\frac{1+2-4}{5\times6-7^8}$. In the MATLAB Command Window, we type

```
>> (1+2-4)/(5*6-7^8)
```

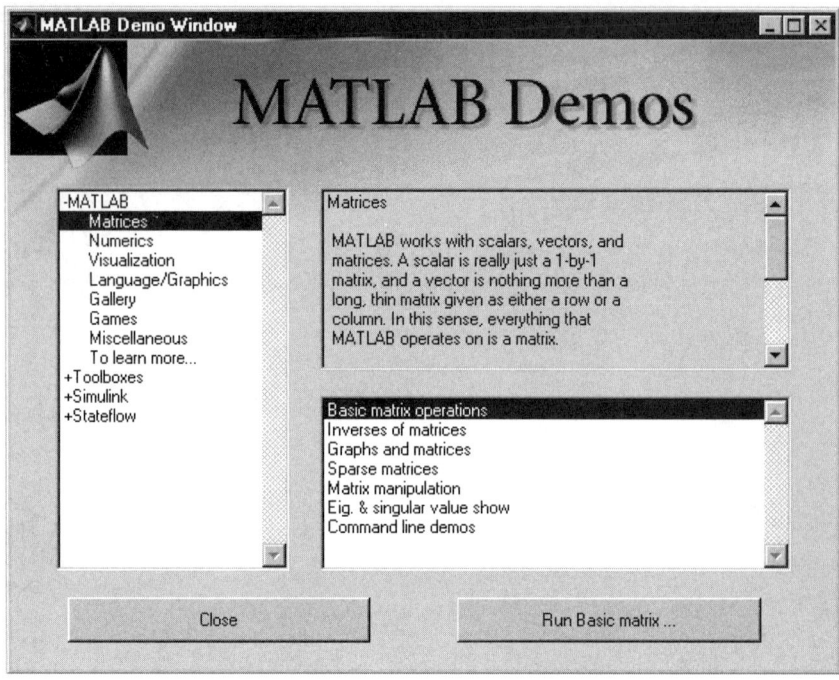

FIGURE 3.1.3.   Matrices MATLAB Demos Window.

and press the Enter key; the answer appears

```
ans =
   1.7347e-007
```

As an alternative, one can use variables; in particular, for the example considered, we type

```
≫a=1;b=2;c=4;d=5;e=6;f=7;g=8;(a+b-c)/(d*e-f^g)
```

then press the Enter key, and the answer is displayed

```
ans =
   1.7347e-007
```

To display the value of variables used, one should enter the variable name at the prompt and press the Enter key; for example,

```
≫ b,g
b =
   2
g =
   8
```

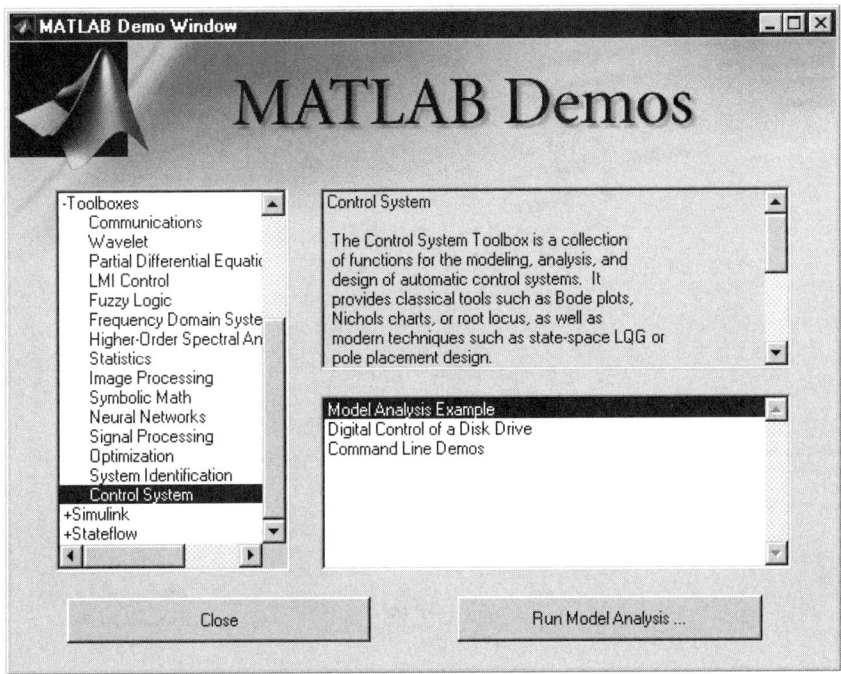

FIGURE 3.1.4. MATLAB Demos Window: Control System Toolbox.

A list of the variables used are obtained using the command who. One has

```
≫ who
Your variables are:
a        b        d        f
ans      c        e        g
```

The basic complex and coordinate transformations, exponential, remainder, rounding, and specialized mathematic and trigonometric functions are supported by MATLAB. For example, to calculate and plot the following function $f(t) = \sin(10t) + \sin(10t)\cos(20t + 1)e^{-5t}$ for $0 \le t \le 0.1$ sec, one has to assign the time interval of interest, calculate $f(t)$, and plot this function using the plot command. We have

```
≫ t=linspace(0,.1,100);
f=sin(10*t)+sin(10*t).*cos(20*t+1). *exp(-5*t); plot(t,f)
```

The plot of $f(t)$ is illustrated in Figure 3.1.5.

Polynomial analysis, curve fitting, and interpolation are easily performed. Consider a polynomial $p_1(x) = x^8 + 2x^7 + 3x^6 + 4x^5 + 5x^4 + 7x^3 + 8x^2 + 9x + 10$.

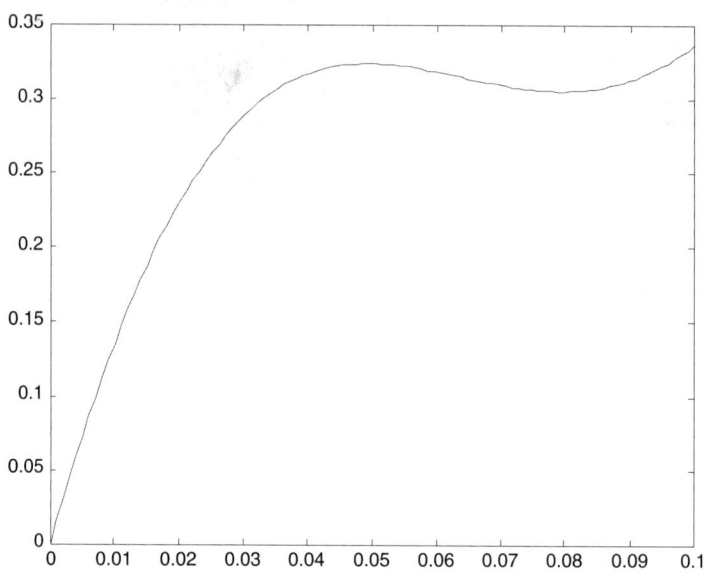

FIGURE 3.1.5.   Plot of function $f(t) = \sin(10t) + \sin(10t)\cos(20t + 1)e^{-5t}$.

To find the roots, we use the `roots` command. In particular,

```
≫ p=[1 2 3 4 5 7 8 9 10]; roots(p)
ans =
    -1.3078 + 0.4022i
    -1.3078 - 0.4022i
    -0.7181 + 1.1938i
    -0.7181 - 1.1938i
     0.9086 + 0.9127i
     0.9086 - 0.9127i
     0.1173 + 1.2828i
     0.1173 - 1.2828i
```

Four different formats are supported by MATLAB. The numerical values are displayed in 15-digit fixed and 15-digit floating point formats if `format long` and `format long e` are used. Five digits are displayed if `format short` and `format short e` are assigned to be used. For example, in `format short e`, we have

```
≫ format short e; p=[1 2 3 4 5 7 8 9 10]; roots(p)
ans =
```

$$-1.3078e + 000 + 4.0224e - 001i$$
$$-1.3078e + 000 - 4.0224e - 001i$$
$$-7.1810e - 001 + 1.1938e + 000i$$
$$-7.1810e - 001 - 1.1938e + 000i$$
$$9.0863e - 001 + 9.1267e - 001i$$
$$9.0863e - 001 - 9.1267e - 001i$$
$$1.1731e - 001 + 1.2828e + 000i$$
$$1.1731e - 001 - 1.2828e + 000i$$

For matrix

$$A = \begin{bmatrix} 1 & 2 & 3 \\ 0 & 4 & 5 \\ 6 & 7 & 8 \end{bmatrix},$$

find the inverse matrix, calculate the eigenvalues, derive $B = 2AA^{-1}$, and find the determinant of B. Using the inv, eig, and det commands, we have

```
>>A=[123;045;678];B=2*A*(A)^(-1);inv(A),eig(A),B,det(B)
ans =
     0.2000   -0.3333    0.1333
    -2.0000    0.6667    0.3333
     1.6000   -0.3333   -0.2667
ans =
    -1.3811
     0.7997
    13.5814
B =
     2.0000   0.0000   0.0000
          0   2.0000   0.0000
    -0.0000   0.0000   2.0000
ans =
     8.0000
```

That is,

$$A^{-1} = \begin{bmatrix} 0.2 & -0.33 & 0.13 \\ -2 & 0.67 & 0.33 \\ 1.6 & -0.33 & -0.27 \end{bmatrix},$$

the eigenvalues are $-1.38, 0.8, 13.58$, and

$$B = \begin{bmatrix} 2 & 0 & 0 \\ 0 & 2 & 0 \\ 0 & 0 & 2 \end{bmatrix}.$$

The determinant is 8.

To solve the linear algebraic equation $Ax = B$, one finds the solution as given by $x = A^{-1}B$.

Using MATLAB, and assigning

$$A = \begin{bmatrix} 1 & 2 & 3 \\ 0 & 4 & 5 \\ 6 & 7 & 8 \end{bmatrix}$$

and

$$B = \begin{bmatrix} 10 \\ 20 \\ 30 \end{bmatrix},$$

we can find x. In particular,

```
>> A= [1 2 3;0 4 5;6 7 8]; B= [10; 20; 30]; x=inv(A)*B
x =
    -0.6667
    3.3333
    1.3333
```

That is, the following solution is obtained:

$$x = \begin{bmatrix} -0.67 \\ 3.33 \\ 1.33 \end{bmatrix}.$$

3.2.   Analysis and Modeling of Dynamic Systems Using MATLAB

Linear and nonlinear differential equations should be solved to perform the analysis of dynamic systems. In the state-space vector form, n-states $x \in \mathbb{R}^n$ and m-controls $u \in \mathbb{R}^m$ are used. In particular, mathematical models of multivariable systems are found as n first-order differential equations. The transient dynamics of linear continuous-time systems is modeled as

$$\frac{dx_1}{dt} = a_{11}x_1 + a_{12}x_2 + \cdots + a_{1n-1}x_{n-1} + a_{1n}x_n + b_{11}u_1 + b_{12}u_2$$
$$+ \cdots + b_{1m-1}u_{m-1} + b_{1m}u_m, \ x_1(t_0) = x_{10},$$

$$\frac{dx_2}{dt} = a_{21}x_1 + a_{22}x_2 + \cdots + a_{2n-1}x_{n-1} + a_{2n}x_n + b_{21}u_1 + b_{22}u_2$$
$$+ \cdots + b_{2m-1}u_{m-1} + b_{2m}u_m, \ x_2(t_0) = x_{20},$$

$$\vdots$$

$$\frac{dx_{n-1}}{dt} = a_{n-11}x_1 + a_{n-12}x_2 + \cdots + a_{n-1n-1}x_{n-1} + a_{n-1n}x_n$$
$$+ b_{n-11}u_1 + b_{n-12}u_2 + \cdots + b_{n-1m-1}u_{m-1}$$
$$+ b_{n-1m}u_m, \ x_{n-1}(t_0) = x_{n-10},$$

$$\frac{dx_n}{dt} = a_{n1}x_1 + a_{n2}x_2 + \cdots + a_{nn-1}x_{n-1} + a_{nn}x_n + b_{n1}u_1 + b_{n2}u_2$$
$$+ \cdots + b_{nm-1}u_{m-1} + b_{nm}u_m, \ x_n(t_0) = x_{n0},$$

which is represented in matrix form as

$$
\frac{dx}{dt} =
\begin{bmatrix}
\dfrac{dx_1}{dt} \\[2mm]
\dfrac{dx_2}{dt} \\[2mm]
\vdots \\[2mm]
\dfrac{dx_{n-1}}{dt} \\[2mm]
\dfrac{dx_n}{dt}
\end{bmatrix}
=
\begin{bmatrix}
a_{11} & a_{12} & \cdots & a_{1n-1} & a_{1n} \\
a_{21} & a_{22} & \cdots & a_{2n-1} & a_{2n} \\
\vdots & \vdots & \ddots & \vdots & \vdots \\
a_{n-11} & a_{n-12} & \cdots & a_{n-1n-1} & a_{n-1n} \\
a_{n1} & a_{n2} & \cdots & a_{nn-1} & a_{nn}
\end{bmatrix}
\begin{bmatrix}
x_1 \\ x_2 \\ \vdots \\ x_{n-1} \\ x_n
\end{bmatrix}
$$

$$
+
\begin{bmatrix}
b_{11} & b_{12} & \cdots & b_{1m-1} & b_{1m} \\
b_{21} & b_{22} & \cdots & b_{2m-1} & b_{2m} \\
\vdots & \vdots & \ddots & \vdots & \vdots \\
b_{n-11} & b_{n-12} & \cdots & b_{n-1m-1} & b_{n-1m} \\
b_{n1} & b_{n2} & \cdots & b_{nm-1} & b_{nm}
\end{bmatrix}
\begin{bmatrix}
u_1 \\ u_2 \\ \vdots \\ u_{m-1} \\ u_m
\end{bmatrix}
= Ax + Bu,
$$

where x is the state vector with initial conditions $x(t_0) = x_0$, u is the control vector, and $A \in \mathbb{R}^{n \times n}$ and $B \in \mathbb{R}^{n \times m}$ are the matrices of constant coefficients.

Nonlinear multivariable dynamic systems are modeled by a set of n first-order nonlinear differential equations

$$
\frac{dx_1}{dt} = F_1(t, x_1, x_2, \ldots, x_{n-1}, x_n, u_1, u_2, \ldots, u_{m-1}, u_m), \; x_1(t_0) = x_{10},
$$

$$
\frac{dx_2}{dt} = F_2(t, x_1, x_2, \ldots, x_{n-1}, x_n, u_1, u_2, \ldots, u_{m-1}, u_m), \; x_2(t_0) = x_{20},
$$

$$
\vdots
$$

$$
\frac{dx_{n-1}}{dt} = F_{n-1}(t, x_1, x_2, \ldots, x_{n-1}, x_n, u_1, u_2, \ldots, u_{m-1}, u_m), \; x_{n-1}(t_0) = x_{n-10},
$$

$$
\frac{dx_n}{dt} = F_n(t, x_1, x_2, \ldots, x_{n-1}, x_n, u_1, u_2, \ldots, u_{m-1}, u_m), \; x_n(t_0) = x_{n0},
$$

or $\quad \dfrac{dx}{dt} = F(t, x, u), \; x(t_0) = x_0,$

where t is the time and $F(t, x, u)$ is the nonlinear map.

Hence, for n–degree-of-freedom dynamic systems, we use:

- n-state variables (for example, current, voltage, angular velocity, displacement, and acceleration)

$$x = \begin{bmatrix} x_1 \\ x_2 \\ \vdots \\ x_{n-1} \\ x_n \end{bmatrix}$$

- m-control inputs (for example, voltages applied to motor windings or deflection of control surfaces)

$$u = \begin{bmatrix} u_1 \\ u_2 \\ \vdots \\ u_{m-1} \\ u_m \end{bmatrix}$$

In addition to the states and controls, the disturbance vector (denoted by d) is considered. The l-dimensional disturbance vector (for example, the load and disturbances torques and forces) is

$$d = \begin{bmatrix} d_1 \\ d_2 \\ \vdots \\ d_{l-1} \\ d_l \end{bmatrix}.$$

Multi-input/multi-output dynamic systems are studied. In Chapter 2, we considered the formulation of the motion control problem for the aircraft. The aircraft outputs are the Euler angles, and the flight vehicle is controlled by deflecting the control surfaces (if the aircraft is studied as a rigid-body mechanical system and flight actuators are not integrated) in the analysis. The multi-input (eight control surfaces)–multi-output (three Euler angles θ, ϕ, and ψ to be controlled) nature is obvious. The pilot assigns the desired Euler angles r_θ, r_ϕ, and r_ψ (reference commands). Using the errors between the reference vector

$$r = \begin{bmatrix} r_\theta \\ r_\phi \\ r_\psi \end{bmatrix}$$

and output vector

$$y = \begin{bmatrix} \theta \\ \phi \\ \psi \end{bmatrix},$$

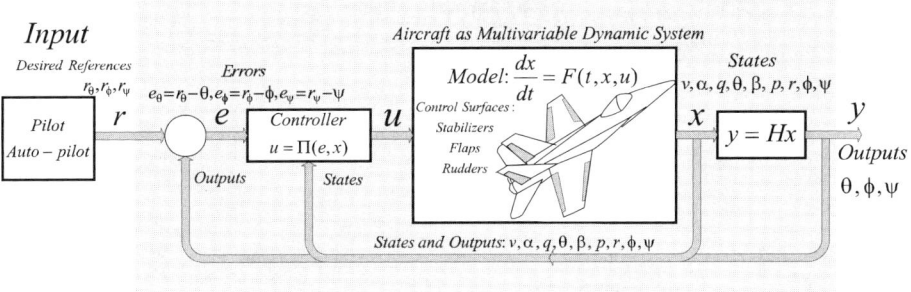

FIGURE 3.2.1.   Block diagram representation of a multi-input/multi-output aircraft.

as defined by

$$
e = r - y = \begin{bmatrix} r_\theta \\ r_\phi \\ r_\psi \end{bmatrix} - \begin{bmatrix} \theta \\ \phi \\ \psi \end{bmatrix},
$$

the controller $u = \Pi(e, x)$ calculates the control inputs (control surface deflections). The aircraft outputs (θ, ϕ, and ψ) can be obtained by using the state variables ($v, \alpha, q, \theta, \beta, p, r, \phi$, and ψ), and the output equation is

$$
y = \begin{bmatrix} \theta \\ \phi \\ \psi \end{bmatrix} = \begin{bmatrix} 0 & 0 & 0 & 1 & 0 & 0 & 0 & 0 & 0 \\ 0 & 0 & 0 & 0 & 0 & 0 & 0 & 1 & 0 \\ 0 & 0 & 0 & 0 & 0 & 0 & 0 & 0 & 1 \end{bmatrix} \begin{bmatrix} v \\ \alpha \\ q \\ \theta \\ \beta \\ p \\ r \\ \phi \\ \psi \end{bmatrix} = Hx.
$$

Figure 3.2.1 illustrates the block diagram representation of the multivariable aircraft with nine states $x \in \mathbb{R}^9 (v, \alpha, q, \theta, \beta, p, r, \phi, \psi)$, eight control surfaces $u \in \mathbb{R}^8$ (right and left horizontal stabilizers, right and left leading and trailing edge flaps, right and left rudders), three outputs $y \in \mathbb{R}^3 (\theta, \phi, \psi)$, and three reference inputs $r \in \mathbb{R}^3 (r_\theta, r_\phi, r_\psi)$.

In electric drives, the output angular velocity is proportional to the rotor angular velocity ω_{motor}, and $y = k_{\text{gear}} \omega_{\text{motor}}$. In servo-systems, the output is ether the linear position or the angular displacement, which is proportional to the angular rotor displacement θ_{rotor}. Hence, the output equation is $y = k_{\text{gear}} \theta_{\text{rotor}}$.

In general, for multi-input/multi-output dynamic systems, one has b reference inputs $r \in \mathbb{R}^b$ and b outputs $y \in \mathbb{R}^b$ (obtained in terms of n states $x \in \mathbb{R}^n$). The

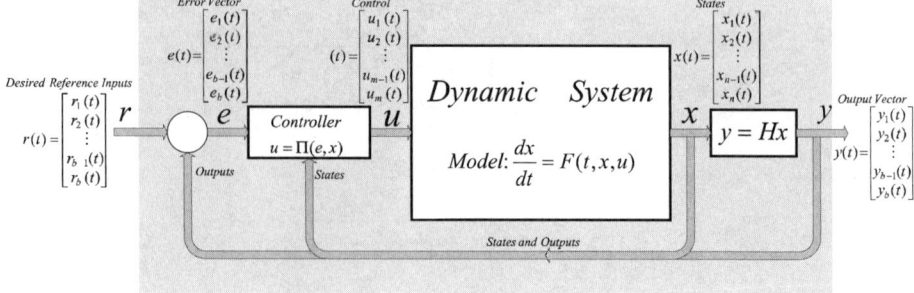

FIGURE 3.2.2.   Functional block diagram of multi-input/multi-output dynamic systems.

output equation is

$$
y_1 = h_{11}x_1 + h_{12}x_2 + \cdots + h_{1n-1}x_{n-1} + h_{1n}x_n,
$$
$$
y_2 = h_{21}x_1 + h_{22}x_2 + \cdots + h_{2n-1}x_{n-1} + h_{2n}x_n,
$$
$$
\vdots
$$
$$
y_{b-1} = h_{b-11}x_1 + h_{b-12}x_2 + \cdots + h_{b-1n-1}x_{n-1} + h_{b-1n}x_n,
$$
$$
y_b = h_{b1}x_1 + h_{b2}x_2 + \cdots + h_{bn-1}x_{n-1} + h_{bn}x_n,
$$

and

$$
y = \begin{bmatrix} y_1 \\ y_2 \\ \vdots \\ y_{b-1} \\ y_b \end{bmatrix} = \begin{bmatrix} h_{11} & h_{12} & \cdots & h_{1n-1} & h_{1n} \\ h_{21} & h_{22} & \cdots & h_{2n-1} & h_{2n} \\ \vdots & \vdots & \ddots & \vdots & \vdots \\ h_{b-11} & h_{b-12} & \cdots & h_{b-1n-1} & h_{b-1n} \\ h_{b1} & h_{b2} & \cdots & h_{bn-1} & h_{bn} \end{bmatrix} \begin{bmatrix} x_1 \\ x_2 \\ \vdots \\ x_{n-1} \\ x_n \end{bmatrix} = Hx,
$$

where $H \in \mathbb{R}^{b \times n}$ is the matrix of coefficients.

The functional block diagram of nonlinear multivariable dynamic systems (n-states, m-controls, b-reference inputs and b-outputs), which are described by the state-space equation

$$
\frac{dx}{dt} = F(t, x, u), x(t_0) = x_0
$$

with the output equation

$$
y = Hx,
$$

is illustrated in Figure 3.2.2.

Having described the basic notations commonly used, practical application of the MATLAB environment must be illustrated. In the MATLAB environment, the differential equations solvers ode113, ode15s, ode23, ode23s, and ode45 are widely used. These solvers allow one to solve linear and nonlinear differential equations given in Cauchy's form. The commonly applied ode45 solver uses the fourth/fifth-order Runge–Kutta method.

MATLAB and SIMULINK m- and mdl-files are macros that are stored as ordinary files with the extension m or mdl (filename.m or filename.mdl). An m-file can be either a function with input and output variables or a list of commands. These m- and mdl-files must be stored either in the working directory or in a directory that is specified in the MATLAB path list. If the user stores the files in the specified directory DIRECTORY on the j drive, to access to m- and mdl-files, one should go to the directory typing j:\DIRECTORY in the Command Window. Alternatively, the directory can be assigned in the path.

Example 3.2.1.

Solve a system of highly nonlinear differential equations

$$\frac{dx_1(t)}{dt} = -5x_1 + 4|x_2|, \, x_1(t_0) = x_{10},$$

$$\frac{dx_2(t)}{dt} = -3x_1x_2 + 2\sin x_1 - x_2 + x_3, \, x_2(t_0) = x_{20},$$

$$\frac{dx_3(t)}{dt} = -x_1x_2 + 10x_2\cos x_1 - 5x_3, \, x_3(t_0) = x_{30}.$$

Two m-files (c_3_2_1a.m and c_3_2_1b.m) are developed, and modeling was performed assigning the initial conditions to be

$$x_0 = \begin{bmatrix} x_{10} \\ x_{20} \\ x_{30} \end{bmatrix} = \begin{bmatrix} 10 \\ -10 \\ 5 \end{bmatrix}.$$

Two- and three-dimensional graphics illustrate the MATLAB visualization capabilities. These two- and three-dimensional plots are found to be useful in addition to the transient analysis. For the example studied, a three-dimensional plot is obtained using x_1, x_2, and x_3 as the variables to be analyzed; see the plot3 command. Comments, which are not executed, appear after the % symbol. These comments explain particular steps in MATLAB scripts. See, for example, the comments % initial and final time and % initial conditions.

MATLAB ode45 and plot3 script (c_3_2_1a.m)

```
echo on; clear all
tspan=[0 2];            % initial and final time
y0= [10 -10 5]';        % initial conditions
[t,y]=ode45('c_3_2_1b',tspan,y0);       %ode45 MATLAB solver
% Plot of the time history found solving differential
% equations
plot(t,y(:,1),'--',t,y(:,2),'-',t,y(:,3),':');
xlabel('Time (seconds)');
```

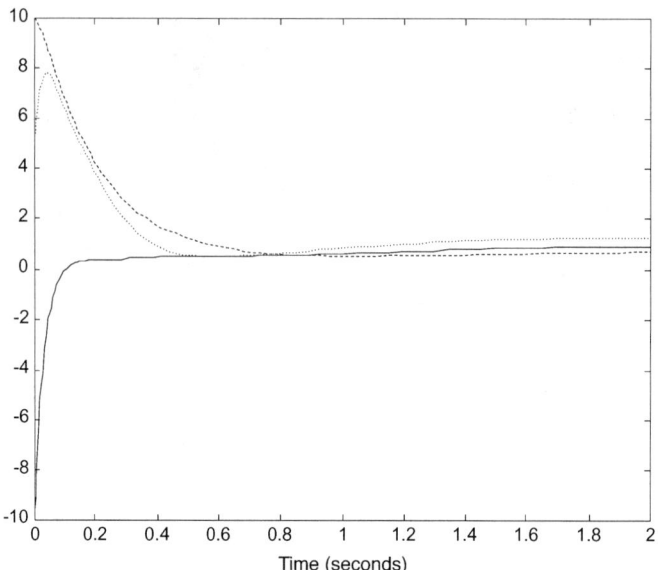

FIGURE 3.2.3.   Evolution of the state variables $x_0 = \begin{bmatrix} x_{10} \\ x_{20} \\ x_{30} \end{bmatrix} = \begin{bmatrix} 10 \\ -10 \\ 5 \end{bmatrix}$.

```
title('Differential equation time history');
pause
% 3-D plot w(y1,y2,y3)
plot3(y(:,1),y(:,2),y(:,3))
xlabel('x1'), ylabel('x2'), zlabel('x3')
text(10,-10,5,'X0 Initial')
text(0,0,0,'0 Origin')
v=axis
pause; disp('END')
```

MATLAB script (c_3_2_1b.m)

```
function yprime = difer(t,y);
a11=-5; a12=4; a21=-3; a22=2; a31=-1; a32=10; a33=-5;
yprime=[a11*y(1,:)+a12*abs(y(2,:));...
a21*y(1,:)*y(2,:)+a22*(sin(y(1,:))-y(2,:)+y(3,:));...
a31*y(1,:)*y(2,:)+a32*cos(y(1,:))*y(2,:)+a33*y(3,:)];
```

To calculate the transient dynamics, type in the Command Window ≫c_3_2_1a and press the Enter key. The resulting transient behavior and three-dimensional plot result; see Figures 3.2.3 and 3.2.4.

The transfer function concept is thoroughly covered in the undergraduate engineering curriculum. The transfer function of a linear, time-invariant (real constant

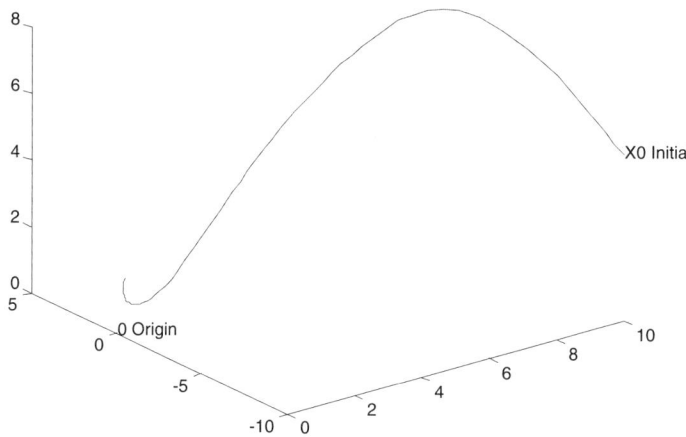

FIGURE 3.2.4.   Three-dimensional plot.

coefficients) n-order differential equation

$$a_n \frac{d^n y(t)}{dt^n} + a_{n-1} \frac{d^{n-1} y(t)}{dt^{n-1}} + \cdots + a_1 \frac{dy(t)}{dt} + a_0 = b_m \frac{d^m u(t)}{dt^m}$$

$$+ b_{m-1} \frac{d^{m-1} u(t)}{dt^{m-1}} + \cdots + b_1 \frac{du(t)}{dt} + b_0, \quad n \geq m,$$

$$\sum_{i=0}^{n} a_i \frac{d^i y(t)}{dt^i} = \sum_{i=0}^{m} b_i \frac{d^i u(t)}{dt^i}, \quad n \geq m,$$

is the ratio of the Laplace transform of the system output $y(t)$ to the Laplace transform of the system input $u(t)$ with initial conditions set to zero.

Assuming that initial conditions are zero, we apply the Laplace transform to both sides of equation $\sum_{i=0}^{n} a_i \frac{d^i y(t)}{dt^i} = \sum_{i=0}^{m} b_i \frac{d^i u(t)}{dt^i}$. Thus, we have

$$\left(\sum_{i=0}^{n} a_i s^i \right) Y(s) = \left(\sum_{i=0}^{m} b_i s^i \right) U(s).$$

Hence, the transfer function is the ratio of polynomials, and

$$G(s) = \frac{Y(s)}{U(s)} = \frac{b_m s^m + b_{m-1} s^{m-1} + \cdots + b_1 s + b_0}{a_n s^n + a_{n-1} s^{n-1} + \cdots + a_1 s + a_0}.$$

By setting the denominator polynomial of the transfer function to zero, one obtains the characteristic equation. The stability of linear systems is guaranteed if all characteristic eigenvalues (characteristic roots), obtained by solving the characteristic equations

$$a_n s^n + a_{n-1} s^{n-1} + \cdots + a_1 s + a_0 = 0,$$

have negative real parts.

The transfer function

$$G(s) = \frac{Y(s)}{U(s)}$$

can be found using the state-space equations.

Consider the linear time-invariant dynamic systems modeled as

$$\frac{dx}{dt} = Ax + Bu \quad \text{(the state-space equation)}$$

and

$$y = Hx \quad \text{(the output equation)}.$$

Taking the Laplace transform of the state-space and output equations, we have

$$sX(s) - x(t_0) = AX(s) + BU(s), \, Y(s) = HX(s).$$

Assuming that $x(t_0) = 0$, one finds

$$X(s) = (sI - A)^{-1} BU(s).$$

Thus, $Y(s) = H(sI - A)^{-1} BU(s)$, and hence the following transfer function results:

$$G(s) = \frac{Y(s)}{U(s)} = H(sI - A)^{-1} B.$$

Transfer functions

$$G(s) = \frac{Y(s)}{U(s)}$$

can be found from

$$\frac{dx}{dt} = Ax + Bu, \, y = Hx$$

using the ss2tf command. The transfer function can be converted to the differential equations using the tf2ss command.

From the output equation $y = Hx + Du$, $D \in \mathbb{R}^{b \times m}$, one concludes that the system output is a function of states and control. In real-world systems, the output usually is not a function of the control, and therefore,

$$D = \begin{bmatrix} 0 & 0 & \cdots & 0 & 0 \\ 0 & 0 & \cdots & 0 & 0 \\ \vdots & \vdots & \ddots & \vdots & \vdots \\ 0 & 0 & \cdots & 0 & 0 \\ 0 & 0 & \cdots & 0 & 0 \end{bmatrix} \in \mathbb{R}^{b \times m}.$$

However, in the MATLAB environment, the general form of the output equation $y = Hx + Du$ is used.

Example 3.2.2.

In Chapter 4, it is shown that the transfer function

$$G(s) = \frac{X(s)}{U(s)} = \frac{1}{s^6 + 3.3s^5\omega_n + 6.6s^4\omega_n^2 + 8.6s^3\omega_n^3 + 7.5s^2\omega_n^4 + 4s\omega_n^5 + \omega_n^6}$$

minimizes the transient responses with regard the integral of time and error.
Let $\omega_n = 5$. Then, we have

$$G(s) = \frac{X(s)}{U(s)} = \frac{1}{s^6 + 16.5s^5 + 165s^4 + 1075s^3 + 4688s^2 + 12500s + 15625}.$$

Let us find the eigenvalues and analyze the stability.
From

$$(s^6 + 16.5s^5 + 165s^4 + 1075s^3 + 4688s^2 + 12500s + 15625)X(s) = U(s),$$

one finds the following sixth-order differential equation

$$\frac{d^6x}{dt^6} + 16.5\frac{d^5x}{dt^5} + 165\frac{d^4x}{dt^4} + 1075\frac{d^3x}{dt^3} + 4688\frac{d^2x}{dt^2} + 12500\frac{dx}{dt} + 15625x = u.$$

The characteristic equation is

$$s^6 + 16.5s^5 + 165s^4 + 1075s^3 + 4688s^2 + 12500s + 15625 = 0.$$

To find the eigenvalues (characteristic roots), we use the `roots` command. Type

```
≫den=[1 16.5 165 1075 4688 12500 15625];eigenvalues=roots(den)
```

and press the Enter key, and the following roots are displayed:

```
eigenvalues=
  -1.3159 + 6.1956i
  -1.3159 - 6.1956i
  -3.2119 + 3.8261i
  -3.2119 - 3.8261i
  -3.7221 + 1.3238i
  -3.7221 - 1.3238I
```

The stability is guaranteed because the real parts of all eigenvalues are negative.
The eigenvalues map in the s-plane is found using the `pznap` command. By typing

```
≫den=[1 16.5 165 1075 4688 12500 15625];num=[1];pzmap(num,den)
```

the eigenvalues map results, as given in Figure 3.2.5.
Our goal is to calculate the impulse and step responses and find the state-space model.

To perform the analysis of the transient dynamics, one uses the commands `impulse` and `step`. In particular, we have

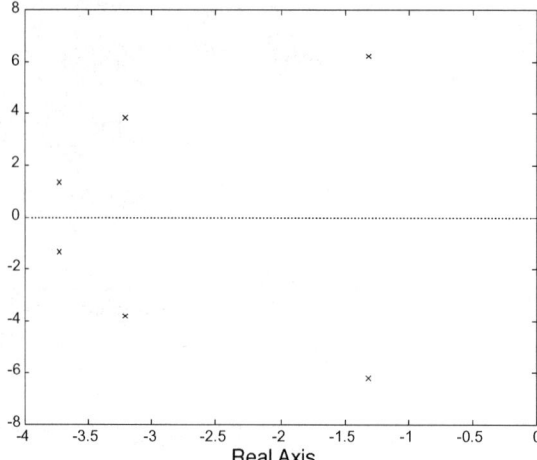

FIGURE 3.2.5.   Eigenvalues map.

```
>>den=[1 16.5 165 1075 4688 12500 15625]; num=[1];
Tfinal=4; impulse(num,den,Tfinal)
```

and

```
>>den=[1 16.5 165 1075 4688 12500 15625]; num=[1];
Tfinal=4; step(num,den,Tfinal)
```

The resulting transient dynamics are illustrated in Figures 3.2.6.

It should be emphasized that the solution can be found analytically applying the residue command. In particular, we have

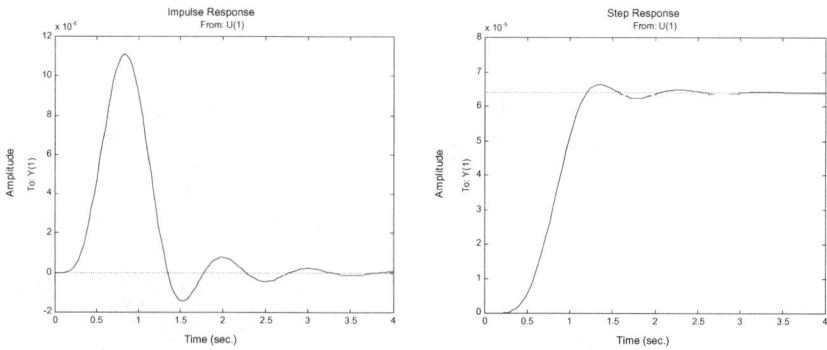

FIGURE 3.2.6.   System dynamics.

```
≫num=[1];den=[1 16.5 165 1075 4688 12500 15625];
[residues,eig,dc]=residue(num,den)
residues =
  1.0e-003  *
    0.0607 + 0.0036i
    0.0607 − 0.0036i
   −0.2264 + 0.2254i
   −0.2264 − 0.2254i
    0.1657 − 0.6452i
    0.1657 + 0.6452i
eig =
  −1.3159 + 6.1956i
  −1.3159 − 6.1956i
  −3.2119 + 3.8261i
  −3.2119 − 3.8261i
  −3.7221 + 1.3238i
  −3.7221 − 1.3238i
dc =
    []
```

and the analytic solution $x(t)$ results.

Let us find the system response if the input signal is

$$u(t) = \begin{cases} 100, \forall t \in [0\ 4)\sec \\ -100, \forall t \in [4\ 8]\sec \end{cases}.$$

The following file is developed using the lsim command:

```
t=0:.01:8; den=[1 16.5 165 1075 4688 12500 15625];
num=[1];
u=100*ones(size(t));
for j=min(find(t>=4)):length(u)u(j)=-100; end;
y=lsim(num,den,u,t); plot(t,y,'-',t,u/100000,':')
```

The resulting dynamics are illustrated in Figure 3.2.7.

To find the state-space model, we apply the tf2ss command. From

```
≫den=[1 16.5 165 1075 4688 12500 15625];num=[1];
[A,B,H,D]=tf2ss(num,den)
A =
  1.0e+004 *
    −0.0017   −0.0165   −0.1075   −0.4688   −1.2500   −1.5625
     0.0001        0         0         0         0         0
          0    0.0001        0         0         0         0
          0         0    0.0001        0         0         0
          0         0         0    0.0001        0         0
          0         0         0         0    0.0001        0
```

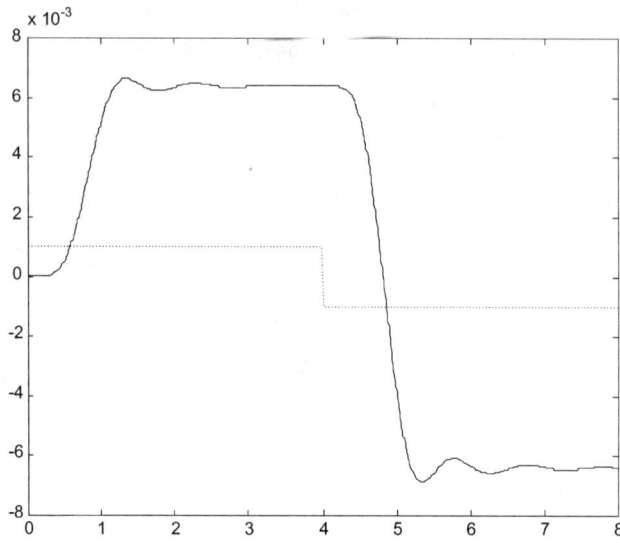

FIGURE 3.2.7.   System response if $u(t) = \begin{cases} 100, \forall t \in [0\ 4)\text{sec} \\ -100, \forall t \in [4\ 8]\text{sec}. \end{cases}$

B =
1
0
0
0
0
0

H =
0   0   0   0   0   1

D =
0

one has

$$A = \begin{bmatrix} -17 & -165 & -1075 & -4688 & -12500 & -15625 \\ 1 & 0 & 0 & 0 & 0 & 0 \\ 0 & 1 & 0 & 0 & 0 & 0 \\ 0 & 0 & 1 & 0 & 0 & 0 \\ 0 & 0 & 0 & 1 & 0 & 0 \\ 0 & 0 & 0 & 0 & 1 & 0 \end{bmatrix},$$

$$B = \begin{bmatrix} 1 \\ 0 \\ 0 \\ 0 \\ 0 \\ 0 \end{bmatrix}, H = \begin{bmatrix} 0 & 0 & 0 & 0 & 0 & 1 \end{bmatrix}, \text{ and } D = [0].$$

That is, we received the following state-space model:

$$\frac{dx}{dt} = \begin{bmatrix} \frac{dx_1}{dt} \\ \frac{dx_2}{dt} \\ \frac{dx_3}{dt} \\ \frac{dx_4}{dt} \\ \frac{dx_5}{dt} \\ \frac{dx_6}{dt} \end{bmatrix} = Ax + Bu = A \begin{bmatrix} x_1 \\ x_2 \\ x_3 \\ x_4 \\ x_5 \\ x_6 \end{bmatrix} + Bu, \ y = Hx + Du = H \begin{bmatrix} x_1 \\ x_2 \\ x_3 \\ x_4 \\ x_5 \\ x_6 \end{bmatrix} + Du.$$

Using the ss2tf command, the results can be verified, and we have

```
≫ [num, den] = ss2tf(A, B, H, D)
num =
    0   -0.0000   -0.0000   -0.0000   -0.0000   -0.0000   1.0000
den =
    1.0e + 004 *
       0.0001   0.0017   0.0165   0.1075   0.4688   1.2500   1.5625
```

Taking note of the matrices A, B, H, and D of the state-space model, the transient responses can be calculated using the lsim command. To find the state variable evolution, we assign $u = 100$ and the initial conditions are $x_{10} = -1$, $x_{20} = 1$, $x_{30} = 0$, $x_{40} = 0$, $x_{50} = 0$, and $x_{60} = 0$. Then, we have

```
≫t=0:.01:4; u=100*ones(size(t)); x0=[-1 1 0 0 0 0];
≫ [y,x]=lsim(A,B,H,D,u,t,x0);plot(t,x)
```

The states evolution are shown in Figure 3.2.8.
If one needs to plot the output transient, it can be done by typing

```
≫ plot(t,y)
```

Figure 3.2.9 documents the output dynamics.
In many practical problems (filters analysis and design), it is critical that the frequency response of dynamic systems is analyzed. The frequency response is obtained by letting $s = j\omega$, where ω is the angular frequency in radians/second. Two Bode plots document how the amplitude of $G(j\omega)$ as well as the phase angle of $G(j\omega)$ vary with the angular frequency (the log scale is used for ω). Usually, the magnitude of $G(j\omega)$ is plotted in decibels, and $|G(j\omega)|_{dB} = 20 \log_{10} |G(j\omega)|$.

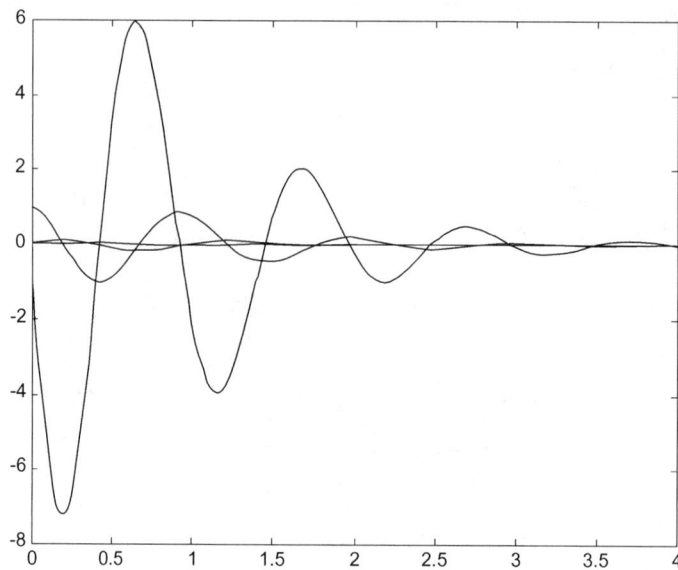

FIGURE 3.2.8.   State variables evolution caused by step input and initial conditions.

Example 3.2.3.

Using MATLAB, let us calculate the Bode plots for

$$G(s) = \frac{X(s)}{U(s)} = \frac{1}{s^6 + 16.5s^5 + 165s^4 + 1075s^3 + 4688s^2 + 12500s + 15625}.$$

Using the bode command, type

```
>>num=[1]; den=[1 16.5 165 1075 4688 12500 15625];
bode(num,den)
```

and the Bode plots are found. Figure 3.2.10 documents the plots for $|G(j\omega)|_{dB}$ and the phase angle of $G(j\omega)$.

Statistical analysis and filtering can be performed in the MATLAB environment. For example, the Butterworth and Chebyshev filters can be designed using the butter and cheby1 or cheby2 commands.

Example 3.2.4.

Calculate the transfer function of the fourth-order Butterworth filter with bandwidth 5 Hz. For the input $u(t) = \sin 2t$, find the filter's output dynamics.

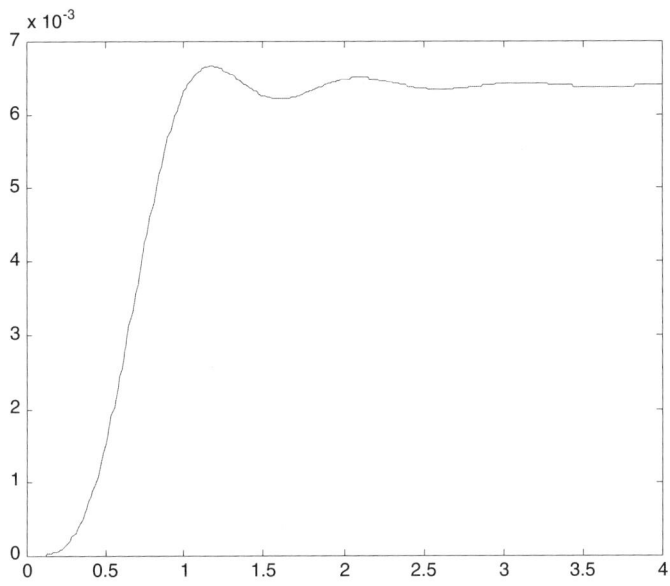

FIGURE 3.2.9.   Output dynamics.

Type in the Command Window

```
>> N=4;Wn=1;W0=2*pi*5;[num,den]=butter(N,W0,'s');
bode(num,den);num,den
```

using `format short e`, one obtains

```
num =
            0            0            0      0 9.7409e+005
den =
   1.0000e+000 8.2094e+001 3.3697e+003 8.1023e+004 9.7409e+005
```

Thus, the transfer function of the Butterworth filter is

$$G(s) = \frac{9.74 \times 10^5}{s^4 + 8.21 \times 10^1 s^3 + 3.37 \times 10^3 s^2 + 8.1 \times 10^4 s + 9.74 \times 10^5}.$$

The Bode plots are plotted in Figure 3.2.11.

The response of the filter for the input signal $u(t) = \sin 2t$ is calculated. We have

```
>> t=0:.01:8; u=sin(2*t); y=lsim(num,den,u,t);
plot (t,y,'-',t,u,':')
```

Bode Diagrams

From: U(1)

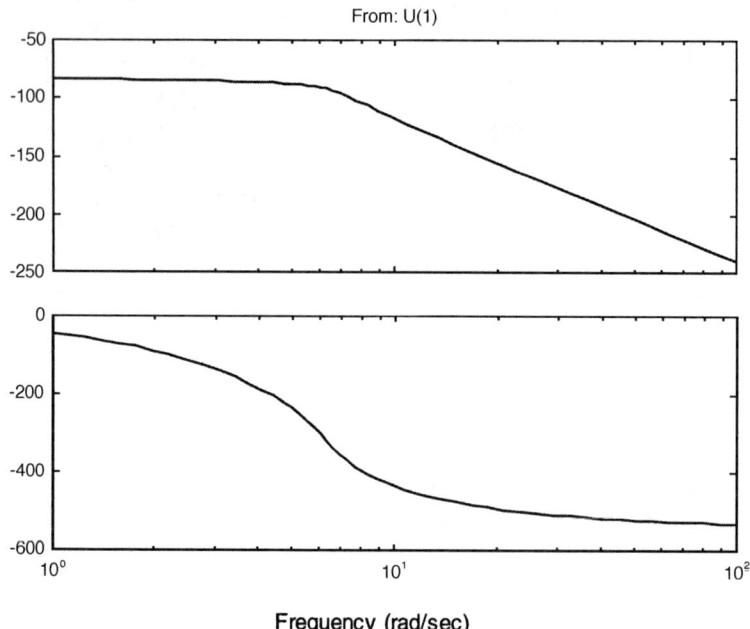

Frequency (rad/sec)

FIGURE 3.2.10.   Bode plots for $G(s) = \frac{1}{s^6+16.5s^5+165s^4+1075s^3+4688s^2+12500s+15625}$.

and the output transient dynamics as well as $u(t)$ are documented in Figure 3.2.12.

Using the Symbolic Math Toolbox, one can perform the symbolic analysis and arithmetic in terms of symbolic expressions. Some class of linear and nonlinear differential and integral equations can be solved.

Example 3.2.5.

Analytically solve the third-order differential equation

$$\frac{d^3x}{dt^3} + 2\frac{dx}{dt} + 3x = d.$$

Using the dsolve command, type in the Command Window

```
≫x=dsolve('D3x+2*Dx+3*x=d')
```

The resulting solution is

```
x =
1/3*d+C1*exp(-t)+C2*exp(1/2*t)*cos(1/2*11^(1/2)*t)
+C3* exp(1/2 *t)* sin(1/2*11^(1/2)*t)
```

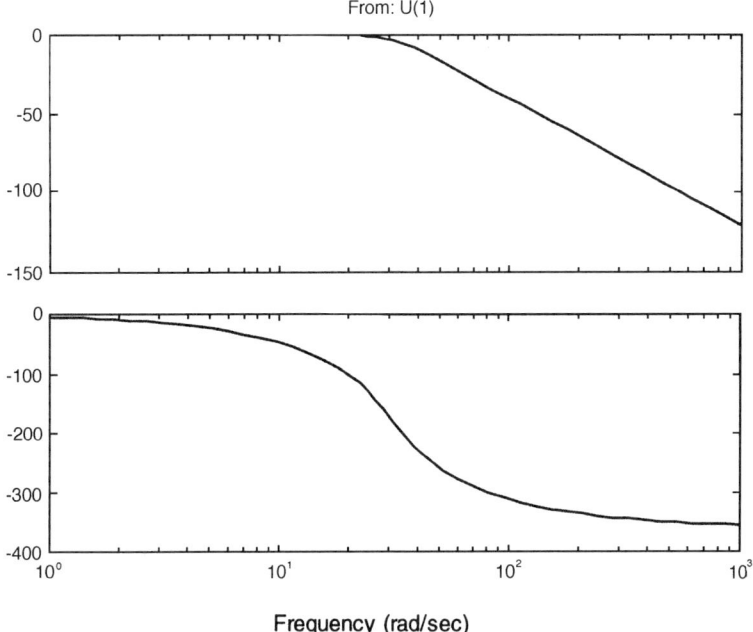

FIGURE 3.2.11.   Bode plots for the fourth-order Butterworth filter.

Using the `pretty` command, find

```
≫pretty(x)
                                             1/2
   1/3 d + C1 exp(-t) + C2 exp(1/2 t) cos(1/2 11    t)
                                  1/2
        + C3 exp(1/2 t) sin(1/2 11    t)
```

Thus, the solution is

$$x(t) = \tfrac{1}{3}d + c_1 e^{-t} + c_2 e^{0.5t} \cos\left(\tfrac{1}{2}\sqrt{11}t\right) + c_3 e^{0.5t} \sin\left(\tfrac{1}{2}\sqrt{11}t\right).$$

Using the initial conditions, the unknown constants are found. As an example, let us assign

$$\left(\frac{d^2x}{dt^2}\right)_0 = 1, \left(\frac{dx}{dt}\right)_0 = 2, \text{ and } x_0 = 3.$$

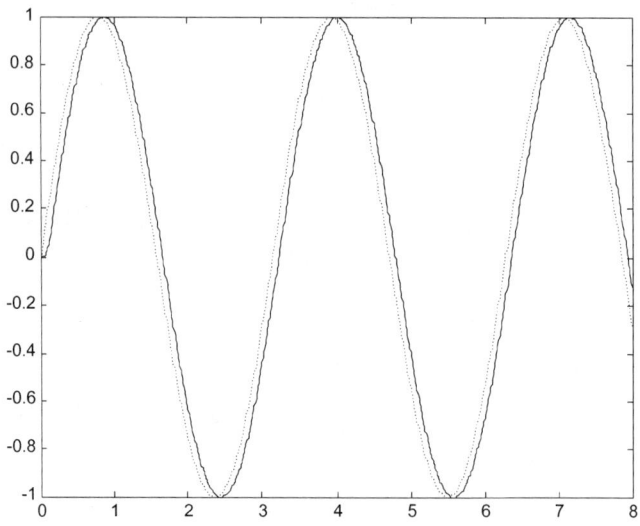

FIGURE 3.2.12.   Evolution of the filter output and plot of $u(t) = \sin 2t$.

We have

```
>>x=dsolve('D3x+2*Dx+3*x=d','D2x(0)=1','Dx(0)=2','x(0)=3');
pretty(x)
   1/3 d + (- 1/5 d + 8/5) exp(-t)

                                                  1/2
           + (- 2/15 d + 7/5) exp(1/2 t) cos(1/2 11    t)
                               1/2                           1/2
           - 1/165 (4 d - 87) 11    exp(1/2 t) sin(1/2
 11    t)
```

Hence, $c_1 = -\frac{1}{5}d + \frac{8}{5}$, $c_2 = -\frac{2}{15}d + \frac{7}{5}$, and $c_3 = -\frac{1}{165}(4d - 87)\sqrt{11}$.
If the forcing function is time-varying, the analytic solution of

$$\frac{d^3x}{dt^3} + 2\frac{dx}{dt} + 3x = d(t)$$

is found as

```
>>x=dsolve('D3x+2*Dx+3*x=d(t)'); pretty(x)
```

```
      /        /
      |        |
1/55  |11      |  exp(t) d(t) dt
      |        |
      \        /
                  /
                  |              1/2
      +           |  -d(t) (3 11       %1 + 11 %2) exp (-1/2 t) dt exp(3/2 t) %2
                  |
                  /
                  /                                                                  \
                  |              1/2                                                  |
      +           |  d(t) (3 %2 11      - 11 %1) exp (-1/2 t) dt exp (3/2 t) %1 |
                  |                                                                  |
                  /                                                                  /
          exp(-t) + C1 exp (-t) + C2 exp (1/2 t) %2 + C3 exp (1/2 t) %1
                      1/2
%1   := sin (1/2 11      t)
                      1/2
%2   := cos (1/2 11      t)
```

To plot the functions, the `ezplot` plotter command is used.

Example 3.2.6.

Plot the function $y = \sin^3 x + \cos^4 x$ using the `ezplot` command.
   To plot the function $y = \sin^3 x + \cos^4 x$, type

```
≫ezplot('sin(x)^3+cos(x)^4')
```

The resulting plot is documented in Figure 3.2.13.
   Continuous- and discrete-time dynamic systems are studied. The discrete-time, n-order, linear, constant-coefficient difference equation is

$$\sum_{i=0}^{n} a_i y_{n-i} = \sum_{i=0}^{m} b_i u_{n-i}, n \geq m.$$

Using the z-transform and assuming that initial conditions are zero, we have

$$\left(\sum_{i=0}^{n} a_i z^i\right) Y(z) = \left(\sum_{i=0}^{m} b_i z^i\right) U(z).$$

Hence, the transfer function is expressed as

$$G(z) = \frac{Y(z)}{U(z)} = \frac{b_m z^m + b_{m-1} z^{m-1} + \cdots + b_1 z + b_0}{a_n z^n + a_{n-1} z^{n-1} + \cdots + a_1 z + a_0}.$$

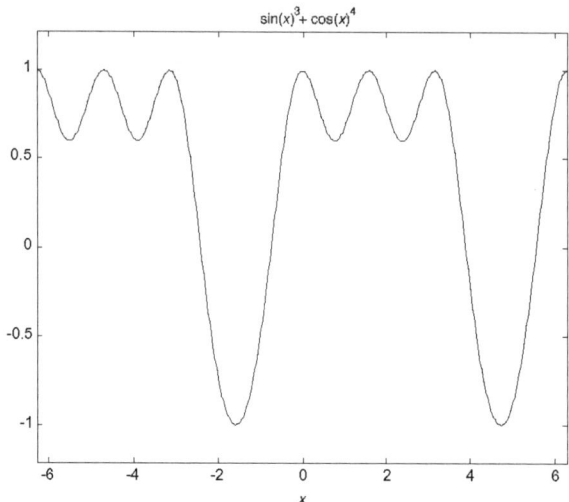

$$\sin(x)^3 + \cos(x)^4$$

FIGURE 3.2.13.   Plot of a nonlinear function $y = \sin^3 x + \cos^4 x$.

If $n = m$, we have

$$G(z) = \frac{Y(z)}{U(z)} = \frac{b_n z^n + b_{n-1} z^{n-1} + \cdots + b_1 z + b_0}{a_n z^n + a_{n-1} z^{n-1} + \cdots + a_1 z + a_0}$$

or

$$G(z) = \frac{Y(z)}{U(z)} = \frac{b_n + b_{n-1} z^{-1} + \cdots + b_1 z^{-n+1} + b_0 z^{-n}}{a_n + a_{n-1} z^{-1} + \cdots + a_1 z^{-n+1} + a_0 z^{-n}}.$$

Example 3.2.7.

Consider a discrete-time system with transfer function

$$G(z) = \frac{Y(z)}{U(z)} = \frac{0.25 + 0.5z^{-1} + 0.75z^{-2}}{1 + 2z^{-1} + 3z^{-3} + 4z^{-4}} = \frac{0.25z^2 + 0.5z + 0.75}{z^4 + 2z^3 + 3z + 4}.$$

Find the system dynamics caused by step input.

The dstep and dimpulse commands are used to find the step response, and the stem command is applied to plot the dynamics.

To find the transient behavior of a discrete-time system for eight instants $k = 0, 1, \ldots, 6, 7$, if the unit step input is applied ($u_k = 1, k \geq 0$), we type in the Command Window

```
≫k=0:1:7;num=[0.25 0.5 0.75];den=[1 2 3 4];
x=dstep(num,den,8);stem(k,x)
```

As one presses the Enter key, the resulting plot is displayed, and the transient behavior is documented in Figure 3.2.14.

The filter command can be used to simulate the discrete-time systems.

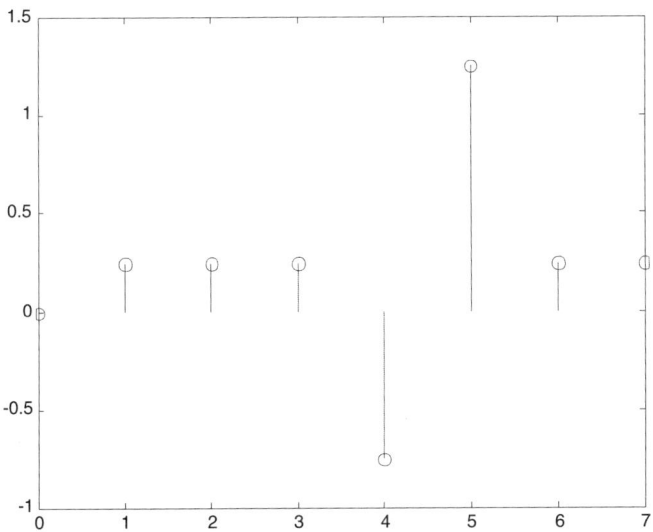

FIGURE 3.2.14.   Transient dynamics for x_k for the unit-step input.

Example 3.2.8.

Simulate the dynamics of the discrete-time systems if

$$G(z) = \frac{X(z)}{U(z)} = \frac{z^2 + 2z + 3}{z^2 - 0.5z + 0.25}.$$

The system input is

$$u_k = \begin{cases} 1, k = 0, \ldots, 9 \\ 0, k = 10, \ldots, 19 \\ -1, k = 20, \ldots, 30 \end{cases}.$$

Perform analysis of stability based on numerical and analytical results.
Type in the Command Window

```
≫k=0:1:30;u=[ones(1,10) zeros(1,10) -ones(1,11)];
num=[1 2 3]; den=[1 -0.5 0.25];
x=filter(num,den,u);plot(k,x,'-',k,u,'+')
```

and the system dynamics is found. We use the "+" symbol to plot the input signal
and the symbol "x" illustrates the system output; see Figure 3.2.15.

The numerical analysis shows that the single-input/single-output discrete system
is stable. The numerical results are easily justified analytically as one finds the

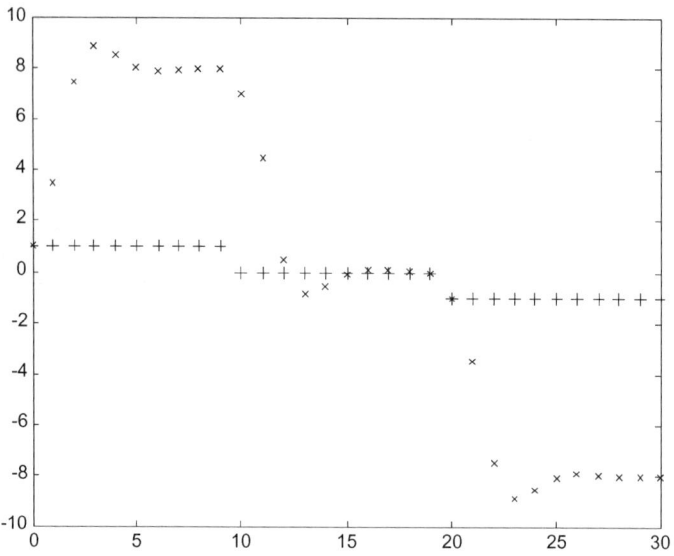

FIGURE 3.2.15.   Transient dynamics of the discrete-time output x_k if

$$u_k = \begin{cases} 1, k = 0, \ldots, 9 \\ 0, k = 10, \ldots, 19 \\ -1, k = 20, \ldots, 30. \end{cases}$$

characteristic eigenvalues, which for stable discrete-time systems must lay in the unit circle. Using the `roots` command, we find that this criterion for stability is guaranteed, and using `format short e`, we have

```
>>format short e; roots(den)
ans =
   2.5000e-001 +4.3301e-001i
   2.5000e-001 -4.3301e-001i
```

We considered n–degree-of-freedom dynamic systems that are modeled using difference equations. In the case of n-dimensional state, m-dimensional control, and b-dimensional output, one has

$$x_k = \begin{bmatrix} x_{k1} \\ x_{k2} \\ \vdots \\ x_{kn-1} \\ x_{kn} \end{bmatrix}, u_k = \begin{bmatrix} u_{k1} \\ u_{k2} \\ \vdots \\ u_{km-1} \\ u_{km} \end{bmatrix}, \text{ and } y_k = \begin{bmatrix} y_{k1} \\ y_{k2} \\ \vdots \\ y_{kb-1} \\ y_{kb} \end{bmatrix}.$$

In matrix form, the state-space linear and nonlinear equations are given as

$$
x_{k+i} = \begin{bmatrix} x_{k+1,1} \\ x_{k+1,2} \\ \vdots \\ x_{k+1,n-1} \\ x_{k+1,n} \end{bmatrix} = \begin{bmatrix} a_{k11} & a_{k12} & \cdots & a_{k1\,n-1} & a_{k1n} \\ a_{k21} & a_{k22} & \cdots & a_{k2\,n-1} & a_{k2n} \\ \vdots & \vdots & \ddots & \vdots & \vdots \\ a_{kn-11} & a_{kn-12} & \cdots & a_{kn-1\,n-1} & a_{kn-1n} \\ a_{kn1} & a_{kn2} & \cdots & a_{kn\,n-1} & a_{knn} \end{bmatrix}
$$

$$
\times \begin{bmatrix} x_{k1} \\ x_{k2} \\ \vdots \\ x_{k\,n-1} \\ x_{kn} \end{bmatrix} + \begin{bmatrix} b_{k11} & b_{k12} & \cdots & b_{k1\,m-1} & b_{k1m} \\ b_{k21} & b_{k22} & \cdots & b_{k2\,m-1} & b_{k2m} \\ \vdots & \vdots & \ddots & \vdots & \vdots \\ b_{kn-11} & b_{kn-12} & \cdots & b_{kn-1\,m-1} & b_{kn-1m} \\ b_{kn1} & b_{kn2} & \cdots & b_{kn\,m-1} & b_{knm} \end{bmatrix} \begin{bmatrix} u_{k1} \\ u_{k2} \\ \vdots \\ u_{k\,m-1} \\ u_{km} \end{bmatrix}
$$

$$
= A_k x_k + B_k u_k,
$$

$$
x_{k=k_0} = x_{k0},
$$

and

$$
x_{k+1} = F_k(k, x_k, u_k), \; x_{k=k_0} = x_{k0},
$$

where $A_k \in \mathbb{R}^{n \times n}$ and $B_k \in \mathbb{R}^{n \times m}$ are the matrices of constant coefficients and $F_k(k, x_k, u_k)$ is the nonlinear map.

The output equations is

$$
y_k = H_k x_k,
$$

where $H_k \in \mathbb{R}^{b \times n}$ is the matrix of constant coefficients.

The state-space and output equations for continuous- and discrete-time systems are

$$
\frac{dx}{dt} = Ax + Bu, \; y = Hx \text{ and } x_{k+1} = A_k x_k + B_k u_k, \; y_k = H_k x_k.
$$

Attacking a great variety of problems in analysis and design of dynamic systems, the digital controller must be synthesized to be implemented using microcontrollers and DSPs. Let us discretize the continuous-time system. From

$$
\frac{dx(t)}{dt} = Ax(t) + Bu(t),
$$

we have

$$
\frac{dx(t)}{dt} - Ax(t) = Bu(t).
$$

Premultiplying both sides of this equation by the inverse matrix exponential e^{-At}, one finds

$$
e^{-At} \left(\frac{dx(t)}{dt} - Ax(t) \right) = \frac{d \left(e^{-At} x(t) \right)}{dt} = e^{-At} Bu(t),
$$

where the matrix exponential is

$$e^{At} = \sum_{k=0}^{\infty} \frac{A^k t^k}{k!}, \quad e^{At} e^{-At} = I,$$

and

$$\frac{de^{At}}{dt} = Ae^{At} = e^{At} A.$$

Then,

$$e^{-At} x(t) = x(0) + \int_0^t e^{-A\tau} Bu(\tau) d\tau.$$

Multiplying both sides by e^{At}, one obtains the solution of the differential equation as

$$x(t) = e^{At} x(0) + \int_0^t e^{A(t-\tau)} Bu(\tau) \, d\tau, \quad t \geq 0,$$

Sampling the input function $u(t)$ and applying it to the zero-order data hold, we have $u(t) = u_k$ if $kT_s \leq t < (k+1)T_s$. From

$$x(t) = e^{At} x(0) + \int_0^t e^{A(t-\tau)} Bu(\tau) d\tau,$$

one obtains

$$x_{k+1} = x((k+1)T_s) = e^{A(k+1)T_s} x(0) + e^{A(k+1)T_s} \int_0^{(k+1)T_s} e^{-A\tau} Bu(\tau)d\tau,$$

$$x_k = x(kT_s) = e^{AkT_s} x(0) + e^{AkT} \int_0^{kT_s} e^{-A\tau} Bu(\tau)d\tau.$$

Subtracting these two equations, the following formula results:

$$x_{k+1} = x((k+1)T_s) = e^{AT_s} x(dT_s) + \int_0^{T_s} e^{A(T_s-t)} Bu(kT_s) d(T_s - t)$$

$$= e^{AT_s} x_k + \int_0^{T_s} e^{A(T_s-t)} Bu_k d(T_s - t).$$

Thus, the matrices of the difference equation are found as

$$A_k = e^{AT_s} \text{ and } B_k = \left(\int_0^{T_s} e^{A(T_s-t)} d(T_s - t) \right) B.$$

If matrix A is nonsingular, the following formula results:

$$B_k = \left(\int_0^{T_s} e^{A(T_s-t)} d(T_s - t) \right) B = A^{-1} \left(e^{AT_s} - I \right) B = \left(e^{AT_s} - I \right) A^{-1} B.$$

It is evident that matrices A_k and B_k depend on the sampling time T_s. The output equations are

$$y_k = H_k x_k, \quad y(kT_s) = H_k x(kT_s) \text{ and } y(t) = Hx(t).$$

That is, $H_k = H$.

Example 3.2.9.

The continuous-time system is given by

$$\frac{dx}{dt} = \begin{bmatrix} \frac{dx_1}{dt} \\ \frac{dx_2}{dt} \\ \frac{dx_3}{dt} \\ \frac{dx_4}{dt} \end{bmatrix} = Ax + Bu = \begin{bmatrix} -1 & 2 & 3 & 4 \\ 5 & -6 & 7 & 8 \\ 9 & 8 & -7 & 6 \\ 5 & 4 & 3 & -2 \end{bmatrix} \begin{bmatrix} x_1 \\ x_2 \\ x_3 \\ x_4 \end{bmatrix} + \begin{bmatrix} 1 \\ 2 \\ 3 \\ 4 \end{bmatrix} u.$$

Find the discrete-time model $x_{k+1} = A_k x_k + B_k u_k$ if the sampling period is 0.1 sec. That is, $T_s = 0.1$ sec.

To solve the problem, the command `c2d` is used. We type

```
≫A=[-1 2 3 4;5 -6 7 8;9 8 -7 6;5 4 3 -2];B=[1; 2; 3; 4];
Ts=0.1;
[Ak,Bk]=c2d(A,B,Ts)
```

Pressing the Enter key, the following matrices of the discrete-time system A_k and B_k result:

```
≫Ak =
   1.2511e+000   3.6137e-001   3.8078e-001   5.9529e-001
   9.1518e-001   9.6928e-001   6.5010e-001   9.6947e-001
   1.1051e+000   7.3373e-001   9.1703e-001   9.2525e-001
   7.6618e-001   4.9576e-001   4.3322e-001   1.2175e+000
Bk =
   2.9005e-001
   4.8830e-001
   5.3987e-001
   5.4780e-001
```

Hence, one has

$$x_{k+1} = \begin{bmatrix} x_{k+11} \\ x_{k+12} \\ x_{k+13} \\ x_{k+14} \end{bmatrix}$$

$$= A_k x_k + B_k u_k = \begin{bmatrix} 1.25 & 0.36 & 0.38 & 0.6 \\ 0.92 & 0.97 & 0.65 & 0.97 \\ 1.1 & 0.73 & 0.92 & 0.93 \\ 0.77 & 0.5 & 0.43 & 1.22 \end{bmatrix} \begin{bmatrix} x_{k1} \\ x_{k2} \\ x_{k3} \\ x_{k4} \end{bmatrix} + \begin{bmatrix} 0.29 \\ 0.49 \\ 0.54 \\ 0.55 \end{bmatrix} u_k.$$

Example 3.2.10.

For the dynamic system, mathematical model of which is

$$\frac{d\theta_r}{dt} = \omega_r,$$

$$\frac{d\omega_r}{dt} = \frac{1}{J}(-B_m \omega_r + T_e),$$

find analytic expressions for matrices A_k and B_k as functions of the sampling period T_s.

Denoting

$$x = \begin{bmatrix} x_1 \\ x_2 \end{bmatrix} = \begin{bmatrix} \theta_r \\ \omega_r \end{bmatrix}$$

and $u = T_e$, we have the following state-space form:

$$\frac{dx(t)}{dt} = Ax + Bu = \begin{bmatrix} 0 & 1 \\ 0 & -\frac{B_m}{J} \end{bmatrix} \begin{bmatrix} x_1 \\ x_2 \end{bmatrix} + \begin{bmatrix} 0 \\ \frac{1}{J} \end{bmatrix} u.$$

To find the matrix exponential, the inverse matrix is derived first, and

$$(sI - A)^{-1} = \begin{bmatrix} s & -1 \\ 0 & s + \frac{B_m}{J} \end{bmatrix} = \frac{1}{s\left(s + \frac{B_m}{J}\right)} \begin{bmatrix} s + \frac{B_m}{J} & 1 \\ 0 & s \end{bmatrix}.$$

Applying the Laplace transforms, one has

$$e^{At} = \begin{bmatrix} 1 & \frac{B_m}{J}\left(1 - e^{-\frac{B_m}{J}t}\right) \\ 0 & e^{-\frac{B_m}{J}t} \end{bmatrix}.$$

Hence, the analytic expressions for matrices A_k and B_k result as

$$A_k = e^{AT_s} = \begin{bmatrix} 1 & \frac{B_m}{J}\left(1 - e^{-\frac{B_m}{J}T_s}\right) \\ 0 & e^{-\frac{B_m}{J}T_s} \end{bmatrix}$$

and

$$B_k = \left(\int_0^{T_s} e^{A(T_s - t)} d(T_s - t)\right) B = \left(\int_0^{T_s} e^{A(T_s - t)} d(T_s - t)\right) \begin{bmatrix} 0 \\ \frac{1}{J} \end{bmatrix}$$

$$= \begin{bmatrix} \frac{T_s}{B_m} - \frac{J}{B_m^2}\left(1 - e^{-\frac{B_m}{J}T_s}\right) \\ \frac{1}{B_m}\left(1 - e^{-\frac{B_m}{J}T_s}\right) \end{bmatrix}.$$

To analyze and model continuous- and discrete-time dynamic systems, the block diagrams are used, and SIMULINK extends the MATLAB environment. SIMULINK offers a large variety of ready-to-use building blocks to build mathematical models. The MATLAB Demos is available, and one lacking enough experience to use the MATLAB environment shall find a great deal of help using the MATLAB Demos. After double clicking Simulink in the MATLAB Demos, the subtopics become available to the user; see Figure 3.2.16.

A good starting point is to start with Simple models, as shown in Figure 3.2.16.

Simple pendulum and spring-mass system simulations, tracking a bouncing ball, Van der Pol equations simulation, as well as other examples are available. Double clicking the simple pendulum simulation, the SIMULINK block diagram appears, which is documented in Figure 3.2.17.

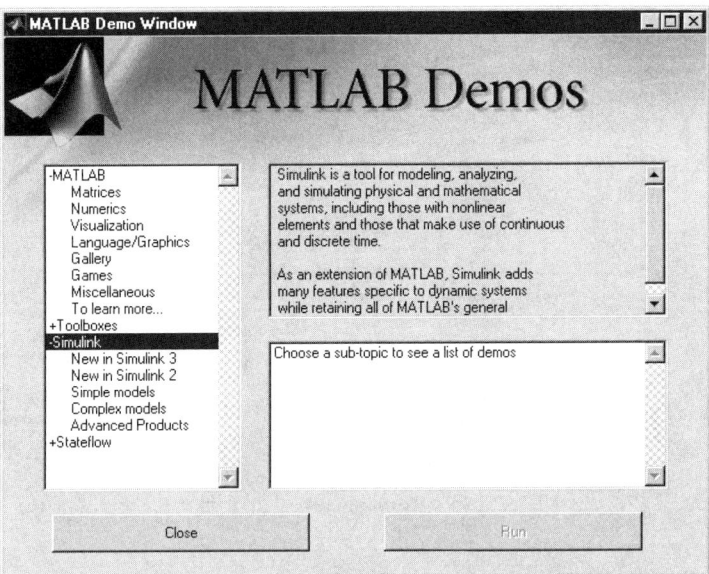

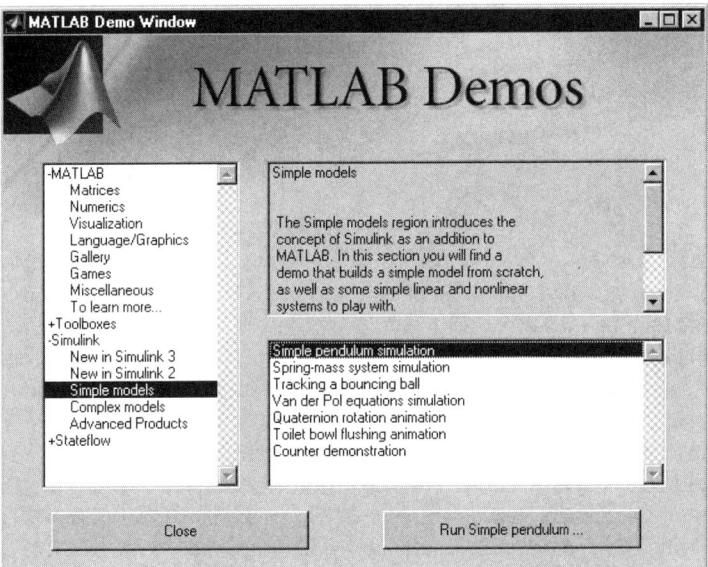

FIGURE 3.2.16.   MATLAB Demos: SIMULINK environment.

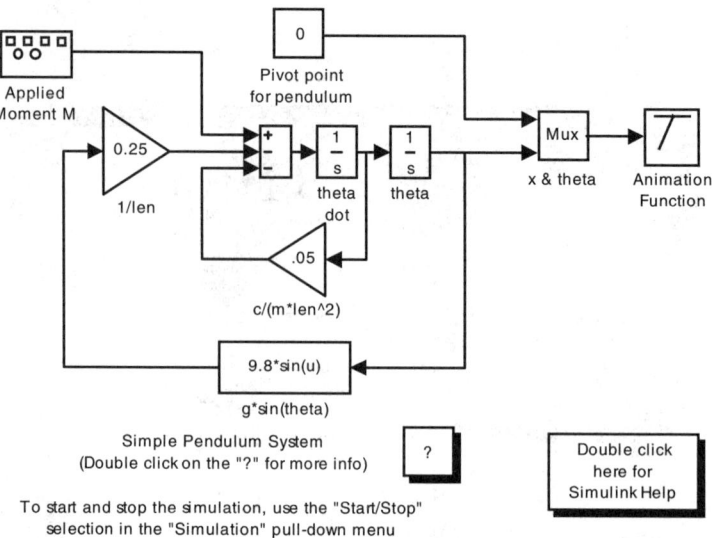

FIGURE 3.2.17.   SIMULINK block-diagram to model and animate the simple pendulum.

The equations of motion for a simple pendulum were derived in Example 2.2.4. In particular, we found that the following two differential equations result:

$$\frac{d\omega}{dt} = \frac{1}{J}(-mgl\sin\theta + T_a - B_m\omega) \text{ and } \frac{d\theta}{dt} = \omega.$$

The moment of inertia is $J = ml^2$, and hence, if $B_m = 0$, we have

$$\frac{d\omega}{dt} = -\frac{g}{l}\sin\theta + \frac{1}{ml^2}T_a \text{ and } \frac{d\theta}{dt} = \omega.$$

These equations are used, and one simulates the pendulum by clicking Simulation, and then by clicking Start.

Clicking the SIMULINK icon, two new windows (SIMULINK Library Browser and untitled empty SIMULINK model window, where new models should be built) appear on the screen, as shown in Figures 3.2.18.

One can open Simulink and Communication Blockset, Control System Toolbox, Fuzzy Logic Toolbox, Neural Network Blockset, Stateflow, Simulink Extra, as well as System ID (identification) Blocks. By clicking on Simulink and Simulink Extra, and then opening the Continuous, one has Simulink Library Browsers, as documented in Figures 3.2.19.

In addition to Continuous, the designer can open the Discrete, Function & Tables, Math, Nonlinear, Signal & Systems, Sinks, and Sources block libraries by double clicking on the corresponding icon. Ready-to-use building blocks, commonly applied in analysis and design of dynamic systems, become available.

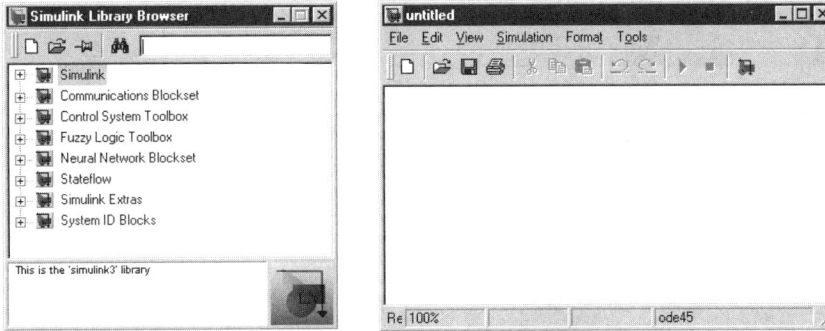

FIGURE 3.2.18.   SIMULINK Library Browser and `untitled` `mdl`-file.

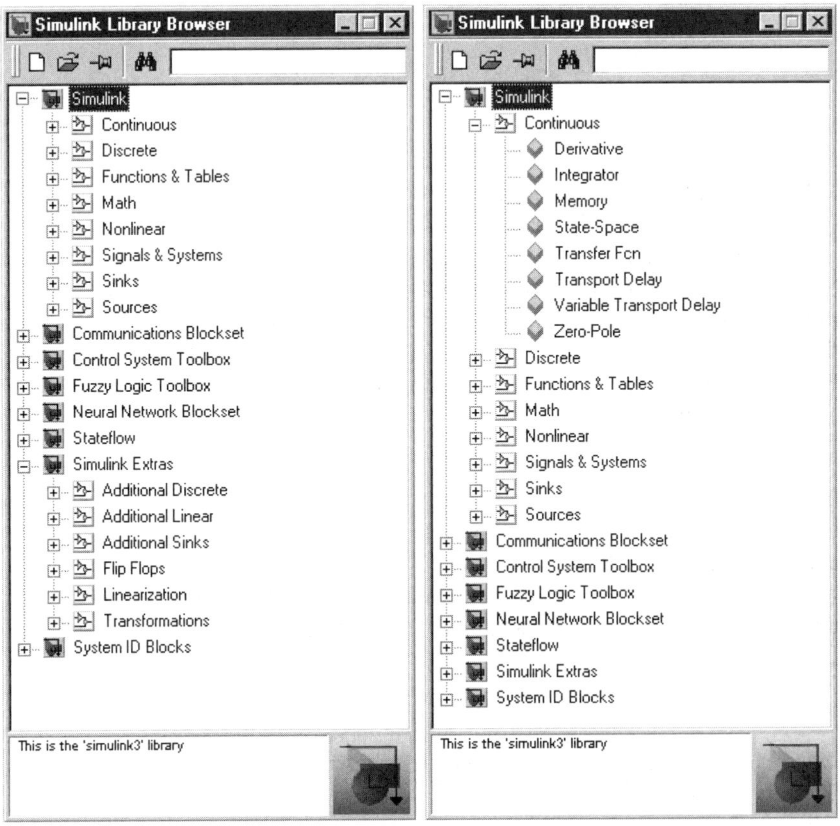

FIGURE 3.2.19.   SIMULINK Library Browsers.

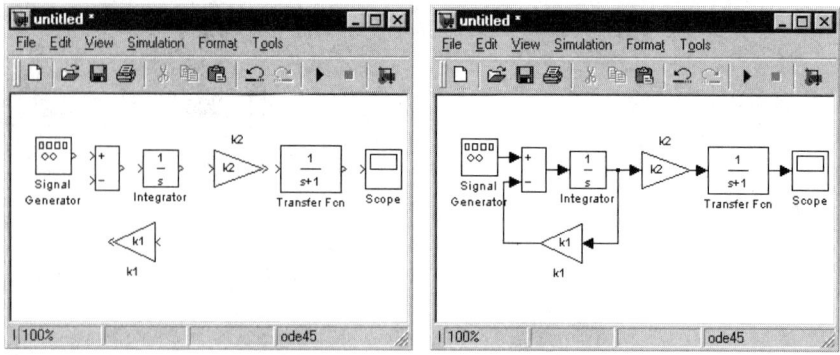

FIGURE 3.2.20.   Untitled window where the SIMULINK block diagram should be built.

To demonstrate how to use SIMULINK, consider the simplest example. Assume that the system dynamics is modeled as

$$\frac{dx_1(t)}{dt} = -k_1 x_1(t) + u(t), x_1(t_0) = x_{10},$$

$$\frac{dx_2(t)}{dt} = k_2 x_1 - x_2, x_2(t_0) = x_{20}.$$

Let the input be a sinusoidal signal with magnitude 50 and frequency 2 Hz. The coefficients and an initial condition are $k_1 = 5$, $k_2 = 10$, $x_{10} = 20$, and $x_{20} = 0$.

We use the Signal Generator, Sum, Gain, Integrator, Transfer Function, and Scope blocks. These blocks were dragged from the above-mentioned SIMULINK block libraries to the untitled window, positioned (placed), and connected using the signal lines, as shown in Figure 3.2.20. That is, by connecting the blocks, the SIMULINK block diagram needed to be used in modeling results.

In the Command Window, type

```
≫k1=5;  k2=10;  x10=20;
```

to download the coefficients and initial condition. The Signal Generator block is used to generate the sinusoidal input. In particular, one specifies the amplitude and frequency as illustrated in Figure 3.2.21.

An initial condition is assigned in the Integrator block. Specifying the modeling time to be 10 sec, the transient behavior of the system output is plotted in the Scope; see Figure 3.2.22.

Example 3.2.11.

Let us simulate the Van der Pol oscillator, as described by the following nonlinear differential equation:

$$\frac{d^2x}{dt^2} - k(1 - x^2)\frac{dx}{dt} + x = d(t),$$

where $d(t)$ is the disturbance (forcing function).

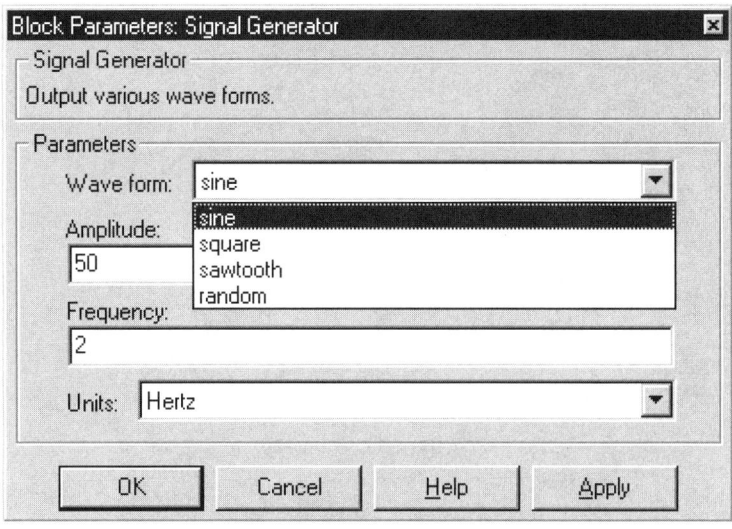

FIGURE 3.2.21.   Block Parameters: Signal Generator.

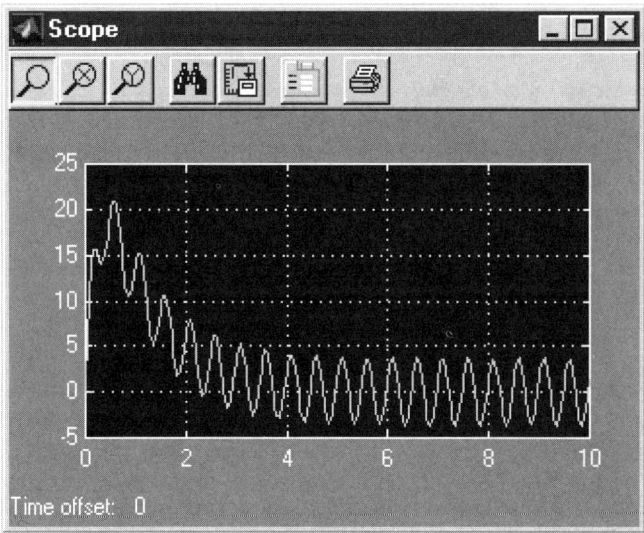

FIGURE 3.2.22.   Modeling results at Scope.

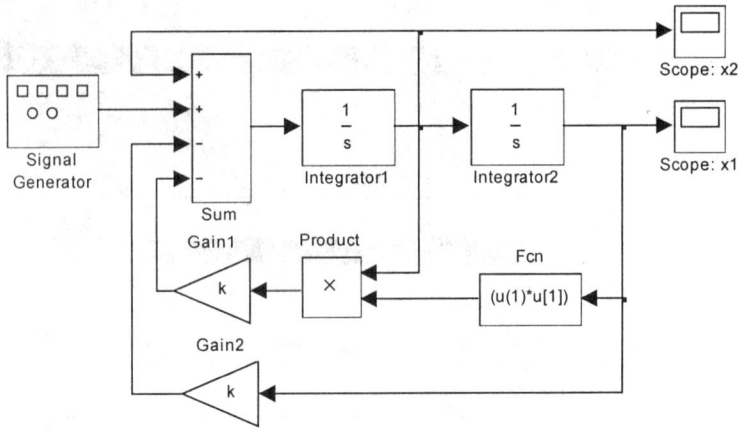

FIGURE 3.2.23.   SIMULINK block diagram (c_3_2_11.mdl).

The second-order Van der Pol differential equation is rewritten as a system of coupled first-order differential equations. In particular,

$$\frac{dx_1(t)}{dt} = x_2, \, x_1(t_0) = x_{10},$$

$$\frac{dx_2(t)}{dt} = -x_1 + kx_2 - kx_1^2 x_2 + d(t), \, x_2(t_0) = x_{20}.$$

The SIMULINK block diagram is built using the following blocks: Signal Generator, Gain, Integrator, Sum, and Scope; see Figure 3.2.23. Modeling of the dynamics was performed assigning $k = 10, d(t) = 10\text{rect}(2t)$ (by double clicking on the Signal Generator block, one selects the square function and assigns the corresponding magnitude 10 and frequency 2 Hz), and

$$x_0 = \begin{bmatrix} x_{10} \\ x_{20} \end{bmatrix} = \begin{bmatrix} -2 \\ 2 \end{bmatrix}$$

(these initial conditions are assigned by double clicking on the Integrator blocks and by typing x10 and x20). It should be emphasized that the coefficient k can be assigned by double clicking on the Gain block and entering the value needed. In the Command Window, type

```
≫k=10; x10=-2; x20=2;
```

The transient dynamics is displayed by Scopes. The plot command can be used, and in the Scopes, one uses the Data History and Scope properties assigning the

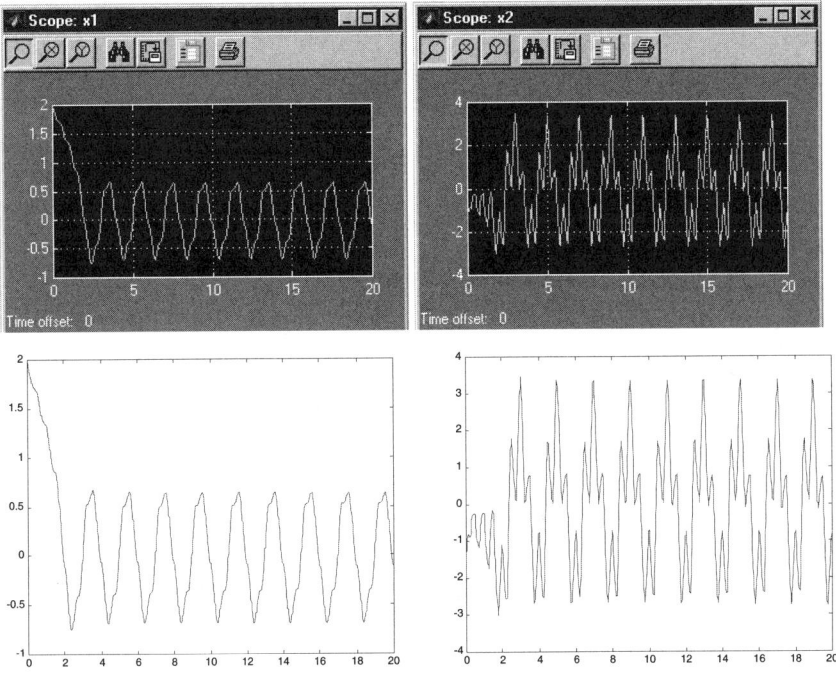

FIGURE 3.2.24.   Dynamics of the state variables.

Variable names. We use the following variables: x1 and x2. Then, the designer types

```
≫plot(x1(:,1),x1(:,2))
≫plot(x2(:,1),x2(:,2))
```

The resulting plots are illustrated in Figure 3.2.24.

It was emphasized that a great deal of viable examples are given in the MATLAB Demos, and the Van der Pol equations simulation is available. By double clicking on the Van der Pol equations simulation, the SIMULINK block diagram appears; see Figure 3.2.25. In particular, one simulates the following differential equations:

$$\frac{dx_1(t)}{dt} = x_2,$$

$$\frac{dx_2(t)}{dt} = -x_1 + x_2 - x_1^2 x_2.$$

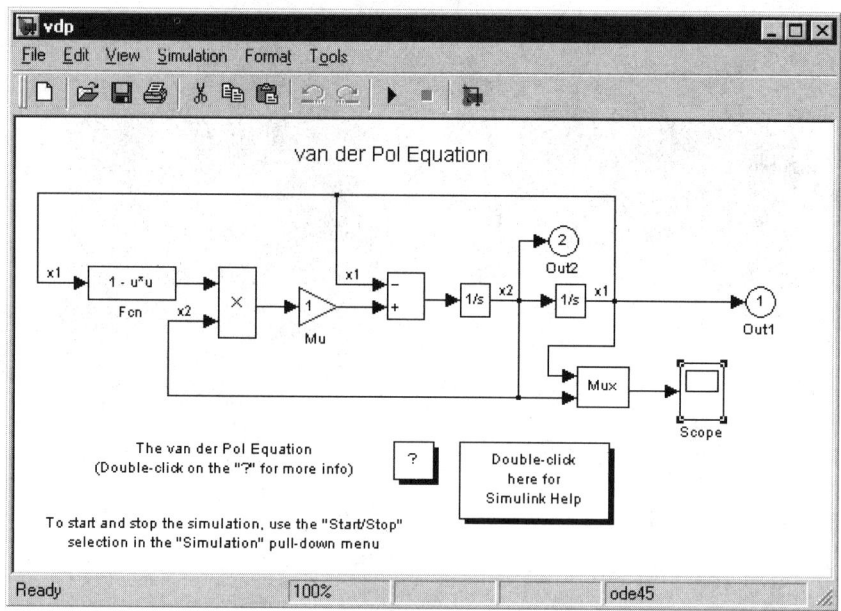

FIGURE 3.2.25.   SIMULINK block diagram to model the Van der Pol equations.

Example 3.2.12. Simulation of a Buck (Step-Down) converter.

The step-down converter is modeled by the following nonlinear differential equations:

$$\frac{du_C}{dt} = \frac{1}{C}(i_L - i_a),$$

$$\frac{di_L}{dt} = \frac{1}{L}(-u_C - (r_L + r_c)i_L + r_c i_a - r_s i_L d_D + V_d d_D),$$

$$\frac{di_a}{dt} = \frac{1}{L_a}(u_C + r_c i_L - (r_a + r_c)i_a - E_a).$$

The parameters are $r_s = 0.05\Omega$, $r_L = 0.03\Omega$, $r_c = 0.2\Omega$, $r_a = 5\Omega$, $C = 0.01$ F, $L = 0.001$ H, $L_a = 0.005$ H, $V_d = 40$ V, and $E_a = 10$ V.

Assume that $t_{on} = 3t_{off}$. The initial conditions are

$$\begin{bmatrix} u_{C0} \\ i_{L0} \\ i_{a0} \end{bmatrix} = \begin{bmatrix} 5 \\ 0.5 \\ 1 \end{bmatrix}.$$

From $t_{on} = 3t_{off}$, one concludes that the duty ratio is $d_D = 0.75$.

Using the differential equations, the following m-files are written to perform the modeling and plot the results.

MATLAB script (c_3_2_12.m)

```
t0=0; tfinal=0.06; tspan=[t0 tfinal];
y0=[5 0.5 1]';
[t,y]=ode45('c_3_2_12b',tspan,y0);
subplot(2,2,1); plot(t,y);
xlabel('Time (seconds)');
title('Dynamics of the state variables');
subplot(2,2,2); plot(t,y(:,1),'-');
xlabel('Time (seconds)'); title('Voltage uc, [V]');
subplot(2,2,3); plot(t,y(:,2),'-');
xlabel('Time (seconds)'); title('Current iL, [A]');
subplot(2,2,4); plot(t,y(:,3),'-');
xlabel('Time (seconds)'); title('Current ia, [A]');
disp('End');
```

MATLAB script (c_3_2_12b.m)

```
% Dynamics of the buck converter
function yprime=difer(t,y);
% Converter parameters
rs=0.05; rl=0.03; rc=0.2; ra=5; C=0.01; L=0.001; La=0.005;
Vd=40; Ea=10; D=0.75;
% differential equations for buck converters
yprime=[(y(2,:)-y(3,:))/C;...
(-y(1,:)-(rl+rc)*y(2,:)+rc*y(3,:)-rs*y(2,:)*D+Vd*D)/L;...
(y(1,:)+rc*y(2,:)-(rc+ra)*y(3,:)-Ea)/La];
```

The transient dynamics for three state variables $u_C(t)$, $i_L(t)$, and $i_a(t)$ are illustrated in Figure 3.2.26, and the settling time is 0.035 sec.

Making note of the output equation $u_a = u_C + r_c i_L - r_c i_a$, the voltage at the load terminal u_a can be plotted. In particular, type in the Command Window

```
>> rc=0.2;plot(t,y(:,1)+rc*y(:,2)-rc*y(:,3),'-');xlabel
('Time (seconds)');title('Voltage ua, [V]');
```

The resulting plot for $u_a(t)$ is documented in Figure 3.2.27.

Example 3.2.13. Simulation of a Boost Converter.

Our goal is to simulate the boost converter, as modeled as

$$\frac{du_C}{dt} = \frac{1}{C}(i_L - i_a - i_L d_D),$$

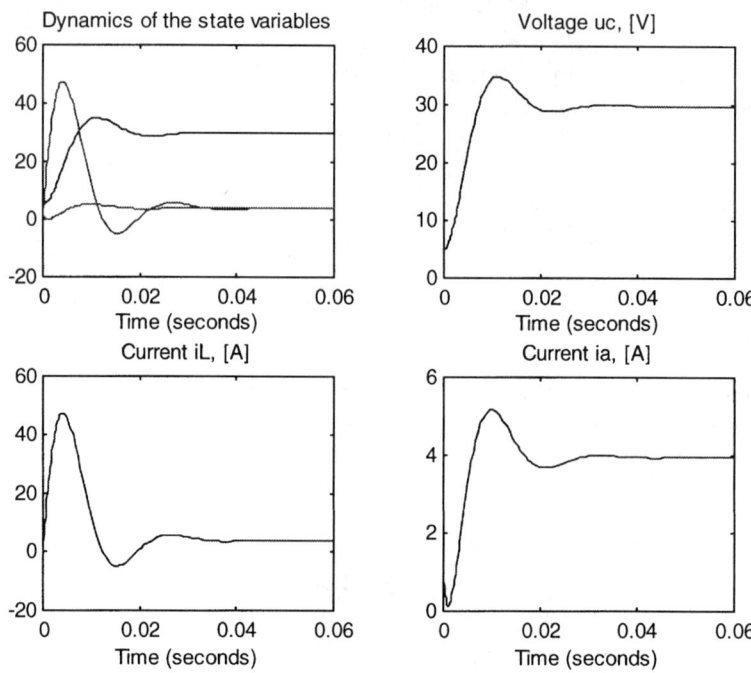

FIGURE 3.2.26.   Transient dynamics of the buck converter.

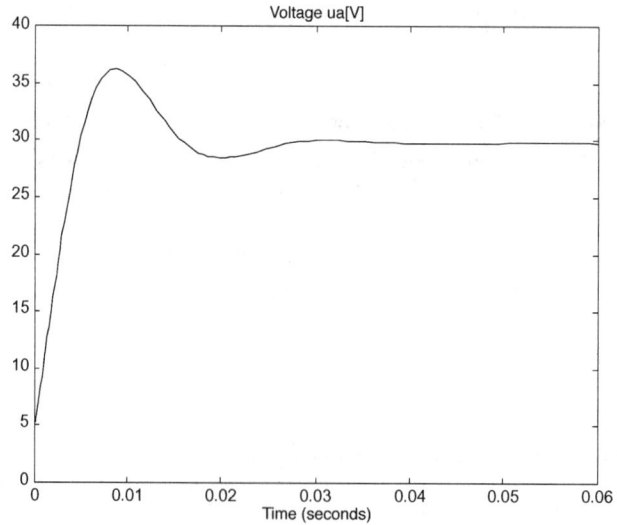

FIGURE 3.2.27.   Transient dynamics of u_a.

$$\frac{di_L}{dt} = \frac{1}{L}(-u_C - (r_L + r_c)i_L + r_c i_a + u_C d_D$$
$$+ (r_c - r_s)i_L d_D - r_c i_a d_D + V_d),$$

$$\frac{di_a}{dt} = \frac{1}{L_a}(u_C + r_c i_L - (r_a + r_c)i_a - r_c i_L d_D - E_a),$$

using the following parameters: $r_s = 0.05\Omega$, $r_L = 0.03\Omega$, $r_c = 0.2\Omega$, $r_a = 5\Omega$, $C = 0.01$ F, and $L = 0.001$ H, $L_a = 0.005$ H, $V_d = 40$ V, and $E_a = 10$ V.
Let $t_{\text{on}} = \frac{1}{3}t_{\text{off}}$. The initial conditions are

$$\begin{bmatrix} u_{c0} \\ i_{L0} \\ i_{a0} \end{bmatrix} = \begin{bmatrix} 5 \\ 2.5 \\ 0 \end{bmatrix}.$$

Taking note of the differential equations, the following m-files are written.

MATLAB script (c_3_2_13a.m)

```
t0=0; tfinal=0.06; tspan=[t0 tfinal];
y0=[5 2.5 0]';
[t,y]=ode45('c_3_2_13b',tspan,y0);
subplot(2,2,1); plot(t,y);
xlabel('Time (seconds)');
title('Dynamics of the state variables');
subplot(2,2,2); plot(t,y(:,1),'-');
xlabel('Time (seconds)'); title('Voltage uc, [V]');
subplot(2,2,3); plot(t,y(:,2),'-');
xlabel('Time (seconds)'); title('Current iL, [A]');
subplot(2,2,4); plot(t,y(:,3),'-');
xlabel('Time (seconds)'); title('Current ia, [A]');
disp('End');
```

MATLAB script (ch_3_2_13b.m)

```
% Dynamics of the boost converter
function yprime=difer(t,y);
% Converter parameters
rs=0.05; rl=0.03; rc=0.2; ra=5; C=0.01; L=0.001;
La=0.005; Vd=40; Ea=10; D=0.25;
% differential equations for boost converters
yprime=[(y(2,:)-y(3,:)-y(2,:)*D)/C;...
(-y(1,:)-(rl+rc)*y(2,:)+rc*y(3,:)+y(1,:)*D+(rc-rs)*y(2,:)*
D-rc*y(3,:)*D+Vd)/L;...
(y(1,:)+rc*y(2,:)-(ra+rc)*y(3,:)-rc*y(2,:)*D-Ea)/La];
```

It should be emphasized that from $t_{\text{on}} = \frac{1}{3}t_{\text{off}}$, one concludes that the duty ratio is 0.25. The transient dynamics of the state variables $u_C(t)$, $i_L(t)$, and $i_a(t)$ is illustrated in Figure 3.2.28.

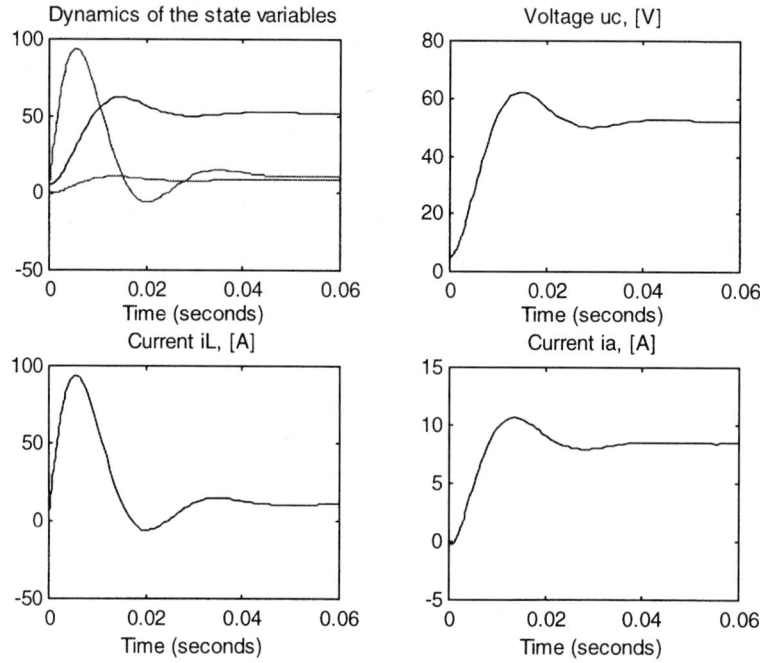

FIGURE 3.2.28.   Transient dynamics of the boost converter.

Example 3.2.14. Simulation of permanent-magnet DC motors.

Let us simulate permanent-magnet DC motors in SIMULINK.

Two linear differential equations must be used to model the motor dynamics (see Example 2.2.11). In particular, we have

$$\frac{di_a}{dt} = -\frac{r_a}{L_a}i_a - \frac{k_a}{L_a}\omega_r + \frac{1}{L_a}u_a,$$

$$\frac{d\omega_r}{dt} = \frac{k_a}{J}i_a - \frac{B_m}{J}\omega_r - \frac{1}{J}T_L.$$

Using an s-domain block diagram documented in Figure 2.2.11, a SIMULINK diagram is built and illustrated in Figure 3.2.29.

The motor parameters (coefficient of differential equations) must be assigned to perform numerical simulations. The motor parameters are assigned to be $r_a = 5\Omega$, $L_a = 0.01$ H, $k_a = 0.2$ V-sec/rad, $k_a = 0.2$ N-m/A, $J = 0.0005$ kg-m^2, and $B_m = 0.00001$ N-m-sec/rad. The applied armature voltage is $u_a = 50\text{rect}(t)$ V, and the load torque is $T_L = 0.2\text{rect}(2t)$ N-m.

Download these motor parameters in the Command Window by typing

```
>> ra=5; La=0.01; ka=0.2; J=0.0005; Bm=0.00001;
```

The transient responses for the state variables (armature current $x_1 = i_a$ and

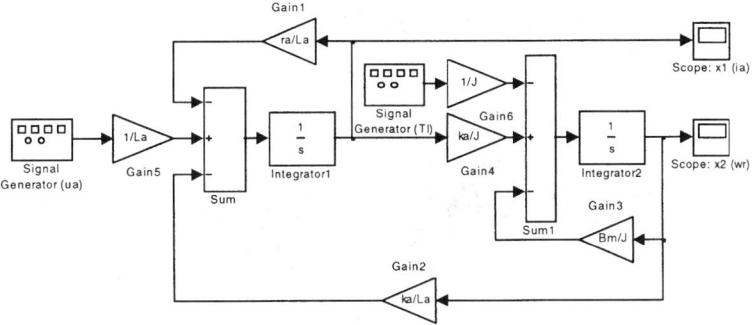

FIGURE 3.2.29. SIMULINK diagram to model permanent-magnet DC motors (c_3_2_14.mdl).

angular velocity $x_2 = \omega_r$) are illustrated in Figure 3.2.30. It should be emphasized that the plot command was used. In particular, to plot the motor dynamics, use

```
plot(x1(:,1),x1(:,2));
xlabel('Time (seconds)'); title('Armature current ia, [A]');
plot(x2(:,1),x2(:,2));
xlabel('Time (seconds)'); title('Velocity wr, [rad/sec]');
```

Example 3.2.15. SIMULINK diagram for stepper motors.

Let us develop the SIMULINK diagram to simulate permanent-magnet stepper motors in the *abc* variables. The mathematical model is given by a set of nonlinear differential equations

$$\frac{di_{as}}{dt} = -\frac{r_s}{L_{ss}}i_{as} + \frac{RT\psi_m}{L_{ss}}\omega_{rm}\sin(RT\theta_{rm}) + \frac{1}{L_{ss}}u_{as},$$

$$\frac{di_{bs}}{dt} = -\frac{r_s}{L_{ss}}i_{bs} - \frac{RT\psi_m}{L_{ss}}\omega_{rm}\cos(RT\theta_{rm}) + \frac{1}{L_{ss}}u_{bs},$$

$$\frac{d\omega_{rm}}{dt} = -\frac{RT\psi_m}{J}\left[i_{as}\sin(RT\theta_{rm}) - i_{bs}\cos(RT\theta_{rm})\right] - \frac{B_m}{J}\omega_{rm} - \frac{1}{J}T_L,$$

$$\frac{d\theta_{rm}}{dt} = \omega_{rm}.$$

The two-phase voltages supplied to the ab stator windings are

$$u_{as} = -\sqrt{2}u_M\sin(RT\theta_{rm}) \text{ and } u_{bs} = \sqrt{2}u_M\cos(RT\theta_{rm}).$$

The SIMULINK diagram is built; see Figure 3.2.31.

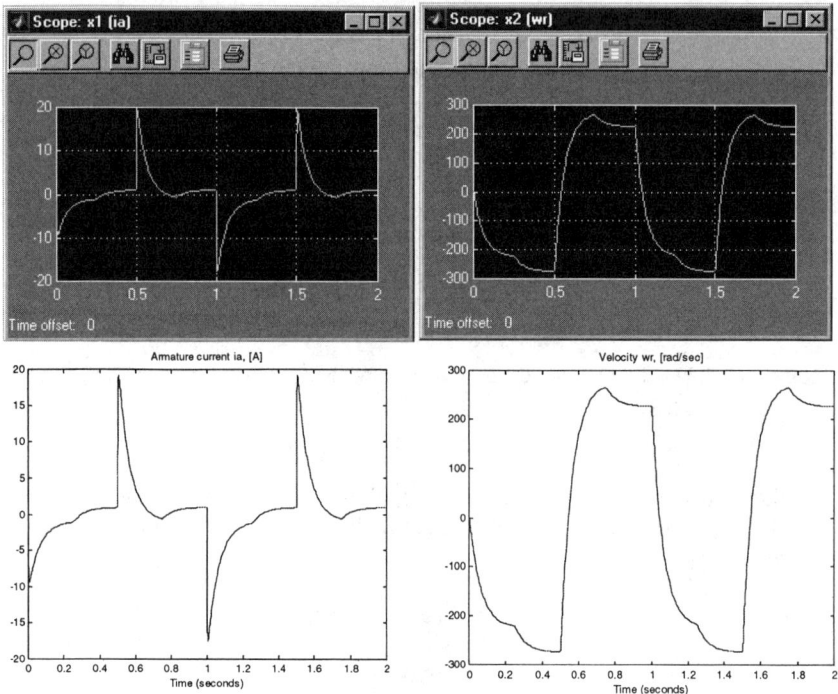

FIGURE 3.2.30.   Permanent-magnet motor dynamics.

Example 3.2.16. Simulation of an elementary reluctance motor.

The nonlinear differential equations to model an elementary synchronous reluctance motor are found in the following form:

$$\frac{di_{as}}{dt} = -\frac{r_s}{L_{ls} + \bar{L}_m - L_{\Delta m}\cos 2\theta_r} i_{as} - \frac{2L_{\Delta m}}{L_{es} + \bar{L}_m - L_{\Delta m}\cos 2\theta_r} i_{as}\omega_r \sin 2\theta_r$$

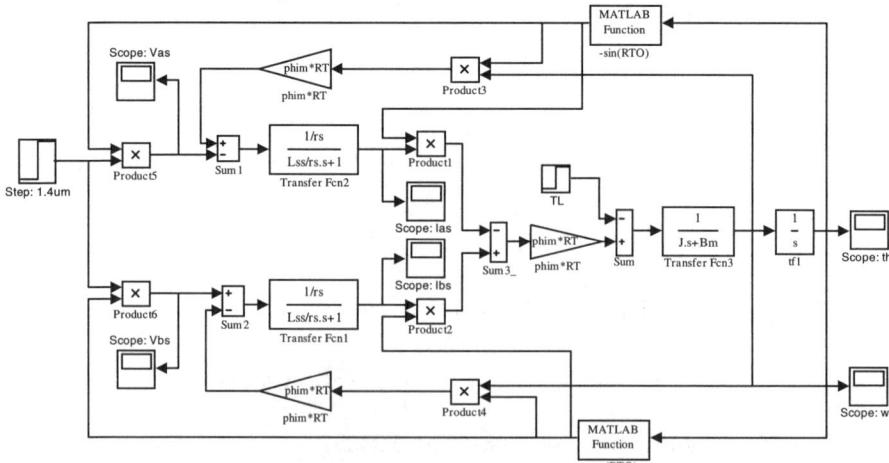

FIGURE 3.2.31.   SIMULINK diagram for simulation of stepper motors.

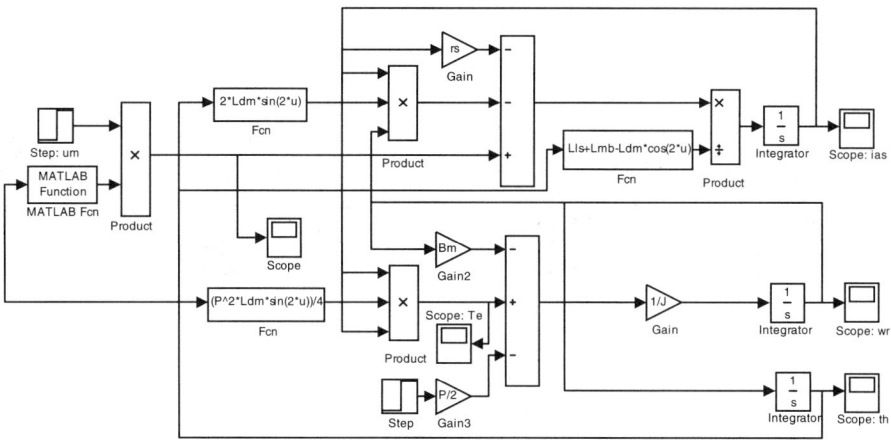

FIGURE 3.2.32. SIMULINK diagram for simulation elementary reluctance motors (c_3_2_16.mdl).

$$+ \frac{1}{L_{ls} + \bar{L}_m - L_{\Delta m}\cos 2\theta_r} u_{as},$$

$$\frac{d\omega_r}{dt} = \frac{1}{J}(L_{\Delta m}i_{as}^2 \sin 2\theta_r - B_m\omega_r - T_L),$$

$$\frac{d\theta_r}{dt} = \omega_r.$$

The motor parameters are $r_s = 2\,\Omega$, $L_{md} = 0.8$ H, $L_{mq} = 0.02$ H, $L_{ls} = 0.04$ H, $J = 0.0006$ kg-m^2, and $B_m = 0.00003$ N-m-sec/rad. The voltage applied to the stator winding is $u_{as} = 220\sin(2\theta_r - 0.62)$. The load torque is $T_L = 0$ N-m.

The motor parameters are downloaded by using the data m-file c_3_2_16d.m.

```
% an elementary synchronous reluctance motor
P=2; rs=2; Lmd=0.8; Lmq=0.02; Lls=0.004; J=0.0006;
Bm=0.00003; Lmb=(Lmq+Lmd)/3; Ldm=(Lmd-Lmq)/3;
%rms value of the voltage
um=220;
%load torque
Tl=0;
```

One must run this data file before the SIMULINK file. The SIMULINK block diagram is documented in Figure 3.2.32.

The transient responses for the state variables must be studied, and the current in the *as* winding as well as the angular velocity are plotted in Figure 3.2.33.

Example 3.2.17. Simulation of three-phase squirrel-cage induction motors.

The mathematical model of a three-phase induction motor is governed by equations

$$u_{as} = r_s i_{as} + (L_{ls} + L_{ms})\frac{di_{as}}{dt} - \frac{1}{2}L_{ms}\frac{di_{bs}}{dt} - \frac{1}{2}L_{ms}\frac{di_{cs}}{dt}$$

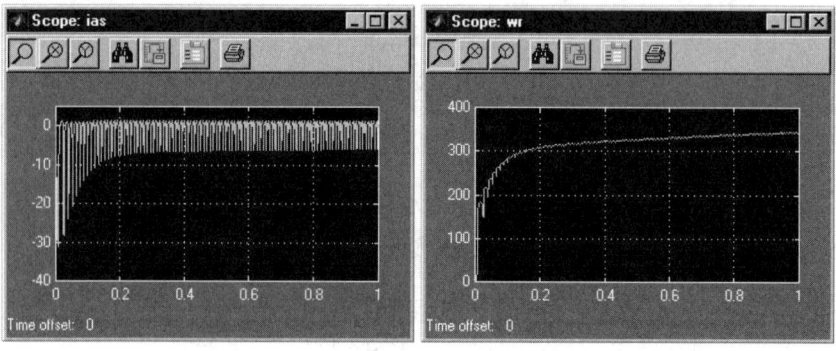

FIGURE 3.2.33.   Transient responses for i_{as} and ω_r.

$$+ L_{ms}\frac{d(i'_{ar}\cos\theta_r)}{dt} + L_{ms}\frac{d(i'_{br}\cos(\theta_r + \frac{2\pi}{3}))}{dt}$$

$$+ L_{ms}\frac{d(i'_{cr}\cos(\theta_r - \frac{2\pi}{3}))}{dt},$$

$$u_{bs} = r_s i_{bs} - \frac{1}{2}L_{ms}\frac{di_{as}}{dt} + (L_{ls} + L_{ms})\frac{di_{bs}}{dt} - \frac{1}{2}L_{ms}\frac{di_{cs}}{dt}$$

$$+ L_{ms}\frac{d(i'_{ar}\cos(\theta_r - \frac{2\pi}{3}))}{dt} + L_{ms}\frac{d(i'_{br}\cos\theta_r)}{dt}$$

$$+ L_{ms}\frac{d(i'_{cr}\cos(\theta_r + \frac{2\pi}{e}))}{dt},$$

$$u_{cs} = r_s i_{cs} - \frac{1}{2}L_{ms}\frac{di_{as}}{dt} - \frac{1}{2}L_{ms}\frac{di_{bs}}{dt} + (L_{ls} + L_{ms})\frac{di'_{cs}}{dt}$$

$$+ L_{ms}\frac{d(i'_{ar}\cos(\theta_r + \frac{2\pi}{3}))}{dt} + L_{ms}\frac{d(i'_{br}\cos(\theta_r - \frac{2\pi}{3}))}{dt}$$

$$+ L_{ms}\frac{d(i'_{cr}\cos\theta_r)}{dt},$$

$$u'_{ar} = r'_r i'_{ar} + L_{ms}\frac{d(i_{as}\cos\theta_r)}{dt} + L_{ms}\frac{d(i_{bs}\cos(\theta_r - \frac{2\pi}{3}))}{dt}$$

$$+ L_{ms}\frac{d(i_{cs}\cos(\theta_r + \frac{2\pi}{3}))}{dt} + (L'_{lr} + L_{ms})\frac{di'_{ar}}{dt}$$

$$- \frac{1}{2}L_{ms}\frac{di'_{br}}{dt} - \frac{1}{2}L_{ms}\frac{di'_{cr}}{dt},$$

$$u'_{br} = r'_r i'_{br} + L_{ms}\frac{d(i_{as}\cos(\theta_r + \frac{2\pi}{3}))}{dt} + L_{ms}\frac{d(i_{bs}\cos\theta_r)}{dt}$$

$$+ L_{ms} \frac{d(i_{cs} \cos(\theta_r - \frac{2\pi}{3}))}{dt} - \frac{1}{2} L_{ms} \frac{di'_{ar}}{dt}$$

$$+ (L'_{lr} + L_{ms}) \frac{di'_{br}}{dt} - \frac{1}{2} L_{ms} \frac{di'_{cr}}{dt},$$

$$u'_{cr} = r'_r i'_{cr} + L_{ms} \frac{d(i_{as} \cos(\theta_r - \frac{2\pi}{3}))}{dt} + L_{ms} \frac{d(i_{bs} \cos(\theta_r + \frac{2\pi}{3}))}{dt}$$

$$+ L_{ms} \frac{d(i_{cs} \cos \theta_r)}{dt} - \frac{1}{2} L_{ms} \frac{di'_{ar}}{dt} - \frac{1}{2} L_{ms} \frac{di'_{br}}{dt} + (L'_{lr} + L_{ms}) \frac{di'_{cr}}{dt},$$

$$\frac{d\omega_r}{dt} = -\frac{P^2}{4J} L_{ms} \left\{ \left[i_{as} (i'_{ar} - \frac{1}{2} i'_{br} - \frac{1}{2} i'_{cr}) \right. \right.$$

$$+ i_{bs} (i'_{br} - \frac{1}{2} i'_{ar} - \frac{1}{2} i'_{cr}) + i_{cs} (i'_{cr} - \frac{1}{2} i'_{br} - \frac{1}{2} i'_{ar}) \right] \sin \theta_r$$

$$+ \frac{\sqrt{3}}{2} \left[i_{as} (i'_{br} - i'_{cr}) + i_{bs} (i'_{cr} - i'_{ar}) \right.$$

$$\left. + i_{cs} (i'_{ar} - i'_{br}) \right] \cos \theta_r \right\} - \frac{B_m}{J} \omega_r - \frac{P}{2J} T_L,$$

$$\frac{d\theta_r}{dt} = \omega_r.$$

Using these equations, which are not given in Cauchy's form, let us build the SIMULINK block diagram to simulate squirrel-cage induction motors. From the above given differential equations, which model induction motor transient dynamics, one obtains

$$\frac{di_{as}}{dt} = \frac{1}{L_{ls} + L_{ms}} \left[-r_s i_{as} + \frac{1}{2} L_{ms} \frac{di_{bs}}{dt} + \frac{1}{2} L_{ms} \frac{di_{cs}}{dt} \right.$$

$$- L_{ms} \frac{d(i'_{ar} \cos \theta_r)}{dt} - L_{ms} \frac{d(i'_{br} \cos(\theta_r + \frac{2\pi}{3}))}{dt}$$

$$\left. - L_{ms} \frac{d(i'_{cr} \cos(\theta_r - \frac{2\pi}{3}))}{dt} + u_{as} \right],$$

$$\frac{di_{bs}}{dt} = \frac{1}{L_{ls} + L_{ms}} \left[-r_s i_{bs} + \frac{1}{2} L_{ms} \frac{di_{as}}{dt} + \frac{1}{2} L_{ms} \frac{di_{cs}}{dt} \right.$$

$$- L_{ms} \frac{d(i'_{ar} \cos(\theta_r - \frac{2\pi}{3}))}{dt} - L_{ms} \frac{d(i'_{br} \cos \theta_r)}{dt}$$

$$\left. - L_{ms} \frac{d(i'_{cr} \cos(\theta_r + \frac{2\pi}{3}))}{dt} + u_{bs} \right],$$

$$\frac{di_{cs}}{dt} = \frac{1}{L_{ls} + L_{ms}} \left[-r_s i_{cs} + \frac{1}{2} L_{ms} \frac{di_{as}}{dt} + \frac{1}{2} L_{ms} \frac{di_{bs}}{dt} \right.$$

$$- L_{ms} \frac{d(i'_{ar} \cos(\theta_r + \frac{2\pi}{3}))}{dt} - L_{ms} \frac{d(i'_{br} \cos(\theta_r - \frac{2\pi}{3}))}{dt}$$

$$-L_{ms}\frac{d(i'_{cr}\cos\theta_r)}{dt}+u_{cs}\Bigg],$$

$$\frac{di'_{ar}}{dt}=\frac{1}{L'_{lr}+L_{ms}}\Bigg[-r'_r i'_{ar}-L_{ms}\frac{d(i_{as}\cos\theta_r)}{dt}$$

$$-L_{ms}\frac{d(i_{bs}\cos(\theta_r-\frac{2\pi}{3}))}{dt}-L_{ms}\frac{d(i_{cs}\cos(\theta_r+\frac{2\pi}{3}))}{dt}$$

$$+\tfrac{1}{2}L_{ms}\frac{di'_{br}}{dt}+\tfrac{1}{2}L_{ms}\frac{di'_{cr}}{dt}+u'_{ar}\Bigg],$$

$$\frac{di'_{br}}{dt}=\frac{1}{L'_{lr}+L_{ms}}\Bigg[-r'_r i'_{br}-L_{ms}\frac{d(i_{as}\cos(\theta_r+\frac{2\pi}{3}))}{dt}-L_{ms}\frac{d(i_{bs}\cos\theta_r)}{dt}$$

$$-L_{ms}\frac{d(i_{cs}\cos(\theta_r-\frac{2\pi}{3}))}{dt}+\tfrac{1}{2}L_{ms}\frac{di'_{ar}}{dt}+\tfrac{1}{2}L_{ms}\frac{di'_{cr}}{dt}+u'_{br}\Bigg],$$

$$\frac{di'_{cr}}{dt}=\frac{1}{L'_{lr}+L_{ms}}\Bigg[-r'_r i'_{cr}-L_{ms}\frac{d(i_{as}\cos(\theta_r-\frac{2\pi}{3}))}{dt}$$

$$-L_{ms}\frac{d(i_{bs}\cos(\theta_r+\frac{2\pi}{3}))}{dt}-L_{ms}\frac{d(i_{cs}\cos\theta_r)}{dt}$$

$$+\tfrac{1}{2}L_{ms}\frac{di'_{ar}}{dt}+\tfrac{1}{2}L_{ms}\frac{di'_{br}}{dt}+u'_{cr}\Bigg],$$

$$\frac{d\omega_r}{dt}=-\frac{P^2}{4J}L_{ms}\Big\{\big[i_{as}(i'_{ar}-\tfrac{1}{2}i'_{br}-\tfrac{1}{2}i'_{cr})+i_{bs}(i'_{br}-\tfrac{1}{2}i'_{ar}-\tfrac{1}{2}i'_{cr})$$

$$+i_{cs}(i'_{cr}-\tfrac{1}{2}i'_{br}-\tfrac{1}{2}i'_{ar})\big]\sin\theta_r+\frac{\sqrt{3}}{2}\big[i_{as}(i'_{br}-i'_{cr})$$

$$+i_{bs}(i'_{cr}-i'_{ar})+i_{cs}(i'_{ar}-i'_{br})\big]\cos\theta_r\Big\}-\frac{B_m}{J}\omega_r-\frac{P}{2J}T_L,$$

$$\frac{d\theta_r}{dt}=\omega_r.$$

To guarantee the balanced operating condition, the following phase voltages should be applied:

$$u_{as}(t)=\sqrt{2}u_M\cos(\omega_f t),$$
$$u_{bs}(t)=\sqrt{2}u_M\cos(\omega_f t-\tfrac{2}{3}\pi),$$

and

$$u_{cs}(t)=\sqrt{2}u_M\cos(\omega_f t+\tfrac{2}{3}\pi).$$

The SIMULINK diagram to simulate three-phase squirrel-cage induction motors is developed (see Figure 3.2.34) using the Derivative block.

A two-pole induction motor has the following parameters: $r_s=0.3\Omega$, $r_r=0.2\Omega$, $L_{ms}=0.035$ H, $L_{ls}=0.001$ H, $L_{lr}=0.001$ H, $J=0.025$ kg-m^2, and $B_m=0.004$ N-m-sec/rad.

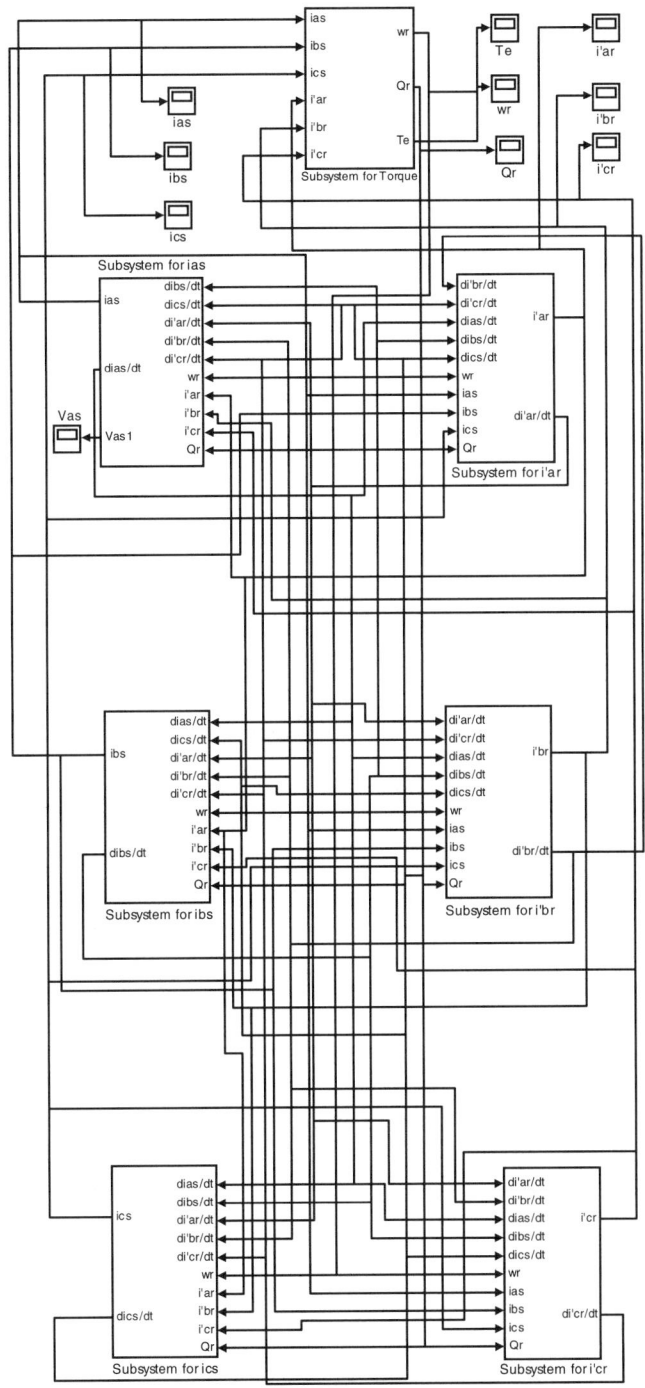

FIGURE 3.2.34. SIMULINK block diagram to simulate squirrel-cage induction motors (c_3_2_17.mdl).

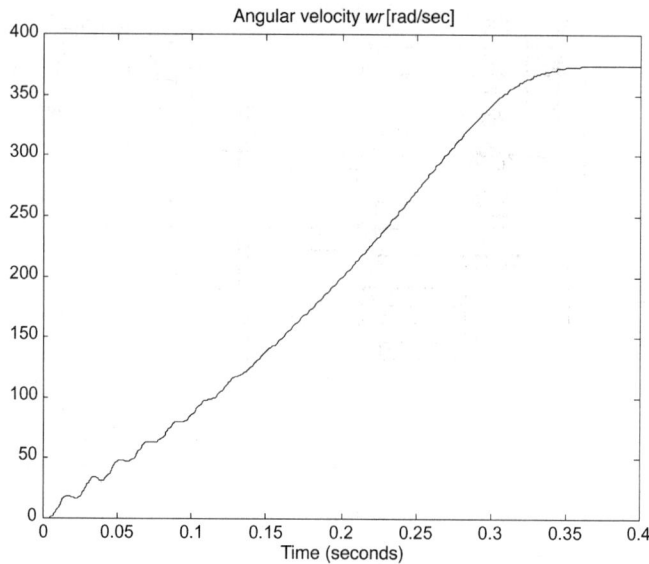

FIGURE 3.2.35.   Transient dynamic for the angular velocity: Squirrel-cage induction motor accelerates from stall.

In the Command Window, the designer downloads the motor parameters as

```
≫P=2; Rs=0.3; Rr=0.2; Lms=0.035; Lls=0.001; Llr=0.001;
Bm=0.004; J=0.025;
```

Nonlinear simulations were performed, and transient dynamics of the stator and rotor currents in the *as*, *bs*, *cs*, *ar*, *br*, and *cr* windings $i_{as}(t)$, $i_{bs}(t)$, $i_{cs}(t)$, $i'_{ar}(t)$, $i'_{br}(t)$, and $i'_{cr}(t)$ as well as the rotor angular velocity $\omega_{rm}(t)$ are documented in Figures 3.2.35 and 3.2.36. The plotting statement to plot the evolution of $\omega_{rm}(t)$ is

```
≫plot(wr(:,1),wr(:,2));
xlabel('Time (seconds)'); title ('Angular velocity wr,
[rad/sec]]');
```

Example 3.2.18. Simulation of three-phase permanent-magnet synchronous motors.

The nonlinear mathematical model of permanent-magnet synchronous motors in Cauchy's form is given by a system of five highly nonlinear differential equations

$$\frac{di_{as}}{dt} = -\frac{r_s(2L_{ss} - \bar{L}_m)}{2L_{ss}^2 - L_{ss}\bar{L}_m - \bar{L}_m^2}i_{as} - \frac{r_s\bar{L}_m}{2L_{ss}^2 - L_{ss}\bar{L}_m - \bar{L}_m^2}i_{bs}$$

$$- \frac{r_s\bar{L}_m}{2L_{ss}^2 - L_{ss}\bar{L}_m - \bar{L}_m^2}i_{cs} - \frac{\psi_m(2L_{ss} - \bar{L}_m)}{2L_{ss}^2 - L_{ss}\bar{L}_m - \bar{L}_m^2}\omega_r \cos\theta_r$$

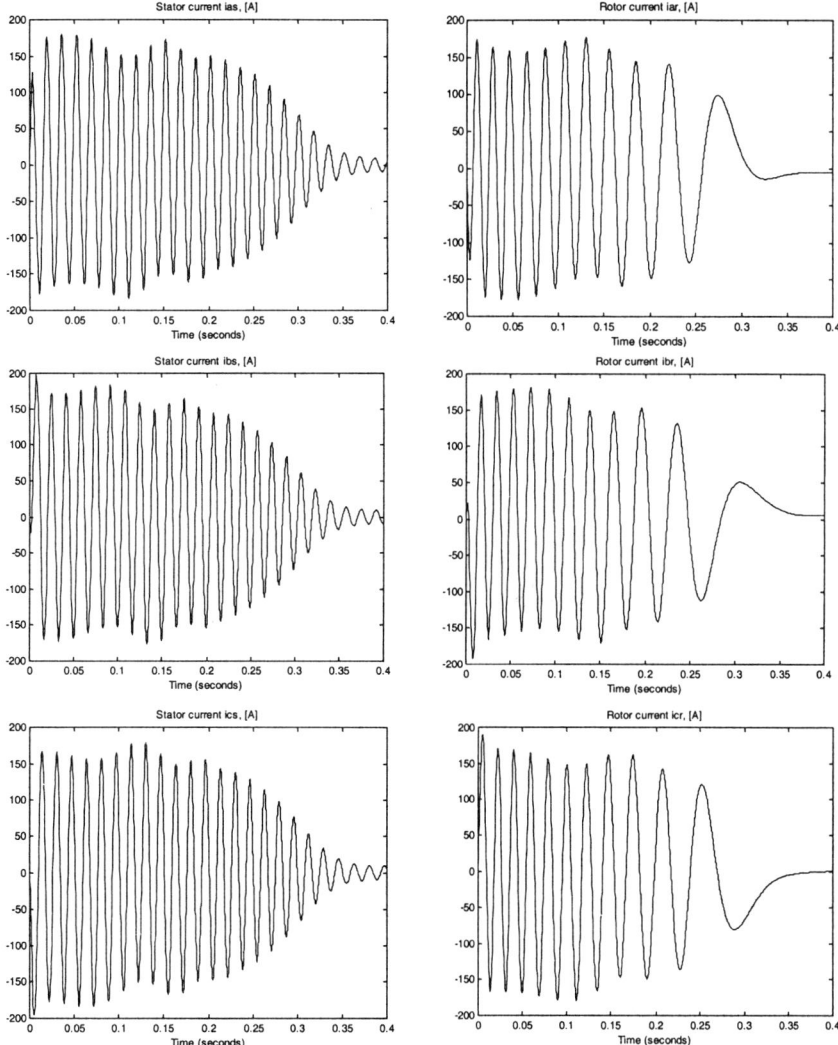

FIGURE 3.2.36.   Transient dynamic of the currents in the stator and rotor windings: Squirrel-cage induction motor accelerates from stall.

$$
-\frac{\psi_m \bar{L}_m}{2L_{ss}^2 - L_{ss}\bar{L}_m - \bar{L}_m^2}\omega_r \cos(\theta_r - \tfrac{2}{3}\pi)
$$

$$
-\frac{\psi_m \bar{L}_m}{2L_{ss}^2 - L_{ss}\bar{L}_m - \bar{L}_m^2}\omega_r \cos(\theta_r + \tfrac{2}{3}\pi) + \frac{2L_{ss} - \bar{L}_m}{2L_{ss}^2 - L_{ss}\bar{L}_m - \bar{L}_m^2}
$$

$$
\times u_{as} + \frac{\bar{L}_m}{2L_{ss}^2 - L_{ss}\bar{L}_m - \bar{L}_m^2}u_{bs} + \frac{\bar{L}_m}{2L_{ss}^2 - L_{ss}\bar{L}_m - \bar{L}_m^2}u_{cs},
$$

$$
\frac{di_{bs}}{dt} = -\frac{r_s \bar{L}_m}{2L_{ss}^2 - L_{ss}\bar{L}_m - \bar{L}_m^2}i_{as} - \frac{r_s(2L_{ss} - \bar{L}_m)}{2L_{ss}^2 - L_{ss}\bar{L}_m - \bar{L}_m^2}i_{bs}
$$

$$
-\frac{r_s \bar{L}_m}{2L_{ss}^2 - L_{ss}\bar{L}_m - \bar{L}_m^2} i_{cs} - \frac{\psi_m \bar{L}_m}{2L_{ss}^2 - L_{ss}\bar{L}_m - \bar{L}_m^2}
$$

$$
\times \omega_r \cos \theta_r - \frac{\psi_m (2L_{ss} - \bar{L}_m)}{2L_{ss}^2 - L_{ss}\bar{L}_m - \bar{L}_m^2} \omega_r \cos(\theta_r - \tfrac{2}{3}\pi)
$$

$$
-\frac{\psi_m \bar{L}_m}{2L_{ss}^2 - L_{ss}\bar{L}_m - \bar{L}_m^2} \omega_r \cos(\theta_r + \tfrac{2}{3}\pi) + \frac{\bar{L}_m}{2L_{ss}^2 - L_{ss}\bar{L}_m - \bar{L}_m^2}
$$

$$
\times u_{as} + \frac{2L_{ss} - \bar{L}_m}{2L_{ss}^2 - L_{ss}\bar{L}_m - \bar{L}_m^2} u_{bs} - \frac{\bar{L}_m}{2L_{ss}^2 - L_{ss}\bar{L}_m - \bar{L}_m^2} u_{cs},
$$

$$
\frac{di_{cs}}{dt} = -\frac{r_s \bar{L}_m}{2L_{ss}^2 - L_{ss}\bar{L}_m - \bar{L}_m^2} i_{as} - \frac{r_s \bar{L}_m}{2L_{ss}^2 - L_{ss}\bar{L}_m - \bar{L}_m^2} i_{bs}
$$

$$
-\frac{r_s (2L_{ss} - \bar{L}_m)}{2L_{ss}^2 - L_{ss}\bar{L}_m - \bar{L}_m^2} i_{cs} - \frac{\psi_m \bar{L}_m}{2L_{ss}^2 - L_{ss}\bar{L}_m - \bar{L}_m^2} \omega_r \cos \theta_r
$$

$$
-\frac{\psi_m \bar{L}_m}{2L_{ss}^2 - L_{ss}\bar{L}_m - \bar{L}_m^2} \omega_r \cos(\theta_r - \tfrac{2}{3}\pi) - \frac{\psi_m (2L_{ss} - \bar{L}_m)}{2L_{ss}^2 - L_{ss}\bar{L}_m - \bar{L}_m^2}
$$

$$
\times \omega_r \cos(\theta_r + \tfrac{2}{3}\pi) + \frac{\bar{L}_m}{2L_{ss}^2 - L_{ss}\bar{L}_m - \bar{L}_m^2} u_{as}
$$

$$
+ \frac{\bar{L}_m}{2L_{ss}^2 - L_{ss}\bar{L}_m - \bar{L}_m^2} u_{bs} + \frac{2L_{ss} - \bar{L}_m}{2L_{ss}^2 - L_{ss}\bar{L}_m - \bar{L}_m^2} u_{cs},
$$

$$
\frac{d\omega_r}{dt} = \frac{P^2 \psi_m}{4J} \left(i_{as} \cos \theta_r + i_{bs} \cos(\theta_r - \tfrac{2}{3}\pi) \right.
$$

$$
\left. + i_{cs} \cos(\theta_r + \tfrac{2}{3}\pi) \right) - \frac{B_m}{j} \omega_r - \frac{P}{2J} T_L,
$$

$$
\frac{d\theta_r}{dt} = \omega_r.
$$

As the differential equations are known, one can develop the SIMULINK diagram to simulate permanent-magnet synchronous motors. To guarantee the balanced operating condition, we apply the following phase voltages:

$$
u_{as}(t) = \sqrt{2} u_M \cos \theta_r, \; u_{bs}(t) = \sqrt{2} u_M \cos(\theta_r - \tfrac{2}{3}\pi),
$$

$$
\text{and} \quad u_{cs}(t) = \sqrt{2} u_M \cos(\theta_r + \tfrac{2}{3}\pi).
$$

The motor parameters are $u_M = 50$ V, $r_s = 0.5\Omega$, $L_{ss} = 0.001$ H, $L_{ls} = 0.0001$ H, $\bar{L}_m = 0.0009$ H, $\psi_m = 0.069$ V-sec/rad or $\psi_m = 0.069$ N-m/A, $B_m = 0.0000115$ N-m-sec/rad, and $J = 0.000017$ kg-m^2.

The SIMULINK diagram is developed and illustrated in Figure 3.2.37. The transient dynamics is studied as the motor accelerates from stall and the rated voltage is supplied to the stator windings

$$
u_{as}(t) = \sqrt{250} \cos \theta_r, \; u_{bs}(t) = \sqrt{250} \cos(\theta_r - \tfrac{2}{3}\pi)
$$

$$
\text{and} \quad u_{cs}(t) = \sqrt{250} \cos(\theta_r + \tfrac{2}{3}\pi).
$$

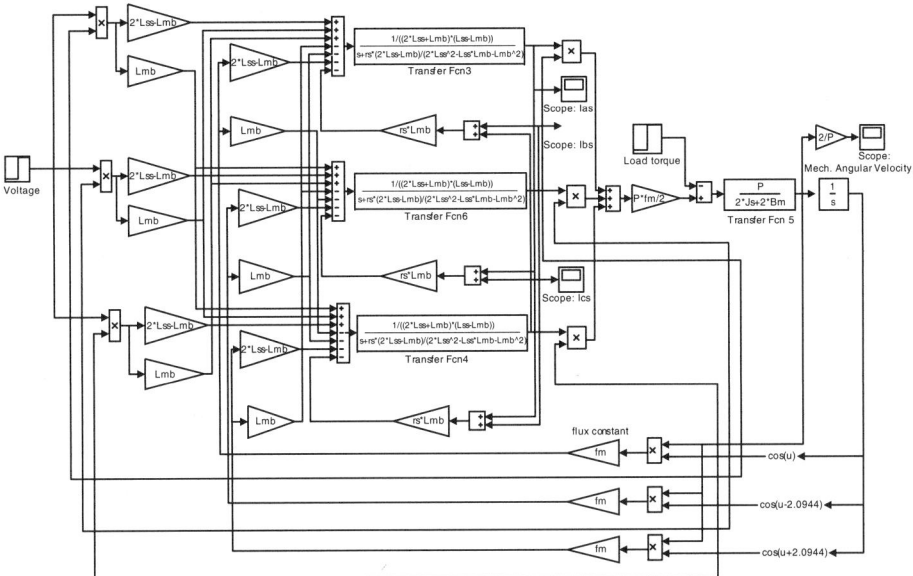

FIGURE 3.2.37. SIMULINK diagram to simulate permanent-magnet synchronous motors (c_3_2_18.mdl).

The motor parameters are downloaded using file c_3_2_18d.m. In particular,

```
% Parameters of the permanent-magnet synchronous motor
P=4; um=50; rs=0.4; Lss=0.0012; Lls=0.0002; fm=0.075;
Bm=0.00002; J=0.00007; Lmb=2*(Lss-Lls)/3;
```

The motor accelerates from stall, and the load torque 0.5 N-m is applied at 0 sec. The acceleration capabilities of the permanent-magnet synchronous motor, documented in Figure 3.2.38, indicate that the motor reaches the steady-state angular velocity within 0.02 sec. It is important to emphasize that permanent-magnet synchronous motors have superior performance compared with DC and induction motors. Because of high starting torque and excellent controllability capabilities in the full torque-speed operating envelope, permanent-magnet synchronous motors are widely used in high-performance electric drives and servo-systems.

The following m-file is developed to plot the transient data (c_3_2_18p.m):

```
% Plots of the transient dynamics of the permanent-magnet
% synchronous motor
plot(Ias(:,1),Ias(:,2)); xlabel('Time (seconds)');
title ('Current Ias, [A]'); pause
plot(Ibs(:,1),Ibs(:,2)); xlabel('Time (seconds)');
title ('Current Ibs, [A]'); pause
plot(Ics(:,1),Ics(:,2)); xlabel('Time (seconds)');
title ('Current Ics, [A]'); pause
plot(wrm(:,1),wrm(:,2)); xlabel('Time (seconds)');
title ('Angular velocity wrm, [rad/sec]'); pause
```

Current, i_{as} [A]   Current, i_{bs} [A]

Current, i_{cs} [A]   Angular velocity,

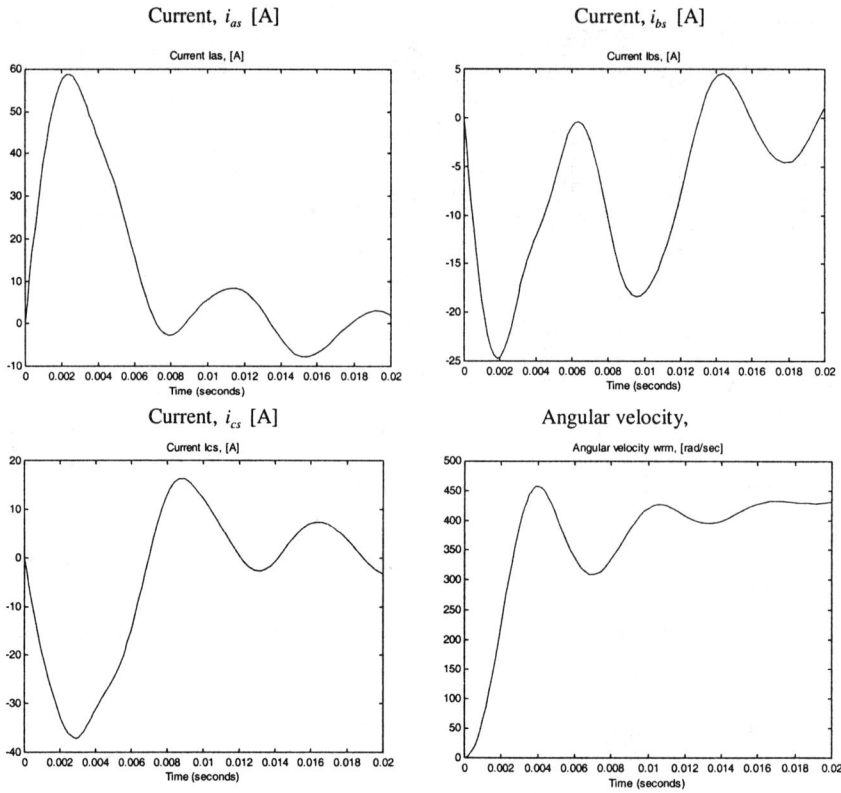

FIGURE 3.2.38.   Acceleration of a three-phase permanent-magnet synchronous motor: $u_{as}(t) = \sqrt{2}50\cos\theta_r$, $u_{bs}(t) = \sqrt{2}50\cos(\theta_r - \frac{2}{3}\pi)$, and $u_{cs}(t) = \sqrt{2}50\cos(\theta_r + \frac{2}{3}\pi)$.

Example 3.2.19. Simulation of permanent-magnet DC motors in the state-space.

Our goal is to simulate permanent-magnet DC motors in SIMULINK using the state-space form. In Example 2.2.11, the differential equations to model permanent-magnet DC motors were found to be

$$\frac{di_a}{dt} = -\frac{r_a}{L_a}i_a - \frac{k_a}{L_a}\omega_r + \frac{1}{L_a}u_a,$$
$$\frac{d\omega_r}{dt} = \frac{k_a}{J}i_a - \frac{B_m}{J}\omega_r - \frac{1}{J}T_L.$$

That is, in matrix form, we have

$$\begin{bmatrix} \dfrac{di_a}{dt} \\ \dfrac{d\omega_r}{dt} \end{bmatrix} = \begin{bmatrix} -\dfrac{r_a}{L_a} & -\dfrac{k_a}{L_a} \\ \dfrac{k_a}{J} & -\dfrac{B_m}{J} \end{bmatrix} \begin{bmatrix} i_a \\ \omega_r \end{bmatrix} + \begin{bmatrix} \dfrac{1}{L_a} \\ 0 \end{bmatrix} u_a.$$

Denoting the state and control variables to be $x_1 = i_a$, $x_2 = \omega_r$ and $u = u_a$, one finds

$$\begin{bmatrix} \dfrac{dx}{dt} \end{bmatrix} = \begin{bmatrix} \dfrac{dx_1}{dt} \\ \dfrac{dx_2}{dt} \end{bmatrix} = \begin{bmatrix} -\dfrac{r_a}{L_a} & -\dfrac{k_a}{L_a} \\ \dfrac{k_a}{J} & -\dfrac{B_m}{J} \end{bmatrix} \begin{bmatrix} x_1 \\ x_2 \end{bmatrix} + \begin{bmatrix} \dfrac{1}{L_a} \\ 0 \end{bmatrix} u,$$

with the initial conditions

$$\begin{bmatrix} i_{a0} \\ \omega_{r0} \end{bmatrix} = \begin{bmatrix} x_{10} \\ x_{20} \end{bmatrix}.$$

The output equation is $y = \omega_r$.
Hence,

$$y = \omega_r = Cx + Du = \begin{bmatrix} 0 & 1 \end{bmatrix} \begin{bmatrix} i_a \\ \omega_r \end{bmatrix}$$

$$+ [0]u_a = \begin{bmatrix} 0 & 1 \end{bmatrix} \begin{bmatrix} x_1 \\ x_2 \end{bmatrix} + [0]u, \; C = \begin{bmatrix} 0 & 1 \end{bmatrix} \text{ and } D = [0].$$

Using the State-Space block, the modeling can be performed. To attain the flexibility, the symbolic notations are used. The State-Space block is illustrated in Figure 3.2.39.

The developed SIMULINK diagram is documented in Figure 3.2.40.

The simulation can be performed assigning the motor parameters and initial conditions.

The following motor parameters are used: $r_a = 5\Omega$, $L_a = 0.01$ H, $k_a = 0.2$ V-sec/rad, $k_a = 0.2$ N-m/A, $J = 0.0005$ kg-m^2, and $B_m = 0.00001$ N-m-sec/rad. The voltage is $u_a = 50\text{rect}(t)$ V.

The initial conditions are

$$\begin{bmatrix} i_{a0} \\ \omega_{r0} \end{bmatrix} = \begin{bmatrix} x_{10} \\ x_{20} \end{bmatrix} = \begin{bmatrix} 5 \\ 100 \end{bmatrix}.$$

To perform the simulations, in the Command Window, we download the motor parameters and initial conditions as

```
≫ra=5; La=0.01; ka=0.2; J=0.0005; Bm=0.00001; x10=5;
x20=100;
```

Running simulation and using the following plotting statement:

```
≫plot(x2(:,1),x2(:,2)); xlabel('Time (seconds)');
title ('Angular velocity wr, [rad/sec]')
```
the dynamics of the motor angular velocity results; see Figures 3.2.41.

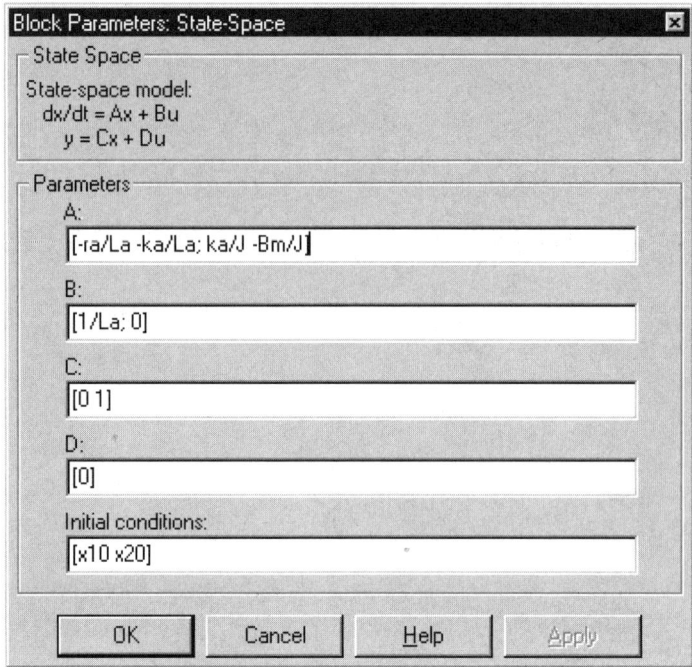

FIGURE 3.2.39.   State-space block with parameters of permanent-magnet DC motors.

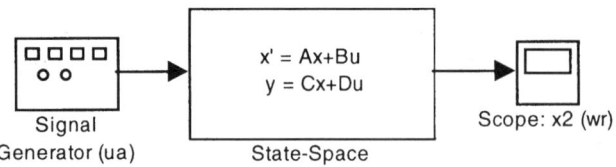

FIGURE 3.2.40.   SIMULINK diagram to simulate the motor dynamics (c_3_2_19.mdl).

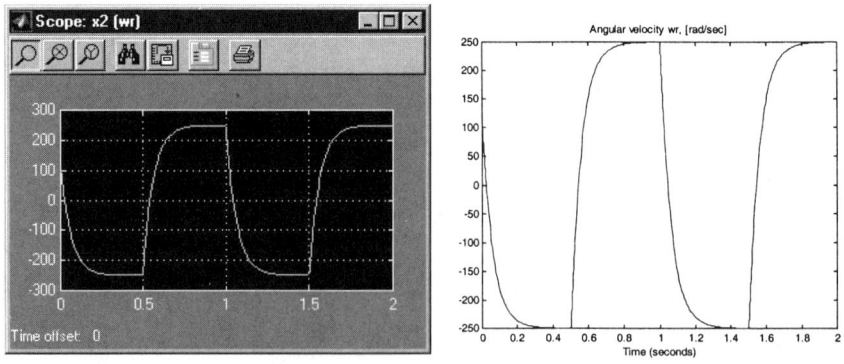

FIGURE 3.2.41.   Motor angular velocity waveforms, $u_a = 50\mathrm{rect}(t)$ V.

4

Analysis and Control of Linear Dynamic Systems

4.1. Introduction: Analysis of Multivariable Continuous- and Discrete-Time Systems

In this chapter, we will develop the basic relationships to solve the motion control problem for dynamic systems using analog and digital proportional-integral-derivative (PID) controllers as well as state-space control algorithms. It should be emphasized that the PID control laws use the tracking error $e(t)$, whereas the state-space controllers use the tracking error $e(t)$ and the state variables $x(t)$. Proportional-integral-derivative and state-space control algorithms are widely used in dynamic systems to stabilize systems, attain tracking and disturbance attenuation, guarantee robustness and accuracy, and so forth. This chapter provides the introduction to nonlinear control and feedback tracking, and single-input/single-output as well as multi-input/multi-output systems are studied. We model the dynamic systems in the s- and z-domains using transfer functions $G_{sys}(s)$ and $G_{sys}(z)$ to synthesize the PID-type controllers. The state-space models are used to design control algorithms applying the Hamilton–Jacobi and Lyapunov theories for multi-input/multi-output systems. The theoretical foundations in analysis and design of dynamic systems modeled using linear differential equations are needed to be covered to fully understand the basic concepts in control of nonlinear systems. It should be emphasized that analysis of linear continuous- and discrete-time closed-loop systems with control constraints will be accomplished. As mathematical models are found in the form of differential or difference equations, system parameters (coefficients of differential or difference equations) are defined, and the analysis can be performed to study stability and stability margins, time response, accuracy, and so on. The system characteristics and performance can be improved and "shaped" using PID-type and state-space stabilizing and tracking controllers studied in this chapter. The general problem approached is the design of PID-type and state-space controllers to ensure the specifications imposed on the desired performance of closed-loop dynamic systems. The synthesis of analog and digital control laws involve the design of controller structures as well as adjusting feedback coefficients to attain certain desired criteria and characteristics. These performance specifications relate stability, robustness, dynamics, accuracy, tracking, disturbance attenuation, as well as other criteria needed to be achieved through the use of control algorithms. The tradeoff between stability and accuracy, robustness and system response, and complexity and implementability is well known. The design procedures are reported to find the structures and feedback coefficients of control laws.

Let us analyze the stability of linear continuos-time systems. In state-space form, we have n-states $x \in \mathbb{R}^n$ and m-controls $u \in \mathbb{R}^m$. The transient dynamics of linear dynamic systems is described by a set of n linear first-order differential equations

$$\frac{dx_1}{dt} = a_{11}x_1 + a_{12}x_2 + \cdots + a_{1n-1}x_{n-1} + a_{1n}x_n + b_{11}u_1 + b_{12}u_2 + \cdots$$
$$+ b_{1m-1}u_{m-1} + b_{1m}u_m, x_1(t_0) = x_{10},$$

$$\frac{dx_2}{dt} = a_{21}x_1 + a_{22}x_2 + \cdots + a_{2n-1}x_{n-1} + a_{2n}x_n + b_{21}u_1 + b_{22}u_2 + \cdots$$
$$+ b_{2m-1}u_{m-1} + b_{2m}u_m, x_2(t_0) = x_{20},$$

$$\vdots$$

$$\frac{dx_{n-1}}{dt} = a_{n-11}x_1 + a_{n-12}x_2 + \cdots + a_{n-1n-1}x_{n-1} + a_{n-1n}x_n + \cdots$$
$$+ b_{n-11}u_1 + b_{n-12}u_2 + \cdots + b_{n-1m-1}u_{m-1} + b_{m-1m}u_m, x_{n-1}(t_0) = x_{n-10},$$

$$\frac{dx_n}{dt} = a_{n1}x_1 + a_{n2}x_2 + \cdots + a_{nn-1}x_{n-1} + a_{nn}x_n + b_{n1}u_1 + b_{n2}u_2 + \cdots$$
$$+ b_{nm-1}u_{m-1} + b_{nm}u_m, x_n(t_0) = x_{n0},$$

which are represented in matrix form as

$$\frac{dx}{dt} = \begin{bmatrix} \dfrac{dx_1}{dt} \\ \dfrac{dx_2}{dt} \\ \vdots \\ \dfrac{dx_{n-1}}{dt} \\ \dfrac{dx_n}{dt} \end{bmatrix} = \begin{bmatrix} a_{11} & a_{12} & \cdots & a_{1n-1} & a_{1n} \\ a_{21} & a_{22} & \cdots & a_{2n-1} & a_{2n} \\ \vdots & \vdots & \ddots & \vdots & \vdots \\ a_{n-11} & a_{n-12} & \cdots & a_{n-1n-1} & a_{n-1n} \\ a_{n1} & a_{n2} & \cdots & a_{nn-1} & a_{nn} \end{bmatrix} \begin{bmatrix} x_1 \\ x_2 \\ \vdots \\ x_{n-1} \\ x_n \end{bmatrix}$$

$$+ \begin{bmatrix} b_{11} & b_{12} & \cdots & b_{1m-1} & b_{1m} \\ b_{21} & b_{22} & \cdots & b_{2m-1} & b_{2m} \\ \vdots & \vdots & \ddots & \vdots & \vdots \\ b_{n-11} & b_{n-12} & \cdots & b_{n-1m-1} & b_{n-1m} \\ b_{n1} & b_{n2} & \cdots & b_{nm-1} & b_{nm} \end{bmatrix} \begin{bmatrix} u_1 \\ u_2 \\ \vdots \\ u_{m-1} \\ u_m \end{bmatrix}$$

$$= Ax + Bu, \quad x(t_0) = x_0.$$

Consider time-invariant dynamic systems. That is, the matrices $A \in \mathbb{R}^{n \times n}$ and $B \in \mathbb{R}^{n \times m}$ are constant. The characteristic equation is given as

$$sI - A = 0 \text{ or } a_n s^n + a_{n-1}s^{n-1} + \cdots + a_1 s + a_0 = 0.$$

Here, $I \in \mathbb{R}^{n \times n}$ is the identity matrix.

Solving the characteristic equation, one finds the eigenvalues (characteristic roots). The system is stable if real parts of all eigenvalues are negative.

The transfer function

$$G(s) = \frac{Y(s)}{U(s)}$$

can be found using the state-space equations. Consider the linear time-invariant dynamic systems as described by

$$\frac{dx}{dt} = Ax + Bu, \quad y = Hx.$$

Taking the Laplace transform of the state-space and output equations, we obtain

$$sX(s) - x(t_0) = AX(s) + BU(s), \quad Y(s) = HX(s).$$

Assuming that $x(t_0) = 0$, we have $X(s) = (sI - A)^{-1}BU(s)$.

Hence, $Y(s) = HX(s) = H(sI - A)^{-1}BU(s)$.

Therefore, the transfer function is

$$G(s) = \frac{Y(s)}{U(s)} = H(sI - A)^{-1}B.$$

For dynamic systems, the transfer function can be found from $dx/dt = Ax + Bu$, $y = Hx$ using the ss2tf MATLAB command. From

$$G(s) = \frac{Y(s)}{U(s)},$$

differential equations $dx/dt = Ax + Bu$, $y = Hx$ are obtained by applying the tf2ss command.

Assuming that the initial conditions are zero, we apply the Laplace transform to both sides of the n-order differential equation

$$\sum_{i=0}^{n} a_i \frac{d^i y(t)}{dt^i} = \sum_{i=0}^{m} b_i \frac{d^i u(t)}{dt^i}.$$

Taking note of

$$\left(\sum_{i=0}^{n} a_i s^i \right) Y(s) = \left(\sum_{i=0}^{m} b_i s^i \right) U(s),$$

one concludes that the transfer function is the ratio of polynomials. In particular,

$$G(s) = \frac{Y(s)}{U(s)} = \frac{b_m s^m + b_{m-1} s^{m-1} + \cdots + b_1 s + b_0}{a_n s^n + a_{n-1} s^{n-1} + \cdots + a_1 s + a_0}.$$

By setting the denominator polynomial of the transfer function to zero, one obtains the characteristic equation. The stability of linear time-invariant dynamic systems is guaranteed if all characteristic eigenvalues, obtained by solving the characteristic equation

$$a_n s^n + a_{n-1} s^{n-1} + \cdots + a_1 s + a_0 = 0,$$

have negative real parts.

Example 4.1.1.

Find the state-space model if the transfer function is expressed as

$$G(s) = \frac{Y(s)}{U(s)} = \frac{200s^2 + 20s + 2}{s^4 + 2s^3 + 3s^2 + 4s + 5}.$$

Obtain the characteristic eigenvalues.

In the Command Window, one downloads the numerator and denominator, and the matrices of the state-space model are found applying the tf2ss command. That is, we have

```
≫num=[200 20 2]; den=[1 2 3 4 5];
[A,B,H,D]=tf2ss(num,den)
```

By pressing the Enter key, the resulting matrices A, B, H and D are displayed as

```
A  =
        -2        -3        -4        -5
         1         0         0         0
         0         1         0         0
         0         0         1         0
B  =
         1
         0
         0
         0
H  =
         0       200        20         2
D  =
         0
```

That is, we have

$$\frac{dx}{dt} = \begin{bmatrix} \dfrac{dx_1}{dt} \\[2mm] \dfrac{dx_2}{dt} \\[2mm] \dfrac{dx_3}{dt} \\[2mm] \dfrac{dx_4}{dt} \end{bmatrix} = Ax + Bu = \begin{bmatrix} -2 & -3 & -4 & -5 \\ 1 & 0 & 0 & 0 \\ 0 & 1 & 0 & 0 \\ 0 & 0 & 1 & 0 \end{bmatrix} \begin{bmatrix} x_1 \\ x_2 \\ x_3 \\ x_4 \end{bmatrix} + \begin{bmatrix} 1 \\ 0 \\ 0 \\ 0 \end{bmatrix} u,$$

$$y = Hx + Du = \begin{bmatrix} 0 & 200 & 20 & 2 \end{bmatrix} \begin{bmatrix} x_1 \\ x_2 \\ x_3 \\ x_4 \end{bmatrix} + [0]u.$$

The eigenvalues are found using the `eig` command. In particular, applying the derived matrix A, we have

```
»eig(A)
ans =
    0.2878  +  1.4161i
    0.2878  -  1.4161i
   -1.2878  +  0.8579i
   -1.2878  -  0.8579i
```

The system is unstable because the real parts of the characteristic eigenvalues are positive.

Consider n–degree-of-freedom dynamic systems modeled using difference equations. In terms of n-dimensional state, m-dimensional control, and b-dimensional output vectors, the system variables (states, controls and outputs) are

$$x_k = \begin{bmatrix} x_{k1} \\ x_{k2} \\ \vdots \\ x_{kn-1} \\ x_{kn} \end{bmatrix}, \quad u_k = \begin{bmatrix} u_{k1} \\ u_{k2} \\ \vdots \\ u_{km-1} \\ u_{km} \end{bmatrix} \quad \text{and} \quad y_k = \begin{bmatrix} y_{k1} \\ y_{k2} \\ \vdots \\ y_{kb-1} \\ y_{kb} \end{bmatrix}.$$

In matrix form, the state-space equations are rewritten as

$$x_{k+1} = \begin{bmatrix} x_{k+1,1} \\ x_{k+1,2} \\ \vdots \\ x_{k+1,n-1} \\ x_{k+1,n} \end{bmatrix} = \begin{bmatrix} a_{k11} & a_{k12} & \cdots & a_{k1n-1} & a_{k1n} \\ a_{k21} & a_{k22} & \cdots & a_{k2n-1} & a_{k2n} \\ \vdots & \vdots & \ddots & \vdots & \vdots \\ a_{kn-11} & a_{kn-12} & \cdots & a_{kn-1n-1} & a_{kn-1n} \\ a_{kn1} & a_{kn2} & \cdots & a_{knn-1} & a_{knn} \end{bmatrix} \begin{bmatrix} x_{k1} \\ x_{k2} \\ \vdots \\ x_{kn-1} \\ x_{kn} \end{bmatrix}$$

$$+ \begin{bmatrix} b_{k11} & b_{k12} & \cdots & b_{k1m-1} & b_{k1m} \\ b_{k21} & b_{k22} & \cdots & b_{k2m-1} & b_{k2m} \\ \vdots & \vdots & \ddots & \vdots & \vdots \\ b_{kn-11} & b_{kn-12} & \cdots & b_{kn-1m-1} & b_{kn-1m} \\ b_{kn1} & b_{kn2} & \cdots & b_{knm-1} & b_{knm} \end{bmatrix} \begin{bmatrix} u_{k1} \\ u_{k2} \\ \vdots \\ u_{km-1} \\ u_{km} \end{bmatrix}$$

$$= A_k x_k + B_k u_k,$$

$$x_{k=k_0} = x_{k0},$$

where $A_k \in \mathbb{R}^{n \times n}$ and $B_k \in \mathbb{R}^{n \times m}$ are the matrices of coefficients. The output equations is

$$y_k = H_k x_k,$$

where $H_k \in \mathbb{R}^{b \times n}$ is the matrix of the constant coefficients.

The n-order linear difference equation can be represented as

$$\sum_{i=0}^{n} a_i y_{n-i} = \sum_{i=0}^{m} b_i u_{n-i}, \quad n \geq m.$$

Assuming that the coefficients are time-invariant (constant-coefficient), using the z-transform, and letting that the initial conditions are zero, one obtains

$$\left(\sum_{i=0}^{n} a_i z^i\right) Y(z) = \left(\sum_{i=0}^{m} b_i z^i\right) U(z).$$

Hence, the transfer function is

$$G(z) = \frac{Y(z)}{U(z)} = \frac{b_m z^m + b_{m-1} z^{m-1} + \cdots + b_1 z + b_0}{a_n z^n + a_{n-1} z^{n-1} + \cdots + a_1 z + a_0}.$$

Example 4.1.2.

Consider the system as described by a system of differential equations in matrix form

$$\frac{dx}{dt} = \begin{bmatrix} \frac{dx_1}{dt} \\ \frac{dx_2}{dt} \\ \frac{dx_3}{dt} \\ \frac{dx_4}{dt} \end{bmatrix} = Ax + Bu = \begin{bmatrix} 10 & 20 & 30 & 40 \\ 1 & 2 & 0 & 0 \\ 0 & 3 & 4 & 0 \\ 0 & 0 & 5 & 6 \end{bmatrix} \begin{bmatrix} x_1 \\ x_2 \\ x_3 \\ x_4 \end{bmatrix} + \begin{bmatrix} 1 \\ 2 \\ 3 \\ 4 \end{bmatrix} u,$$

$$y = Hx + Du = \begin{bmatrix} 1 & 10 & 100 & 1000 \end{bmatrix} \begin{bmatrix} x_1 \\ x_2 \\ x_3 \\ x_4 \end{bmatrix} + [0]u.$$

Let us find the transfer function and calculate the eigenvalues using the matrix A as well as the characteristic equation. Our goal is also to find the discrete-time state-space model (difference equation) if the sampling period is 0.01 sec, and find the eigenvalues.

We use the ss2tf command. The matrices A, B, C, and D must be dowloaded, and the following file is written in the Command Window

```
»A=[10 20 30 40;1 2 0 0;0 3 4 0;0 0 5 6];B=[1; 2; 3; 4];
»H=[1 10 100 1000]; D=[0];
» [num,den]=ss2tf(A,B,H,D)
```

The corresponding numerator and denominator are displayed on the screen.

In particular,

```
» [num,den] =ss2tf (A,B,H,D)
num =
   1.0e+005  *
        0      0.0432      -0.5391       0.5792      -5.7097
den =
      1.0000   -22.0000   144.0000   -378.0000   -60.0000
```

That is, we have

$$G(s) = \frac{Y(s)}{U(s)} = \frac{432s^3 - 5391s^2 + 5792s - 57097}{100000s^4 - 2200000s^3 + 14400000s^2}$$
$$-37800000s - 6000000$$

The eigenvalues can be found using the matrix A. The characteristic equation is obtained by equating the denominator to zero. That is, $sI - A = 0$ or $a_n s^n + a_{n-1}s^{n-1} + \cdots + a_1 s + a_0 = 0$. The following numerical results for eigenvalues are found using the denominator (characteristic equation) as well as matrix A. We have

```
»roots (den)
ans =
   13.3687
    4.3906 + 3.2634i
    4.3906 - 3.2634i
   -0.1500
»eig (A)
ans =
   13.3687
   -0.1500
    4.3906 + 3.2634i
    4.3906 - 3.2634i
```

One concludes that the system is unstable.

The discrete-time state-space model is found by using the c2d command. In particular,

```
»A=[10 20 30 40;1 2 0 0;0 3 4 0;0 0 5 6];B=[1; 2; 3; 4];
Ts=0.01; [Ak,Bk]=c2d(A,B,Ts),  eig(Ak)
```

We have

```
Ak =
      1.1063      0.2173      0.3326      0.4335
      0.0106      1.0213      0.0016      0.0021
      0.0002      0.0309      1.0408      0.0000
      0.0000      0.0008      0.0526      1.0618
```

```
Bk  =
    0.0259
    0.0203
    0.0309
    0.0420
ans =
    1.1430
    0.9985
    1.0443  +  0.0341i
    1.0443  -  0.0341i
```

That is, the matrices of the difference equations are found to be

$$A_k = \begin{bmatrix} 1.106 & 0.217 & 0.333 & 0.434 \\ 0.011 & 1.021 & 0.002 & 0.002 \\ 0 & 0.031 & 1.041 & 0 \\ 0 & 0.001 & 0.053 & 1.062 \end{bmatrix} \quad \text{and} \quad B_k = \begin{bmatrix} 0.026 \\ 0.02 \\ 0.031 \\ 0.042 \end{bmatrix}.$$

The characteristic eigenvalues are 1.143, 0.9985, and $1.0443 \pm 0.0341i$.

4.2. Continuous-Time Dynamic Systems and Analog Controllers

Most real-world systems evolve in continuous-time domain. Therefore, these dynamic systems are in continuous time and modeled using differential equations. Let us study linear systems. The transfer function of the dynamic systems to be controlled is denoted as $G_{sys}(s)$. Whether the designer is involved in the analysis and design of dynamic systems (aircraft, helicopters, missiles, interceptors, pointing systems, robotic manipulators, drives, servos, power electronics, or power converters), the synthesis of PID-type controllers is a critical aspect. In fact, the simplest control algorithms available are the PID-type controllers, and the linear analog PID control law is

$$u(t) = \underset{\text{proportional}}{k_p e(t)} + \underset{\text{integral}}{k_i \frac{e(t)}{s}} + \underset{\text{derivative}}{k_d s e(t)}$$

$$= k_p e(t) + k_i \int e(t)\, dt + k_d \frac{de(t)}{dt}, \quad s = \frac{d}{dt}, \quad (4.2.1)$$

where $e(t)$ is the error between the reference signal and the system output, $e(t) = r(t) - y(t)$; k_p, k_i, and k_d are the proportional, integral, and derivative feedback gains.

The block diagram of the linear analog PID controller (4.2.1) is shown in Figure 4.2.1.

Setting k_d equal to zero, the proportional-integral (PI) controller results. One obtains

$$u(t) = k_p e(t) + k_i \frac{e(t)}{s} = k_p e(t) + k_i \int e(t)\, dt.$$

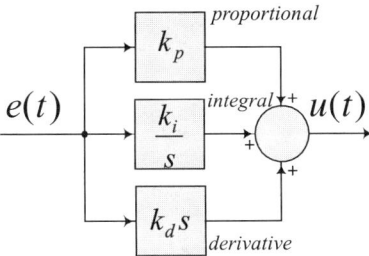

FIGURE 4.2.1.   Linear analog PID controller with k_p, k_i, and k_d feedback gains.

Assigning the integral feedback coefficient k_i to be zero, we have the proportional-derivative (PD) controller as

$$u(t) = k_p e(t) + k_d s e(t) = k_p e(t) + k_d \frac{de(t)}{dt}.$$

If k_i and k_d are set to be equal to zero, the proportional (P) control law results. In particular,

$$u(t) = k_p e(t).$$

Using the Laplace transform, from (4.2.1), we have $U(s) = (k_p + \frac{k_i}{s} + k_d s) E(s)$. One finds the transfer function of the analog PID controllers as

$$G_{\text{PID}}(s) = \frac{U(s)}{E(s)} = \frac{k_d s^2 + k_p s + k_i}{s} = \frac{k_i(T_c^2 s^2 + 2 T_c \xi_c s + 1)}{s}.$$

In general, as will be illustrated later, different PID-type analog control algorithms can be designed, implemented (e.g., P, PI, linear, nonlinear, unbounded and constrained). Therefore, the transfer function of the PID-type controller $G_{\text{PID}}(s) = \frac{U(s)}{E(s)}$ will be used to attain the generality of results. It must be emphasized that the stability analysis using the eigenvalues is valid only for linear dynamic systems.

The closed-loop system, which integrates the dynamic system controlled by the PID-type controller, in the time and s-domains with the negative output loop is represented in Figure 4.2.2.

If the system output $y(t)$ converges to the reference signal $r(t)$ as time approaches infinity (the steady-state value of the system output is equal to the reference input), one concludes that tracking of the reference input is accomplished, and the error vector $e(t) = r(t) - y(t)$ is zero. That is, tracking is achieved if the following is guaranteed:

$$e(t) = r(t) - y(t) = 0 \text{ as } t \to \infty, \text{ or } \lim_{t \to \infty} e(t) = 0.$$

In the time domain, we have the following expression for the tracking error:

$$e(t) = r(t) - y(t).$$

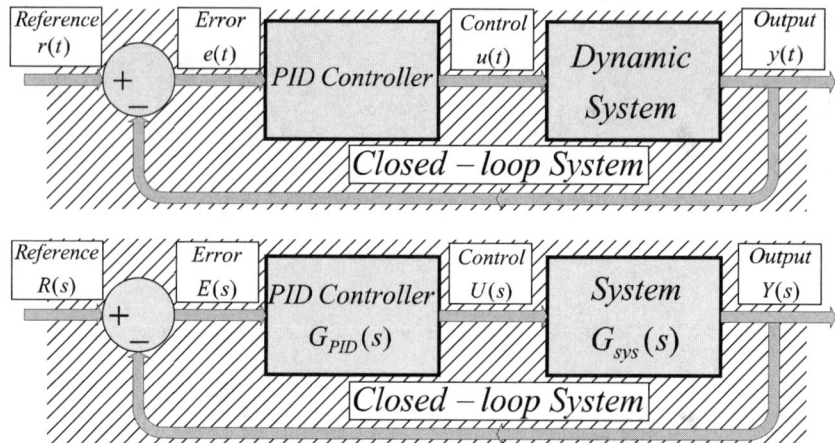

FIGURE 4.2.2.   Time- and s-domain diagrams of closed-loop systems with PID-type controllers.

Hence, the Laplace transform of the error signal is $E(s) = R(s) - Y(s)$. For the closed-loop system, as given in Figure 4.2.2, let us find the Laplace transform of the output $y(t)$. In the s-domain, one finds

$$Y(s) = G_{sys}(s)U(s) = G_{sys}(s)G_{PID}(s)E(s) = G_{sys}(s)G_{PID}(s)[R(s) - Y(s)].$$

Hence, the following transfer function of the closed-loop dynamic system with PID control law results:

$$G(s) = \frac{Y(s)}{R(s)} = \frac{G_{sys}(s)G_{PID}(s)}{1 + G_{sys}(s)G_{PID}(s)}.$$

In the frequency domain, one obtains

$$G(j\omega) = \frac{Y(j\omega)}{R(j\omega)} = \frac{G_{sys}(j\omega)G_{PID}(j\omega)}{1 + G_{sys}(j\omega)G_{PID}(j\omega)}.$$

The characteristic equations of the closed-loop systems can be easily found, and the stability can be guaranteed by adjusting the proportional, integral, and derivative feedback gains (for example, for linear PID control law, k_p, k_i, and k_d must be obtained). In fact, the objective is to design the PID-type algorithms and find the proportional, integral, and derivative feedback coefficients to attain the desired system performance. The closed-loop system performance can be studied using the frequency-domain approach. However, because of the complexity to correlate the transient system dynamics with frequency domain responses, other methods have been widely used.

Example 4.2.1.

Let the transfer function for the open-loop system be

$$G_{sys}(s) = \frac{Y(s)}{U(s)} = \frac{s}{k_d s^2 + k_p s + k_i}.$$

The linear PID controller (4.2.1)

$$G_{PID}(s) = \frac{U(s)}{E(s)} = \frac{k_d s^2 + k_p s + k_i}{s}$$

is used. The closed-loop system, if one implements $G_{PID}(s) = \frac{k_d s^2 + k_p s + k_i}{s}$, is documented in Figure 4.2.3.

The transfer function of the closed-loop system is

$$G(s) = \frac{Y(s)}{R(s)} = \frac{G_{sys}(s)G_{PID}(s)}{1 + G_{sys}(s)G_{PID}(s)} = \frac{1}{1 + 1} = \frac{1}{2}.$$

That is, using the linear PID controller, the system output follows the reference signal, and $y(t) = \frac{1}{2}r(t)$. Furthermore, $e(t) = r(t) - y(t) = \frac{1}{2}r(t)$.

If one implements the controller

$$G_{PID}(s) = \frac{U(s)}{E(s)} = \frac{k(k_d s^2 + k_p s + k_i)}{s},$$

the transfer function of the closed-loop system becomes

$$G(s) = \frac{Y(s)}{R(s)} = \frac{G_{sys}(s)G_{PID}(s)}{1 + G_{sys}(s)G_{PID}(s)} = \frac{k}{1 + k}.$$

Then,

$$y(t) = \frac{k}{1 + k}r(t),$$

and

$$e(t) = r(t) - y(t) = \frac{1}{1 + k}r(t).$$

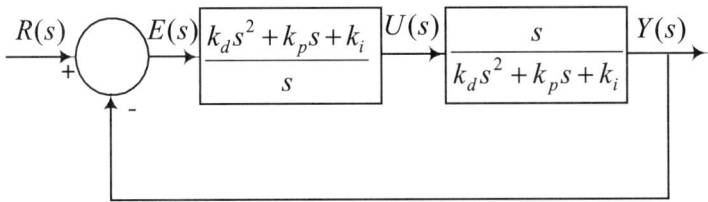

FIGURE 4.2.3.   Closed-loop system with linear PID controller.

The tracking error is equal to zero if $k = \infty$, and the specified accuracy can be guaranteed by assigning the coefficient k, $k \gg 1$ (high-gain coefficient).

It is clear that different controllers can be used, and the relationships between the characteristic eigenvalues and zeros of the open- and closed-loop systems must be studied designing control algorithms. In particular, controllers can be straight-forwardly designed by adjusting the eigenvalues and zeros using the root-locus graphical method introduced by W. R. Evans in 1948 (see W. R. Evans, "Graphi-cal analysis of control systems," *Transactions of AIEE*, vol. 67, pp. 547–551, 1948). The *root loci* are plots of the variations of the eigenvalues of the closed-loop trans-fer function with the gain changes. Changing the characteristic eigenvalues and zeros, the designer is able to adjust simultaneously the transient and frequency re-sponses avoiding tedious relationships between time and frequency domains. This allows one to ensure the logical synthesis pattern, which is based on the following steps:

- Set the specifications (settling, rise time, overshoot, accuracy, as well as other criteria).
- Using these specifications, derive the characteristic eigenvalues and zeros of the closed-loop transfer function.
- Design the controller, and find the feedback gains using open- and closed-loop transfer functions.

Using the closed-loop system configuration, documented in Figure 4.2.4, the closed-loop transfer function is found as

$$G(s) = \frac{Y(s)}{R(s)} = \frac{G_{\text{sys}}(s)G_{\text{controller}}(s)}{1 + G_{\text{sys}}(s)G_{\text{controller}}(s)},$$

where $G_{\text{controller}}(s)$ is the transfer function of the controller.

Augmenting the controller and dynamic system, one finds $G_{\Sigma}(s) = G_{\text{sys}}(s)G_{\text{controller}}(s)$, and we have

$$G(s) = \frac{Y(s)}{R(s)} = \frac{G_{\Sigma}(s)}{1 + G_{\Sigma}(s)}.$$

The closed-loop continuous-time system is illustrated in Figure 4.2.5.

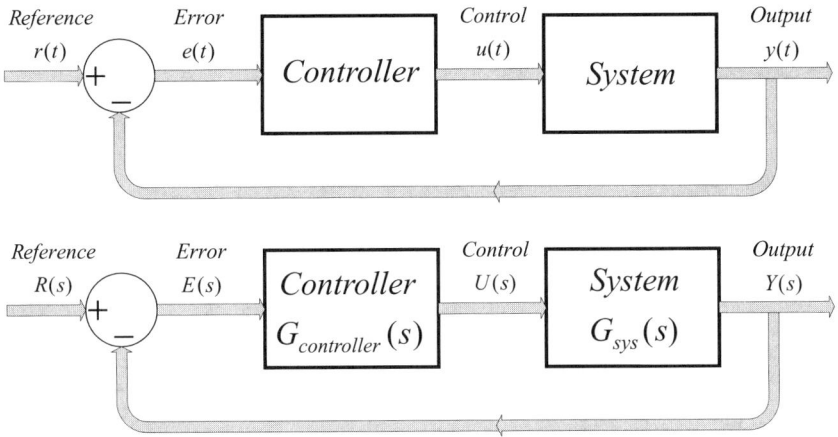

FIGURE 4.2.4.   Time- and s-domain block diagrams of closed-loop systems with controller.

Example 4.2.2.

Let the transfer function for the open-loop system be

$$G_{\text{sys}}(s) = \frac{Y(s)}{U(s)} = \frac{T_0 s + 1}{(T_1^2 s^2 + 2\xi_1 T_1 s + 1)(T_2^2 s^2 + 2\xi_2 T_2 s + 1)(T_3 s + 1)}$$

$$= \frac{2s + 1}{(3s^2 + 4s + 1)(5s^2 + 6s + 1)(7s + 1)}.$$

The controller with transfer function

$$G_{\text{controller}}(s) = \frac{U(s)}{E(s)} = \frac{k(3s^2 + 4s + 1)(5s^2 + 6s + 1)(7s + 1)}{2s + 1}$$

can be used to "compensate" the time constants of the system; see Figure 4.2.6.
The transfer function of the closed-loop system is

$$G(s) = \frac{Y(s)}{R(s)} = \frac{k}{1 + k}.$$

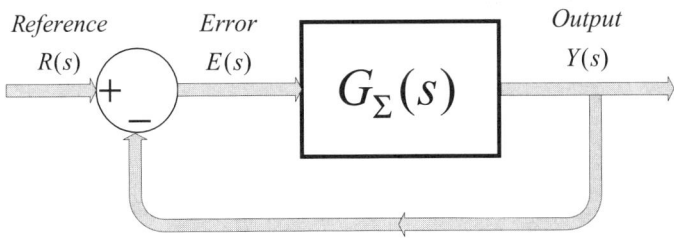

FIGURE 4.2.5. s-domain block diagram of closed-loop dynamic systems.

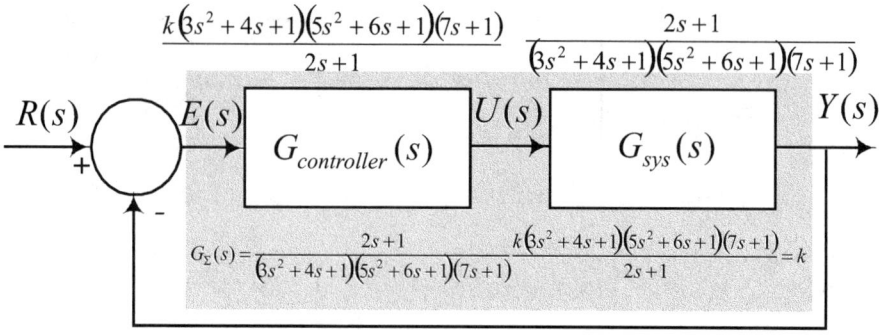

FIGURE 4.2.6.   Closed-loop system with controller.

Therefore,

$$y(t) = \frac{k}{1+k}r(t)$$

and

$$e(t) = r(t) - y(t) = \frac{1}{1+k}r(t).$$

The tracking error is equal to zero if $k = \infty$. The controller with infinity feedback gain $k = \infty$ cannot be implemented. The specified tracking accuracy can be guaranteed by assigning the high value of k, $k \gg 1$.

It must be emphasized that the system parameters vary, and the designer cannot compensate the system time constants as illustrated in Examples 4.2.1 and 4.2.2 because $T_i = $ var. For example, the variations of the armature inductance L_a and the moment of inertia J lead to a change in the time constants (see Example 2.2.11 and Figure 2.2.11). Furthermore, derivative-type controllers are very difficult to implement because of the noise imposed on the measured output $y(t)$ as well as the reference signal $r(t)$. That is, the tracking error $e(t)$ is noisy. In addition, the control is bounded in magnitude. Therefore, the designer must integrate parameter variations and control bounds in the synthesis of robust implementable controllers.

It is required that the system output follow the reference signal, and the desired accuracy depends to a great extent on the control objectives. The steady-state error e_{ss} (steady-state difference between the reference and output) results if the system output $y(t)$ is different compared with the reference signal $r(t)$ as time approaches infinity. Using the final value theorem, from

$$\frac{E(s)}{R(s)} = \frac{1}{1+G_\Sigma(s)},$$

the steady-state error e_{ss} of the closed-loop system is found to be

$$e_{ss} = \lim_{t \to \infty} e(t) = \lim_{s \to 0} s E(s) = \lim_{s \to 0} \frac{s R(s)}{1+G_\Sigma(s)}.$$

That is, the steady-state error e_{ss} depends on the transfer function $G_\Sigma(s) = G_{sys}(s)G_{controller}(s)$ and the reference signal $r(t)$, which is given in the s-domain as $R(s)$. The number of the eigenvalues at the origin defines the error. Using the constant factor k, eigenvalues at the origin, and real and complex-conjugate eigenvalues and zeros, one can write

$$G_\Sigma(s) = \frac{k(T_{n1}s+1)(T_{n2}s+1)\cdots(T_{n,l-1}^2 s^2 + 2\xi_{n,l-1}T_{n,l-1}s + 1) \times (T_{n,l}^2 s^2 + 2\xi_{n,l}T_{n,l}s + 1)}{s^M(T_{d1}s+1)(T_{d2}s+1)\cdots(T_{d,p-1}^2 s^2 + 2\xi_{d,p-1}T_{d,p-1}s + 1) \times (T_{d,p}^2 s^2 + 2\xi_{d,p}T_{d,p}s + 1)},$$

where T_i and ξ_i are the time constants and damping coefficients and M is the order of the eigenvalues at the origin (the system *type* is given by the order M).

We have the following system classes:

- *type* 0 if $G_\Sigma(s) = \frac{k(T_{n1}s+1)(T_{n2}s+1)\cdots(T_{n,l-1}^2 s^2 + 2\xi_{n,l-1}T_{n,l-1}s+1)(T_{n,l}^2 s^2 + 2\xi_{n,l}T_{n,l}s+1)}{(T_{d1}s+1)(T_{d2}s+1)\cdots(T_{d,p-1}^2 s^2 + 2\xi_{d,p-1}T_{d,p-1}s+1)(T_{d,p}^2 s^2 + 2\xi_{d,p}T_{d,p}s+1)}$,

- *type* 1 if $G_\Sigma(s) = \frac{k(T_{n1}s+1)(T_{n2}s+1)\cdots(T_{n,l-1}^2 s^2 + 2\xi_{n,l-1}T_{n,l-1}s+1)(T_{n,l}^2 s^2 + 2\xi_{n,l}T_{n,l}s+1)}{s(T_{d1}s+1)(T_{d2}s+1)\cdots(T_{d,p-1}^2 s^2 + 2\xi_{d,p-1}T_{d,p-1}s+1)(T_{d,p}^2 s^2 + 2\xi_{d,p}T_{d,p}s+1)}$,

- *type* 2 if $G_\Sigma(s) = \frac{k(T_{n1}s+1)(T_{n2}s+1)\cdots(T_{n,l-1}^2 s^2 + 2\xi_{n,l-1}T_{n,l-1}s+1)(T_{n,l}^2 s^2 + 2\xi_{n,l}T_{n,l}s+1)}{s^2(T_{d1}s+1)(T_{d2}s+1)\cdots(T_{d,p-1}^2 s^2 + 2\xi_{d,p-1}T_{d,p-1}s+1)(T_{d,p}^2 s^2 + 2\xi_{d,p}T_{d,p}s+1)}$,

- $\ldots$

- *type* M if $G_\Sigma(s) = \frac{k(T_{n1}s+1)(T_{n2}s+1)\cdots(T_{n,l-1}^2 s^2 + 2\xi_{n,l-1}T_{n,l-1}s+1)(T_{n,l}^2 s^2 + 2\xi_{n,l}T_{n,l}s+1)}{s^M(T_{d1}s+1)(T_{d3}s+1)\cdots(T_{d,p-1}^2 s^2 + 2\xi_{d,p-1}T_{d,p-1}s+1)(T_{d,p}^2 s^2 + 2\xi_{d,p}T_{d,p}s+1)}$.

The reference signal $r(t)$ should be studied as well.

Unit-step input. For the unit step, $r(t) = 1(t)$ and $R(s) = \frac{1}{s}$.

We have the following expression for the steady-state error if $r(t) = 1(t)$:

$$e_{ss} = \lim_{t\to\infty} e(t) = \lim_{s\to 0} sE(s) = \lim_{s\to 0} \frac{sR(s)}{1+G_\Sigma(s)},$$

$$e_{ss} = \lim_{s\to 0} \frac{1}{1+G_\Sigma(s)} = \frac{1}{1+\lim_{s\to 0} G_\Sigma(s)}.$$

For the *type* 0 systems, $\lim_{s\to 0} G_\Sigma(s)$ is finite, and denoting $k_0 = \lim_{s\to 0} G_\Sigma(s)$, the steady-state error is found as

$$e_{ss} = \frac{1}{1+\lim_{s\to 0} G_\Sigma(s)} = \frac{1}{1+k_0}.$$

For the systems of *type* 1 or higher, $k_0 = \lim_{s\to 0} G_\Sigma(s) = \infty$. Thus,

$$e_{ss} = \frac{1}{1+\lim_{s\to 0} G_\Sigma(s)} = 0.$$

Hence, the steady-state error is zero.

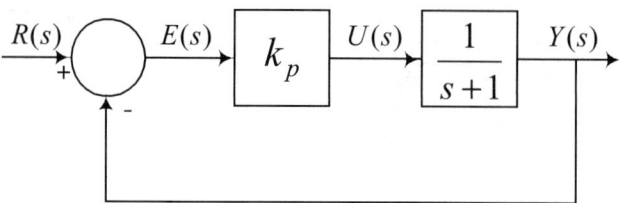

FIGURE 4.2.7.   Closed-loop system with proportional controller.

Example 4.2.3.

Let

$$G_\Sigma(s) = \frac{10(s+1)(2s+1)(3s^2+4s+1)}{s(5s+1)(6s+1)(7s^2+8s+1)}.$$

Find the steady-state error if $r(t) = 1(t)$.
  We have $\lim_{s \to 0} G_\Sigma(s) = \infty$.
  Thus, $e_{ss} = 0$.

Ramp input. If the reference is the ramp function $r(t) = t$, using

$$R(s) = \frac{1}{s^2},$$

one finds

$$e_{ss} = \lim_{t \to \infty} e(t) = \lim_{s \to 0} s E(s) = \lim_{s \to 0} \frac{s R(s)}{1 + G_\Sigma(s)}$$

$$e_{ss} = \lim_{s \to 0} \frac{1/s}{1 + G_\Sigma(s)} = \frac{1}{\lim_{s \to 0} s(1 + G_\Sigma(s))} = \frac{1}{\lim_{s \to 0} s G_\Sigma(s)}.$$

For the *type* 0 systems, one has $\lim_{s \to 0} s G_\Sigma(s) = 0$, and $e_{ss} = \infty$.
For the *type* 1 systems, the following results: $\lim_{s \to 0} s G_\Sigma(s) = k_1$ and $e_{ss} = \frac{1}{k_1}$.
For the systems of *type* 2 or higher, we have $\lim_{s \to 0} s G_\Sigma(s) = \infty$ and $e_{ss} = 0$.

Example 4.2.4.

Let us find the steady-state errors for the closed-loop system if the reference signal is the unit step and ramp reference. The transfer function for the open-loop system is

$$G(s) = \frac{Y(s)}{U(s)} = \frac{1}{s+1}.$$

The P controller is used. The block diagram is given in Figure 4.2.7.
  One finds the following transfer function

$$G_\Sigma(s) = \frac{k_p}{s+1}.$$

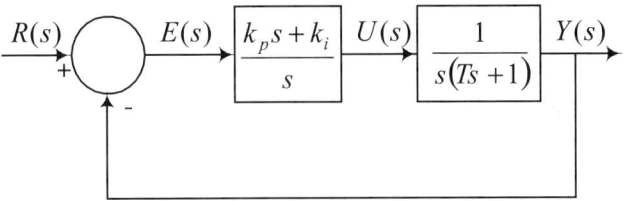

FIGURE 4.2.8.   Closed-loop system with PI control law.

That is, the *type* 0 system is under our consideration.
For the step input, we compute $k_0 = \lim_{s \to 0} G_\Sigma(s) = k_p$.
Using the following formula:

$$e_{ss} = \lim_{s \to 0} \frac{1}{1 + G_\Sigma(s)} = \frac{1}{1 + \lim_{s \to 0} G_\Sigma(s)},$$

the steady-state error is found to be

$$e_{ss} = \frac{1}{1 + k_p}.$$

If $r(t) = t$, we have $\lim_{s \to 0} s G_\Sigma(s) = 0$.
Therefore, from

$$e_{ss} = \frac{1}{\lim_{s \to 0} s G_\Sigma(s)},$$

one has

$$e_{ss} = \infty.$$

Example 4.2.5.

The transfer function of the open-loop system is

$$G(s) = \frac{Y(s)}{U(s)} = \frac{1}{s(Ts + 1)}.$$

Find the steady-state errors if the input signal is the unit step and ramp function.
The PI controller is used. Figure 4.2.8 documents the block diagram.
The *type* 2 system is studied because

$$G_\Sigma(s) = \frac{k_p s + k_i}{s^2(Ts + 1)}.$$

For the step input, we have

$$k_0 = \lim_{s \to 0} G_\Sigma(s) = \infty.$$

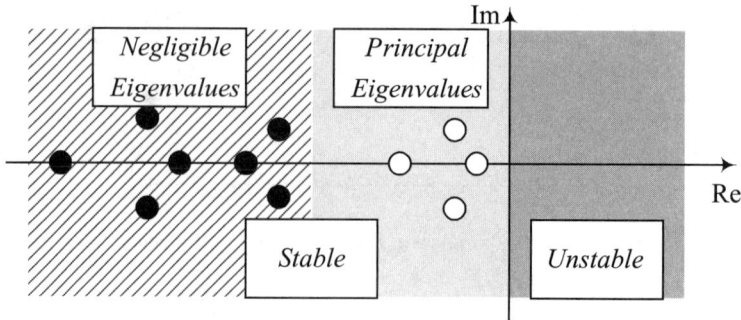

FIGURE 4.2.9.   Eigenvalues in the complex plane.

Taking note of

$$e_{ss} = \frac{1}{1 + \lim_{s \to 0} G_{\Sigma}(s)},$$

one has

$$e_{ss} = 0.$$

That is, the steady-state errors for the step input is zero.
Having found $\lim_{s \to 0} s G_{\Sigma}(s) = \infty$, and making use of

$$e_{ss} = \frac{1}{\lim_{s \to 0} s G_{\Sigma}(s)},$$

for the ramp input, the following analytic result for the steady-state error is obtained:

$$e_{ss} = \infty.$$

Thus, the steady-state error for the ramp input is infinity.
For dynamic systems, the controller transfer function (structure of control laws) can be found assigning the desired location of the characteristic eigenvalues and zeros. For example, if the linear PID controllers are used, the feedback gains k_p, k_i, and k_d can be derived to attain the desired location of the principal characteristic eigenvalues because other eigenvalues can be located far out left in the complex plane; see Figure 4.2.9.

Undergraduate textbooks in automatic control, listed in the references, cover the controller design using the root-locus methods.

To attain the desired steady-state and dynamic performance, as well as to ensure accuracy and stability, advanced control algorithms should be designed and implemented. The most general structure (configuration) of analog PID-type controllers is given as

$$u(t) = \sum_{j=1}^{2N_p-1} k_{p(2j-1)} e^{zj-1}(t) + \sum_{j=1}^{2N_i-1} k_{i(2j-1)} \frac{e(t)}{s^{2j-1}}$$

proportional   integral

$$+ \sum_{j=1}^{2N_d-1} k_{d(2j-1)} \frac{d^{2j-1}e(t)}{dt^{2j-1}},$$

derivative

$$(4.2.2)$$

where N_p, N_i, and N_d are the positive integers assigned by the designer and $k_{p(2j-1)}, k_{i(2j-1)}$, and $k_{d(2j-1)}$ are the proportional, integral, and derivative feedback coefficients.

Letting $N_p = 1$, $N_i = 1$, and $N_d = 1$, we have the linear PID control law as given by (4.2.1).

If $N_p = 2$, $N_i = 2$, and $N_d = 1$, from (4.2.2), we have the $PP^3P^5PII^3I^5D$ control law as

$$u(t) = k_{p1}e(t) + k_{p3}e^3(t) + k_{p5}e^5(t) + k_{i1}\frac{e(t)}{s} + k_{i3}\frac{e(t)}{s^3} + k_{i5}\frac{e(t)}{s^5} + k_{d1}\frac{de(t)}{dt}.$$

In addition to PID-type controllers (4.2.2), nonlinear control laws with nonlinear error mappings can be used to improve system performance. The family of the nonlinear PID- type controllers is obtained as

$$u(t) = \sum_{j=0}^{\varsigma} k_{pj} e^{\frac{2j+1}{2\beta+1}}(t) + \sum_{j=0}^{\sigma} k_{ij} \frac{1}{s} e^{\frac{2j+1}{2\mu+1}}(t) + \sum_{j=0}^{\eta} k_{dj} se(t)^{\frac{2j+1}{2\gamma+1}}, \quad (4.2.3)$$

proportional   integral   derivative

where ς, β, σ, μ, η, and γ are the nonnegative integers assigned by the designer. In (4.2.3), the nonlinear

- proportional $\sum_{j=0}^{\varsigma} k_{pj} e^{\frac{2j+1}{2\beta+1}}(t)$
- integral $\sum_{j=0}^{\sigma} k_{ij} \frac{1}{s} e^{\frac{2j+1}{2\mu+1}}(t)$
- derivative $\sum_{j=0}^{\eta} k_{dj} se(t)^{\frac{2j+1}{2\gamma+1}}$

feedback are given in terms of the error vector.

Nonnegative integers ς, β, σ, μ, η, and γ are assigned by the designer to attain the specified criteria in stability, tracking, accuracy, disturbance attenuation, and so on.

Letting $\varsigma = \beta = \sigma = \mu = \eta = \gamma = 0$, the linear PID controller in (4.2.1) results.

If $\varsigma = \beta = 1$, $\sigma = \mu = 2$, and $\eta = \gamma = 3$, we have the following nonlinear control law:

$$u(t) = k_{p0}e^{\frac{1}{3}} + k_{p1}e + k_{i0}\frac{1}{s}e^{\frac{1}{5}} + k_{i1}\frac{1}{s}e^{\frac{3}{5}} + k_{i2}\frac{1}{s}e + k_{d0}se^{\frac{1}{7}} + k_{d1}se^{\frac{3}{7}}$$

$$+ k_{d2}se^{\frac{5}{7}} + k_{d3}se.$$

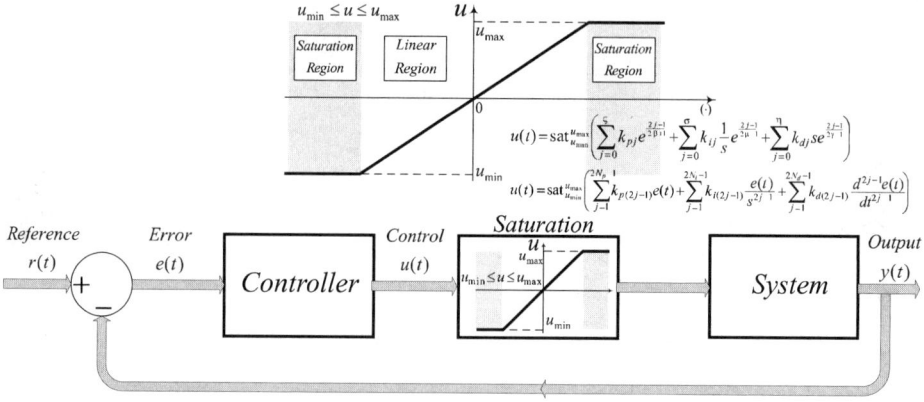

FIGURE 4.2.10.   Closed-loop system with the admissible (saturated) control $u_{\min} \le u \le u_{\max}$.

That is, the tracking error is in the corresponding power.

One cannot calculate the eigenvalues because the closed-loop system with (4.2.3) is nonlinear.

4.2.1.   Control Bounds

In real-world dynamic systems, the control inputs, states, and output variables are bounded. For electric motors, controlled by power converters, the voltages applied to the windings are bounded. Mechanical limits are imposed on the maximum angular or linear velocities of motors. In addition, the voltages and currents in power converters are bounded as well. These rated (maximum allowed) voltages and currents, angular and linear velocities, and displacements are specified. The applied voltages to the motor windings cannot exceed the rated voltage, and the currents in windings cannot be greater than the maximum allowed peak and continuous rated currents. Hence, because of the limits imposed, the allowed control is bounded, and the system variables must be within the maximum admissible (rated) set. Thus, the control authority is constrained, and the typical closed-loop system with saturated control law is shown in Figure 4.2.10.

Using the proportional, integral, and derivative feedback, the saturated control, as a function of the error vector, can be found as

$$u(t) = \mathrm{sat}_{u_{\min}}^{u_{\max}} \left(k_p e(t) + k_i \int e(t)\, dt + k_d \frac{de(t)}{dt} \right), \quad u_{\min} \le u \le u_{\max}. \quad (4.2.4)$$

Compared with the linear unbounded PID controller (4.2.1), algorithm (4.2.4) illustrates that the control signal u is bounded between the minimum and maximum values $u_{\min} \le u \le u_{\max}$, $u_{\min} < 0$ and $u_{\max} > 0$. In the linear region, the control varies between the maximum $u_{\max}$ and minimum $u_{\min}$ values, and one obtains $u(t) = k_p e(t) + k_i \int e(t)\, dt + k_d \frac{de(t)}{dt}$. If $k_p e(t) + k_i \int e(t)\, dt + k_d \frac{de(t)}{dt} > u_{\max}$, the control is saturated, and $u(t) = u_{\max}$. For $k_p e(t) + k_i \int e(t)\, dt + k_d \frac{de(t)}{dt} < u_{\min}$, we have $u(t) = u_{\min}$.

The bounded PID-type control algorithms are found as

$$u(t) = \operatorname{sat}_{u_{\min}}^{u_{\max}} \left(\sum_{j=1}^{2N_p-1} k_{p(2j-1)} e^{2j-1}(t) + \sum_{j=1}^{2N_i-1} k_{i(2j-1)} \frac{e(t)}{s^{2j-1}} \right.$$
$$\left. + \sum_{j=1}^{2N_d-1} k_{d(2j-1)} \frac{d^{2j-1} e(t)}{dt^{2j-1}} \right), \qquad u_{\min} \le u \le u_{\max}.$$

Using nonlinear error mappings, from (4.2.3), one obtains the bounded controller as

$$u(t) = \operatorname{sat}_{u_{\min}}^{u_{\max}} \left(\sum_{j=0}^{\varsigma} k_{pj} e^{\frac{2j+1}{2\beta+1}} + \sum_{j=0}^{\sigma} k_{ij} \frac{1}{s} e^{\frac{2j+1}{2\mu+1}} + \sum_{j=0}^{\eta} k_{dj} se(t)^{\frac{2j+1}{2\gamma+1}} \right),$$

$$u_{\min} \le u \le u_{\max}.$$

Example 4.2.6. PID control of a servo with permanent-magnet DC motor.

Consider the rotating table (servo-mechanism) actuated by a permanent-magnet DC motor; see Figure 4.2.11. The motor is attached to the table through the gear with coefficient k_{gear}. Our goal is to guarantee the assigned angular displacement of the pointer. The reference signal is denoted as $r(t)$. The table angular displacement is a function of the rotor displacement, and $y(t) = k_{\text{gear}} \theta_r(t)$ is the output equation. To change the angular velocity and displacement, one regulates the armature voltage applied to the motor winding u_a (in practice, as was illustrated, power amplifiers are used, and to control u_a, one changes the duty ratio). The linear analog PID control law should be designed, and the feedback coefficients must be found.

The mathematical model of permanent-magnet DC motors was developed in Example 2.2.11, and simulations were presented in Examples 3.2.14 and 3.2.19. We have the following differential equations:

$$\frac{di_a}{dt} = -\frac{r_a}{L_a} i_a - \frac{k_a}{L_a} \omega_r + \frac{1}{L_a} u_a,$$

$$\frac{d\omega_r}{dt} = \frac{1}{J}(T_e - T_{\text{viscous}} - T_L) = \frac{1}{J}(k_a i_a - B_m \omega_r - T_L),$$

$$\frac{d\theta_r}{dt} = \omega_r.$$

Using the Laplace operator $s = \frac{d}{dt}$, one has

$$\left(s + \frac{r_a}{L_a} \right) i_a(s) = -\frac{k_a}{L_a} \omega_r(s) + \frac{1}{L_a} u_a(s),$$

$$\left(s + \frac{B_m}{J} \right) \omega_r(s) = \frac{1}{J} k_a i_a(s) - \frac{1}{J} T_L(s), \quad s\theta_r(s) = \omega_r(s).$$

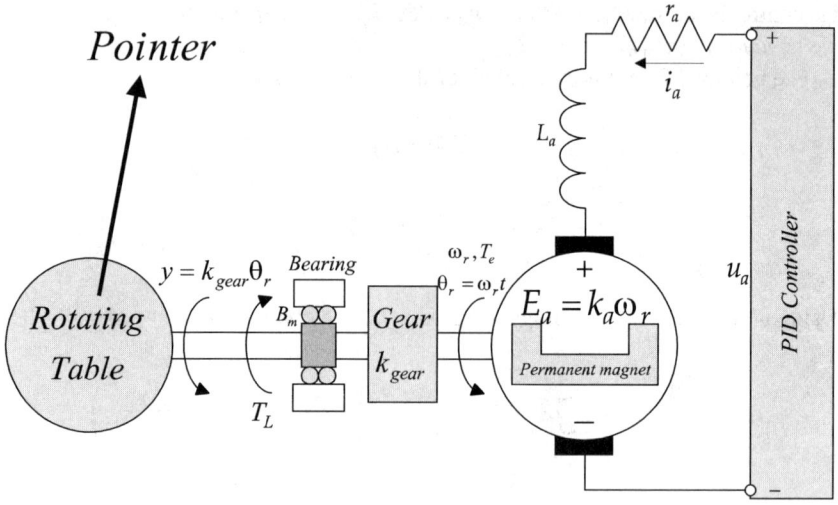

FIGURE 4.2.11.   Schematic diagram of a rotating table (servo) actuated by a DC motor.

Taking note of the output equation $y(t) = k_{\text{gear}}\theta_r(t)$, $y(s) = k_{\text{gear}}\theta_r(s)$, one obtains the block diagram of the open-loop positioning table actuated by the permanent-magnet DC motor; see Figure 4.2.12.

The transfer function for an open-loop system is

$$G_{\text{sys}}(s) = \frac{Y(s)}{U(s)} = \frac{k_{\text{gear}}k_a}{s(L_aJs^2 + (r_aJ + L_aB_m)s + r_aB_m + k_a^2)}.$$

The linear PID controller is $u_a(t) = k_pe(t) + k_i\frac{e(t)}{s} + k_dse(t)$, and therefore, $G_{\text{PID}}(s) = U(s)/E(s) = \frac{k_ds^2 + k_ps + k_i}{s}$.

The closed-loop block diagram is documented in Figure 4.2.13.

The closed-loop transfer function is found as

$$G(s) = \frac{Y(s)}{R(s)} = \frac{G_{\text{sys}}(s)G_{\text{PID}}(s)}{1 + G_{\text{sys}}(s)G_{\text{PID}}(s)}$$

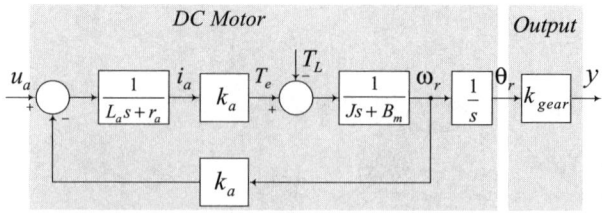

FIGURE 4.2.12.   Block diagram of the open-loop servo-mechanism.

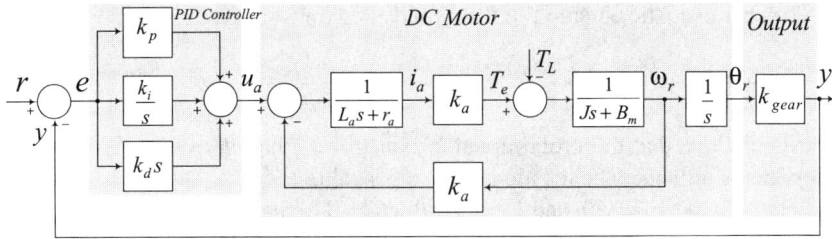

FIGURE 4.2.13.   Block diagram of the closed-loop servo-mechanism.

$$
= \frac{k_{\text{gear}}k_a(k_ds^2 + k_ps + k_i)}{s^2(L_aJs^2 + (r_aJ + L_aB_m)s + r_aB_m + k_a^2) + k_{\text{gear}}k_a(k_ds^2 + k_ps + k_i)}
$$

$$
= \frac{\frac{k_d}{k_i}s^2 + \frac{k_p}{k_i}s + 1}{\frac{L_aJ}{k_{\text{gear}}k_ak_i}s^4 + \frac{(r_aJ+L_aB_m)}{k_{\text{gear}}k_ak_i}s^3 + \frac{(r_aB_m+k_a^2+k_{\text{gear}}k_ak_d)}{k_{\text{gear}}k_ak_i}s^2 + \frac{k_p}{k_i}s + 1}.
$$

In Examples 3.2.14 and 3.2.19, the following motor parameters were used: $r_a = 5\,\Omega$, $L_a = 0.01$ H, $k_a = 0.2$ V-sec/rad, $k_a = 0.2$ N-m/A, $J = 0.0005$ kg-m^2, and $B_m = 0.00001$ N-m-sec/rad. The rated armature voltage for the motor studied is ± 50 V.

The numerical values of the numerator and denominator coefficients in $G_{\text{sys}}(s) = \frac{Y(s)}{U(s)} = \frac{k_{\text{gear}}k_a}{s(L_aJs^2+(r_aJ+L_aB_m)s+r_aB_m+k_a^2)}$ are found using the developed m-file.

MATLAB script (ch_4_2_31.m)

```
% Servo-mechanism parameters
ra=5; La=0.01; ka=0.2; J=0.0005; Bm=0.00001; kgear=0.05;
% Numerator and denominator of the open-loop transfer
% function
num`s=[ka*kgear]; den`s=[La*J ra*J+La*Bm ra*Bm+ka^2 0];
num`s, den`s
```

Using the following numerical results obtained:

```
num`s =
   1.0000e-002
den`s =
   5.0000e-006   2.5001e-003   4.0050e-002          0
```

one finds

$$
G_{\text{sys}}(s) = \frac{Y(s)}{U(s)} = \frac{0.01}{s(0.000005s^2 + 0.0025s + 0.04)}.
$$

The open-loop system is unstable.

The characteristic equation of the closed-loop transfer function is

$$\frac{L_a J}{k_{\text{gear}} k_a k_i} s^4 + \frac{(r_a J + L_a B_m)}{k_{\text{gear}} k_a k_i} s^3 + \frac{(r_a B_m + k_a^2 + k_{\text{gear}} k_a k_d)}{k_{\text{gear}} k_a k_i} s^2 + \frac{k_p}{k_i} s + 1 = 0.$$

It is obvious that the proportional k_p, integral k_i, and derivative k_d feedback coefficients influence the location of the eigenvalues.

Let $k_p = 500$, $k_i = 50$, and $k_d = 5$ (which lead to the particular location of the characteristic eigenvalues). Then, the controller is given as

$$u_a(t) = 500e(t) + 50 \int e(t)\, dt + 5\frac{de(t)}{dt}.$$

Running the m-file (the MATLAB script is ch_4_2_32.m)

```
% Servo-mechanism parameters
ra=5; La=0.01; ka=0.2; J=0.0005; Bm=0.00001; kgear=0.05;
%  Feedback coefficients
kp=500; ki=50; kd=5;
% Denominator of the closed-loop transfer function
den c=[(La*J)/(kgear*ka*ki) (ra*J+La*Bm)/(kgear*ka*ki)
(ra*Bm+ka^2+kgear*ka*kd)/(kgear*ka*ki) kp/ki 1];
format short e; roots(den c)
```

the numerator and denominator are found, and the eigenvalues of the closed-loop system are

```
ans =
 -4.6597e+002
 -1.6973e+001 +4.3059e+001i
 -1.6973e+001 -4.3059e+001i
 -1.0018e-001
```

The MATLAB script (ch_4_2_33.m)

```
% Servo-mechanism parameters
ra=5; La=0.01; ka=0.2; J=0.0005; Bm=0.00001; kgear=0.05;
%  Feedback coefficients
kp=500; ki=50; kd=5;
% Numerator and denominator of the closed-loop transfer
% function
num c=[kd/ki kp/ki 1];
den c=[(La*J)/(kgear*ka*ki) (ra*J+La*Bm)/(kgear*ka*ki)
(ra*Bm+ka^2+kgear*ka*kd)/(kgear*ka*ki) kp/ki 1];
t=0:0.01:0.5;
u=2*ones(size(t));
y=lsim(num c,den c,u,t);
plot(t,y,'-',y,u,':'); xlabel('Time (seconds)');
title('Output and Input Evolution');
axis([0 0.5,0 3])   % axis limits
```

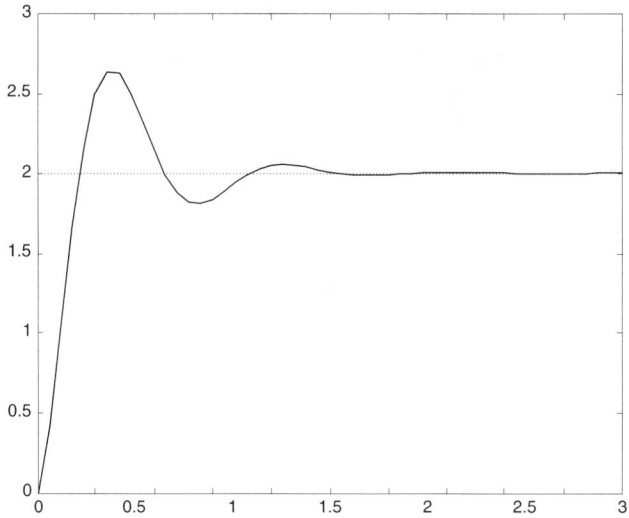

FIGURE 4.2.14.   Closed-loop system with PID controller: evolution of $y(t)$ if $r(t) = 2$ rad.

is used for simulations. The closed-loop system is stable, and the servo-mechanism output (the angular displacement) is illustrated in Figure 4.2.14.

The feedback gains influence the stability. Let us increase the proportional gain k_p and keep the integral and derivative feedback coefficients as before. In particular, $k_p = 5000$, $k_i = 50$, and $k_d = 5$. The following characteristic eigenvalues are found using these gains

```
ans  =
 -5.0368e+002
  1.8350e+000  +1.4089e+002i
  1.8350e+000  -1.4089e+002i
 -1.0000e-002
```

That is, the closed-loop system is unstable. One concludes that the use of the same control algorithm (linear PID controller) does not guarantee the system stability, and one must find the feedback coefficients.

Let $k_p = 500$, $k_i = 25$, and $k_d = 5$ (these feedback coefficients will be used to study the closed-loop servo-mechanism performance in SIMULINK). We have the following eigenvalues:

```
ans  =
 -4.6597e+002
 -1.6998e+001  +4.3072e+001i
 -1.6998e+001  -4.3072e+001i
 -5.0045e-002
```

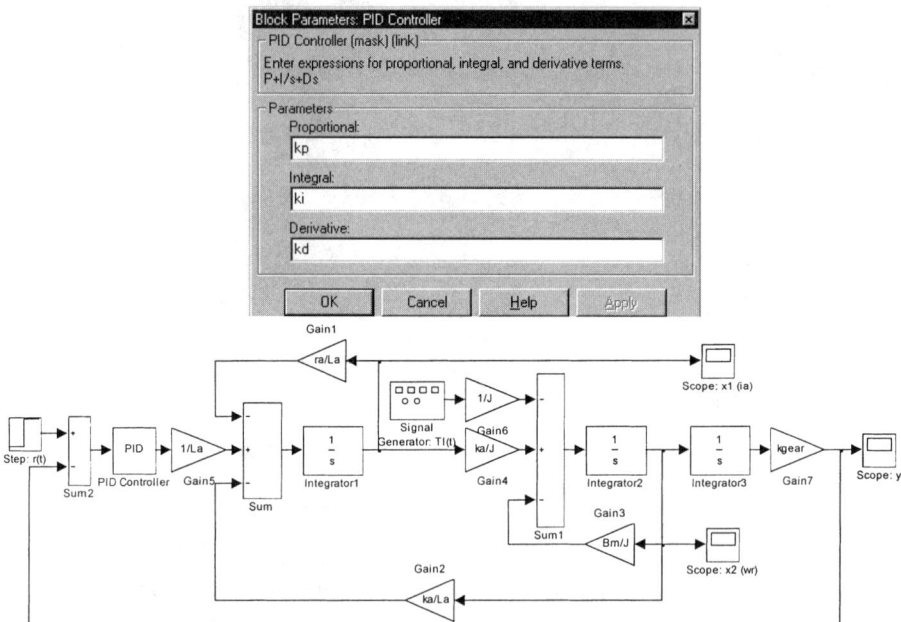

FIGURE 4.2.15.   PID controller and SUMULINK diagram (c_4_2_34.mdl).

The modeling can be performed in SIMULINK, and the resulting diagram is illustrated in Figure 4.2.15. Here, the `PID Controller` building block is used. Downloading the servo-mechanism parameters and feedback gains

```
% Servo-mechanism parameters
ra=5; La=0.01; ka=0.2; J=0.0005; Bm=0.00001; kgear=0.05;
%    Feedback gains
kp=500; ki=25; kd=5;
```

and running the mdl-file, the simulation results become available.

To plot the output angular displacement $y(t) = k_{gear}\theta_r(t)$, one types in the Command Window

```
»plot(y(:,1),y(:,2)); xlabel('Time (seconds)');
title('Output angular displacement [rad]');
```

The transient dynamics of the servo-mechanism output is documented in Figure 4.2.16.

The commonly used performance criteria to be attained are:

- Stability with the desired stability margins in the full operating envelope
- Robustness to parameter variations
- Tracking and disturbance attenuation
- Dynamic and steady-state accuracy

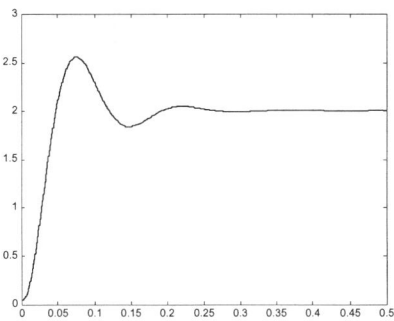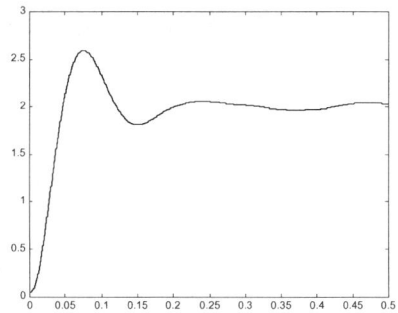

FIGURE 4.2.16.   Output of the servo-system if $r(t) = 2$ rad, $y(t) = k_{\text{gear}}\theta_r$:
(a) $T_L = 0$ N-m and (b) $T_L = 0.5\text{rect}(5t)$ N-m.

- State and output transient responses (the settling time, the maximum overshoot, the rise time, the delay time, as well as the peak time)

It was illustrated that the closed-loop servo-system is stable. The transient
analysis indicates that the settling time is 0.25 sec and the maximum overshoot
is 30%. The control bound must be integrated in the modeling, simulations, and
design. The rated armature voltage is ± 50 V. That is, $-50 \leq u_a \leq 50$ V. The
bounded PID controller is expressed by

$$u_a(t) = \text{sat}_{-50}^{+50}\left(500e(t) + 25\int e(t)\,dt + 5\frac{de(t)}{dt}\right).$$

The modeling can be performed, and the SUMULINK diagram with the Saturation block is illustrated in Figure 4.2.17.

For the desired reference angular displacement $r(t) = 2$ rad, the bounded
armature voltage and the output transient response are illustrated in Figure 4.2.18
if $T_L = 0$ N-m.

It is evident that the settling time and overshoot are different compared with the
linear system (see Figures 4.2.16 and 4.2.18). As the saturation was incorporated
into the modeling and simulation, the settling time is 0.36 sec and the overshoot
is 10%. The saturation effect degrades the system stability, and the eigenvalue
analysis cannot be used to study stability. Other methods should be applied for
dynamic systems with control bounds or other nonlinear phenomena.

4.3. Control of Dynamic Systems Using Digital PID Control Laws

Microprocessors, microcontrollers, and DSPs are widely used to implement control laws, perform diagnostics and filtering, as well as to accomplish data acqui-

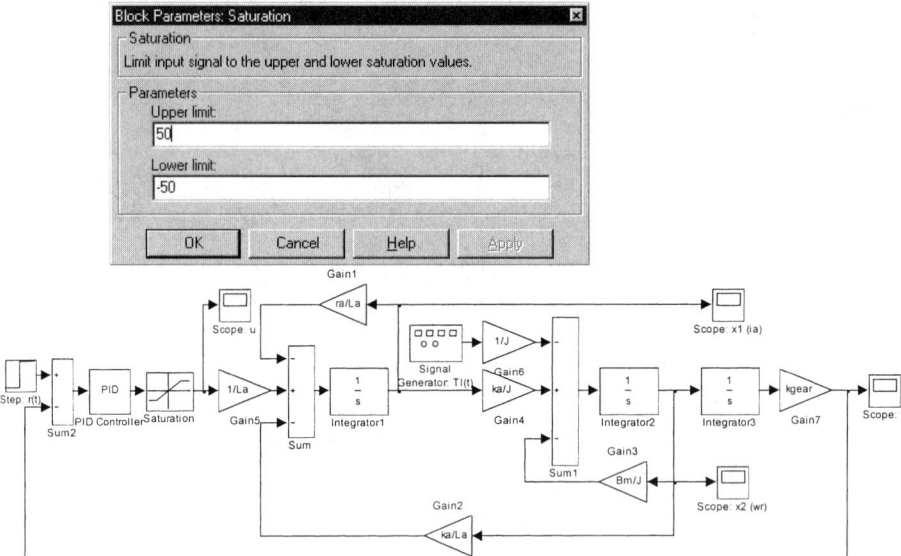

FIGURE 4.2.17.   SIMULINK diagram of the closed-loop servo-mechanism with the saturation (c_4_2_35.mdl).

sition. Therefore, digital control algorithms must be designed, and discrete-time systems must be studied.

To model dynamic systems in the discrete-time domain, the discretization of differential equations must be performed. The continuous-time signal $e(t)$ can be sampled with the sampling period T_s, and the continuous- and discrete-time domains are related as $t = kT_s$, where k is the integer. In addition to the signal sampling, other features must be illustrated. It was shown that by solving differential equations, one finds the evolution of the state variables in the continuous-time domain. In contrast, discrete-time systems are described by difference equations,

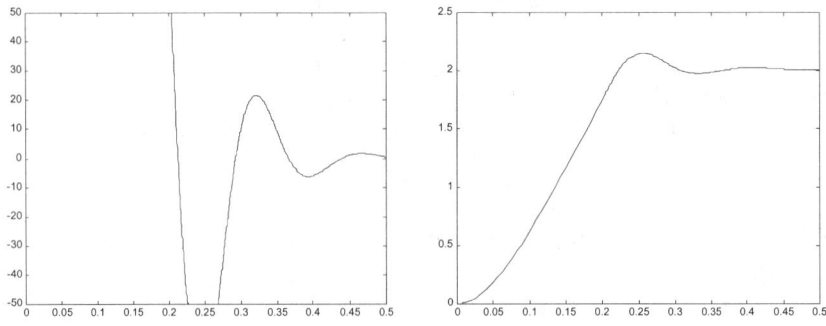

FIGURE 4.2.18.   Armature voltage and servo-mechanism output $y(t)$.

and the solution is found in the discrete-time domain. Furthermore, the differential equations can be represented in the form of difference equations.

Example 4.3.1.

For the first-order, linear, constant-coefficient differential equation $\frac{dx}{dt} = -ax(t) + bu(t)$, let us find the discrete-time model in the form of a difference equation and a transfer function.

From $\frac{dx}{dt} = -ax(t) + bu(t)$, we have $\frac{dx}{dt}\big|_{t=kT_s} = -ax(kT_s) + bu(kT_s)$.

For a small sampling period T_s, the forward rectangular rule, the so-called Euler approximation, gives

$$\frac{dx}{dt} \approx \frac{x(t + T_s) - x(t)}{T_s} \quad \text{and} \quad \frac{dx}{dt}\bigg|_{t=kT_s} = \frac{x(kT_s + T_s) - x(kT_s)}{T_s}.$$

Thus, making use of the forward difference, one obtains

$$\frac{x(kT_s + T_s) - x(kT_s)}{T_s} = -ax(kT_s) + bu(kT_s).$$

We denote $x(t)$ and $u(t)$ at t_k and t_{k+1} as $x_k = x(t)|_{t=kT_s}$, $x_{k+1} = x(t)|_{t=(k+1)T_s}$, and $u_k = u(t)|_{t=kT_s}$.

Therefore,

$$\frac{x_{k+1} - x_k}{T_s} = -ax_k + bu_k,$$

where $x_{k+1} = x[(k+1)T_s]$, $x_k = x(kT_s)$, and $u_k = u(kT_s)$.

Hence, the following difference equation results:

$$x_{k+1} = (1 - aT_s)x_k + bT_s u_k,$$

which can be written as $x_k = (1 - aT_s)x_{k-1} + bT_s u_{k-1}$.

From the derived difference equation, the transfer function is

$$G(z) = \frac{X(z)}{U(z)} = \frac{bT_s z^{-1}}{1 - (1 - aT_s)z^{-1}} = \frac{bT_s}{z - (1 - aT_s)}.$$

That is, the continuous-time system is represented in the discrete-time and z-domains.

Consider the continuous-time signal $e(t)$. Sampling this signal with the sampling period T_s at instances $k = \ldots, -2, -1, 0, 1, 2, \ldots$, we have the values of the signal $\ldots, e(-2T_s), e(-T_s), e(0), e(T_s), e(2T_s), \ldots$. The impulse-sampled signal is expressed as $e_{\text{sampled}}(t) = \sum_{k=-\infty}^{\infty} e(kT_s)\delta(t - kT_s)$, where δ is the Dirac delta function. Modeling causal continuous-time systems and systems, we have $t \geq 0$. Hence, $k \geq 0$ and $e_{\text{sampled}}(t) = \sum_{k=0}^{\infty} e(kT_s)\delta(t - kT_s)$.

The ideal sampler can be viewed as a modulator that multiplies two signals: $e(t)$ and a train of unit impulses $\sum_{k=0}^{\infty} \delta(t - kT_s)$; see Figure 4.3.1.

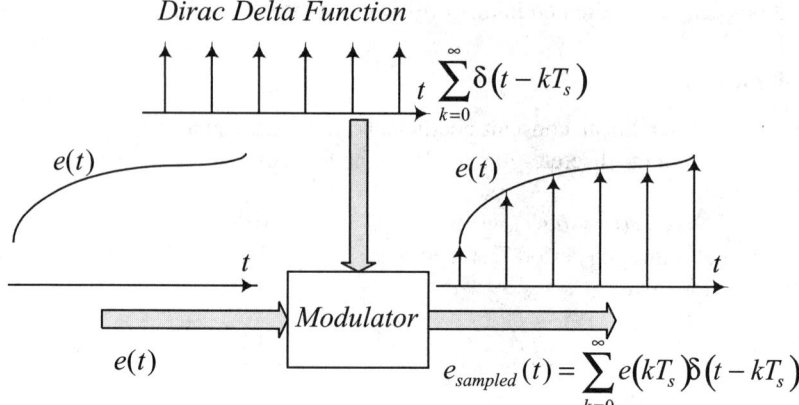

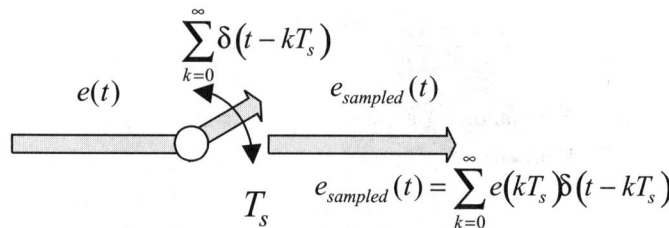

FIGURE 4.3.1.   Sampling of the continuous-time signal $e(t)$.

For $t \geq 0$, $t \in [0, +\infty)$, we have $k = 0, 1, 2, \ldots$ or $k \geq 0$. Hence, the z-transform is found as

$$E(z) = Z[e(t)] = Z[e(kT_s)] = \sum_{k=0}^{\infty} e(kT_s) e^{-kT_s s} = \sum_{k=0}^{\infty} e(kT_s) z^{-k}.$$

The block diagram of closed-loop hybrid dynamic systems, which integrate a continuous-time system to be controlled, A/D and D/A converters, and a digital controller, is shown in Figure 4.3.2.a. It is evident that mathematical models of dynamic systems, as well as data-hold circuits should be studied. Assuming that the system dynamics is modeled by linear constant-coefficient (time-invariant) differential equations, the closed-loop system is documented in Figure 4.3.2.b using the transfer function of system $G_{sys}(s)$, the data-hold circuit $G_H(s)$, and the digital controller $G_C(s)$.

Dynamic systems must be controlled. It has been emphasized and illustrated that a great number of subsystems must be considered and integrated into the analysis. It is evident that the actuators are used (for example, flight actuators drive the control surfaces). The actuators are controlled by power amplifiers, and amplifiers

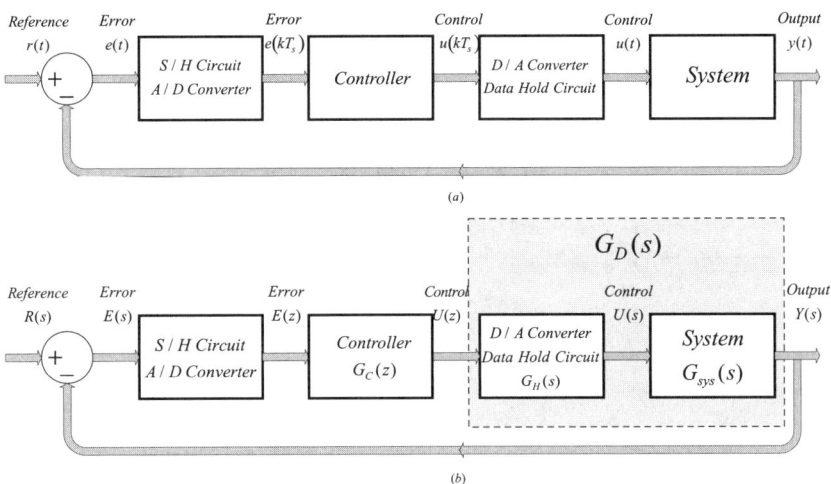

FIGURE 4.3.2.   Block diagrams of dynamic systems with digital controllers.

are controlled by signal-level signals from microcontrollers or DSPs. It is almost impossible to integrate all subsystems and elements. For example, the simplest operational amplifier has more than 50 transistors, and the transistor drivers (integrated circuits) have more than 200 transistors. Mathematically, the radiation and electromagnetic interference, noise, vibration, heat, pressure, as well as other effects can be integrated into the model developments. As a result, the mathematical model of the simplest electric drive, which integrate a power amplifier and a permanent-magnet DC motor, can be developed in the form of thousands of differential equations. However, because of the negligible influence of the secondary effects (very fast dynamics of integrated circuits, high-frequency transistors, slow variations of the motor parameters such as the armature resistance and inductace, *back emf* and *torque* constant, etc.), two differential equations provide a useful mathematical model of an electric drive with a permanent-magnet DC motor to be used in engineering practice to analyze the motor performance and design control algorithms (see Example 4.2.6). In particular, the first equation models the armature circuitry dynamics, whereas the second one models the torsional–mechanical dynamics.

Integrated circuits are widely used in continuous- and discrete-time systems. To convert the discrete-time signals from microcontrollers and DSPs into piecewise continuous signals to be fed to power amplifiers (IGBT, MOSFET, or GTO drivers) data-hold circuits are used. The data-hold circuit can be viewed as a w-order polynomial extrapolator. Zero- and first-order data-hold circuits are typically used to avoid the complexity and time delay associated with the application of high-order data-hold circuits. Consider the sampler and the N-order data-hold circuit, as shown in Figure 4.3.3.

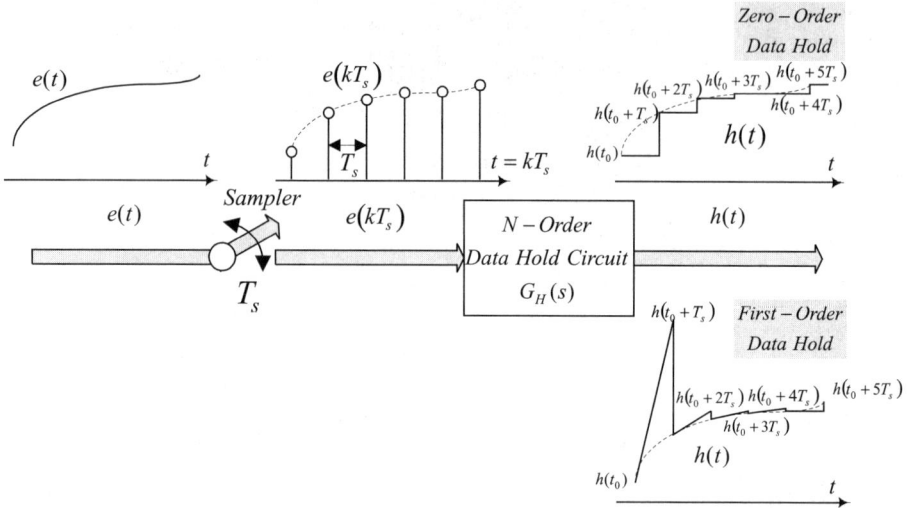

FIGURE 4.3.3.   Sampler and N-order data hold: zero- and first-order data holds.

Assume that the input continuous-time signal $e(t)$ is causal.

For the zero-order data hold, the piecewise, continuous data-hold output is

$$h(t) = \sum_{k=0}^{\infty} e(kT_s)[1(t - kT_s) - 1(t - (k+1)T_s)].$$

The output of the zero-order data hold is the piecewise continuous signal, which is equal to the last sampled value (the value of the continuous-time signal at $t = kT_s$) until the next sampled value is available. The piecewise continuous signal $h(t)$ is illustrated in Figure 4.3.3; see the values $h(t_0), \ldots, h(t_0 + 5T_s)$. In general, we have

$$h(kT_s + t) = h(kT_s) = e(t)|_{t=kT_s} \text{ for } 0 \le t < T_s.$$

Using the Laplace transforms $L[1(t)] = \frac{1}{s}$ and $L[1(t - kT_s)] = \frac{e^{-kT_s s}}{s}$, one finds

$$L[h(t)] = \sum_{k=0}^{\infty} e(kT_s) \frac{e^{-kT_s s} - e^{-(k+1)T_s s}}{s} = \frac{1 - e^{-T_s s}}{s} \sum_{k=0}^{\infty} e(kT_s)e^{-kT_s s}.$$

Let us derive the transfer function for the zero-order data hold. Using the Laplace transform of the signal $e(kT_s)$, as given by $E_{sampled}(s) = \sum_{k=0}^{\infty} e(kT_s)e^{-kT_s s}$, one obtains the transfer function of the zero-order data hold. In particular, from $L[h(t)] = \frac{1 - e^{-T_s s}}{s} \sum_{k=0}^{\infty} e(kT_s)e^{-kT_s s}$, we have

$$G_H(s) = \frac{1 - e^{-T_s s}}{s}.$$

The first-order data hold performs the direct linear extrapolation; see Figure 4.3.3.

The expression for the hold output in time domain is

$$h(t) = 1(t) + \frac{t}{T_s}1(t) - \frac{t - T_s}{T_s}1(t - T_s) - 1(t - T_s).$$

Hence, the following transfer function results:

$$G_H(s) = \frac{1}{s} + \frac{1}{T_s s^2} - \frac{1}{T_s s^2}e^{-T_s s} - \frac{1}{s}e^{-T_s s} = \left(1 - e^{-T_s s}\right)\frac{T_s s + 1}{T_s s^2}.$$

The augmented transfer function $G_D(s)$ of the dynamic system $G_{\text{sys}}(s)$ and the data-hold circuit $G_H(s)$, as illustrated in Figure 4.3.2, is

$$G_D(s) = G_H(s)G_{\text{sys}}(s).$$

Let us find the transfer function of the digital PID controller. The analog PID controller is

$$u(t) = k_p e(t) + k_i \frac{e(t)}{s} + k_d s e(t),$$

and the transfer function is $G_{\text{PID}}(s) = \frac{U(s)}{E(s)} = \frac{k_d s^2 + k_p s + k_i}{s}$.
For the proportional analog controller, we have $u_p(t) = k_p e(t)$ and

$$G_P(s) = \frac{U_p(s)}{E(s)} = k_p.$$

The proportional digital control law is $u_p(kT_s) = k_p e(kT_s)$ and

$$G_P(z) = \frac{U_p(z)}{E(z)} = k_p.$$

The integral

$$u_i(t) = k_i \frac{e(t)}{s}$$

and the derivative $u_d(t) = k_d s e(t)$ terms of the PID controller, with

$$G_I(s) = \frac{U_I(s)}{E(s)} = \frac{k_i}{s}$$

and

$$G_D(s) = \frac{U_D(s)}{E(s)} = k_d s,$$

can be represented in the discrete-time and z-domains. In particular, the z-transforms, as given in Table 4.3.1, are used. For the integral part, using the Euler approximation, the transfer function is

$$G_I(z) = \frac{U_I(z)}{E(z)} = \frac{T_s}{1 - z^{-1}} = \frac{T_s z}{z - 1}.$$

TABLE 4.3.1. z-transform table.

Laplace transform $X(s)$	Time-domain signal $x(t)$	Time-domain signal $x(kT_s)$	z-transform $X(z)$
$\dfrac{1}{s}$	Unit step $1(t)$	$1(kT_s)$	$\dfrac{1}{1-z^{-1}} = \dfrac{z}{z-1}$
$\dfrac{1}{s^2}$	$t1(t)$	$kT_s1(kT_s)$	$\dfrac{T_sz^{-1}}{(1-z^{-1})^2} = \dfrac{T_sz}{(z-1)^2}$
$\dfrac{2}{s^3}$	$t^21(t)$	$(kT_s)^21(kT_s)$	$\dfrac{T_s^2z^{-1}(1+z^{-1})}{(1-z^{-1})^3}$
$\dfrac{6}{s^4}$	$t^31(t)$	$(kT_s)^31(kT_s)$	$\dfrac{T_s^3z^{-1}(1+4z^{-1}+z^{-2})}{(1-z^{-1})^4}$
$\dfrac{24}{s^5}$	$t^41(t)$	$(kT_s)^41(kT_s)$	$\dfrac{T_s^4z^{-1}(1+11z^{-1}+11z^{-2}+z^{-3})}{(1-z^{-1})^5}$
$\dfrac{1}{s+a}$	$e^{at}1(t)$	$e^{-akT_s}1(kT_s)$	$\dfrac{1}{1-e^{-aT_s}z^{-1}}$
$\dfrac{a}{s(s+a)}$	$(1-e^{-at})1(t)$	$(1-e^{-akT_s})1(kT_s)$	$\dfrac{(1-e^{-aT_s})z^{-1}}{(1-z^{-1}) \times (1-e^{-aT_s}z^{-1})}$
$\dfrac{b-a}{(s+a)(s+b)}$	$(e^{-at}-e^{-bt})1(t)$	$(e^{-akT_s}-e^{-bkT_s})1(kT_s)$	$\dfrac{(e^{-aT_s}-e^{-bT_s})z^{-1}}{(1-e^{-aT_s}z^{-1}) \times (1-e^{-bT_s}z^{-1})}$
$\dfrac{1}{(s+a)^2}$	$te^{-at}1(t)$	$kT_se^{-akT_s}1(kT_s)$	$\dfrac{T_se^{-aT_s}z^{-1}}{(1-e^{-aT_s}z^{-1})^2}$
$\dfrac{s}{(s+a)^2}$	$(1-at)e^{-at}1(t)$	$(1-akT_s)e^{-akT_s}1(kT_s)$	$\dfrac{1-(1+aT_s)e^{-aT_s}z^{-1}}{(1-e^{-aT_s}z^{-1})^2}$
$\dfrac{\omega_0}{s^2+\omega_0^2}$	$\sin(\omega_0t)1(t)$	$\sin(\omega_0kT_s)1(kT_s)$	$\dfrac{z^{-1}\sin(\omega_0T_s)}{1-2z^{-1}\cos(\omega_0T_s)+z^{-2}}$
$\dfrac{s}{s^2+\omega_0^2}$	$\cos(\omega_0t)1(t)$	$\cos(\omega_0kT_s)1(kT_s)$	$\dfrac{1-z^{-1}\cos(\omega_0T_s)}{1-2z^{-1}\cos(\omega_0T_s)+z^{-2}}$
$\dfrac{\omega_0}{(s+a)^2+\omega_0^2}$	$e^{-at}\sin(\omega_0t)1(t)$	$e^{-akT_s}\sin(\omega_0kT_s)1(kT_s)$	$\dfrac{e^{-akT_s}z^{-1}\times\sin(\omega_0T_s)}{1-2e^{-aT_s}z^{-1}\times\cos(\omega_0T_s)+e^{-2aT_s}z^{-2}}$
$\dfrac{s+a}{(s+a)^2+\omega_0^2}$	$e^{-at}\cos(\omega_0t)1(t)$	$e^{-akT_s}\cos(\omega_0kT_s)1(kT_s)$	$\dfrac{1-e^{-aT_s}z^{-1}\cos(\omega_0T_s)}{1-2e^{-aT_s}z^{-1}\times\cos(\omega_0T_s)+e^{-2aT_s}z^{-2}}$

To find the derivative term, using the trapezoidal approximation, the first difference results, and one obtains

$$G_D(z) = \frac{U_D(z)}{E(z)} = \frac{1 - z^{-1}}{T_s} = \frac{z - 1}{T_s z}.$$

An infinite number of analog PID-type controllers with transfer functions denoted as $G_{PID}(s)$ exists. For example, PID and PI^3DD^3 controllers

$$u(t) = k_p e(t) + k_i \frac{e(t)}{s} + k_d s e(t)$$

and

$$u(t) = k_p e(t) + k_{i1} \frac{e(t)}{s} + k_{i3} \frac{e(t)}{s^3} + k_{d1} s e(t) + k_{d3} s^3 e(t)$$

have the following transfer functions:

$$G_{PID}(s) = \frac{U(s)}{E(s)} = \frac{k_d s^2 + k_p s + k_i}{s}$$

and

$$G_{PID}(s) = \frac{U(s)}{E(s)} = \frac{k_{d3} s^6 + k_{d1} s^4 + k_p s^3 + k_{i1} s^2 + k_{i3}}{s^3}.$$

One finds the z-domain representation of the control signal $U(z)$ and the transfer functions of digital PID controllers. In the *error* form (the error signal is used to calculate the control), the following expressions result:

$$U(z) = \left(k_{dp} + \frac{k_{di}}{1 - z^{-1}} + k_{dd} \left(1 - z^{-1} \right) \right) E(z),$$

and

$$G_{PID}(z) = \frac{U(z)}{E(z)} = k_{dp} + \frac{k_{di}}{1 - z^{-1}} + k_{dd} \left(1 - z^{-1} \right).$$

Hence,

$$G_{PID}(z) = \frac{(k_{dp} + k_{di} + k_{dd})z^2 - (k_{dp} + 2k_{dd})z + k_{dd}}{z^2 - z}.$$

The *reference-output* form of the digital PID control law can be found. In particular, using the reference input and system output, we have

$$U(z) = -k_{dp} Y(z) - k_{di} \frac{Y(z) - R(z)}{1 - z^{-1}} - k_{dd} \left(1 - z^{-1} \right) Y(z).$$

These results must be visualized. For the error and reference-output forms, the

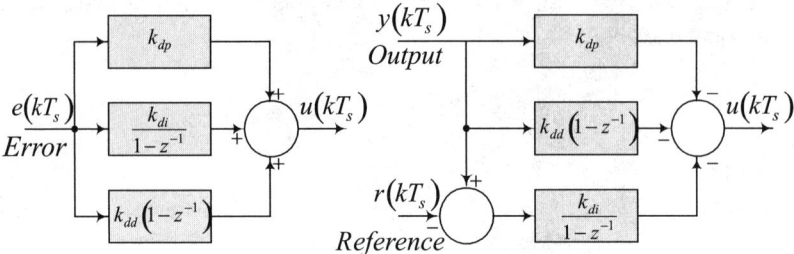

FIGURE 4.3.4.   Representations of error $U(z) = \left[k_{dp} + \frac{k_{di}}{1-z^{-1}} + k_{dd}\left(1 - z^{-1}\right)\right] E(z)$ and reference-output $U(z) = -k_{dp}Y(z) - k_{di}\frac{Y(z)-R(z)}{1-z^{-1}} - k_{dd}\left(1 - z^{-1}\right)Y(z)$ forms of the digital controllers.

block diagrams of the digital PID controllers are documented in Figure 4.3.4.

The feedback gains k_{dp}, k_{di}, and k_{dd} of the digital control law are found using the proportional, integral, and derivative coefficients of the analog PID controller k_p, k_i, and k_d as well as the sampling period. We have

$$k_{dp} = k_p - \tfrac{1}{2}k_{di}, \quad k_{di} = T_s k_i \quad \text{and}$$

$$k_{dd} = \frac{k_d}{T_s}.$$

Using microcontrollers and DSPs, the designer has a wide variety of options to implement digital controllers. For example, the PID controller can be implemented as

$$u(kT_s) = \underbrace{k_p e(kT_s)}_{\text{Proportional}} + \underbrace{\tfrac{1}{2}k_i T_s \sum_{i=1}^{k}[e((i-1)T_s) + e(iT_s)]}_{\text{Integral}}$$

$$\underbrace{+\frac{k_d}{T_s}[e(kT_s) - e((k-1)T_s)]}_{\text{Derivative}}.$$

To find the transfer function for discrete-time systems and controllers applied, the transfer functions in the s-domain are used. In particular, the Tustin substitution method is commonly applied to obtain the approximation of transfer functions in the z-domain. One can approximate the differential equations in the form of the difference equations using the Tustin formula. In particular, from $z = e^{sT_s}$, we have

$$s = \frac{1}{T_s}\ln(z),$$

and the so-called Tustin approximation is given as

$$\ln(z) \approx 2\frac{z-1}{z+1} = 2\frac{1-z^{-1}}{1+z^{-1}}.$$

This approximation is obtained by truncating the series expansion of $\ln(z)$, which is

$$\ln(z) = 2\left[\frac{z-1}{z+1} + \frac{1}{3}\left(\frac{z-1}{z+1}\right)^3 + \frac{1}{5}\left(\frac{z-1}{z+1}\right)^5 + \cdots\right], \quad z > 0.$$

Example 4.3.2.

Using the Tustin approximation, let us approximate the s-domain transfer function of the linear PID controller

$$G_{\text{PID}}(s) = \frac{U(s)}{E(s)} = \frac{k_d s^2 + k_p s + k_i}{s}$$

and find $G_{\text{PID}}(z)$. Also, the expression to implement the digital controller must be obtained. That is, find $u(k)$.

From

$$G_{\text{PID}}(s) = \frac{U(s)}{E(s)} = \frac{k_d s^2 + k_p s + k_i}{s},$$

one finds

$$G_{\text{PID}}(z) = \frac{U(z)}{E(z)}$$

$$= \frac{k_d\left(\frac{2}{T_s}\frac{1-z^{-1}}{1+z^{-1}}\right)^2 + k_p \frac{2}{T_s}\frac{1-z^{-1}}{1+z^{-1}} + k_i}{\frac{2}{T_s}\frac{1-z^{-1}}{1+z^{-1}}}$$

$$= \frac{(2k_p T_s + k_i T_s^2 + 4k_d) + (2k_i T_s^2 - 8k_d)z^{-1}}{+(-2k_p T_s + k_i T_s^2 + 4k_d)z^{-2}}{2T_s(1 - z^{-2})}$$

Thus, the following expression is obtained:

$$U(z) - U(z)z^{-2} = k_{e0}E(z) + k_{e1}E(z)z^{-1} + k_{e2}E(z)z^{-2},$$

$$k_{e0} = k_p + \tfrac{1}{2}k_i T_s + 2\frac{k_d}{T_s},\ k_{e1} = k_i T_s - 4\frac{k_d}{T_s},$$

and

$$k_{e2} = -k_p + \tfrac{1}{2}k_i T_s + 2\frac{k_d}{T_s}.$$

Making use of the z-transform, one finds that to implement the controller, the following control signal must be supplied:

$$u(k) = u(k-2) + k_{e0}e(k) + k_{e1}e(k-1) + k_{e2}e(k-2).$$

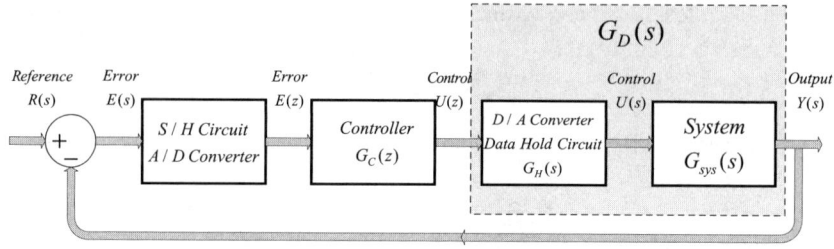

FIGURE 4.3.5.   Block diagrams of the closed-loop dynamic systems with digital controllers.

That is, to implement the PID controller, one uses $e(k)$, $e(k-1)$, $e(k-2)$, and $u(k-2)$.

The Tustin approximation can be performed by applying the c2dm command. Using the help command, we have


```
» help c2dm
C2DM    Conversion of continuous LTI systems to discrete-
time. [Ad,Bd,Cd,Dd]=C2DM(A,B,C,D,Ts,'method') converts
the continuous-time state-space system (A,B,C,D) to
discrete time using 'method':
        'zoh'           Convert to discrete time assuming a
                        zero order hold on the inputs.
        'foh'           Convert to discrete time assuming a
                        first order hold on the inputs.
        'tustin'        Convert to discrete time using the
                        bilinear (Tustin) approximation to
                        the derivative.
        'prewarp'       Convert to discrete time using the
                        bilinear (Tustin) approximation with
                        frequency prewarping. Specify the
                        critical frequency with an additional
                        argument, i.e., C2DM(A,B,C,D,Ts,
                        'prewarp',Wc)
        'matched'       Convert the SISO system to discrete
                        time using the matched pole-zero
                        method.
[NUMd,DENd]=C2DM(NUM,DEN,Ts,'method') converts the
continuous-time polynomial transfer function G(s) =
NUM(s)/DEN(s) to discrete-time, G(z)=NUMd(z)/DENd(z),
using 'method'. See also: C2D, and D2CM.
```


The closed-loop system with digital controller $G_C(s)$ is illustrated in Figure 4.3.5.

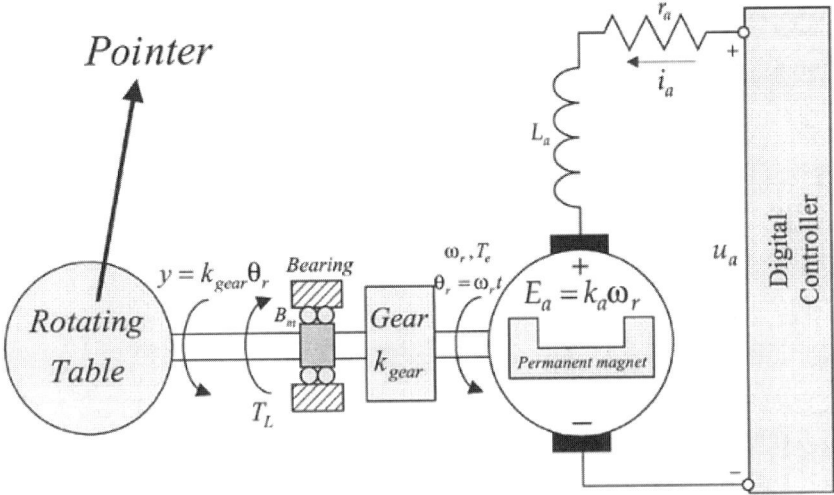

The transfer function of the closed-loop dynamic systems, given in Figure 4.3.5, is

$$G(z) = \frac{Y(z)}{R(z)} = \frac{G_C(z)G_D(z)}{1 + G_C(z)G_D(z)},$$

and if the digital PID controller is used, we have

$$G(z) = \frac{Y(z)}{R(z)} = \frac{G_{\mathrm{PID}}(z)G_D(z)}{1 + G_{\mathrm{PID}}(z)G_D(z)}.$$

Using these transfer functions, the stability and the closed-loop system performance can be studied.

Example 4.3.3. Control of permanent-magnet DC motors using digital controllers.

Consider the servo-system actuated by the permanent-magnet DC motor, as studied in Chapter 4.2 (Example 4.2.6). Our goal is to design the digital PID controller, as well as simulate and analyze the closed-loop system. The servo-system, to be controlled by the digital controller, is shown in Figure 4.3.6.

Taking note of three differential equations

$$\frac{di_a}{dt} = -\frac{r_a}{L_a}i_a - \frac{k_a}{L_a}\omega_r + \frac{1}{L_a}u_a,$$

$$\frac{d\omega_r}{dt} = \frac{1}{J}(T_e - T_{viscous} - T_L) = \frac{1}{J}(k_a i_a - B_m\omega_r - T_L),$$

$$\frac{d\theta_r}{dt} = \omega_r,$$

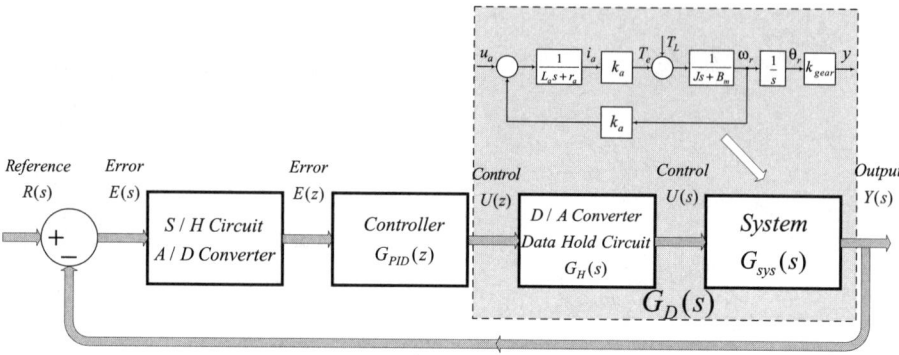

FIGURE 4.3.7.   Block diagrams of the closed-loop systems with the digital PID controller.

one can develop the block diagram of the open-loop servo-system. The closed-loop system is illustrated in Figure 4.3.7.

Augmenting the transfer function of the open-loop servo-systems

$$G_{sys}(s) = \frac{Y(s)}{U(s)} = \frac{k_{gear}k_a}{s(L_aJs^2 + (r_aJ + L_aB_m)s + r_aB_m + k_a^2)},$$

with the transfer function of the zero-order data hold

$$G_H(s) = \frac{1 - e^{-T_ss}}{s},$$

one has

$$G_D(s) = G_H(s)G_{sys}(s) = \frac{1 - e^{-T_ss}}{s} \frac{k_{gear}k_a}{s(L_aJs^2 + (r_aJ + L_aB_m)s + r_aB_m + k_a^2)}.$$

The transfer function $G_D(z)$ is found from $G_D(s)$ using the c2dm command. The filter command is used to model the dynamics.

The following MATLAB m-file was written (c_4_3_21.m):

```
clear all; format short e;
% parameters of the servomechanism with permanent-magnet
% DC motor
ra=5; La=0.01; ka=0.2; J=0.0005; Bm=0.00001; kgear=0.05;
% numerator and denominator of the transfer function of
% open-loop servo H(s)
num`s=[ka*kgear];
den`s=[ra*J (ra*Bm+ka^2) 0];        % armature inductance
                                    % is neglected
den`s=[La*J (ra*J+La*Bm) (ra*Bm+ka^2) 0];% armature in-
                                    % ductance is
                                    % not neglected
```

```
num`s, den`s
pause;

% numerator and denominator of the transfer function
% with zero-order data hold GD(z)
Ts=0.001; % sampling time Ts
[num`dz,den`dz]=c2dm(num`s,den`s,Ts,'zoh');
num`dz, den`dz
pause;
% feedback coefficient gains of the analog PID controller
kp=500;ki=50;kd=5;
% feedback coefficient gains of the digital PID
% controller
kdi=Ts*ki; kdp=kp-kdi/2; kdd=kd/Ts;
% numerator and denominator of the transfer function
% of the digital PID controller
num`pidz=[(kdp+kdi+kdd) -(kdp+2*kdd) kdd];
den`pidz=[1 -1 0];
num`pidz, den`pidz
pause;

% numerator and denominator of the closed-loop transfer
% function G(z)
num`z=conv(num`pidz,num`dz);
den`z=conv(den`pidz,den`dz)+conv(num`pidz,num`dz);
num`z, den`z
pause;

% samples, t=k*Ts
k`final=300; k=0:1:k`final;
% reference input r(t)=2 rad
r=2*ones(1,k`final+1);
% modeling of the servo-system output y(k)
y=filter(num`z,den`z,r);
%plotting statement
plot(k,y,'o',k,y,'--');
title('Output (angular displacement) y(k), [rad]');
xlabel('Discrete time k, t=kTs [seconds]');
disp('End')
```

Having found the numerator and denominator of $G_{sys}(s)$ to be

```
num`s =
  1.0000e-002
den`s =
  5.0000e-006   2.5001e-003   4.0050e-002                 0
```

one concludes that neglecting the armature inductance, the following transfer function can be written for the open-loop system:

$$G_{sys}(s) = \frac{Y(s)}{U(s)} = \frac{0.01}{s(5 \times 10^{-6}s^2 + 2.5 \times 10^{-3}s + 0.04)}.$$

The sampling period is assigned to be 0.001 sec. That is, $T_s = 0.001$ sec. The transfer function $G_D(z)$ is found, and taking note of

```
num˙dz =

        0   2.9539e-007   1.0473e-006   2.3010e-007
den˙dz =
   1.0000e+000  -2.6002e+000   2.2067e+000  -6.0652e-001
```

the following transfer function is found:

$$G_D(z) = \frac{2.95 \times 10^{-7}z^2 + 1.05 \times 10^{-6}z + 2.3 \times 10^{-7}}{z^3 - 2.6z^2 + 2.21z - 0.61}.$$

The transfer function of the digital PID controller is expressed as

$$G_{PID}(z) = \frac{(k_{dp} + k_{di} + k_{dd})z^2 - (k_{dp} + 2k_{dd})z + k_{dp}}{z^2 - z},$$

$$k_{dp} = k_p - \tfrac{1}{2}k_{di}, \, k_{di} = T_s k_i, \, \text{and} \, k_{dd} = \frac{k_d}{T_s}.$$

The feedback gains of the analog PID controller was assigned to be $k_p = 500$, $k_i = 50$, and $k_d = 5$. The fedback coefficients of the digital controller are found using formulas $k_{dp} = k_p - \tfrac{1}{2}k_{di}$, $k_{di} = T_s k_i$, $k_{dd} = \frac{k_d}{T_s}$, and the numerator and denominator of the transfer function

$$G_{PID}(z) = \frac{(k_{dp} + k_{di} + k_{dd})z^2 - (k_{dp} + 2k_{dd})z + k_{dp}}{z^2 - z}$$

are found as

```
num˙pidz =
   5.5000e+003  -1.0500e+004   5.0000e+003
den˙pidz =
       1      -1       0
```

Thus,

$$G_{PID}(z) = \frac{5.5 \times 10^3 z^2 - 1.05 \times 10^4 z + 5 \times 10^3}{z^2 - z}.$$

Hence, for the closed-loop systems

$$G(z) = \frac{Y(z)}{R(z)} = \frac{G_{PID}(z)G_D(z)}{1 + G_{PID}(z)G_D(z)},$$

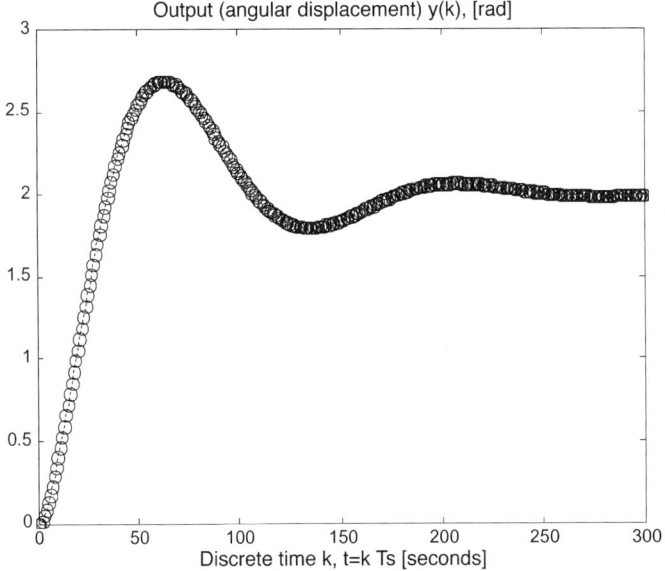

FIGURE 4.3.8.   Output of the servo-system with the digital PID controller.

making use the following numerical results:

```
num z =
          0   1.6247e-003 2.6586e-003 -8.2543e-003 2.8205e-003
          1.1505e-003
den z =
1.0000e+000 -3.5986e+000 4.8096e+000 -2.8215e+000 6.0934e-001
1.1505e-003
```

one obtains

$$G(z) = \frac{1.62 \times 10^{-3}z^4 + 2.66 \times 10^{-3}z^3 - 8.25 \times 10^{-3}z^2}{z^5 - 3.6z^4 + 4.81z^3 - 2.82z^2 + 0.61z + 0.00115}.$$

The output dynamics for the reference input $r(kT_s) = 2$ rad, $k \geq 0$ is shown in Figure 4.3.8.

The settling time is $k_{settling}T_s = 200 \times 0.001 = 0.2$ sec, and the overshoot is 35%.

It must be emphasized that the sampling time significantly influences the system dynamics. Let $T_s = 0.01$ sec. Running the c_4_3_21 m-file, for $T_s = 0.01$ sec,

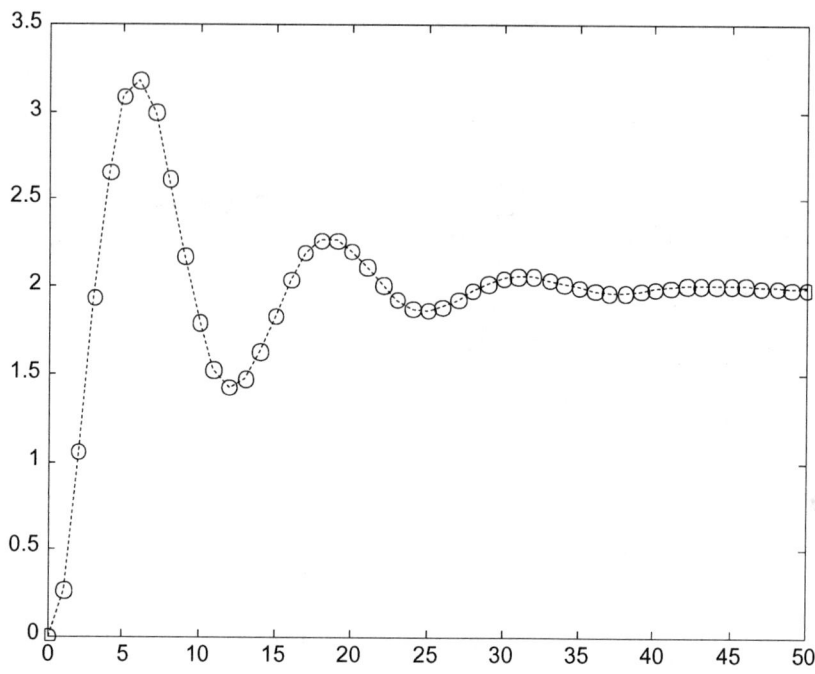

FIGURE 4.3.9.   Output of the servo-system with the digital PID controller, $T_s = 0.01$ sec.

we have

```
num´z =

        0   1.3244e-001 3.3998e-002 -2.6923e-001 9.6337e-002
        6.6416e-003
den´z =
1.0000e+000 -2.7228e+000 2.7513e+000 -1.1380e+000 1.0307e-001
6.6416e-003
```

That is, the closed-loop transfer function is

$$G(z) = \frac{0.13z^4 + 0.034z^3 - 0.27z^2 + 0.096z + 0.00664}{z^5 - 2.72z^4 + 2.75z^3 - 1.14z^2 + 0.103z + 0.00664}.$$

The output $y(kT_s)$ is documented in Figure 4.3.9.

The settling time is $k_{settling}T_s = 30 \times 0.01 = 0.3$ sec, and the overshoot is 60%.

The analytical results are extremely viable in addition to numerical solution and modeling, which were performed in the MATLAB environment. Making note of the transfer functions $G_{\text{sys}}(s)$ and $G_H(s)$, one finds

$$G_D(s) = G_H(s)G_{\text{sys}}(s) = \frac{1 - e^{-T_s s}}{s} \frac{k_{gear}k_a}{s(L_a J s^2 + (r_a J + L_a B_m)s + r_a B_m + k_a^2)}.$$

Assume that the eigenvalues are real and distinct, and the zero-order data hold is used

$$G_H(s) = \frac{1 - e^{-T_s s}}{s}.$$

Thus,

$$G_D(s) = G_H(s)G_{\text{sys}}(s) = \frac{1 - e^{-T_s s}}{s} \frac{k}{s(T_1 s + 1)(T_2 s + 1)}.$$

Here, $k = k_{\text{gear}} k_a$, and the time constants are found from $L_a J s^2 + (r_a J + L_a B_m)s + r_a B_m + k_a^2$.

The partial fraction expansion is performed, letting $T_1 \neq T_2$. That is, the eigenvalues $-\frac{1}{T_1}$ and $-\frac{1}{T_2}$ are distinct (nonrepeated). From Heaviside's expansion formula, one obtains

$$\frac{k}{s^2(T_1 s + 1)(T_2 s + 1)} = \frac{c_1}{s} + \frac{c_2}{s^2} + \frac{c_3}{T_1 s + 1} + \frac{c_4}{T_2 s + 1},$$

where c_1, c_2, c_3, and c_4 are the unknown coefficients that are found as

$$c_1 = \frac{d}{ds}\left(\frac{ks^2}{s^2(T_1 s + 1)(T_2 s + 1)}\right)\Bigg|_{s=0} = \frac{-k(2T_1 T_2 s + T_1 + T_2)}{(T_1 T_2 s^2 + (T_1 + T_2)s + 1)}\Bigg|_{s=0}$$
$$= -k(T_1 + T_2),$$

$$c_2 = \frac{k}{(T_1 s + 1)(T_2 s + 1)}\Bigg|_{s=0} = k, \; c_3 = \frac{k}{s^2(T_2 s + 1)}\Bigg|_{s=-\frac{1}{T_i}} = \frac{kT_1^3}{T_1 - T_2} \; \text{and}$$

$$c_4 = \frac{k}{s^2(T_1 s + 1)}\Bigg|_{s=-\frac{1}{T_2}} = \frac{kT_2^3}{T_2 - T_1}.$$

Let us find $G_D(z)$. We have $G_D(z) = Z[G_D(s)] = Z[G_H(s)G_{\text{sys}}(s)] = \frac{z-1}{z}Z\left[\frac{G_{\text{sys}}(s)}{s}\right]$.

Using the z-transform table, we have the explicit expression for $G_D(z)$. In particular,

$$G_D(z) = \frac{z-1}{z}Z\left[\frac{k}{s^2(T_1 s + 1)(T_2 s + 1)}\right]$$
$$= k\frac{z-1}{z}Z\left[-\frac{T_1 + T_2}{s} + \frac{1}{s^2} + \frac{T_1^3}{T_1 - T_2}\left(\frac{1}{T_1 s + 1}\right)\right.$$
$$\left. + \frac{T_2^3}{T_2 - T_1}\frac{1}{(T_2 s + 1)}\right]$$
$$= k\frac{z-1}{z}\left(-(T_1 + T_2)\frac{z}{z-1} + \frac{T_s z}{(z-1)^2} + \frac{T_1^2}{T_1 - T_2}\frac{z}{z - e^{-T_s/T_1}}\right.$$
$$\left. + \frac{T_2^2}{T_2 - T_1}\frac{z}{z - e^{-T_s/T_2}}\right).$$

Taking note of

$$G_D(z) = k \left(-T_1 - T_2 + T_s \frac{1}{z-1} + \frac{T_1^2}{T_1 - T_2} \frac{z-1}{z - e^{-T_s/T_1}} \right.$$
$$\left. + \frac{T_2^2}{T_2 - T_1} \frac{z-1}{z - e^{-T_s/T_2}} \right)$$

and

$$G_{\text{PID}}(z) = \frac{(k_{dp} + k_{di} + k_{dd})z^2 - (k_{dp} + 2k_{dd})z + d_{dd}}{z^2 - z},$$

the transfer function of the closed-loop system is found as

$$G(z) = \frac{Y(z)}{R(z)} = \frac{G_{\text{PID}}(z)G_D(z)}{1 + G_{\text{PID}}(z)G_D(z)},$$

and the analytic solution $y(k)$ can be straightforwardly derived for different wave-forms of the reference inputs.

4.4.   Hamilton–Jacobi and Lyapunov Methods in Optimal Control of Continuous-Time Systems

4.4.1.   *The Hamilton–Jacobi Theory and Optimal Control*

In Section 4.3, we have considered the application of the PID-type controllers to attain the certain desirable features of closed-loop systems. However, these closed-loop systems, in most cases, do not satisfy the requirements imposed. The designer needs to synthesize control algorithms to minimize or maximize performance indexes that depend on the system variables (states x, outputs y, tracking errors e, and control u), which measure the system performance. In general, the problem is to find the state-feedback control law

$$u = \phi(t, e, x), \tag{4.4.1}$$

such that the performance functional

$$J(x(\cdot), y(\cdot), e(\cdot), u(\cdot)) = \int_{t_0}^{t_f} W_{xyeu}(x, y, e, u)\, dt,$$

is minimized subject to the dynamic system.

Here, $W_{xyeu}(\cdot)\colon \mathbb{R}^n \times \mathbb{R}^b \times \mathbb{R}^b \times \mathbb{R}^m \to \mathbb{R}_{\geq 0}$ is the positive-definite and continuously differentiable integrand function; t_0 and t_f are the initial and final times.

Once the performance functional is chosen, the design involves the straightforward minimization or maximization problem, and the Hamilton–Jacobi concept can be applied in deriving optimal control algorithms.

For example, in a single-input/single-output case, one can wish to minimize the square of the magnitude of the error and control; that is, we have $J(e(\cdot), u(\cdot)) =$

$\int_{t_0}^{t_f} (e^2 + u^2) \, dt$. That is, $W_{eu}(e, u) = e^2 + u^2$. The performance functional must be positive-definite, and one can design and minimize a great variety of functionals to attain better performance. For example, $J(e(\cdot), u(\cdot)) = \int_{t_0}^{t_f} (e^8 + u^{\frac{2}{3}}) \, dt$ or $J(e(\cdot), u(\cdot)) = \int_{t_0}^{t_f} (|e| + u^{\frac{4}{3}}) \, dt$. The question is: Why is the squared quantity commonly used? The answer is very simple: Problems involving squared terms are much more analytically solvable than are other types. In general, the application of nonquadratic integrands results in mathematical complexity. However, the system performance is improved as nonlinear performance functionals are used to find (synthesize) control algorithms.

Illustrative Example 4.4.1

Consider the first-order system $\dot{x}(t) = ax + bu$. Using the squared errors of the state variable x and control input u, as well as the weighting coefficients q and g, we define the quadratic functional to be minimized as $J(x(\cdot), u(\cdot)) = \frac{1}{2} \int_{t_0}^{t_f} (qx^2 + gu^2) \, dt$, $q \geq 0$, $g > 0$. Let us find the controller.

We introduce the Hamiltonian function, which is given as

$$H\left(x, u, \frac{\partial V}{\partial x}\right) = \underbrace{\frac{1}{2}\left(qx^2 + gu^2\right)}_{\substack{\text{Performance Functional} \\ J(x, u) = \int_{t_0}^{t_f} (qx^2 + gu^2)\, dt}} + \frac{\partial V}{\partial x} \underbrace{\frac{dx}{dt}}_{\substack{\text{System Dynamics} \\ \dot{x}(t) = ax + bu}}$$

$$= \frac{1}{2}\left(qx^2 + gu^2\right) + \frac{\partial V}{\partial x}(ax + bu),$$

where $V(x)$ is the return function.

Minimizing the Hamiltonian, that is, making use of

$$\frac{\partial H\left(x, u, \frac{\partial V}{\partial x}\right)}{\partial u} = 0,$$

one obtains

$$gu + \frac{\partial V}{\partial x} b = 0.$$

Thus, the controller is expressed as

$$u = -\frac{b}{g}\frac{\partial V}{\partial x} = -g^{-1} b \frac{\partial V}{\partial x}.$$

Let the return function be given as $V(x) = \frac{1}{2}kx^2$. Then, the controller is

$$u = -g^{-1} bkx.$$

The unknown coefficient k should be found by solving the Riccati equation, as will be illustrated later. From $\dot{x}(t) = ax + bu$ and $u = -g^{-1}bkx$, the closed-loop system is expressed in the form

$$\dot{x}(t) = (a - g^{-1}b^2 k)x.$$

The closed-loop system is stable if $a - g^{-1}b^2 k < 0$.

In general, the system dynamics is modeled by nonlinear differential equations. Let us minimize

$$J(x(\cdot), u(\cdot)) = \int_{t_0}^{t_f} W_{xu}(x, u)\, dt, \qquad (4.4.2)$$

subject to the dynamic system

$$\dot{x}(t) = F(x) + B(x)u, \, x(t_0) = x_0. \qquad (4.4.3)$$

The positive-definite and continuously differentiable integrand function $W_{xu}(\cdot)$: $\mathbb{R}^c \times \mathbb{R}^m \to \mathbb{R}_{\geq 0}$ is used. We assume that $F(x)$ and $B(x)$ are continuous and Lipschitz.

To find an optimal control (4.4.1), the necessary conditions for optimality must be studied.

The Hamiltonian, for (4.4.2) and (4.4.3), is

$$H\left(x, u, \frac{\partial V}{\partial x}\right) = W_{xu}(x, u) + \left(\frac{\partial V}{\partial x}\right)^T (F(x) + B(x)u), \qquad (4.4.4)$$

where $V(\cdot)$: $\mathbb{R}^n \to \mathbb{R}_{\geq 0}$ is the smooth and bounded return function, $V(0) = 0$.

The first- and second-order necessary conditions for optimality are

$$\frac{\partial H\left(x, u, \frac{\partial V}{\partial x}\right)}{\partial u} = 0 \qquad (4.4.5)$$

and

$$\frac{\partial^2 H\left(x, u, \frac{\partial V}{\partial x}\right)}{\partial u \times \partial u^T} > 0. \qquad (4.4.6)$$

Using the first-order necessary condition for optimality (4.4.5), one derives the control function $u(\cdot)$: $[t_0, t_f) \to \mathbb{R}^m$ (4.4.1), which minimizes functional (4.4.2). That is, the goal is to find $u(\cdot)$: $[t_0, t_f) \to \mathbb{R}^m$, which makes

$$H\left(x, u, \frac{\partial V}{\partial x}\right)$$

a minimum.

The Hamilton–Jacobi functional equation is given as

$$-\frac{\partial V}{\partial t} = \min_u \left[W_{xu}(x, u) + \left(\frac{\partial V}{\partial x}\right)^T (F(x) + B(x)u) \right]. \qquad (4.4.7)$$

If the control is constrained, we have

$$-\frac{\partial V}{\partial t} = \min_{u \in U} \left[W_{xu}(x, u) + \left(\frac{\partial V}{\partial x}\right)^T (F(x) + B(x)u) \right].$$

4.4.1.1. Optimal Control of Linear Continuous-Time Systems: Linear Quadratic
Optimal Control

We consider a linear time-invariant system

$$
\frac{dx}{dt} =
\begin{bmatrix}
\dfrac{dx_1}{dt} \\
\dfrac{dx_2}{dt} \\
\vdots \\
\dfrac{dx_{n-1}}{dt} \\
\dfrac{dx_n}{dt}
\end{bmatrix}
=
\begin{bmatrix}
a_{11} & a_{12} & \cdots & a_{1n-1} & a_{1n} \\
a_{21} & a_{22} & \cdots & a_{2n-1} & a_{2n} \\
\vdots & \vdots & \ddots & \vdots & \vdots \\
a_{n-11} & a_{n-12} & \cdots & a_{n-1n-1} & a_{n-1n} \\
a_{n1} & a_{n2} & \cdots & a_{nn-1} & a_{nn}
\end{bmatrix}
\begin{bmatrix}
x_1 \\
x_2 \\
\vdots \\
x_{n-1} \\
x_n
\end{bmatrix}
$$

$$
+
\begin{bmatrix}
b_{11} & b_{12} & \cdots & b_{1m-1} & b_{1m} \\
b_{21} & b_{22} & \cdots & b_{2m-1} & b_{2m} \\
\vdots & \vdots & \ddots & \vdots & \vdots \\
b_{n-11} & b_{n-12} & \cdots & b_{n-1m-1} & b_{n-1m} \\
b_{n1} & b_{n2} & \cdots & b_{nm-1} & b_{nm}
\end{bmatrix}
\begin{bmatrix}
u_1 \\
u_2 \\
\vdots \\
u_{m-1} \\
u_m
\end{bmatrix}
, \quad x(t_0) = x_0,
$$

or

$$
\dot{x}(t) = Ax + Bu, \; x(t_0) = x_0, \tag{4.4.8}
$$

where $A \in \mathbb{R}^{n \times n}$ and $B \in \mathbb{R}^{n \times m}$ are the constant-coefficient matrices.
The quadratic performance functional is given as

$$
J(x(\cdot), u(\cdot)) = \tfrac{1}{2} \int_{t_0}^{t_f} (x^T Q x + u^T G u)\, dt, \quad Q \geq 0, \quad G > 0, \tag{4.4.9}
$$

where $Q \in \mathbb{R}^{n \times n}$ is the positive semidefinite, constant-coefficient weighting matrix
and $G \in \mathbb{R}^{m \times m}$ is the positive-definite constant-coefficient weighting matrix. That
is,

$$
Q =
\begin{bmatrix}
q_{11} & 0 & \cdots & 0 & 0 \\
0 & q_{22} & \cdots & 0 & 0 \\
\vdots & \vdots & \ddots & \vdots & \vdots \\
0 & 0 & \cdots & q_{n-1n-1} & 0 \\
0 & 0 & \cdots & 0 & q_{nn}
\end{bmatrix}
, \quad q_{ii} \geq 0
$$

and

$$
G =
\begin{bmatrix}
g_{11} & 0 & \cdots & 0 & 0 \\
0 & g_{22} & \cdots & 0 & 0 \\
\vdots & \vdots & \ddots & \vdots & \vdots \\
0 & 0 & \cdots & g_{m-1m-1} & 0 \\
0 & 0 & \cdots & 0 & g_{mm}
\end{bmatrix}
, \quad g_{ii} > 0.
$$

The availability of finite state and control "energies" are represented by the terms
$x^T Q x$ and $u^T G u$. The state and control variables must satisfy certain constraints

because the states and control must evolve in admissible envelopes. This situation will be thoroughly studied later.

Then, from (4.4.8) and (4.4.9), we have the expression for the Hamiltonian function as

$$H\left(x, u, \frac{\partial V}{\partial x}\right) = \frac{1}{2}\left(x^T Q x + u^T G u\right) + \left(\frac{\partial V}{\partial x}\right)^T (Ax + Bu), \qquad (4.4.10)$$

and the Hamilton–Jacobi functional equation is

$$-\frac{\partial V}{\partial t} = \min_u \left[\frac{1}{2}\left(x^T Q x + u^T G u\right) + \left(\frac{\partial V}{\partial x}\right)^T (Ax + Bu)\right].$$

The derivative of the Hamiltonian function

$$H\left(x, u, \frac{\partial V}{\partial x}\right)$$

exists, and control function $u(\cdot): [t_0, t_f) \to \mathbb{R}^m$ is found by using the first-order necessary condition (4.4.5). In particular, making use of

$$\frac{\partial H\left(x, u, \frac{\partial V}{\partial x}\right)}{\partial u} = u^T G + \left(\frac{\partial V}{\partial x}\right)^T B,$$

one finds the following optimal controller:

$$u = -G^{-1} B^T \frac{\partial V}{\partial x}. \qquad (4.4.11)$$

This control law is found by minimizing the quadratic performance functional (4.4.9).

It must be emphasized that the second-order necessary condition for optimality (4.4.6) is guaranteed. The weighting matrix G is positive-definite, and we have

$$\frac{\partial^2 H\left(x, u, \frac{\partial V}{\partial x}\right)}{\partial u \times \partial u^T} = G > 0.$$

Plugging the controller (4.4.11) into (4.4.10) or in the Hamilton–Jacobi functional equation

$$-\frac{\partial V}{\partial t} = \min_u \left[\frac{1}{2}\left(x^T Q x + u^T G u\right) + \left(\frac{\partial V}{\partial x}\right)^T (Ax + Bu)\right],$$

one obtains the following Hamilton–Jacobi–Bellman partial differential equation:

$$-\frac{\partial V}{\partial t} = \frac{1}{2}\left(x^T Q x + \left(\frac{\partial V}{\partial x}\right)^T BG^{-1}B^T \frac{\partial V}{\partial x}\right) + \left(\frac{\partial V}{\partial x}\right)^T Ax$$

$$- \left(\frac{\partial V}{\partial x}\right)^T BG^{-1}B^T \frac{\partial V}{\partial x}$$

$$= \frac{1}{2}x^T Q x + \left(\frac{\partial V}{\partial x}\right)^T Ax - \frac{1}{2}\left(\frac{\partial V}{\partial x}\right)^T BG^{-1}B^T \frac{\partial V}{\partial x}. \qquad (4.4.12)$$

This equation must be solved. To find the solution, we assume that (4.4.12) is satisfied by the quadratic return function $V(x)$. That is,

$$V(x) = \tfrac{1}{2} x^T K(t) x, \tag{4.4.13}$$

where $K \in \mathbb{R}^{n \times n}$ is the symmetric matrix,

$$K = \begin{bmatrix} k_{11} & k_{12} & \cdots & k_{1n-1} & k_{1n} \\ k_{21} & k_{22} & \cdots & k_{2n-1} & k_{2n} \\ \vdots & \vdots & \ddots & \vdots & \vdots \\ k_{n-11} & k_{n-12} & \cdots & k_{n-1n-1} & k_{n-1n} \\ k_{n1} & k_{n2} & \cdots & k_{nn-1} & k_{nn} \end{bmatrix}, \qquad k_{ij} = k_{ji}.$$

That is,

$$V(x) = \tfrac{1}{2} \begin{bmatrix} x_1 & x_2 & \cdots & x_{n-1} & x_n \end{bmatrix} \begin{bmatrix} k_{11} & k_{12} & \cdots & k_{1n-1} & k_{1n} \\ k_{21} & k_{22} & \cdots & k_{2n-1} & k_{2n} \\ \vdots & \vdots & \ddots & \vdots & \vdots \\ k_{n-11} & k_{n-12} & \cdots & k_{n-1n-1} & k_{n-1n} \\ k_{n1} & k_{n2} & \cdots & k_{nn-1} & k_{nn} \end{bmatrix}$$

$$\times \begin{bmatrix} x_1 \\ x_2 \\ \vdots \\ x_{n-1} \\ x_n \end{bmatrix}.$$

The unknown matrix K must be positive-definite because positive semidefinite and positive-definite constant-coefficient weighting matrices Q and G have been used in the performance functional (4.4.9). The positive definiteness of the quadratic return function $V(x)$ can be verified using the Sylvester criterion (successive principal minors of K must be positive).

Taking note of (4.4.13) and using the matrix identity $x^T K A x = \tfrac{1}{2} x^T (A^T K + K A) x$, from (4.4.12), one has

$$-\frac{\partial \left(\tfrac{1}{2} x^T K x \right)}{\partial t} = \tfrac{1}{2} x^T Q x + \tfrac{1}{2} x^T A^T K x + \tfrac{1}{2} x^T K A x - \tfrac{1}{2} x^T K B G^{-1} B^T K x. \tag{4.4.14}$$

The boundary condition is

$$V(t_f, x) = \tfrac{1}{2} x^T K(t_f) x = \tfrac{1}{2} x^T K_f x. \tag{4.4.15}$$

From (4.4.14) and (4.4.15), one concludes that the following nonlinear differential equation (the so-called Riccati equation) must be solved to find the unknown symmetric matrix K:

$$-\dot{K} = Q + A^T K + K A - K B G^{-1} B^T K, \; K(t_f) = K_f. \tag{4.4.16}$$

Using (4.4.11) and (4.4.13), the controller is found as

$$u = -G^{-1}B^T Kx. \tag{4.4.17}$$

Here, the matrix B is found using (4.4.8), the weighting matrix G is assigned by the designer by synthesizing the performance functional (4.4.9), and the matrix K is obtained by solving the Ricatti equation (4.4.16).

From (4.4.17), one concludes that the feedback gain matrix is

$$K_F = -G^{-1}B^T K,$$

and the feedback coefficients are derived as one solves (4.4.16). Augmenting (4.4.8) and (4.4.17), we have

$$
\begin{aligned}
\dot{x}(t) &= Ax + Bu = Ax - BG^{-1}B^T Kx = \left(A - BG^{-1}B^T K\right)x \\
&= (A - BK_F)x. \tag{4.4.18}
\end{aligned}
$$

The closed-loop system (4.18) is stable, and the eigenvalues of the matrix $\left(A - BG^{-1}B^T K\right) = (A - BK_F) \in \mathbb{R}^{n \times n}$ have negative real parts.

It is important to emphasize that the solution of the differential nonlinear Riccati equation (4.4.16) can be found by solving the nonlinear algebraic equation if $t_f = \infty$. Thus, matrix K is found from

$$0 = -Q - A^T K - KA + KBG^{-1}B^T K,$$

and the positive-definiteness of the matrix K must be verified.

The Riccati equation solver `lqr` is available in MATLAB. In particular,

```
» help lqr
    LQR   Linear-quadratic regulator design for
    continuous-time systems. [K,S,E] = LQR(A,B,Q,R,N)
    calculates the optimal gain matrix K such that the
    state-feedback law   u = -Kx   minimizes the cost
    function
         J = Integral {x'Qx + u'Ru + 2*x'Nu} dt
    subject to the state dynamics   x = Ax + Bu.
    The matrix N is set to zero when omitted.   Also
    returned are the Riccati equation solution S and
    the closed-loop eigenvalues E:
                           -1
        SA + A'S - (SB+N)R   B'S+N') + Q = 0 ,
        E = EIG(A-B*K)   .
    See also   LQRY, DLQR, LQGREG, CARE, and REG.
```

That is, applying the `lqr` solver, one finds the feedback gain matrix K_F, the return function matrix K, as well as the eigenvalues of the closed-loop system.

Example 4.4.1.

Consider the systems

$$\frac{dx}{dt} = u.$$

Let us find the control law minimizing the performance functional

$$J(x(\cdot), u(\cdot)) = \frac{1}{2} \int_{t_0}^{t_f} \left(x^2 + u^2 \right) dt.$$

Using notations (4.4.8) and (4.4.9), one obtains the following expressions for the matrices A, B, Q, and G:

$$A = [0], \ B = [1], \ Q = [1], \ \text{and} \ G = [1].$$

The matrix K is unknown, and from (4.4.16), we have the following nonlinear differential equation, which has to be solved:

$$-\dot{K}(t) = 1 - K^2(t), \ K(t_f) = 0.$$

The analytic solution of this nonlinear differential equation is

$$K(t) = \frac{1 - e^{-2(t_f - t)}}{1 + e^{-2(t_f - t)}}.$$

Hence, using (4.4.17), one finds an optimal control law that guarantees the minimum of the quadratic functional $J(x(\cdot), u(\cdot)) = \frac{1}{2} \int_{t_0}^{t_f} \left(x^2 + u^2 \right) dt$ with respect to the system dynamics

$$\frac{dx}{dt} = u.$$

The striking application of (4.4.17) and

$$K(t) = \frac{1 - e^{-2(t_f - t)}}{1 + e^{-2(t_f - t)}}$$

gives

$$u(t) = -\frac{1 - e^{-2(t_f - t)}}{1 + e^{-2(t_f - t)}} x.$$

Using the `lqr` MATLAB solver, one finds the feedback and return function coefficients and eigenvalues if $t_f = \infty$. That is, minimizing $J(x(\cdot), u(\cdot)) = \frac{1}{2} \int_0^\infty \left(x^2 + u^2 \right) dt$, we have

```
» [K`feedback,K,Eigenvalues]=lqr(0,1,1,1,0)
K`feedback =
      1
K =
      1
Eigenvalues =
   -1.0000
```

Example 4.4.2.

Consider the first-order continuous-time system

$$\frac{dx}{dt} = -x + u.$$

Minimization of the performance functional

$$J(x(\cdot), u(\cdot)) = \int_0^{t_f} \left(x^2 + u^2\right) dt$$

leads one to the following nonlinear differential equations to find matrix K:

$$-\dot{K}(t) = 1 - 2K(t) - K^2(t), \quad K(t_f) = 0;$$

see (4.4.16).

The analytic solution of this nonlinear differential equation is

$$K(t) = \frac{-e^{\sqrt{2}(t-t_f)} + e^{-\sqrt{2}(t-t_f)}}{(\sqrt{2}-1)e^{\sqrt{2}(t-t_f)} + (\sqrt{2}+1)e^{-\sqrt{2}(t-t_f)}}.$$

From (4.4.17), the optimal controller results as $u(t) = -K(t)x$.
This control law is time-varying.

Let $t_f = \infty$; that is, the performance functional is
$J(x(\cdot), u(\cdot)) = \int_0^\infty (x^2 + u^2)\, dt$.

The solution of the Riccati equations is

$$K(t) = \frac{-e^{\sqrt{2}(t-t_f)} + e^{-\sqrt{2}(t-t_f)}}{(\sqrt{2}-1)e^{\sqrt{2}(t-t_f)} + (\sqrt{2}+1)e^{-\sqrt{2}(t-t_f)}}\Bigg|_{t_f \to \infty}$$

$$= \frac{-e^{-\sqrt{2}t_f} + e^{\sqrt{2}t_f}}{(\sqrt{2}-1)e^{-\sqrt{2}t_f} + (\sqrt{2}+1)e^{\sqrt{2}t_f}}\Bigg|_{t_f \to \infty} = \sqrt{2} - 1.$$

That is, a time-invariant optimal controller is given as $u = -(\sqrt{2}-1)x$.

Example 4.4.3.

Consider the second-order dynamic system modeled as

$$\frac{dx_1}{dt} = x_2, \qquad \frac{dx_2}{dt} = u.$$

The problem is to find a control law minimizing the quadratic functional

$$J(x(\cdot), u(\cdot)) = \frac{1}{2}\int_0^\infty \left(x_1^2 + q_{22}x_2^2 + gu^2\right) dt, \qquad q_{22} \geq 0, \quad g > 0.$$

Using the state-space notations, we have

$$\begin{bmatrix} \dot{x}_1 \\ \dot{x}_2 \end{bmatrix} = \begin{bmatrix} 0 & 1 \\ 0 & 0 \end{bmatrix} \begin{bmatrix} x_1 \\ x_2 \end{bmatrix} + \begin{bmatrix} 0 \\ 1 \end{bmatrix} u, \qquad A = \begin{bmatrix} 0 & 1 \\ 0 & 0 \end{bmatrix}, \qquad B = \begin{bmatrix} 0 \\ 1 \end{bmatrix},$$

and

$$J(x(\cdot), u(\cdot)) = \frac{1}{2} \int_0^\infty \left(\begin{bmatrix} x_1 & x_2 \end{bmatrix} \begin{bmatrix} 1 & 0 \\ 0 & q_{22} \end{bmatrix} \begin{bmatrix} x_1 \\ x_2 \end{bmatrix} + g u^2 \right) dt,$$

$$Q = \begin{bmatrix} 1 & 0 \\ 0 & q_{22} \end{bmatrix}, \, G = g, \quad q_{22} \geq 0, \quad g > 0.$$

Using the quadratic return function (4.4.15)

$$V(x) = \frac{1}{2} k_{11} x_1^2 + k_{12} x_1 x_2 + \frac{1}{2} k_{22} x_2^2 = \frac{1}{2} \begin{bmatrix} x_1 & x_2 \end{bmatrix} \begin{bmatrix} k_{11} & k_{12} \\ k_{21} & k_{22} \end{bmatrix} \begin{bmatrix} x_1 \\ x_2 \end{bmatrix},$$

$$k_{12} = k_{21},$$

the controller (4.4.17) is found as

$$u = -G^{-1} B^T K x = -g^{-1} \begin{bmatrix} 0 & 1 \end{bmatrix} \begin{bmatrix} k_{11} & k_{12} \\ k_{21} & k_{22} \end{bmatrix} \begin{bmatrix} x_1 \\ x_2 \end{bmatrix} = -\frac{1}{g}(k_{21} x_1 + k_{22} x_2).$$

The unknown coefficients of matrix

$$K = \begin{bmatrix} k_{11} & k_{12} \\ k_{21} & k_{22} \end{bmatrix}$$

are found by solving the Riccati equation (4.4.16)

$$-Q - A^T K - K A + K B G^{-1} B^T K$$

$$= -\begin{bmatrix} 1 & 0 \\ 0 & q_{22} \end{bmatrix} - \begin{bmatrix} 0 & 0 \\ 1 & 0 \end{bmatrix} \begin{bmatrix} k_{11} & k_{12} \\ k_{21} & k_{22} \end{bmatrix} - \begin{bmatrix} k_{11} & k_{12} \\ k_{21} & k_{22} \end{bmatrix} \begin{bmatrix} 0 & 1 \\ 0 & 0 \end{bmatrix}$$

$$+ \begin{bmatrix} k_{11} & k_{12} \\ k_{21} & k_{22} \end{bmatrix} \begin{bmatrix} 0 \\ 1 \end{bmatrix} g^{-1} \begin{bmatrix} 0 & 1 \end{bmatrix} \begin{bmatrix} k_{11} & k_{12} \\ k_{21} & k_{22} \end{bmatrix} = \begin{bmatrix} 0 & 0 \\ 0 & 0 \end{bmatrix}.$$

Hence, we have three algebraic equations

$$-\frac{k_{12}^2}{g} - 1 = 0, \quad -k_{11} + \frac{k_{12} k_{22}}{g} = 0, \quad \text{and} \quad -2k_{12} + \frac{k_{22}^2}{g} - q_{22} = 0.$$

The solution is,

$$k_{12} = k_{21} = \pm\sqrt{g}, \quad k_{22} = \pm\sqrt{g(q_{22} + 2k_{12})}, \quad \text{and} \quad k_{11} = \frac{k_{12} k_{22}}{g}.$$

The performance functional $J(x(\cdot), u(\cdot)) = \frac{1}{2} \int_0^\infty \left(x_1^2 + q_{22} x_2^2 + g u^2 \right) dt, q_{22} \geq 0, g > 0$ is positive-definite. Therefore, we have

$$k_{11} = \sqrt{q_{22} + 2\sqrt{g}}, \quad k_{12} = k_{21} = \sqrt{g}, \quad \text{and} \quad k_{22} = \sqrt{g(q_{22} + 2\sqrt{g})}.$$

Thus, the control law is

$$u = -\frac{1}{g}\left(\sqrt{g}x_1 + \sqrt{g(q_{22} + 2\sqrt{g})}x_2\right) = -\frac{1}{\sqrt{g}}x_1 - \sqrt{\frac{q_{22} + 2\sqrt{g}}{g}}x_2.$$

Applying the `lqr` solver, let us obtain the feedback gains and return function coefficients, as well as compute the eigenvalues. Letting $q_{22} = 1$ and $g = 1$, we have

```
» [K`feedback,K,Eigenvalues]=
                        lqr([0 1;0 0],[0;1],[1 0;0 1],[1])
K`feedback =
    1.0000     1.7321
K =
    1.7321     1.0000
    1.0000     1.7321
Eigenvalues =
  -0.8660 + 0.5000i
  -0.8660 - 0.5000i
```

Hence,

$$K = \begin{bmatrix} k_{11} & k_{12} \\ k_{21} & k_{22} \end{bmatrix} = \begin{bmatrix} 1.73 & 1 \\ 1 & 1.73 \end{bmatrix}, \quad k_{11} = 1.73, \ k_{12} = k_{21} = 1, \ k_{22} = 1.73,$$

and $u = -x_1 - \sqrt{3}x_2 = -x_1 - 1.73x_2$.

The analytical and numerical results have been obtained, and the stability of the closed-loop system

$$\frac{dx_1}{dt} = x_2, \qquad \frac{dx_2}{dt} = -x_1 - 1.73x_2$$

is guaranteed because the equivalence have negative real parts.

Example 4.4.4.

Consider the dynamic system as given by

$$\frac{d^2y}{dt^2} + a\frac{dy}{dt} = bu.$$

Hence, one obtains

$$\frac{dx_1}{dt} = x_2, \qquad \frac{dx_2}{dt} = -ax_2 + bu.$$

A controller should be found minimizing the quadratic functional

$$J(x(\cdot), u(\cdot)) = \int_0^\infty \left(q_{11}x_1^2 + q_{22}x_2^2 + gu^2\right)dt, \quad q_{11} \geq 0, \quad q_{22} \geq 0, \quad g > 0,$$

or

$$J(x(\cdot), u(\cdot)) = \frac{1}{2} \int_0^\infty \left(\begin{bmatrix} x_1 & x_2 \end{bmatrix} \begin{bmatrix} q_{11} & 0 \\ 0 & q_{22} \end{bmatrix} \begin{bmatrix} x_1 \\ x_2 \end{bmatrix} + gu^2 \right) dt,$$

$$Q = \begin{bmatrix} q_{11} & 0 \\ 0 & q_{22} \end{bmatrix}, \quad G = g.$$

In the state-space form, one finds

$$\begin{bmatrix} \dot{x}_1 \\ \dot{x}_2 \end{bmatrix} = \begin{bmatrix} 0 & 1 \\ 0 & -a \end{bmatrix} \begin{bmatrix} x_1 \\ x_2 \end{bmatrix} + [b]u, \quad A = \begin{bmatrix} 0 & 1 \\ 0 & -a \end{bmatrix}, \quad B = \begin{bmatrix} 0 \\ b \end{bmatrix}.$$

Using the quadratic return function, the Riccati equation (4.4.16) should be solved. We have

$$-Q - A^T K - K A + K B G^{-1} B^T K$$

$$= - \begin{bmatrix} q_{11} & 0 \\ 0 & q_{22} \end{bmatrix} - \begin{bmatrix} 0 & 0 \\ 1 & -a \end{bmatrix} \begin{bmatrix} k_{11} & k_{12} \\ k_{21} & k_{22} \end{bmatrix} - \begin{bmatrix} k_{11} & k_{12} \\ k_{21} & k_{22} \end{bmatrix} \begin{bmatrix} 0 & 1 \\ 0 & -a \end{bmatrix}$$

$$+ \begin{bmatrix} k_{11} & k_{12} \\ k_{21} & k_{22} \end{bmatrix} \begin{bmatrix} 0 \\ b \end{bmatrix} g^{-1} \begin{bmatrix} 0 & 1 \end{bmatrix} \begin{bmatrix} k_{11} & k_{12} \\ k_{21} & k_{22} \end{bmatrix}$$

$$= - \begin{bmatrix} q_{11} & 0 \\ 0 & q_{22} \end{bmatrix} - \begin{bmatrix} 0 & 0 \\ k_{11} - ak_{12} & k_{12} - ak_{22} \end{bmatrix} - \begin{bmatrix} 0 & k_{11} - ak_{12} \\ 0 & k_{12} - ak_{22} \end{bmatrix}$$

$$+ \frac{b^2}{g} \begin{bmatrix} k_{12}^2 & k_{12}k_{22} \\ k_{12}k_{22} & k_{22}^2 \end{bmatrix} = \begin{bmatrix} 0 & 0 \\ 0 & 0 \end{bmatrix}.$$

The controller (4.4.17) is given by

$$u = -G^{-1} B^T K x = -g^{-1} \begin{bmatrix} 0 & b \end{bmatrix} \begin{bmatrix} k_{11} & k_{12} \\ k_{21} & k_{22} \end{bmatrix} \begin{bmatrix} x_1 \\ x_2 \end{bmatrix}$$

$$= -\frac{b}{g} (k_{21}x_1 + k_{22}x_2).$$

The unknown coefficients k_{11}, k_{21}, and k_{22} can be found solving three nonlinear algebraic equations

$$-q_{11} + \frac{b^2}{g} k_{12}^2 = 0,$$

$$-k_{11} + ak_{12} + \frac{b^2}{g} k_{12}k_{22} = 0,$$

and

$$-q_{22} - 2k_{12} + 2ak_{22} + \frac{b^2}{g} k_{22}^2 = 0,$$

where $k_{12} = k_{21}$.

Example 4.4.5.

Consider the linearized aircraft equations in the longitudinal axis

$$\dot{x} = Ax + Bu = \begin{bmatrix} -0.041 & 5.8 & -0.6 & -9.8 \\ -0.0003 & -0.74 & 1 & -0.0002 \\ -0.00007 & 4.6 & -0.95 & -0.0005 \\ 0 & 0 & 1 & 0 \end{bmatrix} \begin{bmatrix} x_1 \\ x_2 \\ x_3 \\ x_4 \end{bmatrix}$$

$$+ \begin{bmatrix} 0.74 & -0.11 \\ -1.5 & -0.18 \\ -4.3 & -0.62 \\ 0 & 0 \end{bmatrix} \begin{bmatrix} u_1 \\ u_2 \end{bmatrix}.$$

Hence, the constant-coefficient matrices of the aircraft model are

$$A = \begin{bmatrix} -0.041 & 5.8 & -0.6 & -9.8 \\ -0.0003 & -0.74 & 1 & -0.002 \\ -0.00007 & 4.6 & -0.95 & -0.0005 \\ 0 & 0 & 1 & 0 \end{bmatrix}$$

and

$$B = \begin{bmatrix} 0.74 & -0.11 \\ -1.5 & -0.18 \\ -4.3 & -0.62 \\ 0 & 0 \end{bmatrix}.$$

The output equation is

$$y = \begin{bmatrix} 0 & 0 & 0 & 1 \end{bmatrix} \begin{bmatrix} x_1 \\ x_2 \\ x_3 \\ x_4 \end{bmatrix} + \begin{bmatrix} 0 & 0 \end{bmatrix} \begin{bmatrix} u_1 \\ u_2 \end{bmatrix}.$$

Let us find the control law minimizing the quadratic performance functional

$$J(x(\cdot), u(\cdot)) = \tfrac{1}{2} \int_0^\infty \left(x^T Q x + u^T G u \right) dt$$

$$= \tfrac{1}{2} \int_0^\infty \left(\begin{bmatrix} x_1 & x_2 & x_3 & x_4 \end{bmatrix} \begin{bmatrix} 1 & 0 & 0 & 0 \\ 0 & 1 & 0 & 0 \\ 0 & 0 & 1 & 0 \\ 0 & 0 & 0 & 1 \end{bmatrix} \begin{bmatrix} x_1 \\ x_2 \\ x_3 \\ x_4 \end{bmatrix} \right.$$

$$\left. + \begin{bmatrix} u_1 & u_2 \end{bmatrix} \begin{bmatrix} 10 & 0 \\ 0 & 10 \end{bmatrix} \begin{bmatrix} u_1 \\ u_2 \end{bmatrix} \right) dt$$

$$= \tfrac{1}{2} \int_0^\infty \left(x_1^2 + x_2^2 + x_3^2 + x_4^2 + 10u_1^2 + 10u_2^2 \right) dt.$$

That is, the weighting matrices assigned by the designer are

$$Q = \begin{bmatrix} 1 & 0 & 0 & 0 \\ 0 & 1 & 0 & 0 \\ 0 & 0 & 1 & 0 \\ 0 & 0 & 0 & 1 \end{bmatrix}$$

and

$$G = \begin{bmatrix} 10 & 0 \\ 0 & 10 \end{bmatrix}.$$

The feedback control algorithm, as given by (4.4.17), is

$$u = -G^{-1}B^T K x = -K_F x.$$

Let us find the feedback gain matrix K_F, the return function matrix K, and the eigenvalues of the closed-loop system studying $(A - BG^{-1}B^T K) = (A - BK_F) \in \mathbb{R}^{4 \times 4}$. Also, as the control law is derived, the dynamics should be analyzed. The following MATLAB m-file is written:

MATLAB script c_4_4_1.m

```
echo off; clear all; format short e;
% Constant-coefficient matrices A and B
A=[-0.041    5.8  -0.6    -9.6;
   -0.0003  -0.74  1      -0.0002;
   -0.00007  4.6  -0.95   -0.0005
    0        0     1       0];
disp('eigenvalues A'); disp(eig(A));   % Eigenvalues of the
                                       % matrix A
B=[0.74 -0.11;
  -1.5   -0.18;
  -4.3   -0.62;
   0      0];
% Weighting matrices Q and G
Q=[1 0 0 0;
   0 1 0 0;
   0 0 1 0;
   0 0 0 1];
G=[10 0;
   0  10];
% Feedback and return function coefficients, eigenvalues
[K feedback,K,Eigenvalues]=lqr(A,B,Q,G);
disp('K feedback'); disp(K feedback);
disp('K'); disp(K);
disp('eigenvalues A-BK feedback'); disp(Eigenvalues);
% Matrix of the closed-loop system
A closed loop=A-B*K feedback;
```

```
% Longitudinal Aircraft Dynamics
t=0:0.02:4;
% Deflections of control surfaces
uu=[0.5*ones(max(size(t)),4)];
C=[0 0 0 1]; D=[0 0 0 0];
[y,x]=lsim(A closed loop,B*K feedback,C,D,uu,t);
plot(t,x);
title('Aircraft Dynamics, x1, x2, x3, x4');
xlabel('time [seconds]'); pause;
plot(t,y); pause;
plot(t,x(:,1),'-',t,x(:,2),'-',t,x(:,3),'-',t,x(:,4),'-');
pause;
plot(t,x(:,1),'-'); pause;
plot(t,x(:,2),'-'); pause;
plot(t,x(:,3),'-'); pause;
plot(t,x(:,4),'-'); pause;
disp('End')
```

The feedback gain matrix K_F, the return function matrix K, and the eigenvalues of the $(A - BG^{-1}B^T K) = (A - BK_F)$ are found. In particular, using the results displayed

```
eigenvalues A
 -2.9923e+000
  1.3033e+000
 -2.1009e-002 +5.5401e-002i
 -2.1009e-002 -5.5401e-002i
K feedback
  2.9982e-001 -2.0342e-001 -1.2836e+000 -3.5156e+000
  2.6051e-002 -4.9135e-002 -1.6430e-001 -4.1638e-001
K
  1.1373e+000  2.0581e+000 -1.2195e+000 -6.4757e+000
  2.0581e+000  6.8339e+000 -1.5567e+000 -1.3729e+001
 -1.2195e+000 -1.5567e+000  3.3183e+000  1.1851e+001
 -6.4757e+000 -1.3729e+001  1.1851e+001  5.5207e+001
eigenvalues A-BK feedback
 -1.1887e+000 +1.6544e+000i
 -1.1887e+000 -1.6544e+000i
 -3.1399e+000
 -2.3681e+000
```

we have

$$
K = \begin{bmatrix}
1.1 & 2.1 & -1.2 & -6.5 \\
2.1 & 6.8 & -1.6 & -14 \\
-1.2 & -1.6 & 3.3 & 11.8 \\
-6.5 & -14 & 11.8 & 55
\end{bmatrix}
$$

and

$$K_F = \begin{bmatrix} 0.3 & -0.2 & -1.3 & -3.5 \\ 0.026 & -0.049 & -0.16 & -0.42 \end{bmatrix}.$$

The eigenvalues of the closed-loop system are $-1.2 \pm 1.7i$, -3.1, and -2.4. That is, the system is stable (it should be observed that the open-loop system was unstable because the eigenvalues were found to be 2.9, 1.3 and $-0.021 \pm 0.055i$). Using the control algorithm obtained

$$u = -K_F x = -\begin{bmatrix} 0.3 & -0.2 & -1.3 & -3.5 \\ 0.026 & -0.049 & -0.16 & -0.42 \end{bmatrix} \begin{bmatrix} x_1 \\ x_2 \\ x_3 \\ x_4 \end{bmatrix},$$

one has

$$u_1 = -0.3x_1 + 0.2x_2 + 1.3x_3 + 3.5x_4,$$
$$u_2 = -0.026x_1 + 0.049x_2 + 0.16x_3 + 0.42x_4.$$

The transient dynamics of the state variables for

$$\begin{bmatrix} x_{10} \\ x_{20} \\ x_{30} \\ x_{40} \end{bmatrix} = \begin{bmatrix} 0 \\ 0 \\ 0 \\ 0 \end{bmatrix}$$

are plotted in Figure 4.4.1.

4.4.1.2. Tracking Control of Linear Systems

We began the design of optimal control laws by considering the stabilization problem. That is, the reference and tracking error were not studied, and the output equation was not used. Tracking control is a very important problem because the tracking of the reference (command) signal should be accomplished in many real-world systems. The optimal tracking control problem can be formulated as follows: For the dynamic system with output equation $y(t) = Hx(t)$, determine the tracking optimal controller $u = \prod(e, x)$ by minimizing the performance functional.
The tracking error is expressed as

$$e(t) = r(t) - y(t),$$

where $r(t)$ and $y(t)$ are the reference (command) and the output variables.
The output equation is

$$y(t) = Hx(t),$$

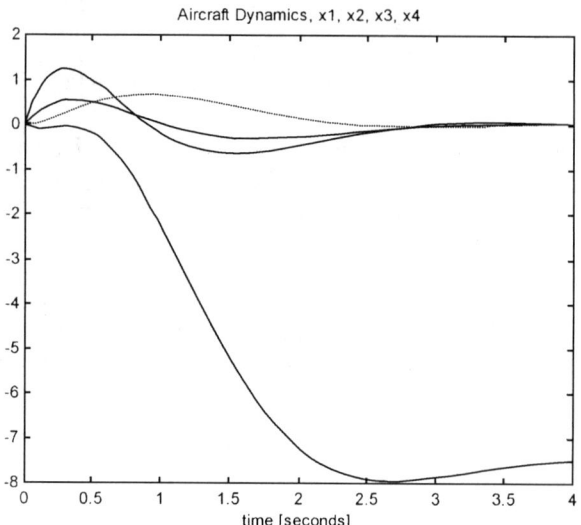

FIGURE 4.4.1.   Evolution of the state variables.

and for multivariable systems, we have

$$e(t) = Nr(t) - y(t) = Nr(t) - Hx(t)$$

$$= \begin{bmatrix} n_{11} & 0 & \cdots & 0 & 0 \\ 0 & n_{22} & \cdots & 0 & 0 \\ \vdots & \vdots & \ddots & \vdots & \vdots \\ 0 & 0 & \cdots & n_{b-1b-1} & 0 \\ 0 & 0 & \cdots & 0 & n_{bb} \end{bmatrix} \begin{bmatrix} r_1 \\ r_2 \\ \vdots \\ r_{b-1} \\ r_b \end{bmatrix}$$

$$+ \begin{bmatrix} h_{11} & h_{12} & \cdots & h_{1n-1} & h_{1n} \\ h_{21} & h_{22} & \cdots & h_{2n-1} & h_{2n} \\ \vdots & \vdots & \ddots & \vdots & \vdots \\ h_{b-11} & h_{b-12} & \cdots & h_{b-1n-1} & h_{b-1n} \\ h_{b1} & h_{b2} & \cdots & h_{bn-1} & h_{bn} \end{bmatrix} \begin{bmatrix} x_1 \\ x_2 \\ \vdots \\ x_{n-1} \\ x_n \end{bmatrix},$$

where $N \in \mathbb{R}^{b \times b}$ and $H \in \mathbb{R}^{b \times n}$ are the constant-coefficient matrices.

Using the output equation $e(t) = Nr(t) - y(t)$, and denoting $e(t) = \dot{x}^{ref}(t)$, we consider the dynamics of the *exogeneous* system

$$\dot{x}^{ref}(t) = Nr - y = Nr - Hx. \tag{4.4.19}$$

Let us augment the system dynamics (4.4.8) $\dot{x}(t) = Ax + Bu$, $x(t_0) = x_0$ with

(4.4.19). Taking note of the output equation $y = Hx$, we have

$$\dot{x}(t) = Ax + Bu, \; y = Hx, \; x_0(t_0) = x_0,$$

$$\dot{x}^{ref}(t) = Nr - y = Nr - Hx.$$

That is,

$$\dot{x}_\Sigma(t) = A_\Sigma x_\Sigma + B_\Sigma u + N_\Sigma r, \; y = Hx, \; x_{\Sigma 0}(t_0) = x_{\Sigma 0}, \qquad (4.4.20)$$

where

$$x_\Sigma = \begin{bmatrix} x \\ x^{ref} \end{bmatrix} \in \mathbb{R}^c \, (c = n + b)$$

is the augmented state vector;

$$A_\Sigma = \begin{bmatrix} A & 0 \\ -H & 0 \end{bmatrix} \in \mathbb{R}^{c \times c}$$

$$B_\Sigma = \begin{bmatrix} B \\ 0 \end{bmatrix} \in \mathbb{R}^{c \times m}$$

and

$$N_\Sigma = \begin{bmatrix} 0 \\ N \end{bmatrix} \in \mathbb{R}^{c \times b}$$

are the time-invariant matrices of coefficients.

The quadratic performance functional is given as

$$J\left(\begin{bmatrix} x(\cdot) \\ x^{ref}(\cdot) \end{bmatrix}, u(\cdot) \right) = \tfrac{1}{2} \int_{t_0}^{t_f} \left(\begin{bmatrix} x \\ x^{ref} \end{bmatrix}^T Q \begin{bmatrix} x \\ x^{ref} \end{bmatrix} + u^T G u \right) dt, \qquad (4.4.21)$$

where $Q \in \mathbb{R}^{c \times c}$ is the positive-semidefinite constant-coefficient matrix and $G \in \mathbb{R}^{m \times m}$ is the positive- definite constant-coefficient matrix.

Making use of the quadratic performance functional (4.4.21), one obtains the Hamiltonian function. In particular,

$$H\left(x_\Sigma, u, r, \frac{\partial V}{\partial x_\Sigma} \right) = \tfrac{1}{2} \left(x_\Sigma^T Q x_\Sigma + u^T G u \right) + \left(\frac{\partial V}{\partial x_\Sigma} \right)^T$$

$$\times (A_\Sigma x_\Sigma + B_\Sigma u + N_\Sigma r). \qquad (4.4.22)$$

From (4.4.22), having found

$$\frac{\partial H\left(x, u, r, \frac{\partial V}{\partial x_\Sigma} \right)}{\partial u} = u^T G + \left(\frac{\partial V}{\partial x_\Sigma} \right)^T B_\Sigma,$$

one finds the control law using the first-order necessary condition for optimality. We have the following form of an optimal control algorithm:

$$u = -G^{-1} B_\Sigma^T \frac{\partial V(x_\Sigma)}{\partial x_\Sigma} = -G^{-1} \begin{bmatrix} B \\ 0 \end{bmatrix}^T \frac{\partial V\left(\begin{bmatrix} x \\ x^{ref} \end{bmatrix} \right)}{\partial \begin{bmatrix} x \\ x^{ref} \end{bmatrix}}. \qquad (4.4.23)$$

The solution of the Hamilton–Jacobi–Bellman partial differential equation

$$-\frac{\partial V}{\partial t} = \tfrac{1}{2}x_\Sigma^T Q x_\Sigma + \left(\frac{\partial V}{\partial x_\Sigma}\right)^T A x_\Sigma - \tfrac{1}{2}\left(\frac{\partial V}{\partial x_\Sigma}\right)^T B_\Sigma G^{-1} B_\Sigma^T \frac{\partial V}{\partial x_\Sigma} \qquad (4.4.24)$$

is satisfied by the quadratic return function

$$V(x_\Sigma) = \tfrac{1}{2}x_\Sigma^T K(t)x_\Sigma, \qquad (4.4.25)$$

where $K \in \mathbb{R}^{c \times c}$ is the symmetric matrix.

Making use of (4.4.24) and (4.4.25), one finds the Riccati equation

$$-\dot{K} = Q + A_\Sigma^T K + K A_\Sigma - K B_\Sigma G^{-1} B_\Sigma^T K, \; K(t_f) = K_f, \qquad (4.4.26)$$

which gives the unknown symmetric matrix K. Equation (4.4.26) needs to be solved.

The controller is found using (4.4.23) and (4.4.25) as

$$u = -G^{-1} B_\Sigma^T K x_\Sigma = -G^{-1} \begin{bmatrix} B \\ 0 \end{bmatrix}^T K \begin{bmatrix} x \\ x^{\text{ref}} \end{bmatrix}. \qquad (4.4.27)$$

From $\dot{x}^{\text{ref}}(t) = e(t)$, one has

$$x^{\text{ref}}(t) = \int e(t)\, dt.$$

Therefore, we obtain the so-called integral control

$$u(t) = -G^{-1} B_\Sigma^T K x_\Sigma(t) = -G^{-1} \begin{bmatrix} B \\ 0 \end{bmatrix}^T K \begin{bmatrix} x(t) \\ \int e(t)\, dt \end{bmatrix}. \qquad (4.4.28)$$

In this control algorithm, the error vector is used in addition to the state feedback.

Example 4.4.6. Tracking control of an aircraft.

We study the tracking control problem for an unstable model of an aircraft. Beginning with the state-space equations (4.4.8), the states are

$$x(t) = [v(t)\; \alpha(t)\; q(t)\; \theta(t)\; \beta(t)\; p(t)\; r(t)\; \phi(t)\; \psi(t)]^T,$$

where v is the forward velocity [m/sec], α is the angle of attack [rad], q is the pitch rate [rad/sec], θ is the pitch angle [rad]; β is the sideslip angle [rad], p is the roll rate [rad/sec], r is the yaw rate [rad/sec], ϕ is the roll angle [rad], and ψ is the yaw angle [rad].

Six control inputs are

$$u(t) = [\delta_{HR}(t)\; \delta_{HL}(t)\; \delta_{FR}(t)\; \delta_{FL}(t)\; \delta_C(t)\; \delta_R(t)]^T,$$

where δ_{HR} and δ_{HL} are the deflections of the right and left horizontal stabilizers [rad], δ_{FR} and δ_{FL} are the deflections of the right and left flaps [rad], and δ_C and δ_R are the canard and rudder deflections [rad].

The longitudinal–lateral dynamics is studied applying the state-space model in

$$\dot{x}(t) = Ax + Bu,$$

where

$$A = \begin{bmatrix}
-0.016 & 8.4 & -0.9 & -9.6 & -1.5 & -0.27 & -0.086 & 0 & 0 \\
-0.003 & -1.2 & 1 & 0 & 0.08 & 0.062 & 0.009 & 0 & 0 \\
-0.0001 & 3.9 & -0.85 & 0 & 0.017 & 0.0038 & 0.04 & 0 & 0 \\
0 & 0 & 1 & 0 & 0 & 0 & 0 & 0 & 0 \\
-0.003 & 0.15 & 0.02 & 0.97 & -0.56 & 0.13 & -0.91 & 0 & 0 \\
-0.00001 & 0.71 & 0.03 & 0.01 & -48 & -3.5 & 0.22 & 0 & 0 \\
0.00001 & -0.94 & 0.06 & 0.005 & 9.2 & -0.028 & -0.51 & 0 & 0 \\
0 & 0 & 0 & 0 & 0 & 1 & 0 & 0 & 0 \\
0 & 0 & 0 & 0 & 0 & 0 & 1 & 0 & 0
\end{bmatrix},$$

$$B = \begin{bmatrix}
0.12 & 0.12 & -0.38 & -0.38 & 0 & 0 \\
-0.16 & -0.16 & -0.27 & -0.27 & 0 & 0 \\
-9.5 & -9.5 & -2.5 & -2.5 & 0 & 0 \\
0 & 0 & 0 & 0 & 0 & 0 \\
0.019 & -0.019 & -0.001 & 0.001 & 0.42 & 0.053 \\
-2.9 & 2.9 & -3.1 & 3.1 & 0.73 & 0.92 \\
3.1 & -3.1 & 0.78 & -0.78 & 0.61 & -0.45 \\
0 & 0 & 0 & 0 & 0 & 0 \\
0 & 0 & 0 & 0 & 0 & 0
\end{bmatrix}.$$

The eigenvalues of A can be found using the characteristic equation. This gives us the following eigenvalues of an open-loop system:

$$-1 \pm 3.52i, -3.16, -2.36, 0.92, -0.0092 \pm 0.204i, 0, 0.$$

Thus, the studied longitudinal–lateral open-loop dynamics of the fighter is unstable.

The tracking problem has to be solved, and the output equation should be used.

One expresses the error vector in terms of the reference inputs and the output vector (Euler angles as the aircraft) as

$$e(t) = r(t) - y(t), r(t) = \begin{bmatrix} r_\theta(t) \\ r_\phi(t) \\ r_\psi(t) \end{bmatrix}$$

and

$$y = Hx = \begin{bmatrix} 0 & 0 & 0 & 1 & 0 & 0 & 0 & 0 & 0 \\ 0 & 0 & 0 & 0 & 0 & 0 & 0 & 1 & 0 \\ 0 & 0 & 0 & 0 & 0 & 0 & 0 & 0 & 1 \end{bmatrix} \begin{bmatrix} v \\ \alpha \\ q \\ \theta \\ \beta \\ p \\ r \\ \phi \\ \psi \end{bmatrix}.$$

Using the longitudinal–lateral aircraft dynamics

$$\dot{x}(t) = Ax + Bu$$

and

$$\dot{x}^{\text{ref}}(t) = \begin{bmatrix} 1 & 0 & 0 \\ 0 & 1 & 0 \\ 0 & 0 & 1 \end{bmatrix} \begin{bmatrix} r_\theta(t) \\ r_\phi(t) \\ r_\psi(t) \end{bmatrix} - Hx \begin{bmatrix} \dot{x}_\theta^{\text{ref}}(t) \\ \dot{x}_\phi^{\text{ref}}(t) \\ \dot{x}_\psi^{\text{ref}}(t) \end{bmatrix}$$

$$= \begin{bmatrix} 1 & 0 & 0 \\ 0 & 1 & 0 \\ 0 & 0 & 1 \end{bmatrix} \begin{bmatrix} r_\theta(t) \\ r_\phi(t) \\ r_\psi(t) \end{bmatrix}$$

$$- \begin{bmatrix} 0 & 0 & 0 & 1 & 0 & 0 & 0 & 0 & 0 \\ 0 & 0 & 0 & 0 & 0 & 0 & 0 & 1 & 0 \\ 0 & 0 & 0 & 0 & 0 & 0 & 0 & 0 & 1 \end{bmatrix} \begin{bmatrix} v \\ \alpha \\ q \\ \theta \\ \beta \\ p \\ r \\ \phi \\ \psi \end{bmatrix}$$

an augmented model (4.4.20) results. The augmented state vector is

$$x_\Sigma(t) = \left[v(t)\ \alpha(t)\ q(t)\ \theta(t)\ \beta(t)\ p(t)\ r(t)\ \phi(t)\ \psi(t)\ x_\theta^{\text{ref}}(t)\ x_\phi^{\text{ref}}(t)\ x_\psi^{\text{ref}}(t) \right]^T,$$

Minimization of the quadratic performance functional (4.4.21) gives the controller

$$u = -G^{-1}B_\Sigma^T K x_\Sigma = -G^{-1} \begin{bmatrix} B \\ 0 \end{bmatrix}^T K x_\Sigma.$$

To find the unknown matrix K, obtain the feedback coefficients of the tracking controller, and model the fighter dynamics. The m-file is written as follows.
    MATLAB script (c_4_4_2.m)

```
%   Tracking control of a fighter
clear all    %This will clear out any previous variables
%   Matrices a and b
a=[-0.016    8.40 -0.90 -9.600 -1.500 -0.270 -0.086   0   0;
   -0.003  -1.20   1.00      0  0.080  0.062  0.009   0   0;
   -0.0001  3.90 -0.85      0  0.017 0.0038  0.040   0   0;
        0      0  1.00      0      0      0      0   0   0;
   -0.003   0.15  0.02  0.970 -0.560  0.130 -0.910   0   0;
  -0.00001  0.71  0.03  0.010 -48.00 -3.500  0.220   0   0;
   0.00001 -0.94  0.06  0.005  9.200 -0.028 -0.510   0   0;
```

```
            0       0       0       0       0  1.000       0  0  0;
            0       0       0       0       0       0  1.000  0  0];
b=[0.120   0.120  -0.380  -0.380       0       0;
  -0.160  -0.160  -0.270  -0.270       0       0;
  -9.500  -9.500  -2.500  -2.500       0       0;
       0       0       0       0       0       0;
   0.019  -0.019  -0.001   0.001    0.42   0.053;
  -2.900   2.900  -3.100   3.100    0.73   0.920;
   3.100  -3.100   0.780  -0.780    0.61  -0.450;
       0       0       0       0       0       0;
       0       0       0       0       0       0];
% Matrices h and d of the output equation
h=[0 0 0 1 0 0 0 0 0;
   0 0 0 0 0 0 0 1 0;
   0 0 0 0 0 0 0 0 1];
d=zeros(3,3);
% Here is a new closed-loop continuous-time model
% dx/dt=[A 0;-H 0]*x+[B; 0]*u+[0; N]
%      *reference Euler angles
% =([A 0;-H 0]+[B; 0]*Kcont)
%      *x+[0; N]*reference Euler angles
A=[a zeros(9,3);-h zeros(3,3)];
B=[b; zeros(3,6)];
% Weighting matrices
q=eye(size(A));
q(10,10)=5000; q(11,11)=5000; q(12,12)=5000;
r=25*eye(size(B,2));
% calculation of feedback and return matrices and
% eigenvalues
[Kcont,k,eigenvalues]=lqr(A,B,q,r);
% reference Euler angles: theta(pitch ra), phi (roll rb)
% and psi (yaw - rc)
ra=1; rb=1; rc=1;
reference Euler angles=[ra; rb; rc];
t=0:0.01:5;
ua=ra*ones(size(t'));
ub=rb*ones(size(t'));
uc=rc*ones(size(t'));
ur=[ua ub uc];
N=eye(3,3);
input matrixB=[zeros(9,3);N]*[0 ra 0; 0 rb 0; 0 0 rc];
H=h; H(3,12)=0;
D=d; D(3,3)=0;
%  simulation is performed using the "lsim" solver
lsim(A-B*Kcont,input matrixB,H,D,ur,t)
```

```
[v,x]=lsim(A-B*Kcont,input matrixB,H,D,ur,t);
% fighter outputs
figure, plot(t,v),title('Fighter outputs: Euler angles'),
grid, xlabel('Time, seconds'), pause,
% longitudinal dynamics
figure, plot(t,0.05*x(:,1),t,x(:,2),t,x(:,3),t,x(:,4)),
grid, title('Longitudinal Dynamics'),
xlabel('Time, seconds'), pause
% lateral dynamics
figure, plot(t,x(:,5),t,x(:,6),t,x(:,7),t,x(:,8),t,x(:,9)),
grid, title('Lateral Dynamics'),xlabel('Time, seconds'),
pause
figure, plot(t,v(:,1)),title('Output: theta'),
xlabel('Time, seconds'),pause
figure, plot(t,v(:,2)),title('Output: phi'),
xlabel('Time, seconds'),pause
figure, plot(t,v(:,3)),title('Output psi'),
xlabel('Time, seconds'),pause
%  plotting statement for the states x
figure, plot(t,x),title('State evolution: x1 to x12'),
grid, xlabel('Time, seconds'), pause,
disp('End')
```

The following matrices $K \in \mathbb{R}^{12 \times 12}$ and $K_F \in \mathbb{R}^{6 \times 12}$ are found

```
» format short e
» k
k =
 Columns 1 through 6
 2.2624e+000  1.0713e+001 -2.7581e-001 -9.4273e+000 -1.0028e+000 -1.1631e-002
 1.0713e+001  6.3340e+001 -8.8656e-001 -5.3723e+001  2.5563e+001 -1.8427e+000
-2.7581e-001 -8.8656e-001  1.5760e+000  1.0881e+001  4.0536e+000 -3.9853e-001
-9.4273e+000 -5.3723e+001  1.0881e+001  1.4346e+002  3.5629e+001 -5.2871e+000
-1.0028e+000  2.5563e+001  4.0536e+000  3.5629e+001  6.7072e+002 -5.7417e+001
-1.1631e-002 -1.8427e+000 -3.9853e-001 -5.2871e+000 -5.7417e+001  7.6338e+000
 2.5003e-002 -3.7498e+000 -7.0960e-001 -8.8651e+000 -8.6346e+001  1.1916e+001
 1.0459e-001 -1.2283e+001 -2.5514e+000 -3.3789e+001 -3.9208e+002  5.3177e+001
-5.5690e-001  1.6539e+001  4.5448e-001 -2.0402e+001  2.5126e+002  6.4248e+000
-1.5547e+001 -6.6623e+001 -2.2713e+001 -2.5523e+002  2.7391e+002 -8.1874e+000
-3.5339e-001  9.6956e+000  3.5937e+000  6.1455e+001  5.7974e+002 -1.0349e+002
-1.5326e-001 -1.5478e+001 -2.4070e+000 -3.5252e+000 -3.5267e+002  3.1549e+000
 Columns 7 through 12
 2.5003e-002  1.0459e-001 -5.5690e-001 -1.5547e+001 -3.5339e-001 -1.5326e-001
-3.7498e+000 -1.2283e+001  1.6539e+001 -6.6623e+001  9.6956e+000 -1.5478e+001
-7.0960e-001 -2.5514e+000  4.5448e-001 -2.2713e+001  3.5937e+000 -2.4070e+000
-8.8651e+000 -3.3789e+001 -2.0402e+001 -2.5523e+002  6.1455e+001 -3.5252e+000
-8.6346e+001 -3.9208e+002  2.5126e+002  2.7391e+002  5.7974e+002 -3.5267e+002
 1.1916e+001  5.3177e+001  6.4248e+000 -8.1874e+000 -1.0349e+002  3.1549e+000
 2.7547e+001  7.9171e+001  4.4184e+001 -1.4619e+001 -1.3385e+002 -6.0784e+001
 7.9171e+001  3.9120e+002  4.6941e+001 -6.1257e+001 -8.5474e+002 -1.1531e+001
 4.4184e+001  4.6941e+001  5.9152e+001  2.6598e+002 -2.5177e+002 -9.9342e+002
-1.4619e+001 -6.1257e+001  2.6598e+002  2.1477e+003  3.7706e+000 -2.4248e+002
-1.3385e+002 -8.5474e+002 -2.5177e+002  3.7706e+000  2.8192e+003  2.7695e+002
-6.0784e+001 -1.1531e+001 -9.9342e+002 -2.4248e+002  2.7695e+002  3.0995e+003
» Kcont
Kcont =
```

Columns 1 through 6
5.0792e-002 -2.4885e-001 -6.3320e-001 -4.2949e+000 -5.2456e+000 7.1159e-001
4.3417e-002 2.1472e-001 -5.5584e-001 -3.3771e+000 1.8280e+000 -3.8523e-001
-1.2025e-001 -6.4778e-001 -1.1671e-001 1.3031e-002 3.7327e+000 -5.1258e-001
-1.2477e-001 -8.6874e-001 -1.7095e-001 -7.4213e-001 -5.0650e+000 6.3245e-001
-1.6576e-002 2.8416e-001 3.9149e-002 2.2787e-001 7.4847e+000 -4.5095e-001
-3.0040e-003 5.3877e-002 6.7003e-003 4.0542e-002 8.6322e-001 -5.5286e-002
Columns 7 through 12
2.2617e+000 4.3993e+000 4.6433e+000 8.3276e+000 -5.5817e+000 -7.1582e+000
-1.6742e+000 -2.3020e+000 -5.2058e+000 9.6374e+000 2.7230e+000 9.1842e+000
-5.0358e-001 -3.7220e+000 3.5623e-001 3.7752e+000 8.1745e+000 -1.8634e+000
7.2573e-001 4.4944e+000 -7.8743e-001 2.6789e+000 -9.0919e+000 2.6838e+000
-4.3053e-001 -3.1023e+000 5.4868e+000 4.0060e+000 3.4518e+000 -7.3158e+000
-2.4039e-001 -2.9935e-001 -2.6221e-002 5.4255e-001 -1.7000e-001 4.6256e-001

The eigenvalues of the closed-loop system are

```
» eigenvalues
eigenvalues =
 -6.2821e+000
 -2.1637e+000  +5.3709e+000i
 -2.1637e+000  -5.3709e+000i
 -3.1484e+000  +4.5962e+000i
 -3.1484e+000  -4.5962e+000i
 -3.3798e+000  +1.9347e+000i
 -3.3798e+000  -1.9347e+000i
 -4.1484e+000
 -1.8578e+000  +1.7648e+000i
 -1.8578e+000  -1.7648e+000i
 -5.7498e-001
 -1.1882e+000
```

The real parts of eigenvalues are negative. Hence, the closed-loop system is stable.

The tracking controller is verified through the simulations. We assign the following references inputs:

$$r(t) = \begin{bmatrix} r_\theta(t) \\ r_\phi(t) \\ r_\psi(t) \end{bmatrix} = \begin{cases} 1 \text{ rad, } \forall t \in \begin{bmatrix} 0 & 5 \end{bmatrix} \text{ sec} \\ 1 \text{ rad, } \forall t \in \begin{bmatrix} 0 & 5 \end{bmatrix} \text{ sec} \\ 1 \text{ rad, } \forall t \in \begin{bmatrix} 0 & 5 \end{bmatrix} \text{ sec} \end{cases}$$

The fighter outputs (Euler angles) are documented in Figure 4.4.2.

The evolution of the state variables in the longitudinal and lateral axes are illustrated in Figure 4.4.3.

4.4.1.3. Optimal Control of Linear Continuous-Time Systems With Control Bounds

Our goal is to find the solution of the constrained optimization problem for dynamic systems described by constant-coefficient differential equations (4.4.8) with control bounds $u_{\min} \leq u \leq u_{\max}$. That is, the bounds are imposed on control inputs.

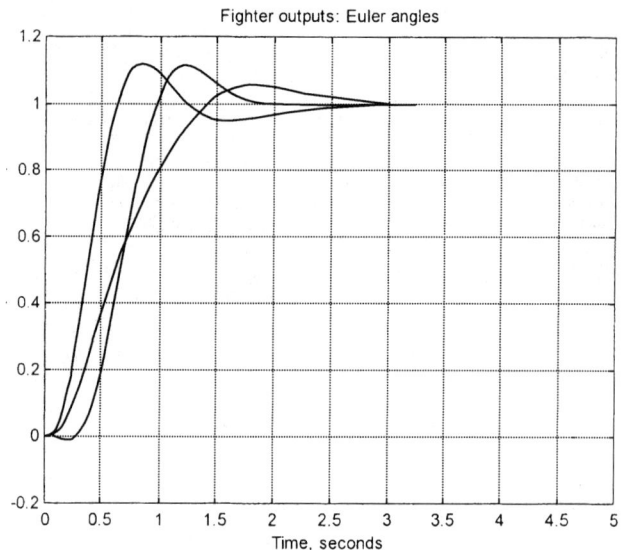

$$\text{FIGURE 4.4.2.} \quad \text{Fighter outputs if } r(t) = \begin{bmatrix} r_\theta(t) \\ r_\phi(t) \\ r_\psi(t) \end{bmatrix} = \begin{cases} 1 \text{ rad, } \forall t \in \begin{bmatrix} 0 & 5 \end{bmatrix} \text{ sec} \\ 1 \text{ rad, } \forall t \in \begin{bmatrix} 0 & 5 \end{bmatrix} \text{ sec} \\ 1 \text{ rad, } \forall t \in \begin{bmatrix} 0 & 5 \end{bmatrix} \text{ sec.} \end{cases}$$

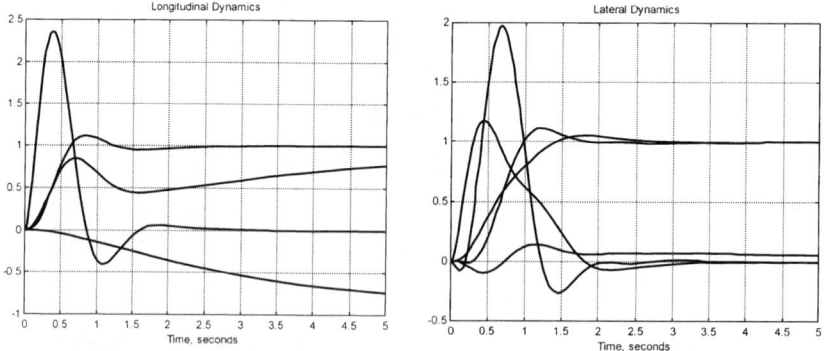

FIGURE 4.4.3.   Longitudinal and lateral aircraft dynamics.

We have the following system:

$$\dot{x}(t) = Ax + Bu, u_{\min} \leq u \leq u_{\max}, x(t_0) = x_0, \qquad (4.4.29)$$

where $u \in U \subset \mathbb{R}^m$ is the bounded control vector and $u_{\min}$ and $u_{\max}$ are finite. A prescribed set of admissible control values is given as $U \subset \mathbb{R}^m$, and the control function $u(\cdot)$: $[t_0, t_f) \rightarrow \mathbb{R}^m$ takes values in the closed set $U \subset \mathbb{R}^m$.

For linear system (4.4.8) $\dot{x}(t) = Ax + Bu$, the quadratic performance functional (4.4.9)

$$J(x(\cdot), u(\cdot)) = \frac{1}{2} \int_{t_0}^{t_f} \left(x^T Q x + u^T G u \right) dt,$$

was minimized. It was shown that the quadratic return function (4.4.13) $V(x) = \frac{1}{2} x^T K x$ is the solution of the Hamilton–Jacobi–Bellman partial differential equation, and the controller is given as (4.4.17). In particular, $u = -G^{-1} B^T K x$.

This controller is not bounded. Our goal is to analytically design the constrained controller for the dynamic system (4.4.29). Let us find the controller as a bounded C^ϵ ($\epsilon \geq 1$) function $\phi(\cdot)$ of the state vector. Using a smooth, integrable one-to-one C^ϵ, function $\phi \in U$, we have

$$u = \phi(x),$$

where $\phi \in U$ such that $\bar{U}_{\min} \leq \phi(x) \leq \bar{U}_{\max}$ for all $x \in X \subset \mathbb{R}^c$ on $[t_0, t_f)$,

$$\bar{U}_{\min} = \begin{bmatrix} u_{1\,\min} \\ \cdots \\ u_{m\,\min} \end{bmatrix} \in \mathbb{R}^m$$

and

$$\bar{U}_{\max} = \begin{bmatrix} u_{1\,\max} \\ \cdots \\ u_{m\,\max} \end{bmatrix} \in \mathbb{R}^m.$$

To solve the tracking control problem, the tracking errors and state variable are used, and one has

$$u = \phi(e, x), \phi \in U.$$

The control inputs are constrained, and the typical closed-loop system with saturated tracking control law is shown in Figure 4.4.4.

Using a smooth, integrable one-to-one C^ϵ, function $\phi \in U$, we design the integrand $\int \left(\phi^{-1}(u) \right)^T G\, du$. This integrand is used instead of commonly applied $\frac{1}{2} u^T G u$; see the quadratic performance functional (4.4.9). That is, the non-quadratic functional is given as

$$J(x(\cdot), u(\cdot)) = \int_{t_0}^{t_f} \left(\frac{1}{2} x^T Q x + \int \left(\phi^{-1}(u) \right)^T G\, du \right) dt. \qquad (4.4.30)$$

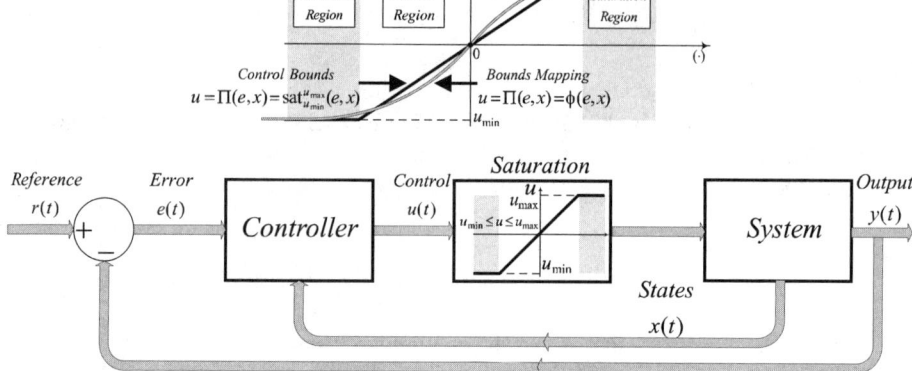

FIGURE 4.4.4.   Closed-loop system with the *admissible* (saturated) tracking control, $u_{min} \le u \le u_{max}$: Saturation is mapped by C^{ϵ}, $(\epsilon \ge 1)$ one-to-one function $\phi(\cdot)$.

Using (4.4.29) and (4.4.30), we have the expression for the Hamiltonian function as

$$H\left(x, u, \frac{\partial V}{\partial x}\right) = \left(\tfrac{1}{2}x^T Q x + \int \left(\phi^{-1}(u)\right)^T G\, du\right)$$
$$+ \left(\frac{\partial V}{\partial x}\right)^T (Ax + Bu). \tag{4.4.31}$$

From the first-order necessary condition for optimality (4.4.5), one finds the controller.

In particular, from

$$\frac{\partial \left(H\left(x, u, \frac{\partial V}{\partial x}\right)\right)}{\partial u} = \left(\phi^{-1}(u)\right)^T G + \left(\frac{\partial V}{\partial x}\right)^T B,$$

we have an admissible control law

$$u = -\phi\left(G^{-1} B^T \frac{\partial V}{\partial x}\right),\ u \in U. \tag{4.4.32}$$

The controller (4.4.32), which was analytically designed minimizing a non-quadratic performance functional (4.4.30), is bounded.

It must be emphasized that the second-order necessary condition for optimality (4.4.6) is guaranteed. In fact,

$$\frac{\partial^2 H\left(x, u, \frac{\partial V}{\partial x}\right)}{\partial u \times \partial u^T} > 0$$

because $G > 0$.

Example 4.4.7.

Consider the dynamic system with the saturation

$$\dot{x}(t) = Ax + Bu, \quad u_{min} \le u \le u_{max}, \quad U = \{|u| \le 1\}.$$

From (4.4.30), mapping the saturation by the integrable one-to-one hyperbolic tangent, one has

$$J(x(\cdot), u(\cdot)) = \int_{t_0}^{t_f} \left(\tfrac{1}{2} x^T Q x + \int \tanh^{-1} uG \, du \right) dt.$$

This functional is positive-definite for all $u \in U$; in particular,

$$\int \tanh^{-1} uG \, du = G \lfloor u \tanh^{-1} u + \tfrac{1}{2} \log \left(1 - u^2 \right) \rfloor.$$

The application of the first-order necessary condition for optimality gives the bounded controller as

$$u = -\tanh \left(G^{-1} B^T \frac{\partial V}{\partial x} \right) \approx -\mathrm{sat} \left(G^{-1} B^T \frac{\partial V}{\partial x} \right). \qquad \square$$

Plugging (4.4.32) in the Hamilton–Jacobi equation we have

$$-\frac{\partial V}{\partial t} = \tfrac{1}{2} x^T Q x + \int \left(\phi^{-1} \left(\phi \left(G^{-1} B^T \frac{\partial V}{\partial x} \right) \right) \right)^T G \, d \left(\phi \left(G^{-1} B^T \frac{\partial V}{\partial x} \right) \right)$$
$$+ \left(\frac{\partial V}{\partial x} \right)^T Ax - \left(\frac{\partial V}{\partial x} \right)^T B\phi \left(G^{-1} B^T \frac{\partial V}{\partial x} \right).$$

Thus,

$$-\frac{\partial V}{\partial t} = \tfrac{1}{2} x^T Q x + \int \frac{\partial V}{\partial x}^T B \, d \left(\phi \left(G^{-1} B^T \frac{\partial V}{\partial x} \right) \right) + \left(\frac{\partial V}{\partial x} \right)^T Ax$$
$$- \left(\frac{\partial V}{\partial x} \right)^T B\phi \left(G^{-1} B^T \frac{\partial V}{\partial x} \right).$$

From

$$\int \frac{\partial V}{\partial x}^T B \, d \left(\phi \left(G^{-1} B^T \frac{\partial V}{\partial x} \right) \right) = \left(\frac{\partial V}{\partial x} \right)^T B\phi \left(G^{-1} B^T \frac{\partial V}{\partial x} \right)$$
$$- \int \left(\phi \left(G^{-1} B^T \frac{\partial V}{\partial x} \right) \right)^T d \left(B^T \frac{\partial V}{\partial x} \right),$$

one finds the following partial differential equation to be solved:

$$-\frac{\partial V}{\partial t} = \tfrac{1}{2} x^T Q x + \left(\frac{\partial V}{\partial x} \right)^T Ax$$
$$- \int \left(\phi \left(G^{-1} B^T \frac{\partial V}{\partial x} \right) \right)^T d \left(B^T \frac{\partial V}{\partial x} \right). \qquad (4.4.33)$$

The following return function can be used to solve (4.4.33):

$$V(x) = \sum_{i=0}^{\varsigma} \frac{2\beta + 1}{2(i + \beta + 1)} \left(x^{\frac{i+\beta+1}{2\beta+1}}\right)^T K_i x^{\frac{i+\beta+1}{2\beta+1}}, \quad K_i \in \mathbb{R}^{c \times c}, \quad (4.4.34)$$

where ς and β are the nonnegative integers.

Example 4.4.8.

For the dynamic system with the saturation

$$\dot{x}(t) = ax + bu, \quad u_{min} \leq u \leq u_{max}, \quad U = \{|u| \leq 1\},$$

by minimizing positive-definite functional

$$J(x(\cdot), u(\cdot)) = \int_{t_0}^{t_f} \left(\tfrac{1}{2}x^2 + \int \tanh^{-1} u \, du\right) dt,$$

the bounded control is found as

$$u = -\tanh\left(\frac{\partial V}{\partial x}\right).$$

To find the feedback coefficients, one must solve (4.4.33). The tanh function was used to map the saturation. From (4.4.33), we have

$$-\frac{\partial V}{\partial t} = \tfrac{1}{2}x^2 + \frac{\partial V}{\partial x}ax - \int \tanh\left(b\frac{\partial V}{\partial x}\right) d\left(b\frac{\partial V}{\partial x}\right),$$

where

$$\int \tanh\left(b\frac{\partial V}{\partial x}\right) d\left(b\frac{\partial V}{\partial x}\right) = \log\cosh\left(b\frac{\partial V}{\partial x}\right).$$

The solution of the partial differential equation

$$-\frac{\partial V}{\partial t} = \tfrac{1}{2}x^2 + \frac{\partial V}{\partial x}ax - \log\cosh\left(b\frac{\partial V}{\partial x}\right)$$

can be found, and the feedback coefficients result.

Example 4.4.9. Control of a servo-system.

Consider a servo-mechanism actuated by a permanent-magnet DC motor. The studied servo is modeled as

$$\frac{di_a}{dt} = -\frac{r_a}{L_a}i_a - \frac{k_a}{L_a}\omega_r - \frac{1}{L_a}u_a,$$

$$\frac{d\omega_r}{dt} = \frac{k_a}{J}i_a - \frac{B_m}{J}\omega_r - \frac{1}{J}T_L$$

$$\frac{d\theta_r}{dt} = \omega_r,$$

and the output equation is given by $y = 0.008\theta_r$.

The applied armature voltage is bounded by $\pm 30V$, and the motor parameters are $r_a = 2.7\Omega$, $L_a = 0.004$ H, $k_a = 0.105$ V-sec/rad (N-m/rad), $J = 0.0001$ kg-m^2 and $B_m = 0.000016$ N-m-sec/rad.

The important feature is the fact that the applied voltage is bounded by

$$u_{a\,min} \le u_a \le u_{a\,max},$$

where $u_{a\,min} = -30$ V and $u_{a\,max} = 30$ V.

Consider the linear model of a servo-system as given by equations with control bounds.

The integral control concept is used in design. The error vector represents the difference between the reference and the actual position of the servo-mechanism; that is,

$$e(t) = \theta_{\text{reference}}(t) - \theta_r(t).$$

The augmented state vector is

$$x = \begin{bmatrix} i_a \\ \omega_r \\ \theta_r \\ \int e\,dt \end{bmatrix}.$$

The straightforward application of the integral control methodology gives the control law (4.4.28). The controller is found as a continuous function, constrained in magnitude and reversible in sign. In particular,

$$u_a = -u_{\max}\tanh\left(k_1 i_a + k_2\omega_r + k_3\theta_r + k_i\int e\,dt\right).$$

If the control is not bounded, minimizing the quadratic functional, one finds the feedback gains by solving the Riccati equation, and the m-file is given as

```
% Unbounded controller design for servos actuated by PM DC
motors format short e;
A=[-675   -26.25   0         0;
    1050  -0.093   0         0;
    0      1       0         0;
    0      0      -0.0008    0];
B=[250; 0; 0; 0];
Q=[0   0   0   0;
   0   0   0   0;
```

```
      0    0    0    0;
      0    0    0    2000000];
G=[1];
[Kfeedback,K,Eigenvalues]=LQR(A,B,Q,G);
disp('Kfeedback'), disp(Kfeedback); pause;
disp('K'), disp(K); pause;
disp('Eigenvalues'), disp(Eigenvalues); pause;
clear; disp('END');
```

The numerical results obtained by running this m-file are

```
Kfeedback
   1.8524e-002   1.1949e-002   5.1495e-001 -1.4142e+003

K
    7.4096e-005   4.7797e-005   2.0598e-003 -5.6569e+000
    4.7797e-005   3.0832e-005   1.3289e-003 -3.6615e+000
    2.0598e-003   1.3289e-003   5.7417e-002 -1.6573e+002
   -5.6569e+000 -3.6615e+000 -1.6573e+002   9.1031e+005

Eigenvalues
-6.3134e+002
-4.3758e+001
-2.3152e+000+ 2.3217e+000i
-2.3152e+000- 2.3217e+000i
```

Hence, one obtains the following feedback gains and eigenvalues:

$$k_1 = 0.019, k_2 = 0.012, k_3 = 0.51, k_i = -1414,$$

and

$$\lambda_1 = -631.3, \lambda_2 = -43.8, \lambda_{3,4} = -2.3 \pm 2.3i.$$

Thus, an integral control is

$$u_a = -0.019i_a - 0.012\omega_r - 0.51\theta_r + 1414 \int e\, dt.$$

Because the bounds are imposed, we have

$$u_a = -\mathrm{sat}_{-30}^{+30}\left(0.019i_a + 0.012\omega_r + 0.51\theta_r - 1414 \int e\, dt\right).$$

The block diagram of the closed-loop system is documented in Figure 4.4.5. The transient dynamics are given in Figure 4.4.6.

The reader can verify that the bounded controller guarantees the robustness to parameter variations. It must be emphasized that in general the designer cannot use the feedback gains obtained solving an unconstrained control problem because

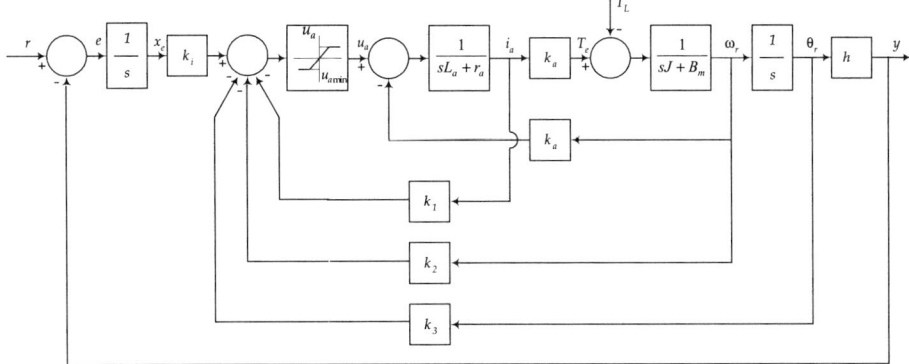

FIGURE 4.4.5.   Block diagram of the servo-system with bounded controller.

the system can be unstable. Specifically, the procedure reported must be used, and the designer must solve the following partial differential equation to obtain the feedback gains:

$$-\frac{\partial V}{\partial t} = \tfrac{1}{2}x^T Q x + \left(\frac{\partial V}{\partial x}\right)^T Ax - \int \tanh\left(\frac{b}{g}\frac{\partial V}{\partial x}\right) d\left(b\frac{\partial V}{\partial x}\right).$$

Using proportional and integral feedback for the tracking error and proportional feedback for the states, the control algorithm is

$$u_a = -\mathrm{sat}_{u_{\min}}^{u_{\max}}\left(k_1 i_a + k_2\omega_r + k_3\theta_r + k_p e + k_i \int e\,dt\right),$$

and the closed-loop system is shown in Figure 4.4.7.
    One obtains the bounded controller as

$$u_a = -\mathrm{sat}_{-30}^{+30}\left(0.019 i_a + 0.012\omega_r + 0.51\theta_r + 2741 e - 1395 \int e\,dt\right),$$

and the transient performance of the resulting closed-loop system if $r = 0.2$ m is documented in Figure 4.4.8. The analysis indicates that the settling time is 0.9 sec with no overshoot. One observes that a favorable angular velocity profile is achieved.
    Control of dynamic systems can be viewed based on the type of control strategies. The continuous and discontinuous constrained and unconstrained control algorithms can be designed for the studied servo. Our goal is to demonstrate the application of the bang-bang (relay-type) control algorithms. A highly refined procedure for nonlinear design of time-optimal controllers can be given using the Hamilton–Jacobi theory. The solution of the Hamilton–Jacobi–Bellman equation can be approximated by applying quadratic and nonquadratic return functions. The

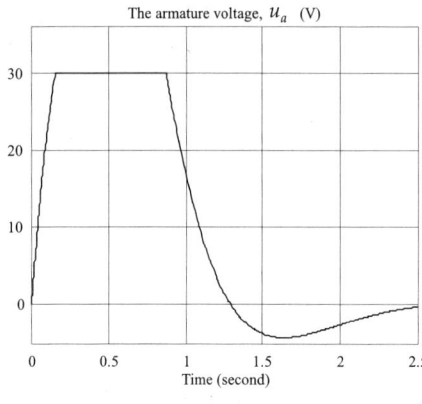

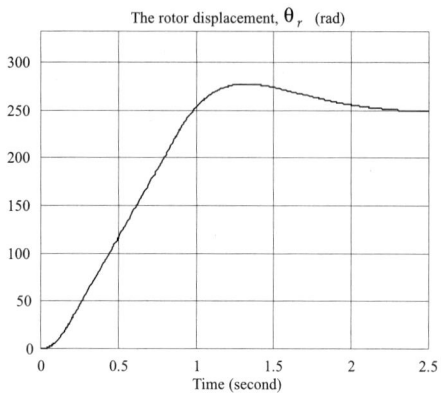

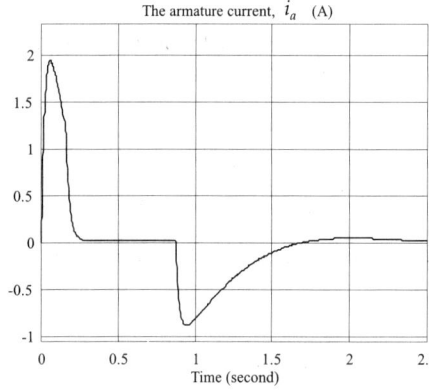

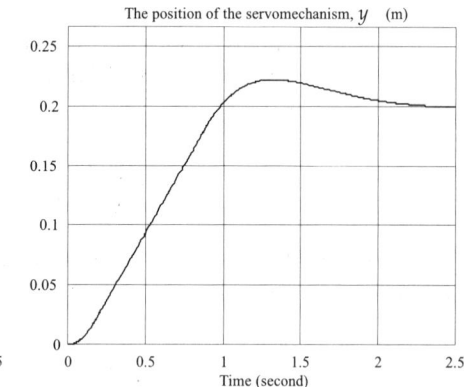

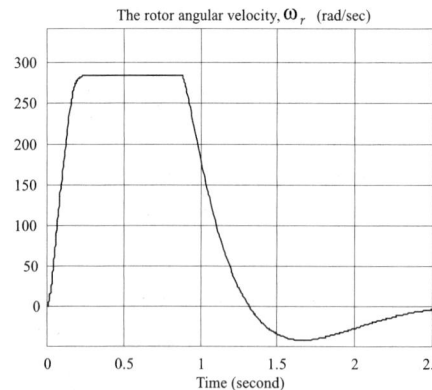

FIGURE 4.4.6.   Servo-system dynamics.

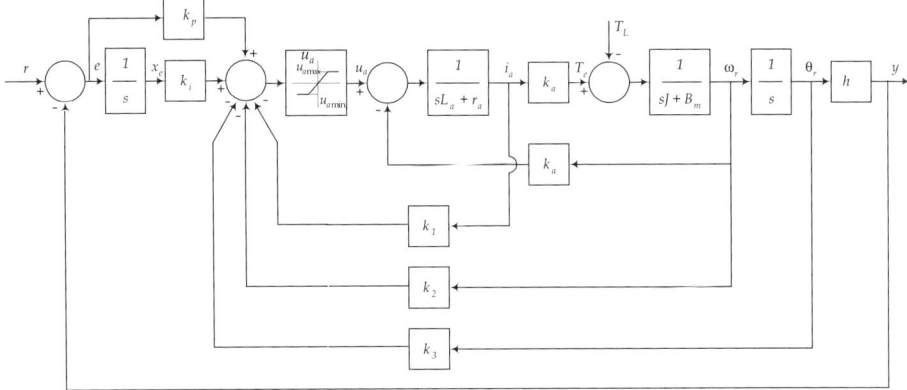

FIGURE 4.4.7.   Block diagram of a servo-system with constrained control law.

quadratic approximation is found to be conservative. However, from illustrative standpoints, viable examples are important, and we lend mathematical credence to the material of Chapter 5 and tackle the minimum-time problem simplifying the mathematical perspectives. Applying the quadratic function $V(x)$, let us attack the time-optimal problem for a servo-mechanism with a relay characteristic with dead zone. By making use of (4.4.32), a relay-type controller structure results, and the resulting servo is shown in Figure 4.4.9.

One has an on–off relay control algorithm; in particular,

$$u_a = - \mathrm{sgn}_{-30}^{+30}\big|_{\text{dead zone} \pm 0.005} (0.000096 i_a + 0.000061 \omega_r$$
$$+ 0.0000140 \theta_r - 3.16e).$$

The SIMULINK diagram is shown in Figure 4.4.10.

The servo-system transient dynamics is illustrated in Figure 4.4.11.

One concludes that the performance of a servo with relay control is improved. Relay- type algorithms have not been implemented widely because the voltage switching (chattering effect) leads to the armature current oscillations. This phenomenon leads to overheating, and the ideal relay cannot be applied. The current chattering varies as the function of the dead zone and motor parameters (armature resistance and inductance). The reader is encouraged to model and analyze the dynamics of a servo by assigning an entire region of parameter variations to check robustness to parameter variations in the full envelope of the parameter changes.

Bounded Integral Control and Experimental Verification

To solve the constrained optimization problem for a servo-system, we map the control bounds as $u = \phi(\cdot) = u_{a \max} \tanh^5(\cdot)$.

The tracking error is $e(t) = \theta_{\text{reference}}(t) - \theta_r(t)$, and the augmented states vector

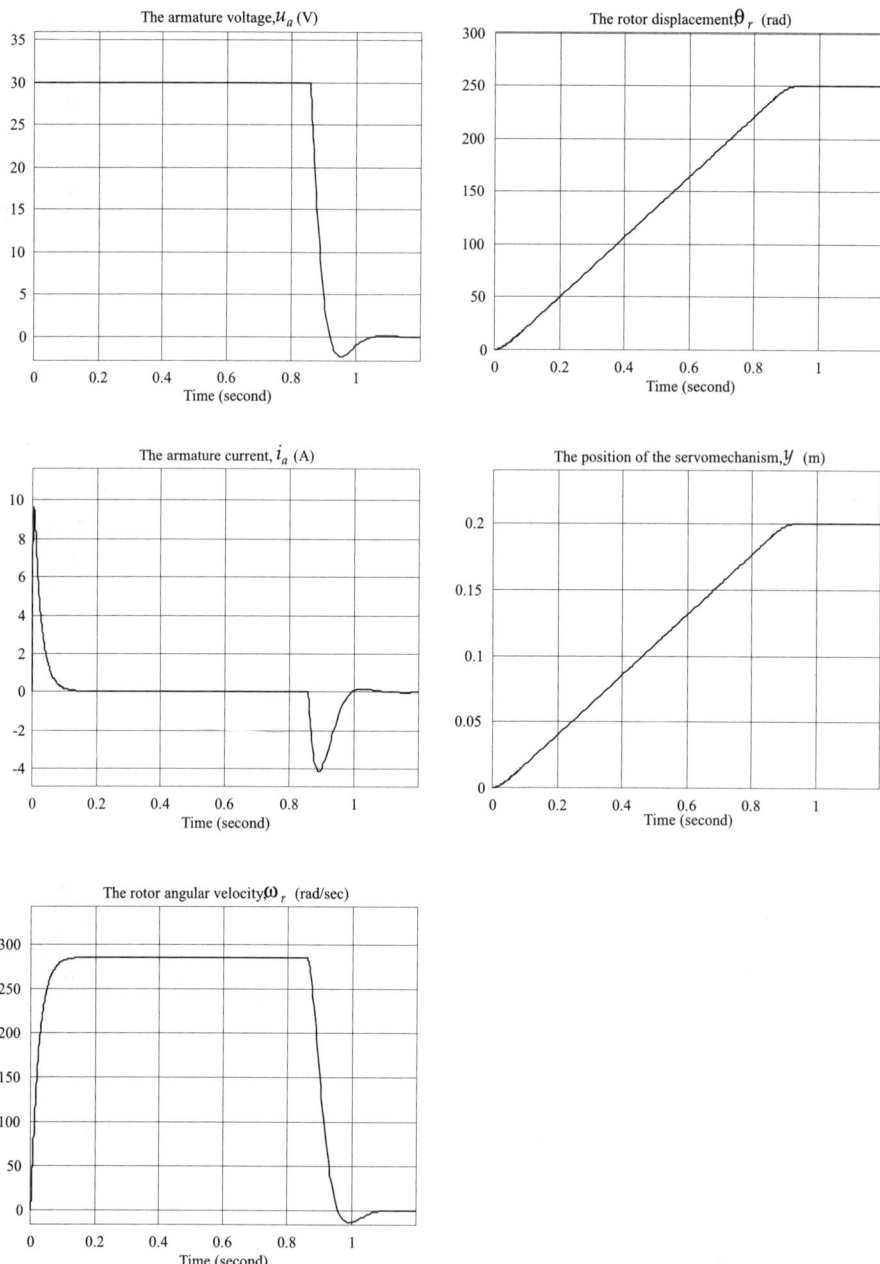

FIGURE 4.4.8.   The servo-mechanism dynamics.

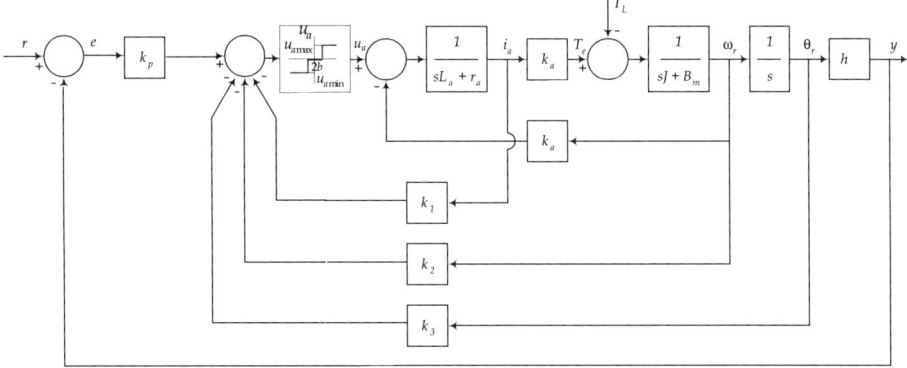

FIGURE 4.4.9.   Block diagram of a servo-mechanism with an on–off relay controller.

is expressed as

$$x = \begin{bmatrix} i_a \\ \omega_r \\ \theta_r \\ \int e\, dt \end{bmatrix}.$$

Minimizing a functional

$$J = \int_0^\infty [W_x(x) + W_u(u)]\, dt,$$

$$W_x(x) = \tfrac{1}{2} x^T Q_0 x + \tfrac{1}{4}(x^T)^2 Q_1 x^2, \quad Q_0 = Q_1 = I \in \mathbb{R}^{4\times 4},$$

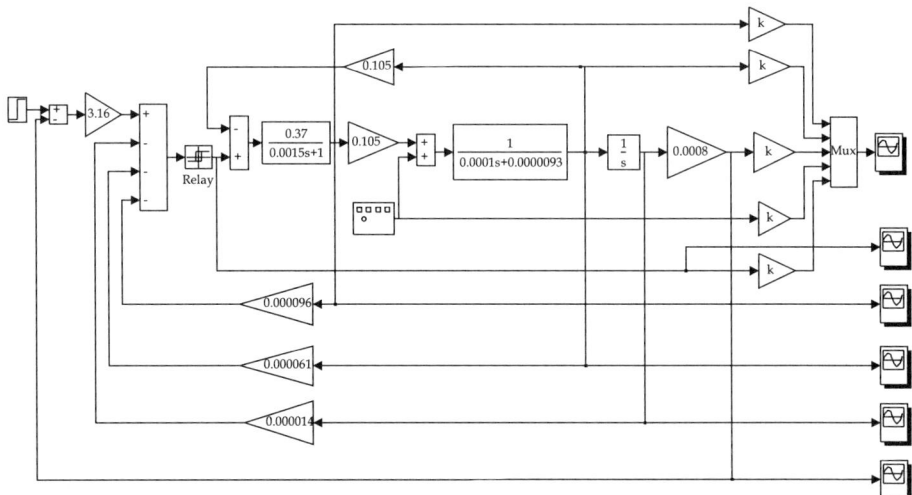

FIGURE 4.4.10.   SIMULINK diagram of a relay servo-mechanism (c_4_4_4.mdl).

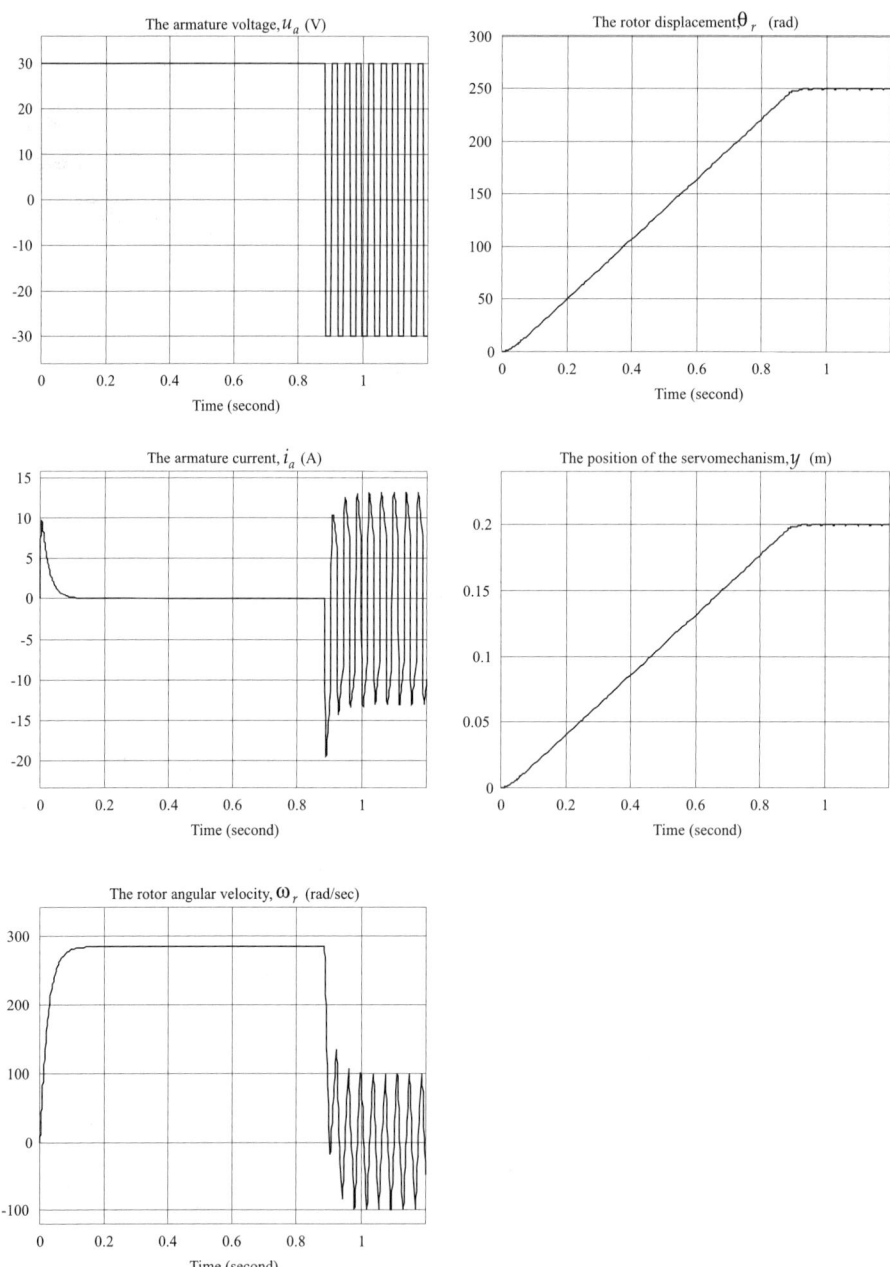

FIGURE 4.4.11.   Transient performance in the relay servo-mechanism.

$$W_u(u) = \int \left(\tanh^5 \left(\frac{u}{u_{a\,\max}} \right) \right)^{-1} G \, du,$$

using the first-order necessary condition for optimality, one finds an admissible control law

$$u = -u_{a\,\max} \tanh^5 \left(G^{-1} B^T \frac{\partial V}{\partial x} \right).$$

Here, the matrix B is

$$B = \begin{bmatrix} -\frac{1}{L_a} \\ 0 \\ 0 \\ 0 \end{bmatrix}.$$

This controller, which renders a minimum, belongs to the admissible closed and bounded set U. To solve the partial differential equation (4.4.33), the following return function is used:

$$V(x) = \tfrac{1}{2} x^T K_0 x + \tfrac{1}{4} \left(x^T \right)^2 K_1 x^2, \ K_0 \in \mathbb{R}^{4 \times 4}, \quad K_1 \in \mathbb{R}^{4 \times 4}.$$

Finally, the following controller is found as

$$u_a = 30 \tanh^5 \Big(-0.096 i_a - 0.2 \omega_r - 0.034 \theta_r - 0.0005 i_a^3 - 0.000001 \omega_r^3$$
$$- 0.0002 \theta_r^3 + 0.84 \int e \, dt + 0.29 \int e^3 \, dt \Big).$$

To verify the applicability of the derived controller, the experiments have been performed. The bounded nonlinear control algorithm was implemented. The servo-mechanism dynamics are shown in Figure 4.4.12 for reference input $\theta_{\text{reference}}(t) = 0.25$ m. The servo-mechanism position follows the specified reference inputs, and the rated armature current, angular velocity, and maximum acceleration are in an admissible operating envelope. The analysis of transient dynamics indicates that good performance has been achieved by using the designed control algorithm. One concludes that the validity and versatility of the offered approach have been justified and proven.

4.4.2. The Lyapunov Theory and Optimal Control

The stability analysis is an extremely important issue. The most powerful stability concept was developed by Lyapunov using the energy analysis. Lyapunov's idea is very simple: If the total energy of a system has a local minimum at a certain equilibrium point, that point is stable.

Consider a system that is described as

$$\dot{x}(t) = F(x), \quad t \geq 0 \text{ with } x(t_0) = x_0. \tag{4.4.35}$$

Our goal is to analyze the stability of (4.4.35).

Let us first consider and study the following example performing simulations.

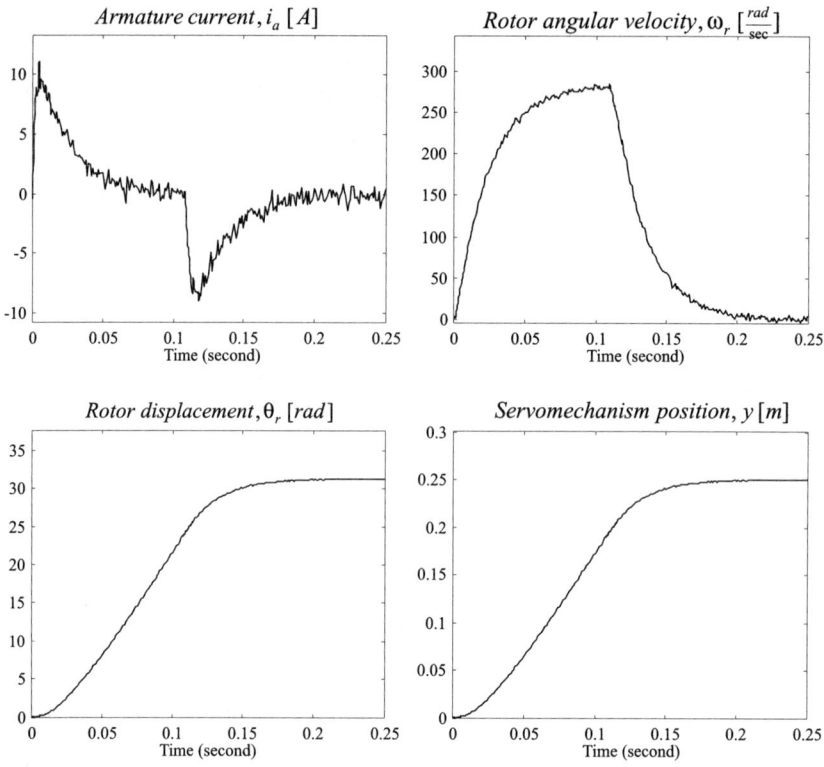

FIGURE 4.4.12.   Experimental results: transient dynamics of a servo-system, $\theta_{\text{reference}}(t) =$ 0.25 m.

Example 4.4.10.

Consider the second-order dynamic system

$$\dot{x}_1(t) \;=\; f_1(x_1, x_2),$$
$$\dot{x}_2(t) \;=\; f_2(x_1, x_2),$$

with $x_1(t_0) = x_{10}$, $x_2(t_0) = x_{20}$.

The evolution of this system is studied. To be specific, consider a nonlinear system modeled as

$$\frac{dx_1(t)}{dt} \;=\; x_2, \; x_1(t_0) = x_{10},$$

$$\frac{dx_2(t)}{dt} \;=\; -x_1 + kx_2 - kk_1 x_1^2 x_2, \; x_2(t_0) = x_{20}.$$

Simulation of this dynamics was performed assigning $k = 5$ and $k_1 = 1$. The

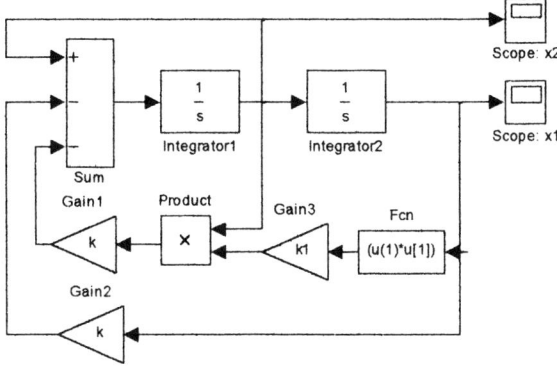

FIGURE 4.4.13.   SIMULINK diagram (c_4_4_21.mdl).

SIMULINK diagram is illustrated in Figure 4.4.13, and the initial conditions are

$$x_0 = \begin{bmatrix} x_{10} \\ x_{20} \end{bmatrix} = \begin{bmatrix} 1 \\ -1 \end{bmatrix}.$$

The transient dynamic waveforms, which are displayed by double clicking the Scope blocks, are shown in Figure 4.4.14.

Assigning $k = 100$ and $k_1 = 1$, the simulated transient responses are plotted in Figure 4.4.15.

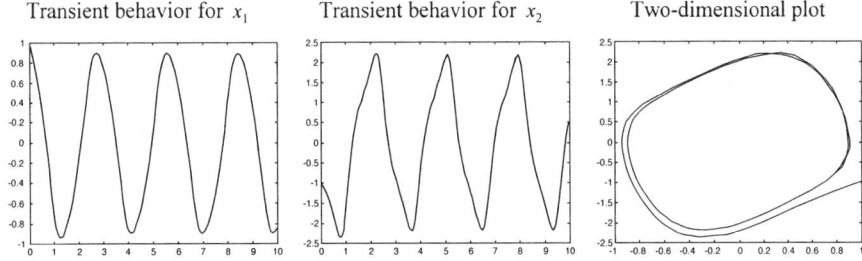

FIGURE 4.4.14.   System dynamics, $k = 5$ and $k_1 = 1$: System is stable with limit cycles.

Transient behavior for x_1 Transient behavior for x_2 Two-dimensional plot

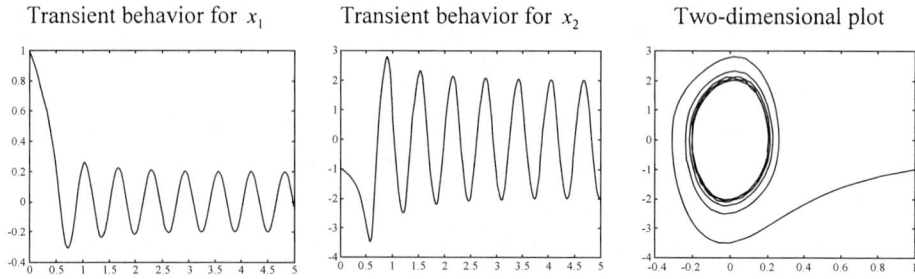

FIGURE 4.4.15.   System dynamics, $k = 100$ and $k_1 = 1$: System is stable with limit cycles.

Let $k = 100$ and $k_1 = 0$. That is, we simulate the system

$$\frac{dx_1(t)}{dt} = x_2, \, x_1(t_0) = 1,$$

$$\frac{dx_2(t)}{dt} = -x_1 + kx_2, \, x_2(t_0) = -1.$$

The system evolution is documented in Figure 4.4.16, and the system is unstable.

The studied example illustrates that dynamic systems can be modeled and analyzed in the time domain, and n-dimensional plots can be studied to analyze the
stability. Figures 4.4.14 to 4.4.16 illustrate that the system is stable if the solution
(system evolution) is bounded. That is, the solution is stable if a constant $\varepsilon > 0$
exists such that $\|x(t)\| < \varepsilon, \, \forall t \in [t_0 \, \infty)$.

Having performed numerical simulations and visualization, let us formulate
more general results using the Lyapunov stability theory.

We represent the system evolution using n-dimensional surfaces (for the second-
order systems, two-dimensional plots result).

Transient behavior for x_1 Transient behavior for x_2 Two-dimensional plot

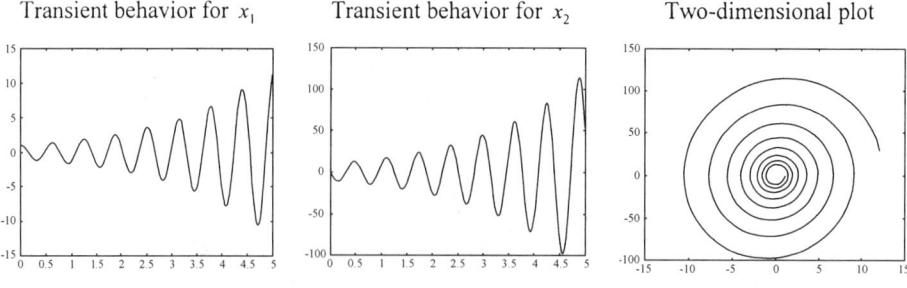

FIGURE 4.4.16.   System dynamics, $k = 100$ and $k_1 = 0$: System is unstable.

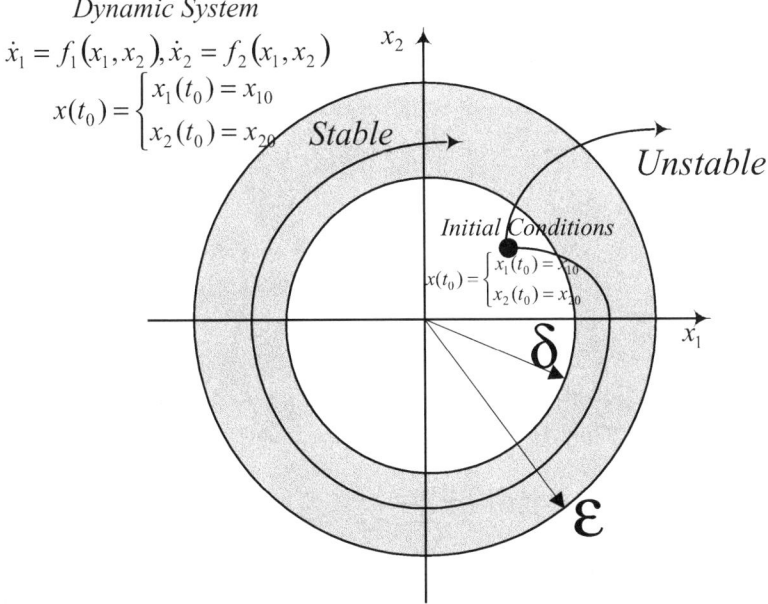

FIGURE 4.4.17.   System trajectories.

In particular, for the second-order dynamic systems, we have the possible system trajectories, as illustrated in Figure 4.4.17.

Two useful definitions are formulated.

Definition 4.4.1. A state x_e is an equilibrium state of the system (4.4.35) if

$$f(x_e) = 0, \quad \forall t \in [t_0 \, \infty). \qquad \Box$$

If one can find a bound on initial conditions such that the trajectories remain within a chosen finite limit, the equilibrium state is said to be stable.

Definition 4.4.2. System (4.4.35) has a stable equilibrium if for any $\varepsilon > 0$ there exists a $\delta(\varepsilon, t_0)$ such that if $\|x(t_0)\| < \delta$, then $\|x(t)\| < \varepsilon$, $\forall t \in [t_0 \, \infty)$. $\qquad \Box$

Lyapunov theorem. Consider the dynamic system described by differential equations

$$\dot{x}(t) = F(x), \quad t \geq 0.$$

If there exists a positive-definite continuous scalar Lyapunov function $V(x)$, $V(0) = 0$ with continuous first-order partial derivatives with respect to x, the equilibrium state is stable if

$$\frac{dV(x)}{dt} = \left(\frac{\partial V}{\partial x}\right)^T \frac{dx}{dt} = \left(\frac{\partial V}{\partial x}\right)^T F(x) \leq 0.$$

If $V(x) > 0$ and

$$\frac{dV(x)}{dt} \leq 0,$$

the equilibrium state of the system $\dot{x}(t) = F(x)$ is asymptotically stable.
Let the system dynamics of the time-varying system be modeled as

$$\dot{x}(t) = F(t, x), \quad t \geq 0.$$

If there exists a positive-definite scalar Lyapunov function $V(t, x)$ with continuous first-order partial derivatives with respect to t and x, the equilibrium state is as follows:

- Stable if the total derivative of the positive-definite function

$$V(t, x) > 0 \text{ is } \frac{dV}{dt} = \frac{\partial V}{\partial t} + \left(\frac{\partial V}{\partial x}\right)^T \frac{dx}{dt} = \frac{\partial V}{\partial t} + \left(\frac{\partial V}{\partial x}\right)^T F(t, x) \leq 0$$

- Uniformly stable if the total derivative of the positive-definite decreasing function $V(t, x) > 0$ is

$$\frac{dV}{dt} = \frac{\partial V}{\partial t} + \left(\frac{\partial V}{\partial x}\right)^T \frac{dx}{dt} = \frac{\partial V}{\partial t} + \left(\frac{\partial V}{\partial x}\right)^T F(t, x) \leq 0$$

- Uniformly asymptotically stable in the large if the total derivative of $V(t, x) > 0$ is negative definite; that is,

$$\frac{dV}{dt} = \frac{\partial V}{\partial t} + \left(\frac{\partial V}{\partial x}\right)^T \frac{dx}{dt} = \frac{\partial V}{\partial t} + \left(\frac{\partial V}{\partial x}\right)^T F(t, x) < 0$$

- Exponentially stable in the large if there exist the K_∞-functions $\rho_1(\cdot)$ and $\rho_2(\cdot)$, and K-function $\rho_3(\cdot)$ such that

$$\rho_1(\|x\|) \leq V(t, x) \leq \rho_2(\|x\|) \quad \text{and} \quad \frac{dV(x)}{dt} \leq -\rho_3(\|x\|) \qquad \square$$

One must perform the analysis of the positive and negative definiteness of $V(x)$, and the following definitions are formulated.

Definition 4.4.3. A scalar time-invariant function $V(x)$ is positive-definite in a region X, which includes the origin, if $V(x) > 0, \forall x \in X, x \neq 0$, and $V(0) = 0$. A scalar time-varying function $V(t, x)$ is positive-definite in X, which includes the origin, if $V(t, x)$ is bounded from below by a time-invariant, positive-definite function $V(x)$ and $V(t, x) > V(x) > 0, \forall t \in [t_0 \infty), \forall x \in X, x \neq 0, V(t, 0) = 0, \forall t \in [t_0 \infty)$.   □

Definition 4.4.4. A scalar time-invariant function $V(x)$ is negative-definite in a region X, which includes the origin, if $V(x) < 0, \forall x \in X, x \neq 0$, and $V(0) = 0$. A scalar time-varying function $V(t, x)$ is negative-definite in X, which includes the origin, if $V(t, x)$ is bounded by a time-invariant negative-definite function $V(x)$ and $V(t, x) < V(x) < 0, \forall t \in [t_0 \infty), \forall x \in X, x \neq 0, V(t, 0) = 0, \forall t \in [t_0, \infty)$.   □

Definition 4.4.5. A scalar time-invariant function $V(x)$ is positive-semidefinite in a region X, which includes the origin, if $V(x) \geq 0, \forall x \in X, x \neq 0$, and $V(0) = 0$.   □

Definition 4.4.6. A scalar time-invariant function $V(x)$ is negative-semidefinite in a region X, which includes the origin, if $V(x) \geq 0, \forall x \in X, x \neq 0$, and $V(0) = 0$.   □

Definition 4.4.7. A scalar time-invariant function $V(x)$ is indefinite in a region X, which includes the origin, if $V(x)$ takes both positive and negative values.   □

For example,

$$V(x) = x_1^2 + x_2^4 + x_3^6 + x_4^2 > 0 \text{ is positive-definite,}$$

$$V(x) = 2x_1^2 + 3x_2^4 + \frac{4x_3^6}{5x_1^2 + 6x_2^2 + 7x_3^6 + 8x_4^2} + 9x_4^2 > 0 \text{ is positive-definite,}$$

$$V(x) = (x_1 + x_2 + x_3 + x_4)^2 \geq 0 \text{ is positive-semidefinite,}$$

$$V(x) = x_1^2 + x_2^4 + x_3^6 + x_4^2 - (x_1^2 + x_2^4 + x_3^6 + x_4^2)^4 < 0 \text{ is negative-definite,}$$

$$V(x) = -x_1^2 + x_2^4 + x_3^6 + x_4^2 - (x_1^2 + x_2^4 + x_3^6 + x_4^2)^2 < 0 \text{ is negative-definite,}$$

$$V(x) = x_1^2 + x_2 x_3^3 + x_4^2 \text{ is indefinite.}$$

Example 4.4.11.

Consider the second-order dynamic system

$$\dot{x}_1(t) = -x_1 x_2^2,$$
$$\dot{x}_2(t) = -x_1^4 x_2^3 - x_2^5.$$

This system is nonlinear, and one cannot analyze the stability using the eigenvalues (which cannot be found) or frequency-domain methods. Using the Lyapunov stability theory, let us apply the simplest quadratic positive-definite function

$$V(x_1, x_2) = \tfrac{1}{2}(x_1^2 + x_2^2), \ V(0, 0) = 0, \forall t \in [0\infty).$$

The total derivative is found as

$$\frac{dV(x_1, x_2)}{dt} = \left(\frac{\partial V}{\partial x}\right)^T \frac{dx}{dt} = \left(\frac{\partial V}{\partial x}\right)^T F(x) = \frac{\partial V}{\partial x_1}(-x_1 x_2^2)$$

$$+ \frac{\partial V}{\partial x_2}(-x_1^4 x_2^3 - x_2^5) = -x_1^2 x_2^2 - x_1^4 x_2^4 - x_2^6.$$

That is, $V(x_1, x_2) = \frac{1}{2}(x_1^2 + x_2^2) > 0$ and

$$\frac{dV(x_1, x_2)}{dt} = -x_1^2 x_2^2 - x_1^4 x_2^4 - x_2^6 < 0.$$

Hence, the system is asymptotically stable.

4.4.2.1. The Relationship Between the Hamilton–Jacobi and Lyapunov Theories in Controller Design

As the basic cornerstone foundations of the application of Lyapunov's stability theory in the stability analysis have been studied, let us apply the Lyapunov concept to design optimal control algorithms. We study the linear system (4.4.8) with controller (4.4.17). That is, we have

$$\dot{x}(t) = Ax + Bu, \quad x(t_0) = x_0$$

with control law

$$u = -G^{-1} B^T K x = -K_F x.$$

The positive-definite performance functional to be minimized is given by (4.4.9), and thus,

$$J(x(\cdot), u(\cdot)) = \frac{1}{2} \int_{t_0}^{\infty} (x^T Q x + u^T G u) \, dt$$

$$= \frac{1}{2} \int_{t_0}^{\infty} (x^T Q x + x^T K_F^T G K_F x) \, dt$$

$$= \frac{1}{2} \int_{t_0}^{\infty} x^T (Q + K_F^T G K_F) x \, dt.$$

Let the total derivative of the $V(x)$ (which must be negative-definite) be expressed as

$$\frac{dV(x)}{dt} = -x^T (Q + K_F^T G K_F) x.$$

The positive-semidefinite and positive-definite constant-coefficient weighting matrices Q and G are used in the quadratic performance functional. That is,

$$\frac{dV(x)}{dt} < 0.$$

Using the quadratic, positive-definite function (4.4.13)

$$V(x) = \tfrac{1}{2} x^T K x,$$

one finds

$$\frac{dV(x)}{dt} = \tfrac{1}{2}(\dot{x}^T K x + x^T K \dot{x}) = \tfrac{1}{2}[(Ax - BK_F x)^T K x + x^T K(Ax - BK_F x)].$$

Recall that we assign

$$\frac{dV(x)}{dt} = -x^T(Q + K_F^T G K_F)x.$$

Then, one has

$$\tfrac{1}{2}\lfloor (Ax - BK_F x)^T K x + x^T K(Ax - BK_F x) \rfloor = -\tfrac{1}{2}x^T(Q + K_F^T G K_F)x$$

Taking note of the expression for the feedback gain matrix

$$K_F = -G^{-1} B^T K,$$

we conclude that the algebraic Riccati equation, which needs to be solved to obtain the unknown matrix K, is given as

$$-Q - A^T K - KA + K^T BG^{-1} B^T K = 0.$$

This equation is related to (4.4.16), and the unknown matrix K is symmetric, $K = K^T$.

From $V(x) = \tfrac{1}{2}x^T K x$, one concludes that the matrix K must be positive-definite to guarantee $V(x) > 0$. Recall that the total derivative was assigned to be negative-definite, and if $K > 0$, we have

$$\frac{dV(x)}{dt} = -x^T\left(Q + K_F^T G K_F\right)x, \quad \frac{dV(x)}{dt} < 0.$$

That is, the closed-loop system

$$\dot{x}(t) = Ax + Bu = Ax - BG^{-1}B^T K x = \left(A - BG^{-1}B^T K\right)x = (A - BK_F)x$$

is stable if the matrix K, obtained by solving the Riccati equation, is positive-definite.

The comparison of the results, obtained using the Hamilton–Jacobi and Lyapunov theories, indicates that the results are closely related, and they complement and support each other.

However, the Lyapunov concept does not support the controller design, whereas using the Hamilton–Jacobi theory, optimal controllers can be straightforwardly analytically designed applying the first-order necessary condition for optimality.

4.5.   Pole Placement Design by Using State Feedback

This section studies the application of the state feedback to ensure the desired pole (eigenvalues) location. One can assign the desired characteristic equation of the closed-loop system to attain the specified dynamics and transient behavior. This avenue of assigning the optimum (desired) form of the transient response in terms of system models and feedback coefficients is equivalent to the specification of the desired solution of differential equations. The eigenvalues can be specified by the designer. Particularly, our goal is to design the linear, time- invariant, full-state control law

$$u = -K_F x$$

for the dynamic systems defined by

$$\dot{x} = Ax + Bu.$$

The characteristic equation of the open-loop system is

$$|sI - A| = 0,$$

where $| \cdot |$ denotes the determinant.

By using the linear controller $u = -K_F x$, the resulting closed-loop system is

$$\dot{x} = Ax - BK_F x = (A - BK_F)x.$$

The cth-order characteristic polynomial of matrix $(A - BK_F) \in \mathbb{R}^{c \times c}$, is given by

$$\begin{aligned} p(s) &= |sI - A + BK_F| = s^c + (a_{c-1} + k_c)s^{c-1} + (a_{c-2} + k_{c-1})s^{c-2} \\ &+ \cdots + (a_0 + k_1). \end{aligned}$$

This characteristic polynomial assigns the essential features of the system dynamics. By equating the found polynomial to zero, one obtains the following characteristic equation:

$$\begin{aligned} p(s) &= |sI - A + BK_F| = s^c + (a_{c-1} + k_c)s^{c-1} \\ &+ (a_{c-2} + k_{c-1})s^{c-2} + \cdots + (a_0 + k_1) \\ &= (s - \lambda_1)(s - \lambda_2) \cdots (s - \lambda_{c-1})(s - \lambda_c) = 0. \end{aligned} \tag{4.5.1}$$

By assigning the desired placement of the characteristic roots (eigenvalues λ_1 through λ_c) for the closed-loop system, the characteristic equation results. The corresponding characteristic equation is

$$s^c + d_{c-1}s^{c-1} + d_{c-2}s^{c-2} + \cdots + d_0 = 0. \tag{4.5.2}$$

Equating coefficients of characteristic equations (4.5.1) and (4.5.2), one obtains the feedback gain coefficients. Hence, the matrix of feedback coefficients

$K_F \in \mathbb{R}^{m \times c}$ is found to meet the specified (desired) dynamics. It is important to emphasize that the eigenvalues are changed by the feedback $u = -K_F x$. However, it is necessary to achieve the balance of bandwidth, damping, overshoot, rise time, sensitivity and robustness to parameter variations, settling time, and other specifications. Furthermore, the designer should choose the pole location with regard to control bounds because control inputs can be saturated if one assigns the desired eigenvalues arbitrarily. Therefore, it is important to achieve a tradeoff between transient dynamics and stability (which is degraded because of control bounds).

The eigenvalues, which are relevant to transient responses, must be assigned. The synthesis of optimum transient performance can be performed by using various concepts. Most effective and widely used frameworks in prototype design are based on the Bessel polynomials and integral transient performance criteria. For example, it is easy to minimize the following integral of time and error:

$$J(t, e) = \min_{t,e} \int_0^\infty t|e|\, dt.$$

The error and settling time are used as criteria of performance to optimize transient responses. The large initial errors $e(0)$ do not lead to a large value of the integral because the transient errors decay to zero for stable systems as time approaches infinity. This integral criterion penalizes the settling time.

Minimizing $J(t, e) = \min_{t,e} \int_0^\infty t|e|\, dt$, one obtains the following transfer functions:

$$H(s) = \frac{1}{s + \omega_n}, \quad H(s) = \frac{1}{s^2 + 1.4s\omega_n + \omega_n^2},$$

$$H(s) = \frac{1}{s^3 + 1.8s^2\omega_n + 2.2s\omega_n^2 + \omega_n^3},$$

$$H(s) = \frac{1}{s^4 + 2.1s^3\omega_n + 3.4s^2\omega_n^2 + 2.7s\omega_n^3 + \omega_n^4},$$

$$H(s) = \frac{1}{s^5 + 2.8s^4\omega_n + 5s^3\omega_n^2 + 5.5s^2\omega_n^3 + 3.4s\omega_n^4 + \omega_n^5},$$

and

$$H(s) = \frac{1}{s^6 + 3.3s^5\omega_n + 6.6s^4\omega_n^2 + 8.6s^3\omega_n^3 + 7.5s^2\omega_n^4 + 4s\omega_n^5 + \omega_n^6}.$$

These transfer functions give the transient responses that minimize the integral of time and error.

If the overdamped dynamics is required, the Bessel polynomials are used to find the desired characteristic equation, as given by (4.5.1). In particular, for the nominal cutoff frequency $\omega_0 = 1$ rad/sec, the resulting eigenvalues locations for the step responses are found for systems with $c = 1$ to $c = 6$; see Table 4.5.1 and Figures 4.5.1.

The m-files to calculate and plot the transient responses are given below.

TABLE 4.5.1.   The pole locations (the numerical values of eigenvalues).

Order	Eigenvalues for criterion $J(t, e)$	Eigenvalues for the Bessel polynomials
$c = 1$	-1	-1
$c = 2$	$-0.71 \pm 0.71j$	$-0.87 \pm 0.5j$
$c = 3$	$-0.71; -0.52 \pm 1.1j$	$-0.94; -0.75 \pm 0.71j$
$c = 4$	$-0.42 \pm 1.26j; -0.62 \pm 0.41j$	$-0.66 \pm 0.83j; -0.9 \pm 0.27j$
$c = 5$	$-0.9; -0.38 \pm 1.3j; -0.58 \pm 0.54j$	$-0.93; -0.59 \pm 0.91j; -0.85 \pm 0.44j$
$c = 6$	$-0.31 \pm 1.27j; -0.58 \pm 0.78j; -0.73 \pm 0.29j$	$-0.54 \pm 0.96j; -0.8 \pm 0.56j; -0.91 \pm 0.19j$

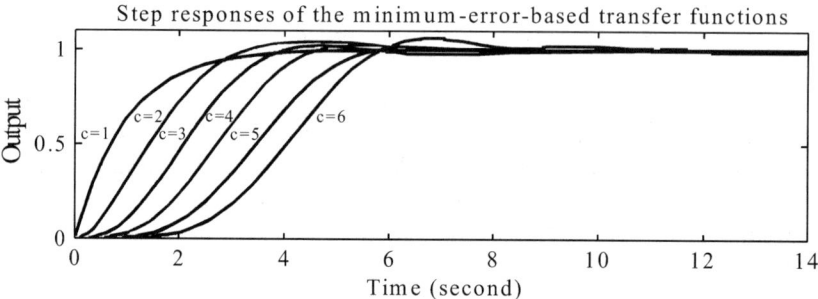

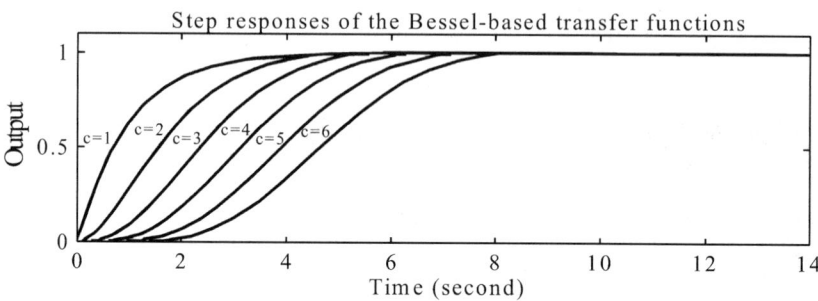

FIGURE 4.5.1.   Unit step responses of the optimum first- through sixth-order systems.

MATLAB script

```
% This m-file shows the step responses
% designed by minimizing the absolute value of J(t,e).
% The step responses are calculated using a m-file step,
y=step(num,den,t)
```

```
t=0:0.001:14;
num=[1]; den=[1 1]; y1=step(num,den,t);
den=[1 1.41 1]; y2=step(num,den,t);
den1=[1 0.71]; den2=[1 1.04 1.41]; den=conv(den1,den2);
y3=step(num,den,t);
den1=[1 0.85 1.77]; dcn2=[1 1.25 0.56]; den=conv(den1,den2);
y4=step(num,den,);
den1=[1 0.9]; den2=[1 0.75 1.81]; den3=[1 1.15 0.62];
den12=conv(den1,den2); den=conv(den12,den3);
y5=step(num,den,t);
den1=[1 0.62 1.69]; den2=[1 1.16 0.95];
den3=[1 1.45 0.62];
den12=conv(den1,den2); den=conv(den12,den3);
y6=step(num,den,t);
subplot(2,1,1); plot(t,y1,t,y2,t,y3,t,y4,t,y5,t,y6);
axis([0 14 0 1.1]);
title ('Step responses of the minimum-error-based
transfer functions');
xlabel('Time (secs)'); ylabel('Output');
text(0.1,0.6,' c=1            c=3         c=5');
text(1.5,0.7,'c=2         c=4            c=6'); pause;
%   These transfer functions can be represented by using
%   the state-space notations. A, B, C and D matrices are
%   found by using a m-file tf2ss,[A,B,C,D]=tf2ss(num,den).
%   The step response can be calculated using a m-file
%   step, y=step(A,B,C,D,1,t).
clear; 'END'
```

MATLAB script

```
% This m-file presents the design of the Bessel analog
% filters using a m-file besself Using the Bessel poly
% -nomials, the characteristic roots are found using a
% m-file eig
w=1; % That is , the cutoff frequency is 1 rad/sec.
t=0:0.001:14;
[num,den]=besself(1,w); [A,B,C,D]=besself(1,1); p1=eig(A);
y1=step(num,den,t);
[num,den]=besself(2,w); [A,B,C,D]=besself(2,1); p2=eig(A);
y2=step(num,den,t);
[num,den]=besself(3,w); [A,B,C,D]=besself(3,1); p3=eig(A);
y3=step(num,den,t);
[num,den]=besself(4,w); [A,B,C,D]=besself(4,1); p4=eig(A);
y4=step(num,den,t);
[num,den]=besself(5,w); [A,B,C,D]=besself(5,1); p5=eig(A);
```

```
y5=step(num,den,t);
[num,den]=besself(6,w);  [A,B,C,D]=besself(6,1);  p6=eig(A);
y6=step(num,den,t);
subplot(2,1,2); plot(t,y1,t,y2,t,y3,t,y4,t,y5,t,y6);
axis([0 14 0 1.1]);
title ('Step responses of the Bessel-based transfer
functions');
xlabel('Time (second)'); ylabel('Output');
text(0.1,0.6,' c=1                    c=3                    c=5');
text(1.8,0.7,'c=2              c=4                    c=6');
pause; clear;  'END'
```

Other criteria can be used for the synthesis of optimum transient behavior, and generally speaking, an infinite number of different indexes might be offered to find the characteristic eigenvalues. For example, the following integrals:

$$J(e) = \min_{e} \int_0^\infty |e|\, dt,$$

$$J(e) = \min_{e} \int_0^\infty e^2\, dt, \quad \text{and}$$

$$J(t, e) = \min_{t,e} \int_0^\infty t^2 |e|\, dt$$

as well as others, have been introduced and applied to determine the characteristic eigenvalues. Using these performance costs, the optimum dynamic responses can be obtained, eigenvalues are found, and the feedback coefficients are calculated. The stability and robustness are the key problem in design. It is clear that the synthesis problem is related to control bounds and parameter variations. The pole placement concept, although guaranteeing the "optimum" location of the characteristic eigenvalues, does not ensure the robustness to parameter variations and constraints because the studied design algorithm is formulated on an assumption that the system is linear and precisely known, and that coefficients are constant (time-invariant). In fact, the stability is prescribed by the desired eigenvalues based on the specified transient dynamics, and the characteristic roots of the designed closed-loop system are largely dependent on certain relations among coefficients of differential equations or transfer functions. Therefore, in many applications, a pole placement framework is found to be conservative because this method does not provide much insight and value into the robust controllers design.

4.5.1. *Control of a Servo-System with Permanent-Magnet DC Motor*

In servo-systems, permanent-magnet DC motors are widely used. Different control algorithms to control servo-mechanisms were studied in Example 4.4.9, and

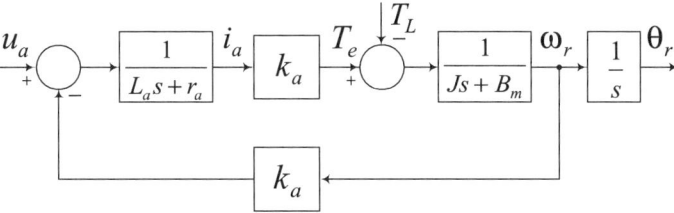

FIGURE 4.5.2. Block diagram of the open-loop servo-systems with a DC motor.

valuable results to analyze different design methods and perspectives were reported. The author feels that all concepts, commonly used, must be thoroughly
studied and examined by using stability, robustness, and performance criteria. Emphasis on the control of servo-systems and servo-drives have been largely based
on qualitative aspects, and quantitative features have been much less studied. As a
result, the designer might try to attain optimum performance and not approach the
robustness of closed-loop systems analyzing stability under parameter variations,
nonlinearities, and disturbance. Dynamics cannot be studied, assuming that the
systems model is accurate and parameters are constant and known precisely. It
appears that for a large class of dynamic systems with control limits, the elegant
pole placement methodology does not show considerable promise, and limitations
of this approach become clear if one considers control bounds and parameter variations. Because of industrial demands, major emphasis on control of dynamic
systems should be concentrated on design of robust controllers, and mathematically derived quantities must be reduced to practical design to satisfy many criteria
and justify the results. The state-space representation of servo-systems actuated
by permanent-magnet motors was developed. In particular, we have the following
state-space equation:

$$
\begin{bmatrix} \frac{di_a}{dt} \\ \frac{d\omega_r}{dt} \\ \frac{d\theta_r}{dt} \end{bmatrix} = \begin{bmatrix} -\frac{r_a}{L_a} & -\frac{k_a}{L_a} & 0 \\ \frac{k_a}{J} & -\frac{B_m}{J} & 0 \\ 0 & 1 & 0 \end{bmatrix} \begin{bmatrix} i_a \\ \omega_r \\ \theta_r \end{bmatrix} + \begin{bmatrix} \frac{1}{L_a} \\ 0 \\ 0 \end{bmatrix} u_a - \begin{bmatrix} 0 \\ \frac{1}{J} \\ 0 \end{bmatrix} T_L. \qquad (4.5.3)
$$

A block diagram of a servo-system is illustrated in Figure 4.5.2.

Let us design control algorithms using the pole-placement method. An integral
control concept is applied to design the closed-loop system. The output of the
studied servo is the position of the servo-mechanism, and $y = h\theta_r$. Hence, by
taking note of the permanent- magnet motor dynamics, one incorporates the output
equation. From (4.5.3), we have

$$
\begin{bmatrix} \frac{di_a}{dt} \\ \frac{d\omega_r}{dt} \\ \frac{d\theta_r}{dt} \end{bmatrix} = \begin{bmatrix} -\frac{r_a}{L_a} & -\frac{k_a}{L_a} & 0 \\ \frac{k_a}{J} & -\frac{B_m}{J} & 0 \\ 0 & 1 & 0 \end{bmatrix} \begin{bmatrix} i_a \\ \omega_r \\ \theta_r \end{bmatrix} + \begin{bmatrix} \frac{1}{L_a} \\ 0 \\ 0 \end{bmatrix} u_a - \begin{bmatrix} 0 \\ \frac{1}{J} \\ 0 \end{bmatrix} T_L,
$$

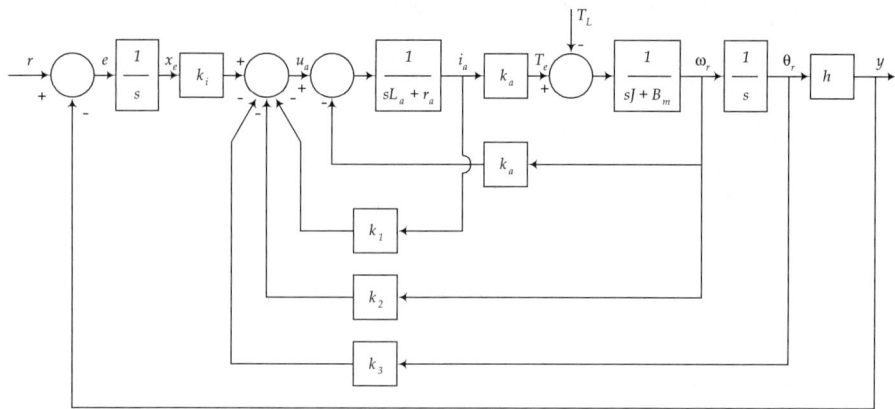

FIGURE 4.5.3.   Block diagram of a servo-system with state and integral feedback.

$$y = Hx = \begin{bmatrix} 0 & 0 & h \end{bmatrix} \begin{bmatrix} i_a \\ \omega_r \\ \theta_r \end{bmatrix}. \tag{4.5.4}$$

The control law is expressed by

$$u_a = -k_1 i_a - k_2 \omega_r - k_3 \theta_r + k_i x_e = -k_1 i_a - k_2 \omega_r - k_3 \theta_r + k_i \int e \, dt,$$

where the error vector is found as $e = r - y = r - h\theta$.

The feedback coefficients k_1, k_2, k_3, and k_i should be determined by comparing the eigenvalues of the closed-loop system with the desired eigenvalues, which can be found by using the Bessel polynomial concept.

The block diagram of a servo-mechanism is illustrated in Figure 4.5.3.

The parameters of the permanent-magnet motor are $r_a = 2.7\,\Omega$, $L_a = 0.004$ H, $B_m = 0.0000093$ N-m-sec/rad, $k_a = 0.105$ V-sec/rad, and $J = 0.0001$ kg-m^2. Let $h = 0.0008$ m/rad.

Making use of the considered notations and parameters, the state-space and output equations of the positioning system have the following form:

$$\begin{bmatrix} \frac{di_a}{dt} \\ \frac{d\omega_r}{dt} \\ \frac{d\theta_r}{dt} \\ \frac{dx_e}{dt} \end{bmatrix} = \begin{bmatrix} -675 - 250k_1 & -26.25 - 250k_2 & -250k_3 & 250k_i \\ 1050 & -0.093 & 0 & 0 \\ 0 & 1 & 0 & 0 \\ 0 & 0 & -0.0008 & 0 \end{bmatrix} \begin{bmatrix} i_a \\ \omega_r \\ \theta_r \\ x_e \end{bmatrix}$$

$$+ \begin{bmatrix} 0 \\ 0 \\ 0 \\ 1 \end{bmatrix} r - \begin{bmatrix} 0 \\ 10000 \\ 0 \\ 0 \end{bmatrix} T_L,$$

$$y = \begin{bmatrix} 0 & 0 & 0.0008 & 0 \end{bmatrix} \begin{bmatrix} i_a \\ \omega_r \\ \theta_r \\ x_e \end{bmatrix}.$$

Then, the characteristic equation for a servo-system is found as

$$\lambda(A) = s^4 + (675 + 250k_1)s^3 + (27563 + 23.3k_1 + 262500k_2)s^2$$
$$+ 262500k_3 s + 210k_i.$$

The characteristic roots are specified based on the desired servo-system performance. We assign the following specification:

- The tracking should be accomplished with zero steady-state error
- The settling time is 1 sec if $r = 0.2$ m; that is, the output reaches the reference input and remains within 5% of its within 1 sec
- The allowed overshoot is less than or equal to 10%

By applying the Bessel polynomials, it is straightforward to find that the cutoff frequency is 8 Hz, and the desired eigenvalues are

$$\lambda_{1,2} = -7.24 \pm 2.17i \text{ and } \lambda_{3,4} = -5.26 \pm 6.64i.$$

The desired characteristic equation is found to be

$$s^4 + 25s^3 + 281s^2 + 1639s + 4096.$$

By equating the like coefficients of the characteristic equations

$$\lambda(A) = s^4 + (675 + 250k_1)s^3 + (27563 + 23.3k_1 + 262500k_2)s^2$$
$$+ 262500k_3 s + 210k_i,$$
$$s^4 + 25s^3 + 281s^2 + 1639s + 4096,$$

one obtains

$$675 + 250k_1 = 25, 27563 + 23.3k_1 + 262500k_2 = 281, 262500k_3 = 1639,$$
$$210k_i = 4096.$$

The feedback gains are
$k_1 = -2.61$, $k_2 = -0.104$, $k_3 = 0.0062$, and $k_i = +19.5$.
Hence, the controller is

$$u_a = 2.61i_a + 0.104\omega_r - 0.0062\theta_r + 19.5 \int e \, dt.$$

MATLAB offers great possibilities to find the unknown feedback coefficients and analyze the dynamics. The m-file for design of controllers by applying a pole-placement methodology is given below.

```
% Controller design for a servo-system actuated by a PM
% DC motor
% A pole-placement technique
format short e;
order=4; cutoff frequency=8; t=0:0.0005:2;
disp('Matrices Ab, Bb, Hb and Db for the given order and
cutoff frequency');
[Ab,Bb,Hb,Db]=besself(order,cutoff frequency), pause;
disp('The desired characteristic roots');
desired roots=eig(Ab), pause;
disp('The specified characteristic polynomial');
poly(desired roots), pause;
disp('Numerator and denumerator for the given order and
cutoff frequency');
[numerator,denumerator]=besself(order,cutoff frequency),
pause;
y=step(numerator,denumerator,t); plot(t,y); pause;
A=[-675   -26.25   0         0;
    1050 -0.093    0         0;
    0     1        0         0;
    0     0       -0.0008    0];
B=[250; 0; 0; 0];
disp('The eigenvalues of matrix A');
eigenvalues A=eig(A), pause;
disp('The feedback coefficients for the specified
desired roots');
Kfeedback=place(A,B,desired roots), pause;
disp('Matrix (A-B*Kfeedback) for the resulting
closed-loop system');
Aa=A-B*Kfeedback, pause;
disp('The characteristic roots of matrix
(A - B*Kfeedback)');
roots Aa=eig(Aa), pause;
disp('The characteristic polynomial for the resulting
closed-loop system');
poly(roots Aa), pause;
clear; disp('END');
```

The following numerical results are obtained:

```
Matrices Ab, Bb, Hb, and Db for the given order and
cutoff frequency
Ab =
  -1.4476e+001 -7.5556e+000          0          0
   7.5556e+000          0            0          0
          0    8.4705e+000 -1.0515e+001 -8.4705e+000
```

```
              0            0  8.4705e+000                0
Bb =
      8
      0
      0
      0
Hb =
      0            0            0  9.4445e-001
Db =
      0
```

The desired characteristic roots
desired roots =
-7.2381e+000+ 2.1673e+000i
-7.2381e+000- 2.1673e+000i
-5.2577e+000+ 6.6413e+000i
-5.2577e+000- 6.6413e+000i

The specified characteristic polynomial
ans =
1.0000e+000 2.4992e+001 2.8106e+002 1.6390e+003 4.0960e+003

Numerator and denumerator for the given order and
cutoff frequency
numerator =
0   0   0   0   4.0960e+003

denumerator =
1.0000e+000 2.4992e+001 2.8106e+002 1.6390e+003 4.0960e+003

The eigenvalues of matrix A
eigenvalues A =
              0
              0
  -4.3757e+001
  -6.3134e+002

The feedback coefficients for the specified desired
roots
place: ndigits= 21
Kfeedback =
  -2.6004e+000 -1.0394e-001 6.2436e-003 -1.9505e+001

Matrix (A-B*Kfeedback) for the resulting closed-loop
system
Aa =

```
-2.4899e+001 -2.6547e-001 -1.5609e+000   4.8762e+003
 1.0500e+003 -9.3000e-002            0            0
           0  1.0000e+000            0            0
           0            0 -8.0000e-004            0
```

```
The characteristic roots of matrix (A - B*Kfeedback)
roots Aa =
-5.2577e+000+ 6.6413e+000i
-5.2577e+000- 6.6413e+000i
-7.2381e+000+ 2.1673e+000i
-7.2381e+000- 2.1673e+000i
```

```
The characteristic polynomial for the resulting closed-
loop system
ans =
1.0000e+000 2.4992e+001 2.8106e+002 1.6390e+003 4.0960e+003
END
```

Hence, for the chosen cutoff frequency, the eigenvalues of the desired characteristic equation are $\lambda_{1,2} = -7.24 \pm 2.17i$ and $\lambda_{3,4} = -5.26 \pm 6.64i$. Thus, the feedback gains and eigenvalues agree with the calculated early, and therefore, the specified requirements shall be satisfied. The servo- system dynamics is shown in Figure 4.5.4.

One might conclude that the design is successfully performed and the assigned criteria are met. However, the designer must analyze the control and state limits. In fact, the applied voltage, armature current, and angular velocity of the rotor are bounded. In particular, the armature voltage and angular velocity are bounded as $u_{a\,max} \le 30$ V, $u_{a\,min} \ge -30$ V, $\omega_{max} \le 300$ rad/sec, and $\omega_{min} \ge -300$ rad/sec. Analysis of dynamics indicates that $u_{a\,max} = 63$ V and $\omega_{max} = 580$ rad/sec. One concludes that the applied voltage must be constrained, and the block diagram with bounded control is given in Figure 4.5.5.

It is of great interest to perform modeling of a servo-mechanism with constrained control. A SIMULINK block diagram is represented in Figure 4.5.6.

The system dynamics is shown in Figure 4.5.7.

The pole-placement technique can be applied only for linear systems. The dynamics is significantly deteriorated by the control limits, and the feedback coefficients should be found by relaxing the design specifications. In fact, electromechanical systems possess nonlinearities, and inherent bounds (maximum applied voltage, current, angular velocity, and displacement) cannot be treated from the linear control theory standpoint.

Let us specify the settling time by 1.5 sec. A new value of the cutoff frequency is assigned to be 4.25 Hz. By using the m-file given earlier, the results are given below.

```
Matrices Ab, Bb, Hb, and Db for the given order and
cutoff frequency
```

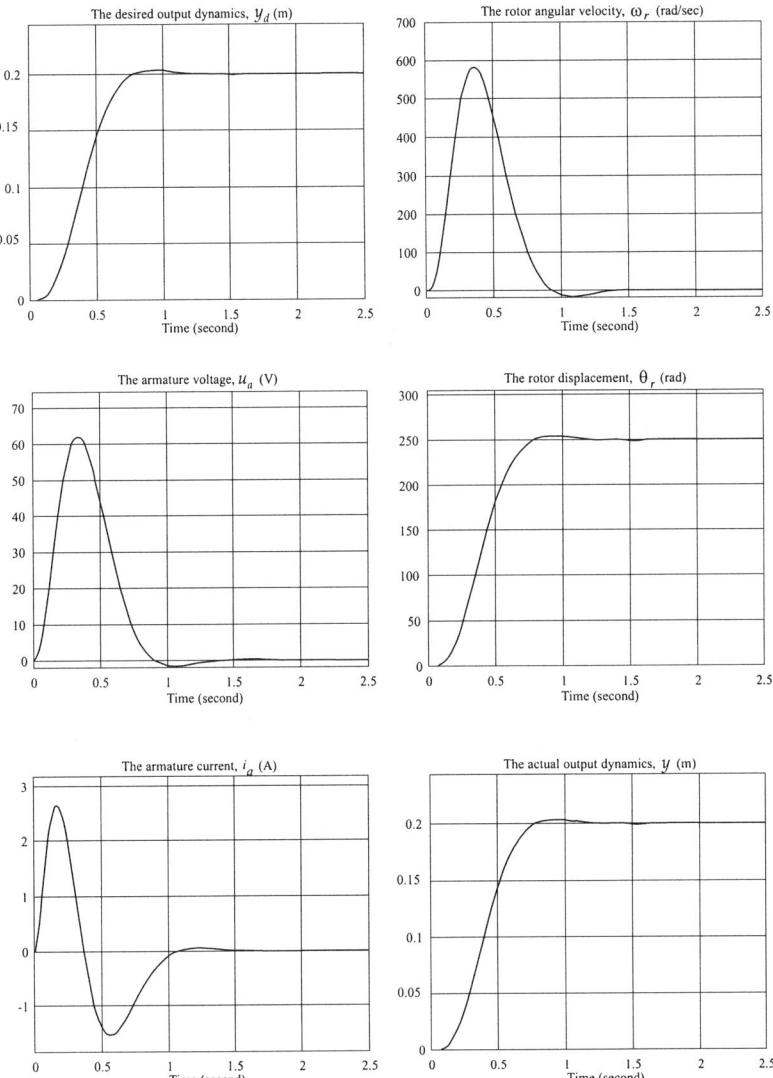

FIGURE 4.5.4.   Transient responses in a servo-system.

```
Ab  =
-7.6904e+000  -4.0139e+000             0             0
 4.0139e+000             0             0             0
            0   4.5000e+000  -5.5863e+000  -4.5000e+000
            0             0   4.5000e+000             0
Bb  =
 4.2500e+000
            0
```

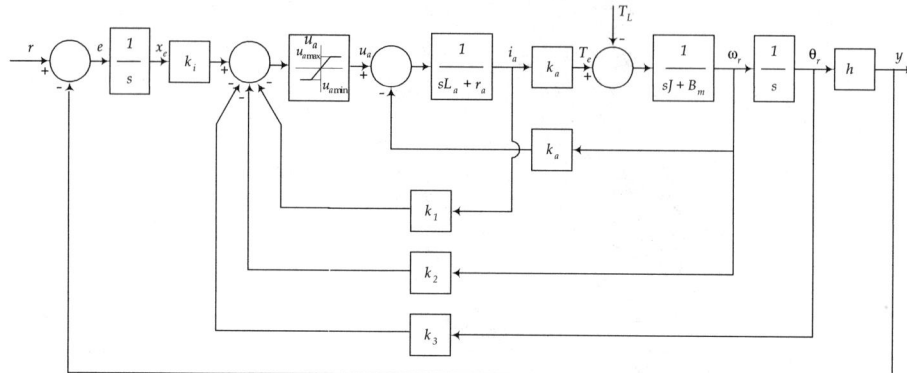

FIGURE 4.5.5.   Block diagram of a servo-system with bounded armature voltage.

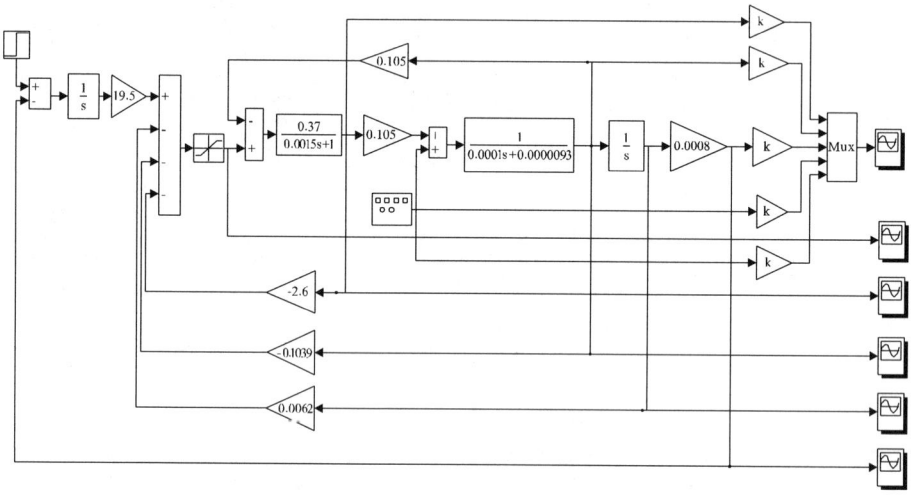

FIGURE 4.5.6.   The block diagram of a servo with the integral control law.

```
                      0
                      0
Hb  =

                      0              0          0   9.4445e-001
Db  =
          0
The desired characteristic roots
desired roots =
-3.8452e+000+ 1.1514e+000i
-3.8452e+000- 1.1514e+000i
-2.7931e+000+ 3.5282e+000i
-2.7931e+000- 3.5282e+000i
```

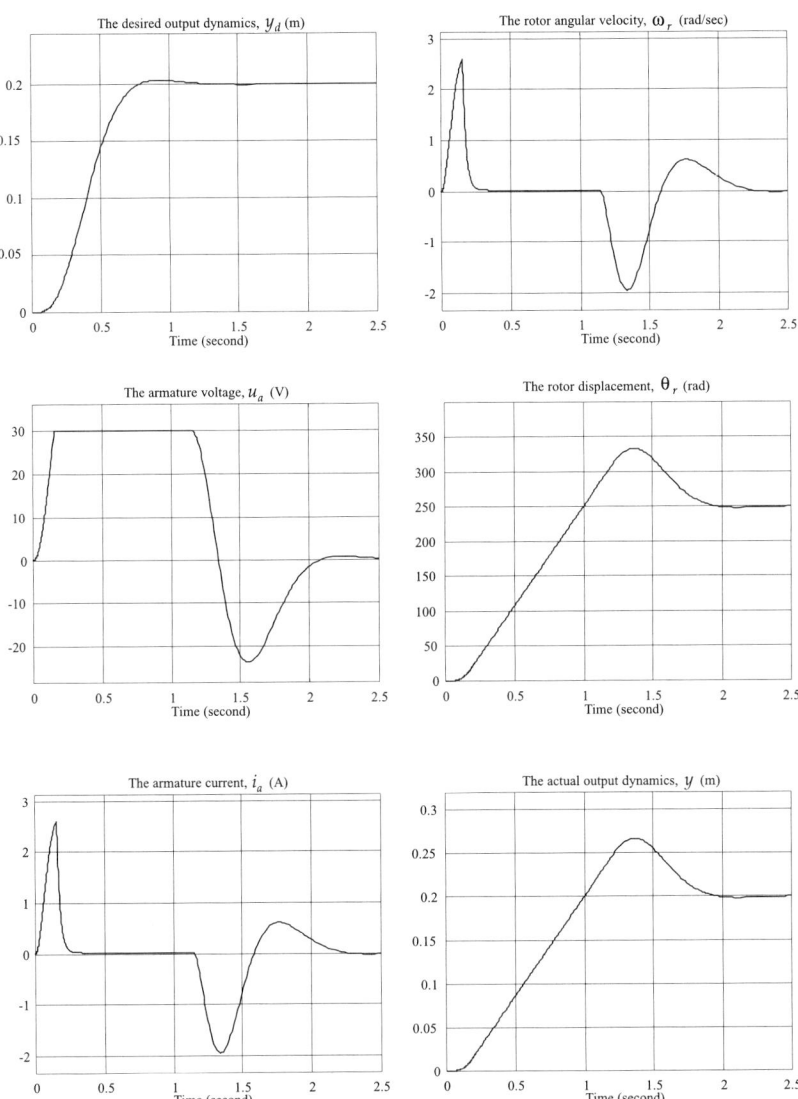

FIGURE 4.5.7.   Transient responses in a servo-system with bounded control.

The specified characteristic polynomial

ans =
1.0000e+000 1.3277e+001 7.9322e+001 2.4573e+002 3.2625e+002

Numerator and denumerator for the given order and
cutoff frequency

```
numerator =
0            0            0            0            3.2625e+002

denumerator =
1.0000e+000 1.3277e+001 7.9322e+001 2.4573e+002 3.2625e+002

The eigenvalues of matrix A
eigenvalues A =
             0
             0
 -4.3757e+001
 -6.3134e+002

The feedback coefficients for the specified desired
roots
place: ndigits= 20
Kfeedback =
 -2.6473e+000 -1.0470e-001  9.3613e-004 -1.5536e+000

Matrix (A-B*Kfeedback) for the resulting closed-loop
system
Aa =
 -1.3184e+001 -7.4377e-002 -2.3403e-001  3.8840e+002
  1.0500e+003 -9.3000e-002            0            0
            0  1.0000e+000            0            0
            0             0 -8.0000e-004            0
The characteristic roots of matrix (A - B*Kfeedback)
roots Aa =
-2.7931e+000+ 3.5282e+000i
-2.7931e+000- 3.5282e+000i
-3.8452e+000+ 1.1514e+000i
-3.8452e+000- 1.1514e+000i
The characteristic polynomial for the resulting closed-
loop system

ans =
1.0000e+000 1.3277e+001 7.9322e+001 2.4573e+002 3.2625e+002

END
```

Hence, with the chosen cutoff frequency 4.25 Hz, the eigenvalues of the desired characteristic equation

$$s^4 + 13.3s^3 + 79.3s^2 + 245.7K_3s + 326.3$$

are

$$\lambda_{1,2} = -3.85 \pm 1.15i \text{ and } \lambda_{3,4} = -2.79 \pm 3.53i.$$

The feedback coefficients are found as

$$k_1 = -2.65, \, k_2 = -0.105, \, k_3 = 0.00094, \text{ and } k_i = +1.55,$$

and the following control law results:

$$u_a = 2.65 i_a + 0.105\omega_r - 0.00094\theta_r + 1.55 \int e \, dt.$$

For completeness, let us study the dynamics, as given in Figure 4.5.8.

Analysis indicates that the specified requirements have been achieved if the settling time is 1.5 sec. The most important criteria are the stability and the robustness to parameter variations. The Routh–Hurwitz criterion can be efficiently applied for linear systems, and the Lyapunov theory gives the sufficient conditions for stability and robustness of linear and nonlinear systems with control bounds and parameter variations. The pole-placement technique may lead to very dangerous results. In particular, the system can be unstable even if very small variations of parameters appear (the motor armature resistance and the back emf constant vary in the region ±40%). To illustrate this feature, let us analyze the feedback gains designed.

The controllers were found as

$$u_a = 2.61 i_a + 0.104\omega_r - 0.0062\theta_r + 19.5 \int e \, dt$$

and

$$u_a = 2.65 i_a + 0.105\omega_r - 0.00094\theta_r + 1.55 \int e \, dt.$$

Because of positive current and angular velocity feedback gains, the resulting closed-loop system is not robust and is very sensitive to parameter variations. The reader is encouraged to model the designed servo-mechanism and calculate the eigenvalues to verify that the closed-loop system is unstable if the motor parameters are slightly changed (back emf constant, rotor resistance, armature inductance, and moment of inertia vary).

It is very difficult (or even impossible) to formulate the general specifications for the optimal location of the characteristic eigenvalues within parameter variations envelope. One concludes that the application of the Bessel polynomial might not guide one to the practical result, and the designer can find that by using the pole-placement concept, the robust design cannot be achieved. Although a suitable value of the characteristic eigenvalues can be found, this trial-and-error process does not provide much insight into the synthesis. The stability region and robustness must be explicitly studied in the entire envelope of parameter variations. Therefore, it is of practical interest to design the robust control algorithms, and other design methods, such as the Hamilton–Jacobi theory and Lyapunov's concept, can be applied to ensure a unitary figure of merit in robustness and system performance.

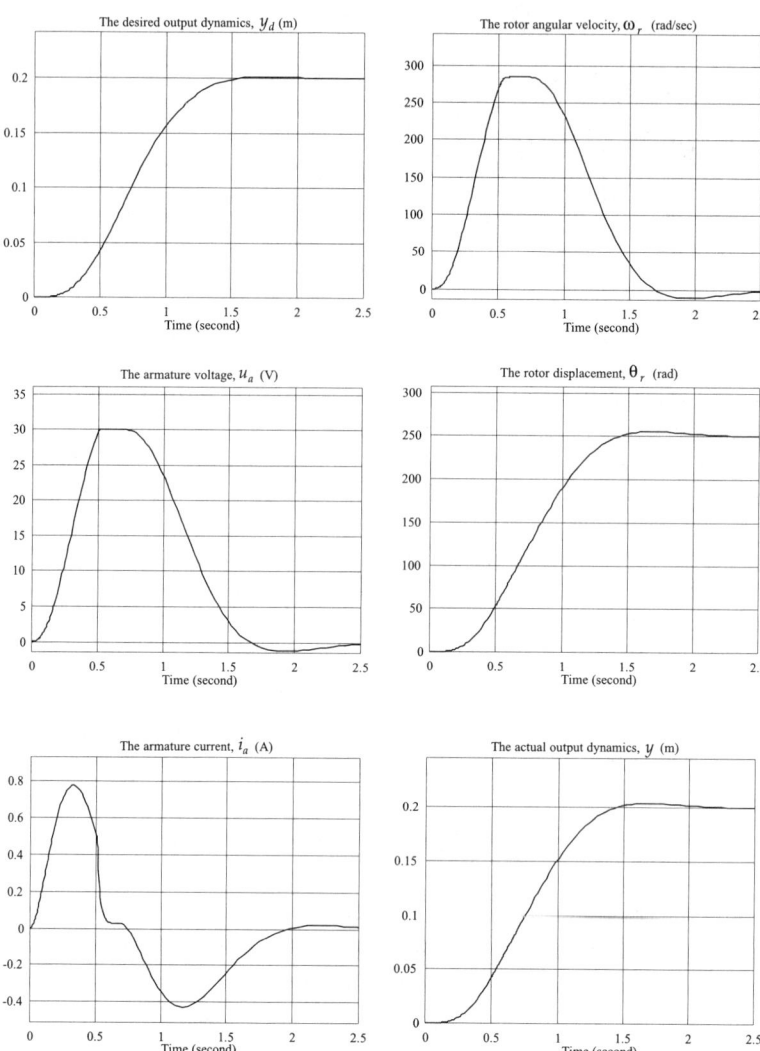

FIGURE 4.5.8.   Transient performance of a servo-mechanism.

4.6.   Control of Discrete-Time Systems Using the Hamilton–Jacobi Theory

4.6.1.   *Liner Quadratic Regulator Problem*

We consider discrete-time systems modeled using the state-space difference equations. In particular, the system dynamics is expressed as

$$x_{n+1} = A_n x_n + B_n u_n, \quad n \geq 0. \tag{4.6.1}$$

The quadratic performance index to be minimized is

$$J = \sum_{n=0}^{N-1} \left[x_n^T Q_n x_n + u_n^T G_n u_n \right], \quad Q_n \geq 0, \quad G_n > 0. \tag{4.6.2}$$

Our goal is to find the control law that guarantees that the value of the performance cost (4.6.2) is minimum or maximum. The minimization and maximization problems can be straightforwardly solved using the Hamilton–Jacobi concept.

For linear dynamic systems and quadratic performance indexes, the solution of the Hamilton–Jacobi–Bellman equations is satisfied by the quadratic return functions. That is, the solution of the Hamilton–Jacobi–Bellman recursive equation

$$V(x_n) = \min_{u_n} \left[x_n^T Q_n x_n + u_n^T G_n u_n + V(x_{n+1}) \right] \tag{4.6.3}$$

is given by

$$V(x_n) = x_n^T K_n x_n. \tag{4.6.4}$$

From (4.6.3) and (4.6.4), we have

$$\begin{aligned}
V(x_n) &= \min_{u_n} \lfloor x_n^T Q_n x_n + u_n^T G_n u_n + (A_n x_n + B_n u_n)^T K_{n+1}(A_n x_n + B_n u_n) \rfloor \\
&= \min_{u_n} \left[x_n^T Q_n x_n + u_n^T G_n u_n + x_n^T A_n^T K_{n+1} A_n x_n + x_n^T A_n^T K_{n+1} B_n u_n \right. \\
&\quad \left. + u_n^T B_n^T K_{n+1} A_n x_n + u_n^T B_n^T K_{n+1} B_n u_n \right].
\end{aligned} \tag{4.6.5}$$

The first-order necessary condition for optimality gives

$$u_n^T G_n + x_n^T A_n^T K_{n+1} B_n + u_n^T B_n^T K_{n+1} B_n = 0,$$

and hence, the optimal controller is expressed as

$$u_n = - \left(G_n + B_n^T K_{n+1} B_n \right)^{-1} B_n^T K_{n+1} A_n x_n. \tag{4.6.6}$$

The second-order necessary condition for optimality is guaranteed because

$$\begin{aligned}
\frac{\partial^2 H(x_n, u_n, V(x_{n+1}))}{\partial u_n \times \partial u_n^T} &= \frac{\partial^2 (u_n^T G_n u_n + u_n^T B_n^T K_{n+1} B_n u_n)}{\partial u_n \times \partial u_n^T} \\
&= 2G_n + 2B_n^T K_{n+1} B_n > 0.
\end{aligned}$$

Plugging the expression for the controller (4.6.6) into (4.6.5), one finds

$$\begin{aligned}
x_n^T K_n x_n &= x_n^T Q_n x_n + x_n^T A_n^T K_{n+1} A_n x_n - x_n^T A_n^T K_{n+1} B_n \left(G_n + B_n^T K_{n+1} B_n \right)^{-1} \\
&\quad \times B_n K_{n+1} A_n x.
\end{aligned}$$

Hence, the difference equation to find the unknown matrix of the quadratic return function (4.6.4) is

$$K_n = Q_n + A_n^T K_{n+1} A_n - A_n^T K_{n+1} B_n \left(G_n + B_n^T K_{n+1} B_n \right)^{-1} B_n K_{n+1} A_n. \tag{4.6.7}$$

If in index (4.6.2) $N = \infty$, we have

$$J = \sum_{n=0}^{\infty} \left[x_n^T Q_n x_n + u_n^T G_n u_n \right], \qquad Q_n \geq 0, \quad G_n > 0.$$

The control law (4.6.6) is expressed as

$$u_n = - \left(G_n + B_n^T K_n B_n \right)^{-1} B_n^T K_n A_n x_n,$$

where the unknown symmetric matrix K_n is found by solving the following non-linear equation:

$$-K_n + Q_n + A_n^T K_n A_n - A_n^T K_n B_n \left(G_n + B_n^T K_n B_n \right)^{-1} B_n K_n A_n = 0, \quad K_n = K_n^T.$$

Matrix K_n is positive-definite, and the MATLAB $dlqr$ solver is used to find the feedback matrix $\left(G_n + B_n^T K_n B_n \right)^{-1} B_n^T K_n A_n$, return function matrix K_n, and eigenvalues. In particular,

```
» help dlqr
  DLQR  Linear-quadratic regulator design for discrete-
      time systems.
      [K,S,E] = DLQR(A,B,Q,R,N)   calculates the optimal
      gain matrix K such that the state-feedback law
      u[n] = -Kx[n]   minimizes the cost function
          J = Sum {x'Qx + u'Ru + 2*x'Nu}
      subject to the state dynamics x[n+1] = Ax[n] + Bu[n].
      The matrix N is set to zero when omitted.   Also
      returned are the Riccati equation solution S and
      the closed-loop eigenvalues E:
                                  -1
      A'SA - S - (A'SB+N)(R+B'SB) (B'SA+N') + Q = 0,
      E = EIG(A-B*K).
```

The closed-loop systems is expressed as

$$x_{n+1} = A_n x_n + B_n u_n \text{ with } u_n = - \left(G_n + B_n^T K_{n+1} B_n \right)^{-1} B_n^T K_{n+1} A_n x_n.$$

That is, $x_{n+1} = \left[A_n - B_n \left(G_n + B_n^T K_{n+1} B_n \right)^{-1} B_n^T K_{n+1} A_n \right] x_n.$

Example 4.6.1.

Consider the second-order discrete-time system

$$x_{n+1} = \begin{bmatrix} x_{1n+1} \\ x_{2n+1} \end{bmatrix} = A_n x_n + B_n u_n = \begin{bmatrix} 1 & 2 \\ 3 & 4 \end{bmatrix} \begin{bmatrix} x_{1n} \\ x_{2n} \end{bmatrix} + \begin{bmatrix} 5 & 6 \\ 7 & 8 \end{bmatrix} \begin{bmatrix} u_{1n} \\ u_{2n} \end{bmatrix}.$$

Find the controller minimizing the performance index

$$
\begin{aligned}
J &= \sum_{n=0}^{\infty} \left[x_n^T Q_n x_n + u_n^T G_n u_n \right] \\
&= \sum_{n=0}^{\infty} \left[\begin{bmatrix} x_{1n} & x_{2n} \end{bmatrix} \begin{bmatrix} 1 & 0 \\ 0 & 1 \end{bmatrix} \begin{bmatrix} x_{1n} \\ x_{2n} \end{bmatrix} + \begin{bmatrix} u_{1n} & u_{2n} \end{bmatrix} \begin{bmatrix} 1 & 0 \\ 0 & 1 \end{bmatrix} \begin{bmatrix} u_{1n} \\ u_{2n} \end{bmatrix} \right] \\
&= \sum_{n=0}^{\infty} \left(x_{1n}^2 + x_{2n}^2 + u_{1n}^2 + u_{2n}^2 \right).
\end{aligned}
$$

We have the following MATLAB script, and the numerical results are displayed:

```
» A=[1 2;3 4];B=[5 6;7 8];Q=eye(size(A));G=eye(size(B));
[KF,Kn,E]=dlqr(A,B,Q,G)
KF =
        0.2557      0.3008
        0.0658      0.1697
Kn =
        2.0154      0.8255
        0.8255      1.6943
E =
        -0.1630
        0.0265
```

Thus, the unknown matrix K_n is

$$
K_n = \begin{bmatrix} 2 & 0.83 \\ 0.83 & 1.7 \end{bmatrix}.
$$

Hence, the controller is found as

$$
u_n = -\left(G_n + B_n^T K_{n+1} B_n \right)^{-1} B_n^T K_{n+1} A_n x = -\begin{bmatrix} 0.26 & 0.3 \\ 0.066 & 0.17 \end{bmatrix} \begin{bmatrix} x_{1n} \\ x_{2n} \end{bmatrix},
$$

$$
u_{1n} = -0.26x_{1n} - 0.3x_{2n}, \quad u_{2n} = -0.066x_{1n} - 0.17x_{2n}.
$$

The system is stable because the eigenvalues -0.16 and 0.02 are within the unit circle.

Example 4.6.2.

Let the matrix difference equation is

$$
x_{n+1} = \begin{bmatrix} x_{1n+1} \\ x_{2n+1} \\ x_{3n+1} \end{bmatrix} = A_n x_n + B_n u_n = \begin{bmatrix} 1 & 2 & 3 \\ 4 & 5 & 6 \\ 7 & 8 & 9 \end{bmatrix} \begin{bmatrix} x_{1n} \\ x_{2n} \\ x_{2n} \end{bmatrix} + \begin{bmatrix} 1 \\ 2 \\ 3 \end{bmatrix} u_n.
$$

Find an optimal controller minimizing the quadratic performance index

$$
\begin{aligned}
J &= \sum_{n=0}^{\infty} \left[x_n^T Q_n x_n + u_n^T G_n u_n \right] \\
&= \sum_{n=0}^{\infty} \left[\begin{bmatrix} x_{1n} & x_{2n} & x_{3n} \end{bmatrix} \begin{bmatrix} 1 & 0 & 0 \\ 0 & 1 & 0 \\ 0 & 0 & 1 \end{bmatrix} \begin{bmatrix} x_{1n} \\ x_{2n} \\ x_{3n} \end{bmatrix} + u_n^2 \right] \\
&= \sum_{n=0}^{\infty} \left(x_{1n}^2 + x_{2n}^2 + x_{3n}^2 + u_n^2 \right).
\end{aligned}
$$

Typing in the Command Window

```
»A=[1 2 3; 4 5 6;7 8 9];B=[1; 2; 3];Q=eye(size(A));G=1;
[KF,Kn,E]=dlqr(A,B,Q,G)
```

the following results are displayed:

```
KF =
    1.9021    2.4470    2.9919
Kn =
   31.4414   18.0738    5.7063
   18.0738   13.7074    7.3410
    5.7063    7.3410    9.9758
E =
   -0.8302
    0.0583
    0.0000
```

Thus,

$$
K_n = \begin{bmatrix} 31.4 & 18.1 & 5.7 \\ 18.1 & 13.7 & 7.3 \\ 5.7 & 7.3 & 10 \end{bmatrix},
$$

and the controller is

$$
u_n = -1.9x_{1n} - 2.4x_{2n} - 3x_{3n}.
$$

The dynamic performance of the closed-loop system, which is stable, is simulated using the `filter` command. In particular, having derived the closed-loop system dynamics in the form of the linear difference equation $x_{n+1} = [A_n - B_n (G_n + B_n^T K_{n+1} B_n)^{-1} B_n^T K_{n+1} A_n] x_n$, one finds the numerator and denominator of the transfer function in the z-domain. We have

```
»A'c=A-B*KF; C=[1 0 0]; H=[0]; [num,den]=ss2tf(A'c,B,C,H);
»k=0:1:20; r=[ones(1,21)]; x=filter(num,den,r);
plot(k,x,'-',k,x,'o',k,r,'+')
```

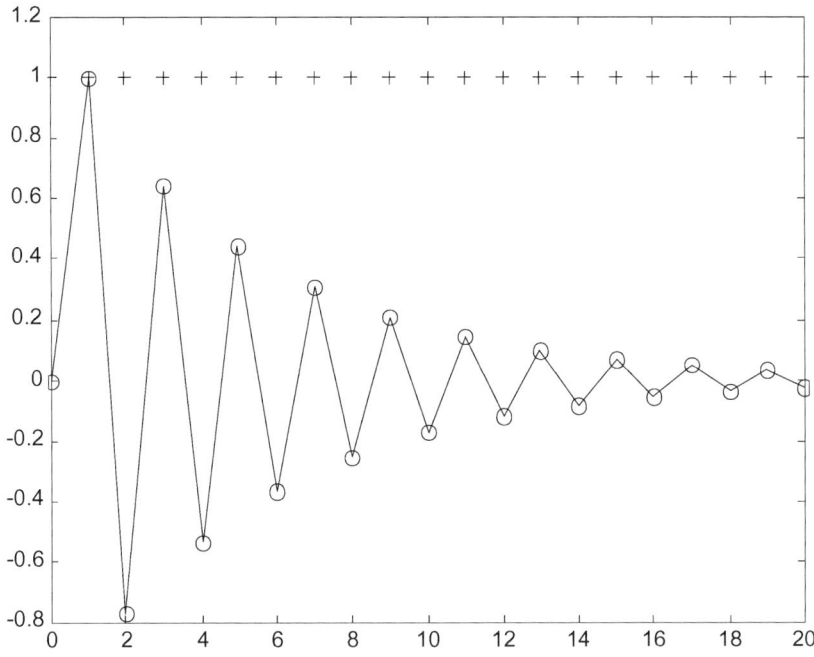

FIGURE 4.6.1.   Output dynamics.

The modeling results (system output "o" and reference r "+") are documented in Figure 4.6.1.

It must be emphasized that the system output does not follow the reference input $r(k)$ because the stabilization problem was solved. It was illustrated that the tracking control problem must be approached and solved to guarantee that the system output follows the reference command.

4.6.2.   *Constrained Optimization of Discrete-Time Systems*

Because of the constraints imposed on control inputs, the designer must synthesize bounded control laws. In this section, the constrained optimization problem for multivariable discrete-time systems is studied, and the constrained digital controllers are synthesized. Our goal is to illustrate that a bounded control law can be analytically designed using the Hamilton–Jacobi theory. The admissibility concept, which is based on Lyapunov's second method, is used to study stability of the resulting closed-loop system. Consider linear discrete-time systems with bounded control, as modeled by

$$x_{n+1} = A_n x_n + B_n u_n, \quad x_{n0} \in X_0, \quad u_n \in U, \quad n \geq 0. \tag{4.6.8}$$

The performance index to be minimized is expressed as

$$J = \sum_{n=0}^{N-1} \left[x_n^T Q_n x_n + 2 \int \left(\phi^{-1}(u_n) \right)^T G_n \, du_n - u_n^T B_n^T K_{n+1} B_n u_n \right]. \quad (4.6.9)$$

Performance indexes must be positive-definite. Therefore, the following inequality must be satisfied $\lfloor x_n^T Q_n x_n + 2 \int \left(\phi^{-1}(u_n) \right)^T G_n \, du_n \rfloor > u_n^T B_n^T K_{n+1} B_n u_n$ for all $x_n \in X$ and $u_n \in U$.

The Hamilton–Jacobi–Bellman recursive equation is

$$V(x_n) = \min_{u_n \in U} \lfloor x_n^T Q_n x_n + 2 \int \left(\phi^{-1}(u_n) \right)^T G_n \, du_n - u_n^T B_n^T K_{n+1} B_n u_n$$

$$+ V(x_{n+1}) \rfloor.$$

Using the quadratic return function (4.6.4) $V(x_n) = x_n^T K_n x_n$, one finds

$$x_n^T K_n x_n = \min_{u_n \in U} \left[x_n^T Q_n x_n + 2 \int \left(\phi^{-1}(u_n) \right)^T G_n \, du_n - u_n^T B_n^T K_{n+1} B_n u_n \right.$$

$$\left. + (A_n x_n + B_n u_n)^T K_{n+1} (A_n x_n + B_n u_n) \right]. \quad (4.6.10)$$

The first- and second-order necessary conditions for optimality are used. From the first-order necessary condition for optimality, we have the bounded controller as

$$u_n = -\phi \left(G_n^{-1} B_n^T K_{n+1} A_n x_n \right), \quad u_n \in U. \quad (4.6.11)$$

It is evident that

$$\frac{\partial^2 \left(2 \int \left(\phi^{-1}(u_n) \right)^T G_n \, du_n \right)}{\partial u_n \times \partial u_n^T}$$

is positive-definite because ϕ and ϕ^{-1} lie in the first and third quadrants, and weighting matrix G_n is positive-definite. One concludes that the second-order necessary condition for optimality is satisfied because

$$\frac{\partial \left(\phi^{-1}(u_n) \right)^T}{\partial u_n} G_n > 0.$$

Using (4.6.10) and (4.6.11), the following equation results:

$$x_n^T K_n x_n = x_n^T Q_n x_n + 2 \int x_n^T A_n^T K_{n+1} B_n \, d \left(\phi \left(G_n^{-1} B_n^T K_{n+1} A_n x_n \right) \right)$$

$$+ x_n^T A_n^T K_{n+1} A_n x_n - 2 x_n^T A_n^T K_{n+1} B_n \phi \left(G_n^{-1} B_n^T K_{n+1} A_n x_n \right),$$

where

$$2 \int x_n^T A_n^T K_{n+1} B_n \, d \left(\phi \left(G_n^{-1} B_n^T K_{n+1} A_n x_n \right) \right)$$

$$= 2 x_n^T A_n^T K_{n+1} B_n \phi \left(G_n^{-1} B_n^T K_{n+1} A_n x_n \right)$$

$$- 2 \int \left(\phi \left(G_n^{-1} B_n^T K_{n+1} A_n x_n \right) \right)^T d \left(B_n^T K_{n+1} A_n x_n \right).$$

Hence, the unknown matrix $K_{n+1} \in \mathbb{R}^{c \times c}$ is found by solving

$$x_n^T K_n x_n = x_n^T Q_n x_n + x_n^T A_n^T K_{n+1} A_n x_n$$
$$-2 \int \left(\phi \left(G_n^{-1} B_n^T K_{n+1} A_n x_n \right) \right)^T d \left(B_n^T K_{n+1} A_n x_n \right). \quad (4.6.12)$$

Mapping the control bounds imposed by the continuous, integrable one-to-one bounded functions $\phi \in U$, one finds the term $2 \int \left(\phi \left(G_n^{-1} B_n^T K_{n+1} A_n x_n \right) \right)^T$ $d \left(B_n^T K_{n+1} A_n x_n \right)$. For example, using the hyperbolic tangent to map the saturation-type constraints, we have

$$\int \tanh z \, dz = \log \cosh z$$

$$\text{and} \int \tanh^g z \, dz = -\frac{\tanh^{g-1} z}{g - 1} + \int \tanh^{g-2} z \, dz, \quad g \neq 1.$$

Solving equation (4.6.12), matrix K_{n+1} is found, and the feedback gains of controller (4.6.11) result.

Minimizing the performance index

$$J = \sum_{n=0}^{\infty} \left[x_n^T Q_n x_n + 2 \int \left(\phi^{-1}(u_n) \right)^T G_n \, du_n - u_n^T B_n^T K_{n+1} B_n u_n \right],$$

one finds the following bounded control law:

$$u_n = -\phi \left(G_n^{-1} B_n^T K_n A_n x_n \right), \quad u_n \in U,$$

where the unknown symmetric matrix K_n is found from

$$x_n^T K_n x_n = x_n^T Q_n x_n + x_n^T A_n^T K_n A_n x_n$$
$$-2 \int \left(\phi \left(G_n^{-1} B_n^T K_n A_n x_n \right) \right)^T d \left(B_n^T K_n A_n x_n \right).$$

The admissibility concept is applied to verify the stability of the resulting closed-loop system if open-loop system (4.6.8) is unstable. For open-loop unstable linear systems (4.6.8) $x_{n+1} = A_n x_n + B_n u_n, x_{n0} \in X_0$ with control bounds $u_n \in U$ bounded controllers do not exist that guarantee stability in the large.

The resulting closed-loop system (4.6.8) with (4.6.11) evolves in X, and

$$\left\{ x_{n+1} = A_n x_n - B_n \phi \left(G_n^{-1} B_n^T K_{n+1} A_n x_n \right), x_{n0} \in X_0 \right\} \in X(X_0, U) \subset \mathbb{R}^c.$$

That is, set X is found by using equations of motion (4.6.8), bounded control law (4.6.11), and the *auxiliary* (initial) conditions $x_{n0} \in X_0$.

A subset of the admissible domain of stability $S \subset \mathbb{R}^c$ is found by using the Lyapunov stability theory. In particular, by applying the criteria imposed on the Lyapunov pair, we have

$$S = \left\{ x_n \in R^c \colon x_{n0} \in X_0, u_n \in U | V(0) = 0, V(x_n) > 0, \Delta V(x_n) < 0, \right.$$
$$\left. \forall x_n \in X(X_0, U) \right\} \subset \mathbb{R}^c.$$

The region of attraction can be studied, and S is an *invariant* domain.

It should be emphasized that the quadratic Lyapunov function is applied, and $V(x_n) = x_n^T K_n x_n$ is positive-definite if $K_n > 0$.

Hence, the first difference, as given by

$$
\begin{aligned}
\Delta V(x_n) = V(x_{n+1}) - V(x_n) = & \, x_n^T A_n^T K_{n+1} A_n x_n \\
& - 2 x_n^T A_n^T K_{n+1} B_n \phi \left(G_n^{-1} B_n^T K_{n+1} A_n x_n \right) \\
& + \phi (G_n^{-1} B_n^T K_{n+1} A_n x_n)^T B_n^T K_{n+1} B_n \phi (G_n^{-1} B_n^T K_{n+1} A_n x_n) - x_n^T K_n x_n,
\end{aligned}
$$

must be negative-definite for all $x_n \in X$ to ensure the stability.

The evolution of the closed-loop system depends on the initial conditions and constraints. In particular, it was illustrated that $X(X_0, U)$. The *sufficiency* analysis of stability is performed studying $S \subset \mathbb{R}^c$ and $X(X_0, U) \subset \mathbb{R}^c$. Stability is guaranteed if $X \subseteq S$.

4.6.3. *Tracking Control of Discrete-Time Systems*

We study the dynamic systems modeled by the following difference equation in matrix form:

$$
x_{n+1}^{\text{system}} = A_n x_n^{\text{system}} + B_n u_n, \quad x_{n0}^{\text{system}} \in X_0, \quad u_n \in U, \quad n \geq 0.
$$

The output equation is

$$
y_n = H_n x_n^{\text{system}}.
$$

Taking note of the *exogeneous* system, we have

$$
x_n^{\text{ref}} = x_{n-1}^{\text{ref}} + r_n - y_n.
$$

Thus, one finds

$$
x_{n+1}^{\text{ref}} = x_n^{\text{ref}} + r_{n+1} - y_{n+1} = x_n^{\text{ref}} + r_{n+1} - H_n \left(A_n x_n^{\text{system}} + B_n u_n \right).
$$

Hence, an augmented model is given by

$$
x_{n+1} = \begin{bmatrix} x_{n+1}^{\text{system}} \\ x_{n+1}^{\text{ref}} \end{bmatrix} = \begin{bmatrix} A_n & 0 \\ -H_n A_n & I_n \end{bmatrix} x_n + \begin{bmatrix} B_n \\ -H_n B_n \end{bmatrix} u_n + \begin{bmatrix} 0 \\ I_n \end{bmatrix} r_{n+1}.
$$

Here, the augmented state vector is

$$
x_n = \begin{bmatrix} x_n^{\text{system}} \\ x_n^{\text{ref}} \end{bmatrix},
$$

and $I_n \in \mathbb{R}^{b \times b}$ is the identity matrix.

To synthesize the bounded controller, the nonquadratic performance index to be minimized is found as

$$
J = \sum_{n=0}^{N-1} \left[x_n^T Q_n x_n + 2 \int \left(\phi^{-1}(u_n) \right)^T G_n \, du_n - u_n^T \begin{bmatrix} B_n \\ -H_n B_n \end{bmatrix}^T \right.
$$

$$
\left. \times K_{n+1} \begin{bmatrix} B_n \\ -H_n B_n \end{bmatrix} u_n \right].
$$

Using the quadratic return function $V(x_n) = x_n^T K_n x_n$, from the Hamilton–Jacobi equation

$$x_n^T K_n x_n = \min_{u_n \in U} \left[x_n^T Q_n x_n + 2 \int \left(\phi^{-1}(u_n) \right)^T G_n \, du_n - u_n^T \begin{bmatrix} B_n \\ -H_n B_n \end{bmatrix}^T \right.$$

$$\times K_{n+1} \begin{bmatrix} B_n \\ -H_n B_n \end{bmatrix} u_n + \left(\begin{bmatrix} A_n & 0 \\ -H_n A_n & I_n \end{bmatrix} x_n + \begin{bmatrix} B_n \\ -H_n B_n \end{bmatrix} u_n \right)^T$$

$$\left. \times K_{n+1} \left(\begin{bmatrix} A_n & 0 \\ -H_n A_n & I_n \end{bmatrix} x_n + \begin{bmatrix} B_n \\ -H_n B_n \end{bmatrix} u_n \right) \right],$$

using the first-order necessary condition for optimality, one obtains the following bounded tracking controller:

$$u_n = -\phi \left(G_n^{-1} \begin{bmatrix} B_n \\ -H_n B_n \end{bmatrix}^T K_{n+1} \begin{bmatrix} A_n & 0 \\ -H_n A_n & I_n \end{bmatrix} x_n \right), \quad u_n \in U.$$

The unknown matrix K_{n+1} is found by solving the following equation:

$$x_n^T K_n x_n = x_n^T Q_n x_n + x_n^T \begin{bmatrix} A_n & 0 \\ -H_n A_n & I_n \end{bmatrix} K_{n+1} \begin{bmatrix} A_n & 0 \\ -H_n A_n & I_n \end{bmatrix} x_n$$

$$-2 \int \left(\phi \left(G_n^{-1} \begin{bmatrix} B_n \\ -H_n B_n \end{bmatrix}^T K_{n+1} \begin{bmatrix} A_n & 0 \\ -H_n A_n & I_n \end{bmatrix} x_n \right) \right)^T$$

$$\times d \left(\begin{bmatrix} B_n \\ -H_n B_n \end{bmatrix}^T K_{n+1} \begin{bmatrix} A_n & 0 \\ -H_n A_n & I_n \end{bmatrix} x_n \right).$$

Example 4.6.3. Tracking control of an aircraft.

The reported constrained optimization technique is applied to an unstable multi-input/multi-output model of a twin-tail fighter, which is described by

$$x_{n+1}^{air} = A_n x_n^{air} + B_n u_n.$$

The aircraft state variables are

$$x_n^{air} = \begin{bmatrix} v & \alpha & q & \theta & \beta & p & r & \phi & \psi \end{bmatrix}^T,$$

where v is the forward velocity, α is the angle-of-attack, q is the pitch rate, θ is the pitch angle, β is the sideslip angle, p is the roll rate, r is the yaw rate, ϕ is the roll angle, and ψ is the yaw angle.

Six control inputs are bounded because of the mechanical limits imposed, and the aircraft control vector is

$$u_n = \begin{bmatrix} \delta_{HR} & \delta_{HL} & \delta_{FR} & \delta_{FL} & \delta_C & \delta_R \end{bmatrix}^T,$$

where δ_{HR} and δ_{HL} are the deflections of the right and left horizontal stabilizers, δ_{FR} and δ_{FL} are the deflections of the right and left flaps, and δ_C and δ_R are the canard and rudder deflections.

The following hard bounds (mechanical limits) on the deflections of control surfaces are imposed $|\delta_{HR}, \delta_{HL}| \leq 0.44$ rad, $|\delta_{FR}, \delta_{FL}| \leq 0.35$ rad, $|\delta_C| \leq 0.47$ rad, and $|\delta_R| \leq 0.52$ rad.

For the aircraft studied, we have the following constant-coefficient matrices:

$$A_n = \begin{bmatrix}
0.9995 & 0.2457 & -0.0273 & -0.2885 & -0.0391 & -0.0075 & -0.002 & 0 & 0 \\
-0.0001 & 0.9663 & 0.0291 & 0 & 0.0011 & 0.0017 & 0.0003 & 0 & 0 \\
0 & 0.1135 & 0.9765 & 0 & 0.0007 & 0.0002 & 0.0012 & 0 & 0 \\
0 & 0.0017 & 0.0296 & 1 & 0 & 0 & 0 & 0 & 0 \\
-0.0001 & 0.0048 & 0.0011 & 0.0288 & 0.977 & 0.0037 & -0.0268 & 0 & 0 \\
0.0001 & 0.0165 & 0.0005 & -0.0198 & -1.3515 & 0.8977 & 0.025 & 0 & 0 \\
0 & -0.0268 & 0.0015 & 0.0041 & 0.2716 & -0.0003 & 0.9811 & 0 & 0 \\
0 & 0.0003 & 0 & -0.0002 & -0.0207 & 0.0285 & 0.0003 & 1 & 0 \\
0 & -0.0004 & 0 & 0 & 0.0041 & 0 & 0.0297 & 0 & 1
\end{bmatrix},$$

$$B_n = \begin{bmatrix}
0.071 & 0.0067 & -0.011 & -0.0117 & -0.0004 & -0.0001 \\
-0.009 & -0.0088 & -0.0091 & -0.009 & 0 & 0 \\
-0.2818 & -0.2819 & -0.0746 & -0.0746 & 0 & 0 \\
-0.0042 & -0.0042 & -0.0011 & -0.0011 & 0 & 0 \\
-0.001 & 0.0007 & -0.0006 & 0.0005 & 0.0123 & 0.0018 \\
-0.0822 & 0.0819 & -0.0881 & 0.0879 & 0.0122 & 0.025 \\
0.0921 & -0.0924 & 0.0233 & -0.0232 & 0.02 & -0.0132 \\
-0.0013 & 0.013 & -0.013 & 0.013 & 0.0002 & 0.0004 \\
0.0014 & -0.0014 & 0.0003 & -0.0003 & 0.0003 & -0.0002
\end{bmatrix}.$$

The corresponding eigenvalues of A_n are

$$1, 1, 0.965 \pm 0.1i, 0.91, 1.03, 1 \pm 0.006i, 0.93.$$

The open-loop system is unstable. The tracking control problem has to be solved, and the output equation should be introduced. The aircraft outputs are the Euler angles θ, ϕ, and ψ. That is,

$$y_n = H_n x_n^{air} = \begin{bmatrix}
0 & 0 & 0 & 1 & 0 & 0 & 0 & 0 & 0 \\
0 & 0 & 0 & 0 & 0 & 0 & 0 & 1 & 0 \\
0 & 0 & 0 & 0 & 0 & 0 & 0 & 0 & 1
\end{bmatrix} \begin{bmatrix} v \\ \alpha \\ q \\ \theta \\ \beta \\ p \\ r \\ \phi \\ \psi \end{bmatrix}.$$

We can express the tracking error vector in terms of the reference inputs (desired Euler's angles are denoted by $r_n = \begin{bmatrix} r_\theta \\ r_\phi \\ r_\psi \end{bmatrix}$) and the aircraft output vector $y_n = H_n x_n^{air}$.

Hence, defining

$$x_n^{ref} = x_{n-1}^{ref} + r_n - y_n,$$

we have

$$x_{n+1}^{ref} = x_n^{ref} + r_{n+1} - y_{n+1} = x_n^{ref} + r_{n+1} - H_n \left(A_n x_n^{air} + B_n u_n \right).$$

The augmented model

$$x_{n+1} = \begin{bmatrix} x_{n+1}^{air} \\ x_{n+1}^{ref} \end{bmatrix} = \begin{bmatrix} A_n & 0 \\ -H_n A_n & I_n \end{bmatrix} x_n + \begin{bmatrix} B_n \\ -H_n B_n \end{bmatrix} u_n + \begin{bmatrix} 0 \\ I_n \end{bmatrix} r_{n+1}.$$

is used to solve the tracking control problem.

Here, the augmented state vector is

$$x_n = \begin{bmatrix} x_n^{air} \\ x_n^{ref} \end{bmatrix};$$

$I_n \in \mathbb{R}^{3 \times 3}$ is the identity matrix.

A bounded controller is synthesized by minimizing the following nonquadratic performance index:

$$J = \sum_{n=0}^{\infty} \left[x_n^T Q_n x_n + 2 \int \left(\tanh^{-1} \left(U_{max}^{-1} u_n \right) \right)^T G_n \, du_n - u_n^T \begin{bmatrix} B_n \\ -H_n B_n \end{bmatrix}^T \right.$$

$$\left. \times K_{n+1} \begin{bmatrix} B_n \\ -H_n B_n \end{bmatrix} u_n \right],$$

where

$$U_{max} = \begin{bmatrix} \delta_{HR\,max} & 0 & 0 & 0 & 0 & 0 \\ 0 & \delta_{HL\,max} & 0 & 0 & 0 & 0 \\ 0 & 0 & \delta_{FR\,max} & 0 & 0 & 0 \\ 0 & 0 & 0 & \delta_{FL\,max} & 0 & 0 \\ 0 & 0 & 0 & 0 & \delta_{C\,max} & 0 \\ 0 & 0 & 0 & 0 & 0 & \delta_{R\,max} \end{bmatrix},$$

$\delta_{HR\,max} = \delta_{HL\,max} = 0.44$ rad, $\delta_{FR\,max} = \delta_{FL\,max} = 0.35$ rad, $\delta_{C\,max} = 0.47$ rad, and $\delta_{R\,max} = 0.52$ rad.

It is easy to see that the imposed control bounds (mechanical limits on the deflections of control surfaces) are mapped by the hyperbolic tangent.

Minimizing the Hamilton–Jacobi equation, one has

$$u_n = -U_{max} \tanh \left(G_n^{-1} \begin{bmatrix} B_n \\ -H_n B_n \end{bmatrix}^T K_n \begin{bmatrix} A_n & 0 \\ -H_n A_n & I_n \end{bmatrix} x_n \right).$$

Using identity matrices $I_{q1n} \in \mathbb{R}^{9 \times 9}$, $I_{q2n} \in \mathbb{R}^{3 \times 3}$, and $I_{gn} \in \mathbb{R}^{6 \times 6}$, we assign the following weighting matrices:

$$Q_n = \begin{bmatrix} I_{q1n} & 0 \\ 0 & 500 I_{q2n} \end{bmatrix} \quad \text{and} \quad G_n = I_{gn}.$$

The equation

$$
\begin{aligned}
x_n^T K_n x_n = {} & x_n^T Q_n x_n + x_n^T \begin{bmatrix} A_n & 0 \\ -H_n A_n & I_n \end{bmatrix}^T K_n \begin{bmatrix} A_n & 0 \\ -H_n A_n & I_n \end{bmatrix} x_n \\
& -2 \int \left(\tanh \left(G_n^{-1} \begin{bmatrix} B_n \\ -H_n B_n \end{bmatrix}^T K_n \begin{bmatrix} A_n & 0 \\ -H_n A_n & I_n \end{bmatrix} x_n \right) \right)^T \\
& \times d \left(\begin{bmatrix} B_n \\ -H_n B_n \end{bmatrix}^T K_n \begin{bmatrix} A_n & 0 \\ -H_n A_n & I_n \end{bmatrix} x_n \right)
\end{aligned}
$$

was solved to find $K_n \in \mathbb{R}^{12 \times 12}$.

The nonquadratic index minimized is positive-definite because

$$
\left[x_n^T Q_n x_n + 2 \int \left(\tanh^{-1} \left(U_{\max}^{-1} u_n \right) \right)^T G_n \, du_n \right] > u_n^T \begin{bmatrix} B_n \\ -H_n B_n \end{bmatrix}^T K_n \begin{bmatrix} B_n \\ -H_n B_n \end{bmatrix} u_n
$$

holds for all $x_n \in X$ and $u_n \in U$.

Using the formula

$$
\frac{\partial \left(\tanh^{-1} u_n \right)}{\partial u_n} = \frac{1}{1 - u_n^2},
$$

one concludes that the second-order necessary condition for optimality is guaranteed.

The sufficient conditions, as given by

$$
V(x_n) > 0 \text{ and } \Delta V(x_n) < 0,
$$

are satisfied in the studied flight operating envelope, $X \subseteq S$ for all $x_n \in X$ and $u_n \in U$.

The aircraft dynamics is studied because the capabilities of a bounded controller must be verified via simulations. Figure 4.6.2 illustrates the fighter output dynamics when the reference inputs (Euler's angles in radians) are assigned to be

$$
r_n = \begin{bmatrix} r_\theta \\ r_\phi \\ r_\psi \end{bmatrix} = \begin{cases} 0.6, & n \geq 0 \\ 0.2, & n \geq 0 \\ 1.5, & n \geq 0 \end{cases} \text{ or } r_t = \begin{bmatrix} r_\theta \\ r_\phi \\ r_\psi \end{bmatrix} = \begin{cases} 0.6, & t \geq 0 \\ 0.2, & t \geq 0 \\ 1.5, & t \geq 0. \end{cases}
$$

The initial conditions are

$$
y_0 = \begin{bmatrix} \theta_0 \\ \phi_0 \\ \psi_0 \end{bmatrix} = \begin{cases} 0, & t = 0 \\ 1, & t = 0 \\ 0, & t = 0. \end{cases}
$$

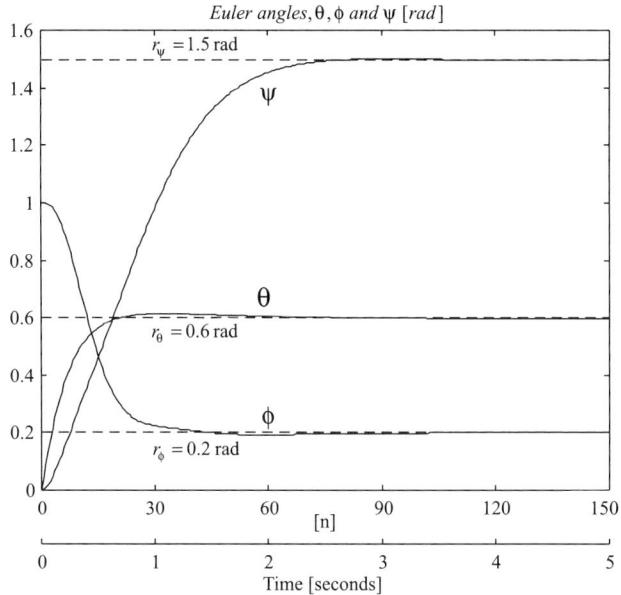

FIGURE 4.6.2. Fighter's output dynamics: $34°$, $11°$, and $86°$ maneuver.

5

Analysis, Identification, and Control of Nonlinear Dynamic Systems

5.1. Nonlinear Analysis of Dynamic Systems

The general results in stability theory were developed by the Ukrainian scientist A. Lyapunov (he was a Professor of Mechanical Engineering at the Kharkov University). Lyapunov formulated a basic concept of stability and derived sufficient stability conditions for dynamic systems described by ordinary differential equations. In recent years, there has been considerable activity in the application of the second method, and innovative methods were developed. The application of efficient and flexible software, such as MATLAB, has changed the aim of the Lyapunov stability theory, and the second method is rarely used for derivation of solutions. Nonlinear differential equations can be solved, and the value of Lyapunov's theory is shifted from the descriptive theory of solutions of differential equations to convergence and stability analysis as well as to design of stabilizing controllers. The Lyapunov concept is a viable tool in solution of stabilization and optimization problems, encountered in the wide areas of control. In particular, the second method has been applied and extensively used in control, diagnostics, identification, and model reduction of nonlinear systems to analyze different algorithms, as well as to study and visualize the qualitative features of the theoretical developments. For example, control and optimization methods employ the sufficient conditions for stability, as given by the second method, and the design techniques are based on the solution of Lyapunov-type equations or inequalities. Further studies are needed to make the so-called Lyapunov second method available for a wide range of applications and to demonstrate additional capabilities and practical benefits of the Lyapunov theory. The main reason and basic motivation of this section is to apply the Lyapunov concept to nonlinear dynamic systems. It is demonstrated that the application of nonquadratic Lyapunov candidates is an important step to using the full potential of the second method. These allow one to cope with more difficult problems in analysis of nonlinear systems. Nonlinear time-invariant systems were considered in Chapter 4.4 (see Section 4.4.2), where the definitions and the Lyapunov theorem were stated.

We consider a nonlinear time-varying system

$$\dot{x}(t) = F(t, x), t \geq 0 \text{ with } x(t_0) = x_0. \qquad (5.1.1)$$

For time-varying systems, we have the following theorem to study the stability.

Theorem 5.1.1. *Consider the dynamic system described by the nonlinear differential equation (5.1.1). If a positive-definite scalar Lyapunov function $V(t, x)$ exists with continuous first-order partial derivatives with respect to t and x, the equilibrium state is:*

- *Stable if the total derivative of the positive-definite function $V(t, x) > 0$ is*

$$\frac{dV}{dt} = \frac{\partial V}{\partial t} + \left(\frac{\partial V}{\partial x}\right)^T \frac{dx}{dt} = \frac{\partial V}{\partial t} + \left(\frac{\partial V}{\partial x}\right)^T F(t, x) \leq 0.$$

- *Uniformly stable if the total derivative of the positive-definite decreasing function $V(t, x) > 0$ is*

$$\frac{dV}{dt} = \frac{\partial V}{\partial t} + \left(\frac{\partial V}{\partial x}\right)^T \frac{dx}{dt} = \frac{\partial V}{\partial t} + \left(\frac{\partial V}{\partial x}\right)^T F(t, x) \leq 0.$$

- *Uniformly asymptotically stable in the large if the total derivative of $V(t, x) > 0$ is negative definite; that is,*

$$\frac{dV}{dt} = \frac{\partial V}{\partial t} + \left(\frac{\partial V}{\partial x}\right)^T \frac{dx}{dt} = \frac{\partial V}{\partial t} + \left(\frac{\partial V}{\partial x}\right)^T F(t, x) < 0.$$

- *Exponentially stable in the large if exist the K_∞-functions $\rho_1(\cdot)$ and $\rho_2(\cdot)$, and the K-function $\rho_3(\cdot)$ such that*

$$\rho_1(\|x\|) \leq V(t, x) \leq \rho_2(\|x\|) \quad and \quad \frac{dV(x)}{dt} \leq -\rho_3(\|x\|).$$

We will use the K-, KL-, and K_∞-functions.

Definition 5.1.1. Function $\rho_K(\cdot)$: $\mathbb{R}_{\geq 0} \to \mathbb{R}_{\geq 0}$, $\rho_K(0) = 0$ is said to be a K-function if it is continuous and strictly increasing; $\rho_{K\infty}(\cdot)$: $\mathbb{R}_{\geq 0} \to \mathbb{R}_{\geq 0}$, $\rho_{K\infty}(0) = 0$, is a K_∞-function if $\rho_{K\infty}$ is of class K, and $\rho_{K\infty}(q) \to \infty$ as $q \to \infty$; $\rho_{KL}(\cdot)$: $\mathbb{R}_{\geq 0} \times \mathbb{R}_{\geq 0} \to \mathbb{R}_{\geq 0}$ is called a KL-function if for fixed t, mapping $\rho_{KL}(\cdot, t)$ is of class K, and $\rho_{KL}(q, t) \to 0$ as $t \to \infty$.

Example 5.1.1.

Consider the second-order time-invariant system

$$\dot{x}_1(t) = -x_1^5 - x_1 x_2^2,$$
$$\dot{x}_2(t) = -x_1 - x_2^5, \quad t \geq 0.$$

A scalar positive-definite function is chosen in the quadratic form. That is,

$$V(x_1, x_2) = \tfrac{1}{2} \left(x_1^2 + x_2^2\right).$$

One must find the expression for the total derivative, and we have

$$\frac{dV(x_1, x_2)}{dt} = \left(\frac{\partial V}{\partial x}\right)^T \frac{dx}{dt} = \left(\frac{\partial V}{\partial x}\right)^T F(x) = \frac{\partial V}{\partial x_1}\left(-x_1^5 - x_1 x_2^2\right)$$

$$+ \frac{\partial V}{\partial x_2}\left(-x_1 - x_2^5\right) = -x_1^4 - x_1^2 x_2^2 - x_1 x_2 - x_2^6.$$

Thus, the total derivative of the $V(x_1, x_2) = \frac{1}{2}\left(x_1^2 + x_2^2\right)$ is negative definite, and

$$V(x_1, x_2) = \frac{1}{2}\left(x_1^2 + x_2^2\right) > 0, \quad \frac{dV(x_1, x_2)}{dt} < 0.$$

These allow one to conclude that the equilibrium state is uniformly asymptotically stable.

Example 5.1.2.

Consider the time-varying system given as

$$\dot{x}_1(t) = -x_1 + x_2^3,$$
$$\dot{x}_2(t) = -e^{-10t} x_1 x_2^2 - 5x_2 - x_2^3, \quad t \geq 0.$$

A scalar positive-definite function is designed as

$$V(t, x_1, x_2) = \frac{1}{2}\left(x_1^2 + e^{10t} x_2^2\right).$$

The total derivative is found as

$$\frac{dV(t, x_1, x_2)}{dt} = \frac{\partial V}{\partial t} + \left(\frac{\partial V}{\partial x}\right)^T \frac{dx}{dt} = \frac{\partial V}{\partial t} + \left(\frac{\partial V}{\partial x}\right)^T F(t, x)$$

$$= \frac{\partial V}{\partial t} + \frac{\partial V}{\partial x_1}\left(-x_1 + x_2^3\right) + \frac{\partial V}{\partial x_2}\left(-e^{-10t} x_1 x_2^2 - 5x_2 - x_2^3\right)$$

$$= -x_1^2 - e^{10t} x_2^4.$$

That is, $V(t, x_1, x_2) = \frac{1}{2}\left(x_1^2 + e^{10t} x_2^2\right) > 0$ and

$$\frac{dV(x_1, x_2)}{dt} = -x_1^2 - e^{10t} x_2^4 < 0.$$

Thus, the equilibrium state is uniformly asymptotically stable.

The Lyapunov stability theory is widely applied in analysis and design of nonlinear systems. It was emphasized that stability plays a major role in control system design, stabilization, and optimization. Innovative Lyapunov candidates can be applied in stability analysis and in design of stabilizing controllers.

Consider dynamic systems given by the nonlinear constant-coefficient equation

$$\dot{x}(t) = F(x), \, x(t_0) = x_0, \, t \geq 0. \tag{5.1.2}$$

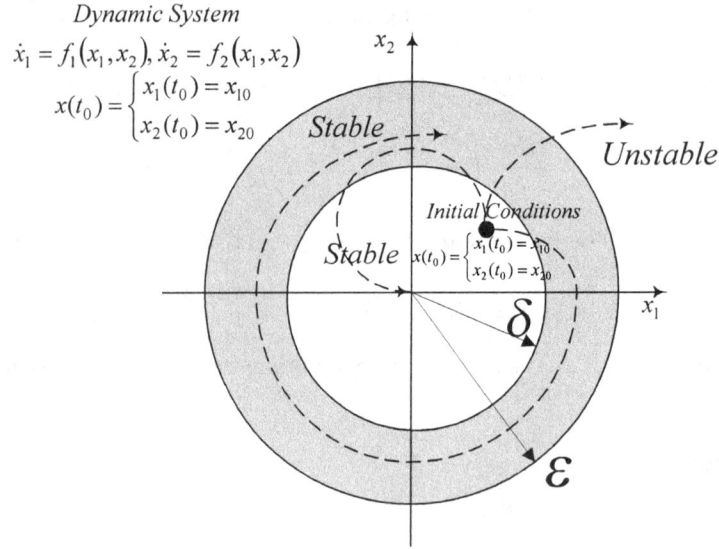

FIGURE 5.1.1.   System trajectories.

Stability or instability of the origin may be global or local. The concept of asymptotic stability should be extended to apply to a specified region $X \subset \mathbb{R}^n$ about the origin rather than to the origin. The system evolution can be illustrated using n-dimensional surfaces or plots. For second-order dynamic systems, the system trajectories are shown in Figure 5.1.1.

Consider the dynamic systems with unstable equilibrium (origin) and asymptotically stable periodic solutions (limit cycles). Our goal is to establish the sufficient conditions that allow us to study the system dynamics. In analysis of stability, *the set-based concept* can be used, and it is essential to formulate the following lemma.

Lemma 5.1.1. *For differential equation (5.1.2), assume existence, uniqueness, and continuous dependence of its solution $x(\cdot)$: $[t_0, \infty) \to \mathbb{R}^n$ from initial conditions in $X_0 = \{x_0 \in \mathbb{R}^n\} \subset \mathbb{R}^n$. A sufficient condition for asymptotic stability is the existence of a continuous real-valued Lipschitzian scalar C^κ ($\kappa \geq 1$) function of the state vector $V(\cdot): \mathbb{R}^n \to \mathbb{R}_{\geq 0}$ with continuous first partial derivatives such that for all $x \in S \subset \mathbb{R}^n$, except $x \in Z \subset \mathbb{R}^n$, the following criteria:*

$$V(x) = 0, \forall x = 0; \; V(x) \to \infty, \forall x \to \infty; \; V(x) > 0$$

$$\text{and } \frac{dV(x)}{dt} < 0, \forall x \neq 0 \qquad (5.1.3)$$

are guaranteed. If $V(\cdot): \mathbb{R}^n \to \mathbb{R}_{\geq 0}$ does not satisfy stability conditions (5.1.3) in the set Z, the equilibrium point is unstable in Z. Let S be the largest invariant set, S is bounded, and $Z \subset S$, $X_0 \subset S$. Then, the dynamic system (5.1.2) is unstable *in the small and asymptotically stable in the large. If $S \subseteq Z$ or $S \subset Z$, then the system is unstable.*

Example 5.1.3.

Study the trajectories of the system

$$\dot{x}_1(t) = x_1 + x_2 - x_1(x_1^2 + x_2^2),$$
$$\dot{x}_2(t) = -x_1 + x_2 - x_2(x_1^2 + x_2^2) + u, \quad -2 \le u \le 2. \quad (5.1.4)$$

By using the positive-definite function, as given by

$$V(x_1, x_2) = \tfrac{1}{2}(x_1^2 + x_2^2),$$

which is defined and differentiable, for $u = 0$ (open-loop system), we find

$$\frac{dV(x_1, x_2)}{dt} = (x_1^2 + x_2^2 - 1)(-x_1^2 - x_2^2).$$

Equation $x_1^2 + x_2^2 = 1$ is an equation of the unit circle centered at the origin with center $C(0, 0)$. One concludes that

$$\frac{dV(x_1, x_2)}{dt} > 0$$

inside the circle (the interior domain of the unit circle satisfies $x_1^2 + x_2^2 < 1$);

$$\frac{dV(x_1, x_2)}{dt} = 0$$

at the set of all points in a plane whose distance from center $C(0, 0)$ is 1;

$$\frac{dV(x_1, x_2)}{dt} < 0$$

outside the circle (the exterior domain of the unit circle is described by $x_1^2 + x_2^2 > 1$). Therefore, the open-loop system has a unique closed trajectory; that is, a stable periodic solution toward a unique circle exists. The open-loop system (5.1.4) is stable.

Let us verify the existence of a limit cycle by solving (5.1.4). The polar coordinates are used. The rectangular coordinates (x_1, x_2) and polar coordinates (r, θ) are related by

$$x_1 = r \cos \theta, \, x_2 = r \sin \theta, \text{ and } r^2 = x_1^2 + x_2^2, \theta = \tan^{-1}\left(\frac{x_2}{x_1}\right).$$

By using (5.1.4), it is easy to find two differential equations in polar coordinates. In particular,

$$\dot{r}(t) = r(1 - r^2), \dot{\theta}(t) = -1.$$

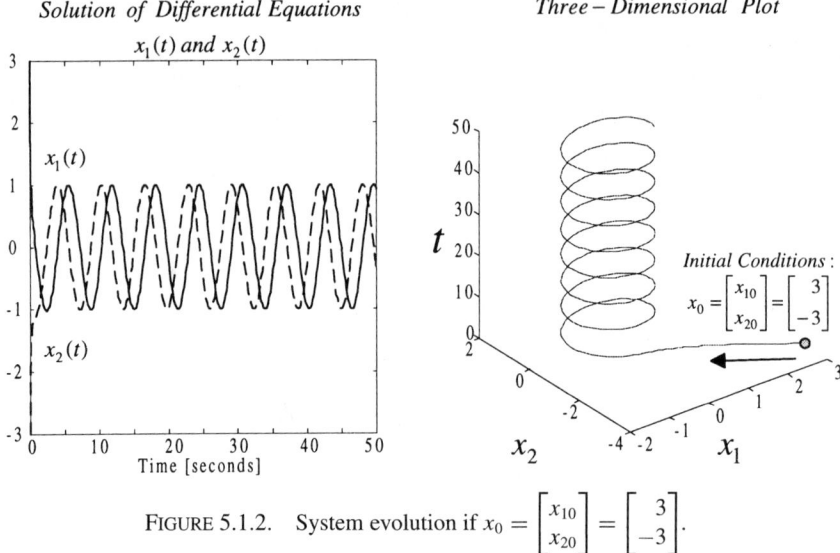

FIGURE 5.1.2.   System evolution if $x_0 = \begin{bmatrix} x_{10} \\ x_{20} \end{bmatrix} = \begin{bmatrix} 3 \\ -3 \end{bmatrix}$.

By separating variables, one obtains

$$r(t) = \frac{1}{\sqrt{1 - ae^{-2t}}}$$

and $\theta(t) = -t + t_0$. Constant a depends on initial conditions. Hence, the corresponding trajectories spiral toward the unit circle from its interior domain or exterior domain as $t \to \infty$ if $a > 0$ or $a < 0$.

Numerical results have been performed. Simulations for different initial conditions are shown in Figures 5.1.2 and 5.1.3.

Bounded control algorithms will be designed in this chapter in the following sections using the Hamilton–Jacobi and Lyapunov theories. For the system under consideration, the control is bounded, and $-2 \leq u \leq 2$. Using the results of Chapter 4, we have

$$u = -\phi \left(B^T \frac{\partial V}{\partial x} \right).$$

Making use of $V(x_1, x_2) = \frac{1}{2}(k_{11}x_1^2 + k_{22}x_2^2)$, we apply

$$u = -\text{sat}_{-2}^{+2} \left(\begin{bmatrix} 0 \\ 1 \end{bmatrix}^T \frac{\partial V}{\partial x} \right) = -\text{sat}_{-2}^{+2}(5x_1 + 5x_2).$$

The evolution of the closed-loop system is illustrated in Figure 5.1.4.

With all formulations and properties thus stated, the scheme to solve the problem of stability analysis is in place. The application of the second method of Lyapunov

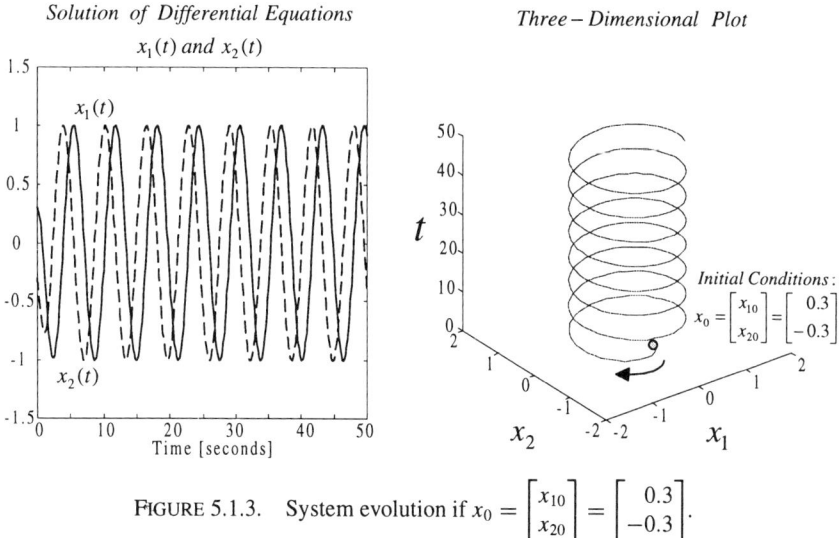

FIGURE 5.1.3.   System evolution if $x_0 = \begin{bmatrix} x_{10} \\ x_{20} \end{bmatrix} = \begin{bmatrix} 0.3 \\ -0.3 \end{bmatrix}$.

involves the design of $V(x)$. Theorem 5.1.1 and Lemma 5.1.1 do not contain the procedure for how to find the appropriate Lyapunov candidates. Though, it is not easy to design a suitable function, it is not a serious disadvantage because various methods are available for synthesis of Lyapunov functions using nonlinear dynamics, and a great deal of candidates are applied *a priori*, assuming the specified form of $V(x)$. One can express Lyapunov candidates as continuous C^κ functions using

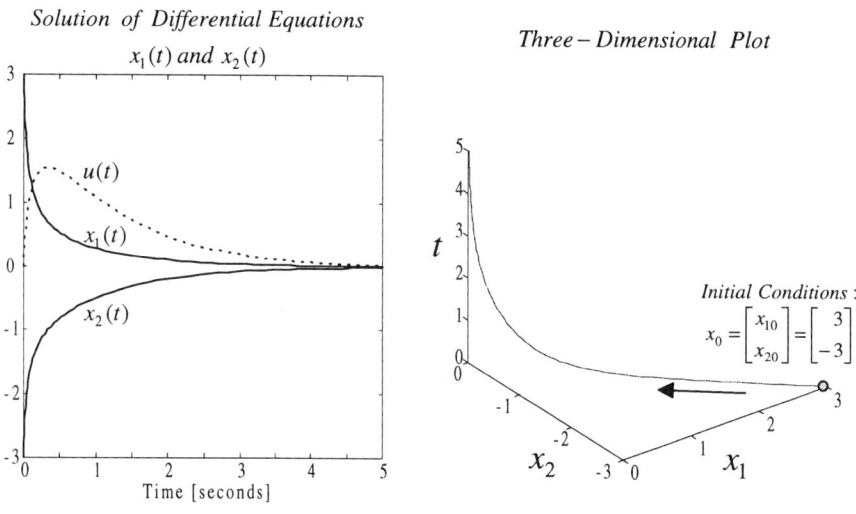

FIGURE 5.1.4.   Closed-loop system dynamics if $x_0 = \begin{bmatrix} x_{10} \\ x_{20} \end{bmatrix} = \begin{bmatrix} 3 \\ -3 \end{bmatrix}$.

quadratic and nonquadratic matrix-product forms. For example, the Lyapunov candidates can be found as

$$V(x) = F^T(x)F(x).$$

We introduce

$$V(x) = \sum_{i=0}^{\eta} \frac{2\gamma + 1}{2(i + \gamma + 1)} \left(x^{\frac{i+\gamma+1}{2\gamma+1}} \right)^T K_i(t) x^{\frac{i+\gamma+1}{2\gamma+1}}, \qquad (5.1.5)$$

where $K_i(\cdot) \in \mathbb{R}^{c \times c}$, $\eta = 0, 1, 2, \ldots$, and $\gamma = 0, 1, 2, \ldots$.

By making the use of (5.1.5), one can find, for example,

$$V(x) = \tfrac{1}{2} x^T K_0 x, \ K_0(\cdot) \in \mathbb{R}^{c \times c} (\text{for } \eta = 0 \text{ and } \gamma = 0)$$

or

$$V(x) = \tfrac{1}{2} x^T K_0 x + \tfrac{1}{4} (x^2)^T K_1 x^2, \ K_0(\cdot) \in \mathbb{R}^{c \times c}$$
$$\text{and } K_1(\cdot) \in \mathbb{R}^{c \times c} \ (\text{for } \eta = 1 \text{ and } \gamma = 1).$$

It was emphasized that the application of the Lyapunov second method involves the design of a continuously differentiable and positive-definite scalar functions specifically chosen to study the stability of dynamic systems. Nonquadratic functions

$$V(x) = \tfrac{1}{3} |x^T| K_2 x^T K_0 x,$$

or

$$V(x) = \tfrac{1}{3} |x^T| K_2 x^T K_0 x + \tfrac{1}{5} |x^T| K_2 (x^2)^T K_1 x^2, \ K_2(\cdot) \in \mathbb{R}^{c \times 1},$$

are suitable and manageable in stability analysis of nonlinear systems.

In general, we have

$$V(x) = f_{|x|}(|x|) \sum_{i=0}^{\eta} \frac{2\gamma + 1}{2(i + \gamma + 1)} \left(x^{\frac{i+\gamma+1}{2\gamma+1}} \right)^T K_i(t) x^{\frac{i+\gamma+1}{2\gamma+1}}. \qquad (5.1.6)$$

Positive-definite scalar C^κ functions $V(x)$, as defined by (5.1.5) and (5.1.6), are continuously differentiable. For analysis of stability, it is necessary to examine the total derivative of the Lyapunov candidate along the trajectories of (5.1.2). Quadratic functions $V(x)$ are commonly used, whereas nonquadratic candidates (5.1.5) and (5.1.6) have not been widely applied. Let us demonstrate the application of nonquadratic and quadratic Lyapunov candidates.

Example 5.1.4.

For a dynamic system

$$\dot{x}_1(t) = -x_1 + x_2,$$
$$\dot{x}_2(t) = -x_1 - x_2 - x_2|x_2|,$$

we apply the positive-definite scalar Lyapunov candidate as given by

$$V(x_1, x_2) = x_1^2 - x_1 x_2 + 2x_2^2 + \tfrac{1}{3} x_2^2 |x_2|.$$

This real-valued, real-analytic function is continuously differentiable. Then, one obtains

$$\frac{dV(x_1, x_2)}{dt} = 2x_1 \dot{x}_1 - x_2 \dot{x}_1 - x_1 \dot{x}_2 + 4x_2 \dot{x}_2 + x_2 |x_2| \dot{x}_2$$

$$= -x_1^2 - x_2^2 (5 + 5|x_2| + x_2^2).$$

Hence,

$$\frac{dV(x_1, x_2)}{dt} < 0$$

for all $x \in \mathbb{R}^2$. By applying the sufficient conditions for stability, one concludes that the equilibrium state is uniformly asymptotically stable, and $V(x_1, x_2) = x_1^2 - x_1 x_2 + 2x_2^2 + \tfrac{1}{3} x_2^2 |x_2|$ is a Lyapunov function.

One can us apply the quadratic Lyapunov candidate

$$V(x_1, x_2) = \tfrac{1}{2}(x_1^2 + x_2^2).$$

The total derivative is found as

$$\frac{dV(x_1, x_2)}{dt} = x_1 \dot{x}_1 + x_2 \dot{x}_2 = -x_1^2 - x_2^2 (1 + |x_2|).$$

That is, $V(x_1, x_2) > 0$ and

$$\frac{dV(x_1, x_2)}{dt} < 0.$$

The equilibrium state is uniformly asymptotically stable, and the quadratic function $V(x_1, x_2) = \tfrac{1}{2}(x_1^2 + x_2^2)$ is the Lyapunov function.

The present status of computer resources and available software indicates that nonlinear differential equations can be easily solved. However, many problems in analysis and control of nonlinear dynamic systems are formulated in terms of stability, optimality, dynamic performance, accuracy, as well as other descriptive properties. Therefore, the application of cornerstone theories, such as the Lyapunov stability theory, is extremely important.

5.2.   State-Space Nonlinear Identification

Advances in computers and digital signal processors, as well as state-of-the-art software development tools and hardware platforms, have stimulated basic research in real-time nonlinear identification of dynamic systems. There is a need to develop computationally efficient and robust identification and design methods to ensure real-time implementation capabilities. The major goal of this section is

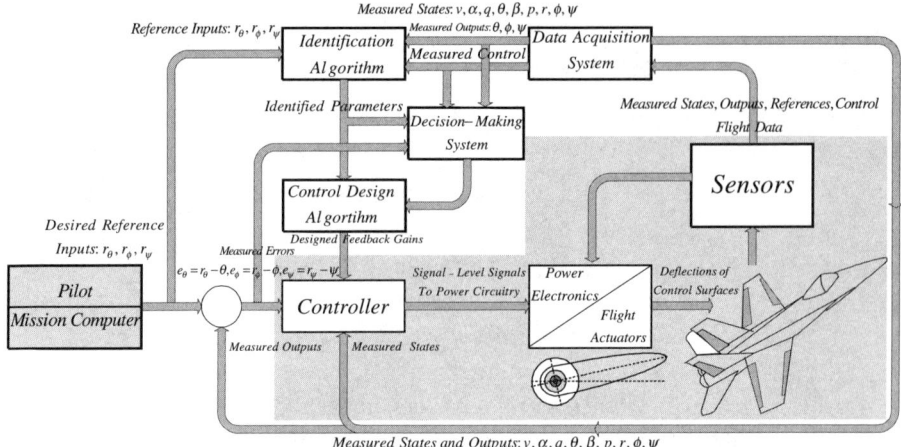

FIGURE 5.2.1.   Real-time identification and control of aircraft: closed-loop configuration.

to approach and solve the nonlinear identification problem to attack and solve a spectrum of extremely important problems in nonlinear analysis and optimization of dynamic systems. These problems are encountered in a wide range of dynamic systems. For example, flight dynamics and control problems need to be solved to control current and next-generation advanced aircraft. High-performance aircraft must satisfy the required performances in the specified operating flight envelope at various attitudes and velocities, and high-angle-of-attack regimes. To improve mission effectiveness, ensure the specified flying and handling qualities (agility, maneuverability, controllability, and other pilotage quantities), guarantee surviv-ability, damage adaptation and recovery, completely automated flight management systems must be designed with diagnostics and failure accommodation features (the so-called self-repairing flight control). That is, a real-time motion control problem must be solved to attain the desired aircraft performance in the expand flight envelopes. To solve the motion control problem, nonlinear identification must be performed. The flight management system must guarantee the required aircraft performance specifications, as tailored as necessary to perform the speci-fied mission and other objectives for a fully functional and damaged/crippled/failed aircraft, by solving the motion control problem which integrates real-time iden-tification, control design (with real-time redesign capabilities), and controller re-configuration. By using this concept, airframe damage and surface and hardware failures can be identified and accommodated through controller redesign and re-configuration. The high-level closed-loop flight management system, described in Chapter 2, is shown in Figure 5.2.1. For rigid-body aircraft, the nine state vari-ables are v (forward velocity), α (angle of attack), q (pitch rate), θ (pitch angle), β (sideslip angle), p (roll rate), r (yaw rate), ϕ (roll angle), and ψ (yaw angle). The reference inputs are the desired Euler angles (denoted as r_θ, r_ϕ, and r_ψ), and correspondingly, the outputs are the Euler angles θ, ϕ, and ψ.

Real-time identification (performed periodically within allowable rate), control redesign, and controller reconfiguration (which can be integrated in advanced flight management systems) enable the aircraft to (among other things)

- Execute perfectly coordinated longitudinal and lateral directional maneuvers
- Improve flying and handling qualities
- Increase agility and maneuverability
- Ensure reliability and survivability
- Guarantee controllability and failure accommodation
- Attain damage adaptation and recovery via adaptation and redesign

The aircraft equations of motion are derived using Newton's law or the Lagrange equations of motion. That is, the mathematical model in the form of nonlinear differential equations is available, and the problem is to identify the unknown parameters (coefficients of differential equations) to perform diagnostics, optimization, and control using updated (periodically identified) model parameters.

5.2.1. Least-Squares Identification

5.2.1.1. Identification of Linear Systems

Linear single-input dynamic systems. As the starting point, we consider the identification problem for a linear single-input dynamic system as described by the following difference equation in matrix form:

$$x(i) = F_m x(i - 1) + G_m u(i - 1). \tag{5.2.1}$$

Our goal is to identify the unknown matrices $F_m \in \mathbb{R}^{k \times k}$ and $G_m \in \mathbb{R}^{k \times 1}$.

The identification process uses a least-squares fit on recent measurements to obtain the unknown parameters in (5.2.1). That is, the least-squares estimation is used to obtain the vector of estimated parameters p in F_m and G_m. We calculate

$$p_n = \left(\Phi_{mn}^T \Phi_{mn} \right)^{-1} \Phi_{mn}^T x_m(i) \hat{=} \Psi_n x_m(i)$$

from measurements in

$$x_m(i) = \Phi p, \; x_m \in R^{nk \times 1}.$$

Here, Φ is the regression matrix and $x_m(i)$ is the vector of measured states.

When $(k + 1)$ measurements are incorporated into the identification, the measurement vector becomes

$$x_m(i, k) = \Phi_x f_m + \Phi_u g_m,$$

where

$$x_m(i, k) = \begin{bmatrix} x_1(i, k) \\ \vdots \\ x_n(i, k) \end{bmatrix}, \quad x_1(i, k) = \begin{bmatrix} x_1(i + 1) \\ \vdots \\ x_1(i + k) \end{bmatrix}, \quad x_n(i, k) = \begin{bmatrix} x_n(i + 1) \\ \vdots \\ x_n(i + k) \end{bmatrix},$$

$$\Phi_x = \begin{bmatrix} \Phi_{x1} & \cdots & 0 \\ \vdots & \vdots & \vdots \\ 0 & \cdots & \Phi_{xn} \end{bmatrix} \in R^{nk \times nn}, \quad \Phi_u = \begin{bmatrix} \Phi_{u1} & \cdots & 0 \\ \vdots & \vdots & \vdots \\ 0 & \cdots & \Phi_{un} \end{bmatrix} \in R^{nk \times nl}$$

$$\Phi_{x1} = \Phi_{xn} = [x_1(i, k-1) \cdots x_n(i, k-1)] \in R^{k \times n},$$
$$\Phi_{u1} = \Phi_{un} = [u_1(i, k-1) \cdots u_1(i, k-1)] \in R^{k \times l},$$
$$f_m = [f_1 \cdots f_n]^T \in R^{nn \times 1}, f_1 = [F_{11} \cdots F_{1n}]^T, f_n = [F_{n1} \cdots F_{nn}]^T,$$
$$g_m = [g_1 \cdots g_n]^T \in R^{nl \times 1}, g_1 = [G_{11} \cdots G_{1l}]^T, g_n = [G_{n1} \cdots G_{nl}]^T.$$

The unknown parameters are found as

$$p_1 = \Psi_1 \begin{bmatrix} f_1 \\ g_1 \end{bmatrix} \in R^{(n+l) \times 1}, \ldots, p_n = \Psi_n \begin{bmatrix} f_n \\ g_n \end{bmatrix} \in R^{(n+l) \times 1}, \Psi_1 = \Psi_n.$$

With this reordering, the regression matrix Φ becomes matrix diagonal. Rows in F_m and G_m are identified independently by Ψ_n. One obtains F_m and G_m, and identification is completed.

Example 5.2.1. Identification of the fighter model in the longitudinal axis

A third-order model of the aircraft is given by the following equation:

$$\begin{bmatrix} \dfrac{dv}{dt} \\ \dfrac{d\alpha}{dt} \\ \dfrac{dq}{dt} \end{bmatrix} = \begin{bmatrix} -0.041 & 18 & -35 \\ -0.0003 & -0.74 & 1 \\ -0.0002 & 4.6 & -0.95 \end{bmatrix} \begin{bmatrix} v \\ \alpha \\ q \end{bmatrix} + \begin{bmatrix} 0.86 \\ -0.21 \\ -19 \end{bmatrix} \delta.$$

Using the sampling period 0.01 sec, let us find the difference equations in the form of $x(i) = F_m x(i-1) + G_m u(i-1)$, and using the modeling results obtained, identify the "unknown" parameters. As the system parameters will be identified, the linear quadratic regulator is designed and the eigenvalues of the closed-loop system are found.

The matrices of the difference equations are found using the MATLAB command c2d, and the script is given below.

MATLAB script c_5_2_11.m

```
% Matrices A and B of the F-18 fighter in the
% longitudinal axis
% Continuous-time model
A=[-0.041   18     -35;
    -0.0003 -0.74   1;
    -0.0002  4.6   -0.95 ]
B=[0.86; -0.21; -19]
```

```
% Sampling period
Ts=0.01
% Matrices Fm (An) and Gm (Bn) of the discrete-time model
[Fm,Gm]=c2d(A,B,Ts)
disp('End')
```

The numerical results are

```
» format short e
» c\`5\`2\`11
A =
  -4.1000e-002   1.8000e+001  -3.5000e+001
  -3.0000e-004  -7.4000e-001   1.0000e+000
  -2.0000e-004   4.6000e+000  -9.5000e-001
B =
   8.6000e-001
  -2.1000e-001
  -1.9000e+001
Ts =
   1.0000e-002
Fm =
   9.9959e-001   1.7131e-001  -3.4740e-001
  -2.9985e-006   9.9286e-001   9.9171e-003
  -2.0589e-006   4.5616e-002   9.9077e-001
Gm =
   4.1500e-002
  -3.0372e-003
  -1.8916e-001
```

That is, we have the following discrete-time model:

$$
\begin{bmatrix} v(i) \\ \alpha(i) \\ q(i) \end{bmatrix} = \begin{bmatrix} 0.999 & 0.171 & -0.347 \\ 0 & 0.993 & 0.0099 \\ 0 & 0.046 & 0.991 \end{bmatrix} \begin{bmatrix} v(i-1) \\ \alpha(i-1) \\ q(i-1) \end{bmatrix}
$$
$$
+ \begin{bmatrix} 0.0415 \\ -0.003 \\ -0.189 \end{bmatrix} \delta(i-1).
$$

This discrete-time model of the aircraft is used to develop the SIMULINK diagram; see Figure 5.2.2. An input (forcing function) is

$$
\delta(i) = Kx(i) + \xi(i)
$$

where $\xi(i)$ consists of five normally distributed random values with a standard deviation of 0.1.

The evolution of the state variables and forcing function (control input) are shown in Figure 5.2.3.

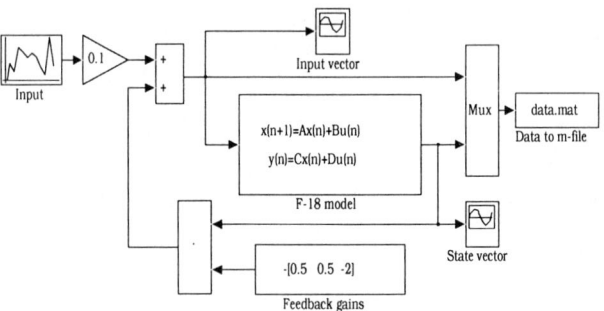

FIGURE 5.2.2.   SIMULINK diagram (c_5_2_12.mdl).

Let us identify the "unknown" parameters using the simulation results.   In particular, the parameters to be identified are

$$F_m = \begin{bmatrix} f_{vv} & f_{v\alpha} & f_{vq} \\ f_{\alpha v} & f_{\alpha\alpha} & f_{\alpha q} \\ f_{qv} & f_{q\alpha} & f_{qq} \end{bmatrix}, \qquad G_m = \begin{bmatrix} g_v \\ g_\alpha \\ g_q \end{bmatrix}$$

Hence, $f_m = [f_{vv} \; f_{v\alpha} \; f_{vq} \; f_{\alpha v} \; f_{\alpha\alpha} \; f_{\alpha q} \; f_{qv} \; f_{q\alpha} \; f_{qq}]^T$ and $g_m = [g_v \; g_\alpha \; g_q]^T$. For five samples, we have

$$x_{mv}(i+1) = \begin{bmatrix} v(1) \\ v(2) \\ v(3) \\ v(4) \\ v(5) \end{bmatrix} = \begin{bmatrix} v(0) & \alpha(0) & q(0) \\ v(1) & \alpha(1) & q(1) \\ v(2) & \alpha(2) & q(2) \\ v(3) & \alpha(3) & q(3) \\ v(4) & \alpha(4) & q(4) \end{bmatrix} \begin{bmatrix} f_{vv} \\ f_{v\alpha} \\ f_{vq} \end{bmatrix} + \begin{bmatrix} \delta(0) \\ \delta(1) \\ \delta(2) \\ \delta(3) \\ \delta(4) \end{bmatrix} g_v$$

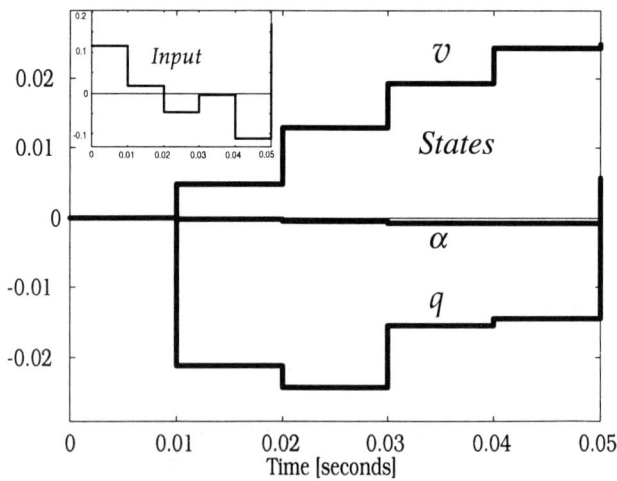

FIGURE 5.2.3.   Aircraft longitudinal dynamics.

Therefore,

$$\begin{bmatrix} v(1) \\ v(2) \\ v(3) \\ v(4) \\ v(5) \end{bmatrix} = \begin{bmatrix} v(0) & \alpha(0) & q(0) & \delta(0) \\ v(1) & \alpha(1) & q(1) & \delta(1) \\ v(2) & \alpha(2) & q(2) & \delta(2) \\ v(3) & \alpha(3) & q(3) & \delta(3) \\ v(4) & \alpha(4) & q(4) & \delta(4) \end{bmatrix} \begin{bmatrix} f_{vv} \\ f_{v\alpha} \\ f_{vq} \\ g_v \end{bmatrix} = \Phi_v \begin{bmatrix} f_{vv} \\ f_{v\alpha} \\ f_{vq} \\ g_v \end{bmatrix}.$$

Similar formulas may be written for other rows. The values of Φ are the same for each row of F_m and G_m. However, the values of x_m, f_m, and g_m are changed. Thus, Ψ_n is calculated only once for each identification. The following MATLAB m-file was used to determine the model parameters and perform the linear-quadratic-regulator design.

MATLAB script c_5_2_13.m

```
% Simulink data are used for identification of the
% "unknown" parameters of the discrete-time F-18 model
% We use data.mat
% Data arrays are t (time), u (input), x(state variables)
% Identification
load data; data=data';
u=data(:,2); x=data(:,3:5);
n=length(u)-1;
phi=[x(1:(n-1),1) x(1:(n-1),2) x(1:(n-1),3) u(1:(n-1))];
psi=inv(phi'*phi)*phi';
fm1=psi*x(2:n,1); fm2=psi*x(2:n,2); fm3=psi*x(2:n,3);
% Identified matrices of the difference equation
fm=[fm1(1:3)';fm2(1:3)';fm3(1:3)']
gm=[fm1(5.2.24) fm2(5.2.24) fm3(5.2.24)]'
% Design of the LQR controller
% Weighting matrices
Qn=[1 0 0;0 1 0;0 0 1]; Gn=1;
% Feedback gains kf, K-matrix, and eigenvalues eig
[kfeedback,K,eig]=dlqr(fm,gm,Qn,Gn)
disp('End')
```

The following numerical results are displayed:

```
» c˙5˙2˙13
fm =
    9.9960e-001   1.7130e-001  -3.4740e-001
   -7.1886e-011   9.9290e-001   9.9300e-003
    3.1373e-009   4.5600e-002   9.9080e-001
gm =
    4.1500e-002
   -3.0000e-003
```

```
 -1.8190e-001
kfeedback =
  8.2171e-001   7.1147e-001 -1.8526e+000
K =
  6.9257e+000   1.1334e+001 -5.2628e+000
  1.1334e+001   2.5665e+002 -6.8910e+000
 -5.2628e+000  -6.8910e+000  1.1771e+001
eig =
  8.0849e-001 +1.2439e-001i
  8.0849e-001 -1.2439e-001i
  9.9738e-001
End
```

That is, the parameters of the fighter model, identified by the algorithm de-scribed, are identical to the original system parameters. In particular, we obtained

$$F_m = \begin{bmatrix} 0.999 & 0.171 & -0.347 \\ 0 & 0.993 & 0.0099 \\ 0 & 0.046 & 0.991 \end{bmatrix} \quad \text{and} \quad G_m = \begin{bmatrix} 0.0415 \\ -0.003 \\ -0.189 \end{bmatrix}.$$

It is evident that the identified parameters are identical to the aircraft parameters used in simulation.

The feedback control law (optimal linear-quadratic-regulator) is found as

$$u(i) = -0.82v(i) - 0.71\alpha(i) + 1.8q(i).$$

The closed-loop system is stable, and eigenvalues are

$$0.808 \pm 0.1244i \quad \text{and} \quad 0.99738.$$

Nonlinear multi-input/multi-output dynamic systems. Let us outline the state-space identification algorithm to find the unknown coefficients of the discrete-time system

$$x_{k+1} = Fx_k + Gu_k, k > 0. \tag{5.2.2}$$

The identification uses a least-squares fit on recent measurements of inputs and states to obtain the coefficients of matrices F and G. These coefficients are calculated row-by-row to achieve efficiency. For an n-order system with m-inputs, a single row of (5.2.2) for x_j is defined as

$$x_{j,k+1} = \begin{bmatrix} f_{j,1} \cdots f_{j,n} \end{bmatrix} x_k + \begin{bmatrix} g_{j,1} \cdots g_{j,m} \end{bmatrix} u_k \equiv F_j x_k + G_j u_k = x_k^T F_1^T + u_k^T G_1^T$$

$$= \begin{bmatrix} x_k^T u_k^T \end{bmatrix} \begin{bmatrix} F_1 \\ G_1 \end{bmatrix} = \begin{bmatrix} x_k^T u_k^T \end{bmatrix} \hat{p}_j, \quad \hat{p}_j = \begin{bmatrix} F_1 \\ G_1 \end{bmatrix}. \tag{5.2.3}$$

The $(z + 1)$ input and states measurements are written as

$$x_{j,z} = \begin{bmatrix} x_{j,1} \\ \vdots \\ x_{j,z} \end{bmatrix} = \begin{bmatrix} x_0^T & u_0^T \\ \vdots & \vdots \\ x_{z-1}^T & u_{z-1}^T \end{bmatrix} \hat{p}_j \equiv \Phi_m \hat{p}_j. \tag{5.2.4}$$

Note that the regression matrix Φ_m remains the same for all rows. The coefficients in the jth row are identified using the following equation:

$$\hat{p}_j = \left(\Phi_m^T \Phi_m\right)^{-1} \Phi_m^T x_{j,z}. \tag{5.2.5}$$

Example 5.2.2. Identification of the multi-input/multi-output fighter

An augmented, longitudinal–lateral aircraft dynamics is modeled by a set of nine linear differential equations, as given in the matrix state-space form, as

$$\dot{x}(t) = Ax + Bu,$$

where the matrices A and B are unknown.

The states are

$$x(t) = [v(t)\ \alpha(t)\ q(t)\ \theta(t)\ \beta(t)\ p(t)\ r(t)\ \phi(t)\ \psi(t)]^T,$$

where v is the forward velocity, α is the angle of attack, q is the pitch rate, θ is the pitch angle, β is the sideslip angle, p is the roll rate, r is the yaw rate, ϕ is the roll angle; and ψ is the yaw angle.

Six control inputs are defined as

$$u(t) = [\delta_{HR}(t)\ \delta_{HL}(t)\ \delta_{FR}(t)\ \delta_{FL}(t)\ \delta_C(t)\ \delta_R(t)]^T,$$

where δ_{HR} and δ_{HL} are the deflections of the right and left horizontal stabilizers, δ_{FR} and δ_{FL} are the deflections of the right and left flaps, and δ_C and δ_R are the canard and rudder deflections.

The corresponding discrete-time model in

$$x_{k+1} = Fx_k + Gu_k$$

results.

One does not have coefficients of matrices A and B, as well as F and G, which must be identified using the flight data. These data can be easily downloaded.

To demonstrate and visualize the analysis, modeling, discretization, control, and identification, we perform the modeling to obtain the data used in identification. That is, the fighter is simulated to perform and validate the identification algorithm. We assign

$$A = \begin{bmatrix}
-0.016 & 8.4 & -0.9 & -9.6 & -1.5 & -0.27 & -0.086 & 0 & 0 \\
-0.003 & -1.2 & 1 & 0 & 0.08 & 0.062 & 0.009 & 0 & 0 \\
-0.0001 & 3.9 & -0.85 & 0 & 0.017 & 0.0038 & 0.04 & 0 & 0 \\
0 & 0 & 1 & 0 & 0 & 0 & 0 & 0 & 0 \\
-0.003 & 0.15 & 0.02 & 0.97 & -0.56 & 0.13 & -0.91 & 0 & 0 \\
0.00001 & 0.71 & 0.03 & 0.01 & -48 & -3.5 & 0.22 & 0 & 0 \\
0.00001 & -0.94 & 0.06 & 0.005 & 9.2 & -0.028 & -0.51 & 0 & 0 \\
0 & 0 & 0 & 0 & 0 & 1 & 0 & 0 & 0 \\
0 & 0 & 0 & 0 & 0 & 0 & 1 & 0 & 0
\end{bmatrix},$$

$$
B = \begin{bmatrix}
0.12 & 0.12 & -0.38 & -0.38 & 0 & 0 \\
-0.16 & -0.16 & -0.27 & -0.27 & 0 & 0 \\
-9.5 & -9.5 & -2.5 & -2.5 & 0 & 0 \\
0 & 0 & 0 & 0 & 0 & 0 \\
0.019 & -0.019 & -0.001 & 0.001 & 0.42 & 0.053 \\
-2.9 & 2.9 & -3.1 & 3.1 & 0.73 & 0.92 \\
3.1 & -3.1 & 0.78 & -0.78 & 0.61 & -0.45 \\
0 & 0 & 0 & 0 & 0 & 0 \\
0 & 0 & 0 & 0 & 0 & 0
\end{bmatrix}.
$$

Then, for the sampling period 0.03 sec, one obtains matrices F and G of (5.2.2) as

$$
F = \begin{bmatrix}
0.9995 & 0.2457 & -0.0273 & -0.2885 & -0.0391 & -0.0075 & -0.002 & 0 & 0 \\
-0.0001 & 0.9663 & 0.0291 & 0 & 0.0011 & 0.0017 & 0.0003 & 0 & 0 \\
0 & 0.1135 & 0.9765 & 0 & 0.0007 & 0.0002 & 0.0012 & 0 & 0 \\
0 & 0.0017 & 0.0296 & 1 & 0 & 0 & 0 & 0 & 0 \\
-0.0001 & 0.0048 & 0.0011 & 0.0288 & 0.977 & 0.0037 & -0.0268 & 0 & 0 \\
0.0001 & 0.0165 & 0.0005 & -0.0198 & -1.3515 & 0.8977 & 0.025 & 0 & 0 \\
0 & -0.0268 & 0.0015 & 0.0041 & 0.2716 & -0.0003 & 0.9811 & 0 & 0 \\
0 & 0.0003 & 0 & -0.002 & -0.0207 & 0.0285 & 0.0003 & 1 & 0 \\
0 & -0.0004 & 0 & 0 & 0.0041 & 0 & 0.0297 & 0 & 1
\end{bmatrix},
$$

$$
G = \begin{bmatrix}
0.071 & 0.0067 & -0.011 & -0.0117 & -0.0004 & -0.0001 \\
-0.009 & -0.0088 & -0.0091 & -0.009 & 0 & 0 \\
-0.2818 & -0.2819 & -0.0746 & -0.0746 & 0 & 0 \\
-0.0042 & -0.0042 & -0.0011 & -0.0011 & 0 & 0 \\
-0.001 & 0.0007 & -0.0006 & 0.0005 & 0.0123 & 0.0018 \\
-0.0822 & 0.0819 & -0.0881 & 0.0879 & 0.0122 & 0.025 \\
0.0921 & -0.0924 & 0.0233 & -0.0232 & 0.02 & -0.0132 \\
-0.0013 & 0.013 & -0.013 & 0.013 & 0.0002 & 0.0004 \\
0.0014 & -0.0014 & 0.0003 & -0.0003 & 0.0003 & -0.0002
\end{bmatrix}.
$$

The controller is designed, minimizing the quadratic performance cost, and the following m-file is used:

MATLAB script c_5_2_21.m

```
% Discretize open-loop continuous model of the fighter
% Find feedback gains by solving the LQR problem
% Generate the random input vector
% A, B, C, and D matrices are
a=[-0.016   8.40 -0.90 -9.600 -1.500 -0.270 -0.086  0   0;
   -0.003 -1.20  1.00      0  0.080  0.062  0.009  0   0;
   -0.0001  3.90 -0.85      0  0.017 0.0038  0.040  0   0;
        0     0  1.00      0      0      0      0   0   0;
   -0.003  0.15  0.02  0.970 -0.560  0.130 -0.910  0   0;
  -0.00001  0.71  0.03  0.010 -48.00 -3.500  0.220  0   0;
   0.00001 -0.94  0.06  0.005  9.200 -0.028 -0.510  0   0;
        0     0     0      0      0  1.000      0   0   0;
        0     0     0      0      0      0  1.000   0   0]
b=[0.120   0.120 -0.380 -0.380      0      0;
  -0.160  -0.160 -0.270 -0.270      0      0;
  -9.500  -9.500 -2.500 -2.500      0      0;
       0       0      0      0      0      0;
   0.019  -0.019 -0.001  0.001   0.42  0.053;
  -2.900   2.900 -3.100  3.100   0.73  0.920;
```

```
       3.100 -3.100  0.780 -0.780  0.61 -0.450;
           0      0      0      0      0      0;
           0      0      0      0      0      0]
c=eye(size(a))
d=zeros(9,6)
% Sampling time
ts=0.03;
% F and G matrices of the discretize fighter model
[f,g]=c2d(a,b,ts)
% Weighting matrices
Q=eye(size(f)); R=0.1*eye(size(g,2));
% LQR design to find the feedback gain matrix
k=dlqr(f,g,Q,R)
% Random input vector
ns=0.01*randn(30,6);
t=(0:ts:29*ts)';
nst=[t ns]
save f g c d ts k nst
```

By running this file, the following data are displayed in the Command Window:

```
» c`5`2`21
a =
Columns 1 through 7
 -0.0160   8.4000  -0.9000  -9.6000  -1.5000  -0.2700  -0.0860
 -0.0030  -1.2000   1.0000        0   0.0800   0.0620   0.0090
 -0.0001   3.9000  -0.8500        0   0.0170   0.0038   0.0400
       0        0   1.0000        0        0        0        0
 -0.0030   0.1500   0.0200   0.9700  -0.5600   0.1300  -0.9100
 -0.0000   0.7100   0.0300   0.0100 -48.0000  -3.5000   0.2200
  0.0000  -0.9400   0.0600   0.0050   9.2000  -0.0280  -0.5100
       0        0        0        0        0   1.0000        0
       0        0        0        0        0        0   1.0000
  Columns 8 through 9
       0        0
       0        0
       0        0
       0        0
       0        0
       0        0
       0        0
       0        0
       0        0
b =
  0.1200   0.1200  -0.3800  -0.3800        0        0
 -0.1600  -0.1600  -0.2700  -0.2700        0        0
 -9.5000  -9.5000  -2.5000  -2.5000        0        0
       0        0        0        0        0        0
```

```
   0.0190   -0.0190   -0.0010    0.0010    0.4200    0.0530
  -2.9000    2.9000   -3.1000    3.1000    0.7300    0.9200
   3.1000   -3.1000    0.7800   -0.7800    0.6100   -0.4500
        0         0         0         0         0         0
        0         0         0         0         0         0
```

c =

```
   1    0    0    0    0    0    0    0    0
   0    1    0    0    0    0    0    0    0
   0    0    1    0    0    0    0    0    0
   0    0    0    1    0    0    0    0    0
   0    0    0    0    1    0    0    0    0
   0    0    0    0    0    1    0    0    0
   0    0    0    0    0    0    1    0    0
   0    0    0    0    0    0    0    1    0
   0    0    0    0    0    0    0    0    1
```

d =

```
   0    0    0    0    0    0
   0    0    0    0    0    0
   0    0    0    0    0    0
   0    0    0    0    0    0
   0    0    0    0    0    0
   0    0    0    0    0    0
   0    0    0    0    0    0
   0    0    0    0    0    0
   0    0    0    0    0    0
```

f =

Columns 1 through 7

```
   0.9995    0.2457   -0.0273   -0.2885   -0.0391   -0.0075   -0.0020
  -0.0001    0.9663    0.0291    0.0000    0.0011    0.0017    0.0003
  -0.0000    0.1135    0.9765    0.0000    0.0007    0.0002    0.0012
  -0.0000    0.0017    0.0296    1.0000    0.0000    0.0000    0.0000
  -0.0001    0.0048    0.0011    0.0288    0.9770    0.0037   -0.0268
   0.0001    0.0165    0.0005   -0.0198   -1.3515    0.8977    0.0250
  -0.0000   -0.0268    0.0015    0.0041    0.2716   -0.0003    0.9811
   0.0000    0.0003    0.0000   -0.0002   -0.0207    0.0285    0.0003
  -0.0000   -0.0004    0.0000    0.0000    0.0041   -0.0000    0.0297
```

Columns 8 through 9

```
        0         0
        0         0
        0         0
        0         0
        0         0
        0         0
        0         0
   1.0000         0
        0    1.0000
```

g =

```
   0.0071    0.0067   -0.0110   -0.0117   -0.0004   -0.0001
  -0.0090   -0.0088   -0.0091   -0.0090    0.0000    0.0000
```

```
-0.2818   -0.2819   -0.0746   -0.0746    0.0000   -0.0000
-0.0042   -0.0042   -0.0011   -0.0011    0.0000   -0.0000
-0.0010    0.0007   -0.0006    0.0005    0.0123    0.0018
-0.0822    0.0819   -0.0881    0.0879    0.0122    0.0250
 0.0921   -0.0924    0.0233   -0.0232    0.0199   -0.0132
-0.0013    0.0013   -0.0013    0.0013    0.0002    0.0004
 0.0014   -0.0014    0.0003   -0.0003    0.0003   -0.0002
k=
  Columns 1 through 7
 0.8396    1.2504   -1.2709   -4.7677   -3.8662   -0.0376    2.3431
 0.8017    1.7038   -1.2351   -3.2587    2.5403    0.0677   -2.2410
-1.7648   -4.0645   -0.1646    4.8461    5.6774   -1.3692   -0.8052
-1.9362   -4.3957   -0.1782    4.0928   -5.9501    1.4482    0.9161
-0.1595    0.0396    0.0530    1.9041    7.7161   -0.2409   -0.4246
-0.0355    0.0133    0.0085    0.3023    0.6529    0.1699   -0.2498
  Columns 8 through 9
    0.4225    1.3482
   -0.2125   -1.6201
   -1.8057   -0.4750
    1.8635    0.1293
   -0.9027    1.4192
    0.0908    0.0267
nst =
      0   -0.0046   -0.0063    0.0049   -0.0103   -0.0090    0.0081
 0.0300    0.0037   -0.0233   -0.0001    0.0024    0.0014    0.0064
 0.0600    0.0073   -0.0123   -0.0028   -0.0126   -0.0014    0.0131
 0.0900    0.0211    0.0106    0.0128   -0.0035   -0.0116    0.0033
 0.1200   -0.0136   -0.0011    0.0186   -0.0094    0.0118   -0.0067
 0.1500   -0.0102    0.0038   -0.0052   -0.0117   -0.0002   -0.0015
 0.1800    0.0104    0.0094    0.0010   -0.0102    0.0054   -0.0245
 0.2100   -0.0039   -0.0212   -0.0081   -0.0040   -0.0072    0.0047
 0.2400   -0.0138   -0.0064    0.0068    0.0017   -0.0066    0.0012
 0.2700    0.0032   -0.0070   -0.0236   -0.0012    0.0031   -0.0059
 0.3000    0.0155   -0.0102    0.0099    0.0106    0.0011   -0.0065
 0.3300    0.0071   -0.0018    0.0022   -0.0025    0.0185   -0.0108
 0.3600    0.0196    0.0152    0.0026   -0.0152   -0.0028   -0.0005
 0.3900    0.0050   -0.0004    0.0121    0.0001    0.0221    0.0038
 0.4200    0.0186    0.0123   -0.0027    0.0007    0.0151   -0.0033
 0.4500   -0.0034   -0.0070   -0.0013    0.0032   -0.0195   -0.0050
 0.4800   -0.0114    0.0001   -0.0127    0.0050   -0.0168   -0.0004
 0.5100   -0.0021   -0.0078   -0.0166    0.0128   -0.0057   -0.0017
 0.5400    0.0119    0.0059   -0.0070   -0.0055   -0.0019   -0.0096
 0.5700   -0.0112   -0.0025    0.0028    0.0026    0.0001    0.0129
 0.6000    0.0064    0.0048   -0.0054   -0.0001    0.0084    0.0044
 0.6300   -0.0060    0.0067   -0.0133   -0.0058   -0.0072    0.0128
 0.6600    0.0055   -0.0008    0.0107    0.0214   -0.0072   -0.0050
 0.6900   -0.0110    0.0089   -0.0071   -0.0026   -0.0020   -0.0112
 0.7200    0.0009    0.0231   -0.0001   -0.0141   -0.0002    0.0081
 0.7500   -0.0200    0.0052   -0.0000    0.0177    0.0028    0.0004
```

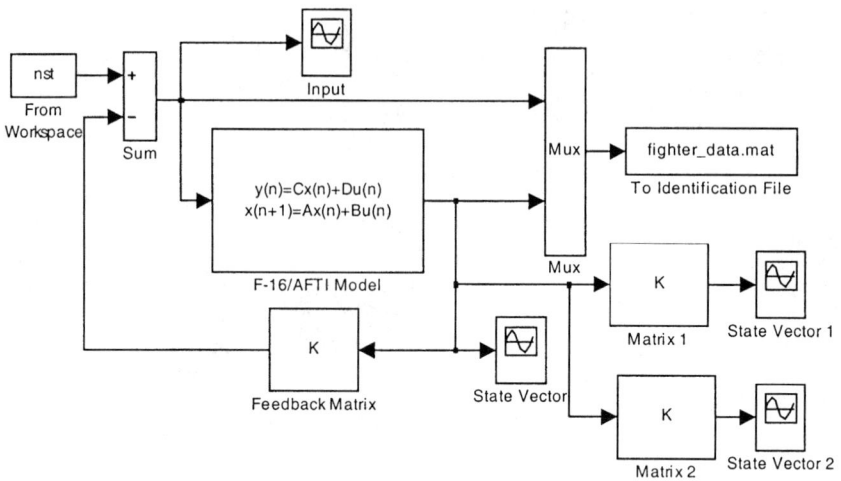

FIGURE 5.2.4.   SIMULINK diagram of an aircraft model (c_5_2_22.mdl).

```
0.7800   -0.0049   -0.0001   -0.0025    0.0033    0.0106   -0.0076
0.8100    0.0046    0.0091    0.0040   -0.0112    0.0062   -0.0009
0.8400   -0.0032    0.0006   -0.0026    0.0062   -0.0175   -0.0201
0.8700    0.0124   -0.0111   -0.0166    0.0127    0.0070    0.0108
```

The aircraft is modeled, and the SIMULINK diagram, as shown in Figure 5.2.4, is developed.

Six inputs $[\delta_{HR}(t)\ \delta_{HL}(t)\ \delta_{FR}(t)\ \delta_{FL}(t)\ \delta_C(t)\ \delta_R(t)]^T$ and nine states $[v(t)\ \alpha(t)\ q(t)\ \theta(t)\ \beta(t)\ p(t)\ r(t)\ \phi(t)\psi(t)]^T$ were sampled, multiplexed, and stored as a matrix in a data file entitled as `fighter_data.mat`. These data are stored in a single sixteen column matrix, in which each row contains sample time, input values, and values of the state variables at that sample time. That is, the $(z+1)$ sets of data for $0 \geq k \geq z$ are given as

$$[\texttt{fighter\_data.mat}] = \begin{bmatrix} t_0 & x_0^T & u_0^T \\ \vdots & \vdots & \vdots \\ t_k & x_k^T & u_k^T \\ \vdots & \vdots & \vdots \\ t_z & x_z^T & u_z^T \end{bmatrix}.$$

The random excitation vector (denoted in Figure 5.2.4 as nst) consists of six normally distributed inputs with mean 0.01. Identification and design are performed in the MATLAB environment using the corresponding m-files, which demultiplex the data, execute the identification algorithm for each row, and sort the results to calculate the estimated arrays of F_m and G_m, and the associated controls. In the demonstration discussed, the discrete-time sampling time (t_s) is chosen to be 0.03 sec. Twenty samples $(z = 20)$ are used to identify the unknown coefficients in F_m and G_m using state variable responses; see Figure 5.2.5.

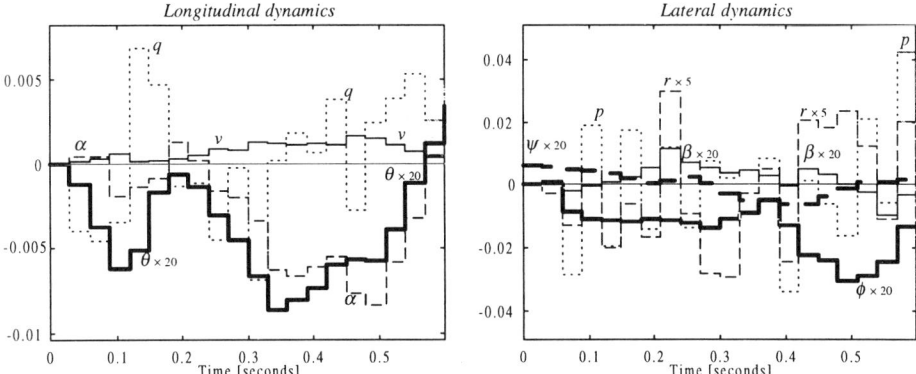

FIGURE 5.2.5.   Longitudinal and lateral aircraft dynamics.

With these settings, using MATLAB on a 450-MHz Pentium computer running Windows98, the identification takes 0.047 sec to calculate F_m and G_m (observe that the design is accomplished within 0.29 sec). The errors between coefficients of matrices F and F_m and G and G_m do not exceed $1 \times 10^{-6}\%$. As the parameters are identified, the tracking control problem can be solved. The eigenvalues of A can be found using the characteristic equation. This gives us the following eigenvalues of an open-loop system: $1 \pm 3.25i$, -3.16, -2.36, 0.92, $-0.0092 \pm 0.204i$, 0, 0. Thus, the studied longitudinal–lateral dynamics of the fighter is unstable. The analysis of the eigenvalues of the closed-loop system shows that the designed controller guarantee stability. The MATLAB script, used in identification, as well as in analysis and control is documented below.

MATLAB script c_5_2_23.m

```
%  Takes the data from simulink simulation c`5`2`22.mdl
%  and identifies the parameters in the f and g matrices
%  of the discrete-time model
%  The complete process is
%  1. Load a continuous model a, b, c, d using c`5`2`21.m
%  2. Discretize the model to find f, g, c, d using
%     c`5`2`21.m
%  3. Find the feedback gains and random input vector
%     from c`5`2`21.m
%  4. Perform simulation using c`5`2`22.mdl to receive
%     fighter`data
%  5. Run this file
% Identification of the unknown parameters
load fighter`data; t0=clock; data=data';
u=data(:,2:7); x=data(:,8:16); n=length(u)-1;
phi=[x(1:(n-1),1:9) u(1:(n-1),1:6)];
psi=inv(phi'*phi)*phi';
```

```
for k=1:9; phat(k,:)=(psi*x(2:n,k))';end;
fm=phat(:,1:9), gm=phat(:,10:15)
identification time=etime(clock,t0)
pause
% The error between the real and the identified matrices
f-fm
g-gm
pause
% Design of LQR and integral LQR controllers
% LQR control
Q=eye(size(fm)); R=eye(size(gm,2));
t0=clock; [kf,k,eig]=dlqr(fm,gm,Q,R)
design time=etime(clock,t0)
pause
% Closed-loop system dynamics
k=200; d=zeros(3,6);
c=[0 0 0 1 0 0 0 0 0; 0 0 0 0 0 0 0 1 0;
   0 0 0 0 0 0 0 0 1];
ref=ones(k,6);    %ref=0.1*randn(k,6);
[y,x]=dlsim(fm-gm*kf,gm,c,d,ref);
plot(0:k-1,y,'-'); pause
plot(0:k-1,x,':'); pause
% Integral LQR control
Fm=[fm zeros(9,3);-c*fm eye(3,3)]; Gm=[gm; -c*gm];
Q=eye(size(Fm)); [kf,k,eig]=dlqr(Fm,Gm,Q,R)
end; disp('END')
```

The following numerical results are obtained:

```
» c 5 2 23
fm =
Columns 1 through 7
  0.9995    0.2457 -0.0273 -0.2885 -0.0391 -0.0075 -0.0020
 -0.0001    0.9663  0.0291  0.0000  0.0011  0.0017  0.0003
 -0.0000    0.1135  0.9765  0.0000  0.0007  0.0002  0.0012
 -0.0000    0.0017  0.0296  1.0000  0.0000  0.0000  0.0000
 -0.0001    0.0048  0.0011  0.0288  0.9770  0.0037 -0.0268
  0.0001    0.0165  0.0005 -0.0198 -1.3515  0.8977  0.0250
 -0.0000   -0.0268  0.0015  0.0041  0.2716 -0.0003  0.9811
  0.0000    0.0003  0.0000 -0.0002 -0.0207  0.0285  0.0003
 -0.0000   -0.0004  0.0000  0.0000  0.0041 -0.0000  0.0297
Columns 8 through 9
  0.0000   -0.0000
 -0.0000    0.0000
 -0.0000    0.0000
 -0.0000    0.0000
  0.0000   -0.0000
```

```
  0.0000   -0.0000
 -0.0000    0.0000
  1.0000    0.0000
 -0.0000    1.0000
gm =
  0.0071    0.0067   -0.0110   -0.0117   -0.0004   -0.0001
 -0.0090   -0.0088   -0.0091   -0.0090    0.0000    0.0000
 -0.2818   -0.2819   -0.0746   -0.0746    0.0000   -0.0000
 -0.0042   -0.0042   -0.0011   -0.0011    0.0000   -0.0000
 -0.0010    0.0007   -0.0006    0.0005    0.0123    0.0018
 -0.0822    0.0819   -0.0881    0.0879    0.0122    0.0250
  0.0921   -0.0924    0.0233   -0.0232    0.0199   -0.0132
 -0.0013    0.0013   -0.0013    0.0013    0.0002    0.0004
  0.0014   -0.0014    0.0003   -0.0003    0.0003   -0.0002
identification time =
  0.0500
difference f =
  1.0e-010 *
Columns 1 through 7
 -0.0067    0.0235   -0.0005   -0.0103    0.0010   -0.0005   -0.0016
  0.0085   -0.0228    0.0006    0.0189   -0.0026    0.0010    0.0020
  0.1968   -0.6658    0.0098    0.6403   -0.0565    0.0252    0.0532
  0.0026   -0.0084    0.0002   -0.0030   -0.0007    0.0003    0.0007
 -0.0012    0.0024   -0.0000   -0.0055    0.0005   -0.0001    0.0000
 -0.0591    0.1475    0.0020   -0.0983    0.0135   -0.0009    0.0014
  0.0489   -0.1109    0.0012    0.1878   -0.0167    0.0052    0.0087
  0.0009   -0.0058    0.0000    0.0010    0.0003   -0.0002   -0.0002
  0.0002   -0.0051    0.0000   -0.0024   -0.0001    0.0001    0.0003
Columns 8 through 9
 -0.0113    0.0086
  0.0097   -0.0120
  0.1436   -0.3956
  0.0058   -0.0014
 -0.0014    0.0038
 -0.0310    0.0946
  0.0283   -0.0916
  0.0030   -0.0005
  0.0013    0.0012
difference g =
  1.0e-012 *
 -0.0008    0.0048    0.0007   -0.0018    0.0000   -0.0003
  0.0012   -0.0072   -0.0010    0.0019   -0.0000    0.0004
  0.0306   -0.1204   -0.0313    0.0294   -0.0024    0.0116
  0.0003   -0.0032   -0.0003    0.0010   -0.0001    0.0001
 -0.0000    0.0001   -0.0000   -0.0000    0.0000   -0.0000
  0.0041   -0.0021    0.0040   -0.0063    0.0017    0.0011
  0.0045   -0.0255   -0.0101    0.0063   -0.0017    0.0026
 -0.0001    0.0010    0.0002    0.0004    0.0001   -0.0001
  0.0000   -0.0008   -0.0002    0.0002   -0.0000    0.0001
```

```
kf =
Columns 1 through 7
  0.4050     0.8672    -0.7405    -3.8007    -3.3367     0.0988     1.4059
  0.5101     1.7979    -0.6688    -2.9200     1.6564    -0.0436    -1.1337
 -0.3917    -1.3876    -0.0982     1.4083     0.9154    -0.2494     0.0393
 -0.4535    -1.5227    -0.0964     1.4023    -1.9936     0.3140     0.1659
  0.0171     0.3136     0.0317     0.3943     3.3124    -0.1534    -0.3018
  0.0052     0.0956     0.0092     0.1117     0.5770     0.0089    -0.2043
Columns 8 through 9
  0.5186     0.3349
 -0.2810    -0.5425
 -0.2746    -0.1687
  0.4821    -0.2737
 -0.4558     0.5954
 -0.0625     0.0115
k =
Columns 1 through 7
  20.2882     46.0801    -2.5217   -58.7260     4.0477    -0.6382    -1.2086
  46.0801    184.1450    -8.3410  -230.9522    40.5469    -2.6291    -7.9715
  -2.5217     -8.3410     3.7207    17.8519     3.5173    -0.1134    -0.5596
 -58.7260   -230.9522    17.8519   393.4845    44.7803    -1.5627    -7.2868
   4.0477     40.5469     3.5173    44.7803   410.0141   -28.8753   -62.1647
  -0.6382     -2.6291    -0.1134    -1.5627   -28.8753     5.1082     5.2940
  -1.2086     -7.9715    -0.5596    -7.2868   -62.1647     5.2940    21.0040
  -0.8730     -9.2194    -0.4508    -3.7791   -67.2140     7.0953    11.8719
   2.7658     23.3115     0.1492   -12.2503    38.3902     0.6709     5.6946
Columns 8 through 9
  -0.8730      2.7658
  -9.2194     23.3115
  -0.4508      0.1492
  -3.7791    -12.2503
 -67.2140     38.3902
   7.0953      0.6709
  11.8719      5.6946
  48.0623     -6.3482
  -6.3482     74.6211
eig =
  0.6570
  0.8522 + 0.1017i
  0.8522 - 0.1017i
  0.8775
  0.9175
  0.9592 + 0.0496i
  0.9592 - 0.0496i
  0.9713
  0.9862
design time =
    0.2900
kf =
```

```
Columns 1 through 7
  0.2543    0.3600   -0.8802   -6.2989   -3.8659    0.5292    2.4401
  0.2207    0.7175   -0.8469   -5.5350    2.3720   -0.3931   -2.2137
 -0.6209   -2.2533   -0.1475    1.1360    6.7582   -1.2405   -1.1316
 -0.6389   -2.5213   -0.1861    0.1801   -7.2796    1.3075    1.2374
 -0.0686    0.1445    0.0255    0.3756    8.5513   -0.4533   -0.3646
 -0.0126    0.0250    0.0030    0.0357    0.5853    0.0532   -0.2020
Columns 8 through 12
  3.1387    7.8144    0.4440   -0.0821   -0.5041
 -2.2407   -8.0777    0.4843    0.0456    0.5452
 -8.5728   -1.2078    0.2315    0.6315   -0.0069
  8.8288    0.8269    0.1919   -0.6345    0.0286
 -2.8822    7.2945    0.1531    0.0488   -0.3848
  0.4540    0.0198    0.0202   -0.0723    0.0232
k =
  1.0e+003 *
Columns 1 through 7
  0.0238    0.0617   -0.0017   -0.0532   -0.0059   -0.0000    0.0002
  0.0617    0.2622   -0.0057   -0.2288    0.0224   -0.0025   -0.0045
 -0.0017   -0.0057    0.0045    0.0304    0.0035   -0.0003   -0.0006
 -0.0532   -0.2288    0.0304    0.6395    0.0665   -0.0085   -0.0133
 -0.0059    0.0224    0.0035    0.0665    1.1382   -0.1156   -0.1549
 -0.0000   -0.0025   -0.0003   -0.0085   -0.1156    0.0217    0.0262
  0.0002   -0.0045   -0.0006   -0.0133   -0.1549    0.0262    0.0544
  0.0036   -0.0099   -0.0023   -0.0578   -0.7607    0.1433    0.1719
 -0.0004    0.0225    0.0002   -0.0276    0.3132    0.0448    0.1415
 -0.0024   -0.0089   -0.0022   -0.0404    0.0139   -0.0004   -0.0006
 -0.0002    0.0000    0.0001    0.0033    0.0389   -0.0102   -0.0105
 -0.0000   -0.0007   -0.0001   -0.0004   -0.0174   -0.0017   -0.0085
Columns 8 through 12
  0.0036   -0.0004   -0.0024   -0.0002   -0.0000
 -0.0099    0.0225   -0.0089    0.0000   -0.0007
 -0.0023    0.0002   -0.0022    0.0001   -0.0001
 -0.0578   -0.0276   -0.0404    0.0033   -0.0004
 -0.7607    0.3132    0.0139    0.0389   -0.0174
  0.1433    0.0448   -0.0004   -0.0102   -0.0017
  0.1719    0.1415   -0.0006   -0.0105   -0.0085
  1.2255    0.3214   -0.0030   -0.1011   -0.0151
  0.3214    1.6245    0.0139   -0.0337   -0.1167
 -0.0030    0.0139    0.0112   -0.0000   -0.0005
 -0.1011   -0.0337   -0.0000    0.0142    0.0019
 -0.0151   -0.1167   -0.0005    0.0019    0.0156
eig =
  0.6533
  0.8755 + 0.1356i
  0.8755 - 0.1356i
  0.8824 + 0.1089i
  0.8824 - 0.1089i
  0.9265 + 0.0776i
```

```
0.9265 - 0.0776i
0.8665 + 0.0469i
0.8665 - 0.0469i
0.9555 + 0.0305i
0.9555 - 0.0305i
0.8551
```

5.2.1.2.   Least-Squares Identification of Nonlinear Systems

Nonlinear multivariable system dynamics is modeled by nonlinear differential equations in state-space form

$$\dot{x}(t) = F(x, d, u), \quad x(t_0) = x_0, \quad t \geq 0, \tag{5.2.6}$$

where $x \in X \subset \mathbb{R}^c$ is the state vector, $u \in U \subset \mathbb{R}^m$ is the control (forcing function) vector, $d \in D \subset \mathbb{R}^l$ is the disturbance vector, and $F(\cdot): \mathbb{R}^c \times \mathbb{R}^l \times \mathbb{R}^m \to \mathbb{R}^c$ is the smooth map.

Our goal is to identify the system parameters by mapping nonlinear systems in $X \times D \times U$. Using experimental results, one obtains $M_{\text{system}} = \{x : x \in X(X_0, D, U)\}$. A parameter set must be identified by applying the incoming measured system data, and the *sufficiency* criterion $M_{\text{model}} \subseteq M_{\text{system}}$ must be guaranteed. That is, using the originally mapped M_{system}, the identified system model can be easily validated.

We design an augmented vector to be used in identification. Using the state variables x, control u (disturbances can be considered as the forcing function), and nonlinear expressions, we have

$$x_a = f(x, u) = \begin{bmatrix} x \\ u \\ f_x(x) \\ f_u(u) \\ f_{xu}(x, u) \end{bmatrix}, \quad x_a \in \mathbb{R}^w. \tag{5.2.7}$$

Denoting the matrix of unknown parameters as $P \in \mathbb{R}^{c \times w}$ and using (5.2.7), the state-space systems model is expressed as

$$\dot{x}(t) = P x_a,$$

and

$$\dot{x}(t)^T = x_a^T P^T. \tag{5.2.8}$$

For identification, we use z-measurements of the state variables, the forcing function (control, noise, and disturbances), the state derivatives, and the nonlinear expressions $f_x(x)$, $f_u(u)$, and $f_{xu}(x, u)$ at instants t_j, $j \in [1, \ldots, z]$. That is, an augmented vector x_a is used. We have

$$\dot{x}_m = \dot{x}(t_j)^T \equiv \begin{bmatrix} \dot{x}_1(t_1) & \cdots & \dot{x}_c(t_1) \\ \vdots & \vdots & \vdots \\ \dot{x}_1(t_z) & \cdots & \dot{x}_c(t_z) \end{bmatrix} = \begin{bmatrix} \dot{x}_a(t_1)^T \\ \vdots \\ \dot{x}_a(t_z)^T \end{bmatrix} \hat{P}^T = \Phi_m \hat{P}^T,$$

where

$$\begin{bmatrix} \dot{x}_1(t_1) & \cdots & \dot{x}_c(t_1) \\ \vdots & \vdots & \vdots \\ \dot{x}_1(t_z) & \cdots & \dot{x}_c(t_z) \end{bmatrix} \in \mathbb{R}^{z \times c} \quad \text{and} \quad \Phi_m \in \mathbb{R}^{z \times w}.$$

Choosing the time instances t_j to guarantee $\det(\Phi_m^T \Phi_m) \neq 0$, a least-squares estimate of $\hat{P}$ can be found. In particular, to find the unknown parameters in (5.2.8), the following expression results to derive the matrix of unknown parameters:

$$\hat{P} = \left[(\Phi_m^T \Phi_m)^{-1} \Phi_m^T \dot{x}_m \right]. \tag{5.2.9}$$

Example 5.2.3. Identification of the unknown parameters of the Volterra equations

To illustrate the identification algorithm, consider the Volterra equations in the state-space form

$$\dot{x}(t) = \begin{bmatrix} \dot{x}_1(t) \\ \dot{x}_2(t) \end{bmatrix} = \begin{bmatrix} p_{11}x_1 + p_{13}x_1x_2 + p_{14}u \\ p_{22}x_2 + p_{23}x_1x_2 + p_{24}u \end{bmatrix}. \tag{5.2.10}$$

From (5.2.8) and (5.2.10), we have

$$\dot{x}(t) = \begin{bmatrix} \dot{x}_1(t) \\ \dot{x}_2(t) \end{bmatrix} = \begin{bmatrix} p_{11}x_1 + p_{13}x_1x_2 + p_{14}u \\ p_{22}x_2 + p_{23}x_1x_2 + p_{24}u \end{bmatrix} = \begin{bmatrix} p_{11} & p_{12} & p_{13} & p_{14} \\ p_{21} & p_{22} & p_{23} & p_{24} \end{bmatrix} \begin{bmatrix} x_1 \\ x_2 \\ x_1x_2 \\ u \end{bmatrix}.$$

Simulation of the nonlinear Volterra equations

$$\dot{x}(t) = 2x_1 - x_1x_2 + u, \quad \dot{x}_2(t) = -8x_2 + 2x_1x_2 + u$$

with zero initial conditions results in the data set for the states, nonlinearity x_1x_2, and control (forcing function) u.

The SIMULINK diagram used to model the Volterra equations is illustrated in Figure 5.2.6.

Making use of the data received, the parameters are identified applying equation (5.2.9). The MATLAB script to perform the identification is documented.

MATLAB script c_5_2_32.m

```
% Simulink data are used for identification of the
% "unknown" parameters of the Voltera equation
% Data are stored in wolves.mat
% Volterra equation
% dx1/dt = 2*x1 - x1*x2 + u
% dx2/dt = -8*x2 + 2*x1*x2 + u
load wolves;
```

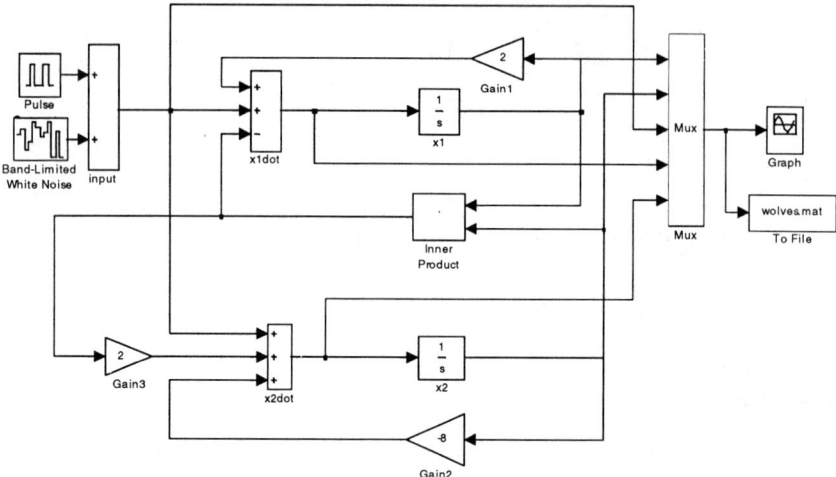

FIGURE 5.2.6.   SIMULINK model (c_5_2_31.mdl)

```
n=6;
t0=clock;
xm=xm';
tt=xm(1:n,1);
u=xm(1:n,4);
xdot=xm(1:n,5:6);
x=xm(1:n,2:3);
%n=length(u);
z3=x(:,1).*x(:,2);
z=[x z3];
phi=[z u];
phihat=inv(phi'*phi)*phi';
% fm=inv(phi)*z(2:5,1)
fm=(phihat*xdot)'
idtime=etime(clock,t0)
disp('End')
```

The resulting "unknown" parameters are identified, and the time needed for identification is 0.05 sec. We have

```
» c'5'2'32
fm =
    2.0000   -0.0000   -1.0000    1.0000
   -0.0000   -8.0000    2.0000    1.0000
idtime =
    0.0500
End
```

That is, solving (5.2.9), one obtains the following unknown coefficients of nonlinear differential equations (5.2.10):

$$p_{11} = 2, \ p_{12} = 0, \ p_{13} = -1, \ p_{14} = 1, \ p_{21} = 0, \ p_{22} = -8, \ p_{23} = 2 \text{ and } p_{24} = 1.$$

Hence, we have

$$\dot{x}(t) \begin{bmatrix} \dot{x}_1(t) \\ \dot{x}_2(t) \end{bmatrix} = \begin{bmatrix} 2x_1 - x_1 x_2 + u \\ -8x_2 + 2x_1 x_2 + u \end{bmatrix}.$$

Example 5.2.4. Identification of an advanced fighter

We study the identification problem for an aircraft, which is modeled as

$$
\dot{x}^{air}(t) = \begin{bmatrix} \dot{v}(t) \\ \dot{\alpha}(t) \\ \dot{q}(t) \\ \dot{\theta}(t) \\ \dot{\beta}(t) \\ \dot{p}(t) \\ \dot{r}(t) \\ \dot{\phi}(t) \\ \dot{\psi}(t) \end{bmatrix} = Ax^{air} + F(x^{air}) + Bu = A \begin{bmatrix} v \\ \alpha \\ q \\ \theta \\ \beta \\ p \\ r \\ \phi \\ \psi \end{bmatrix}
$$

$$
+ \begin{bmatrix} 0 \\ -p\cos\alpha\tan\beta - r\sin\alpha\tan\beta \\ \frac{1}{I_Y}\left[(I_Z - I_X)pr - I_{XZ}p^2 + I_{XZ}r^2\right] \\ q\cos\phi - r\sin\phi \\ p\sin\alpha - r\cos\alpha \\ \frac{1}{I_X I_Z - I_{XZ}^2}\left[I_{XZ}(I_X - I_Y + I_Z)qp + (I_Y I_Z - I_{XZ}^2 - I_Z^2)qr\right] \\ \frac{1}{I_X I_Z - I_{XZ}^2}\left[(I_X^2 - I_X I_Y + I_{XZ}^2)qp - I_{XZ}(I_X - I_Y + I_Z)qr\right] \\ q\tan\theta\sin\phi + r\tan\theta\cos\phi \\ q\cos^{-1}\theta\sin\phi + r\cos^{-1}\theta\cos\phi \end{bmatrix} + B \begin{bmatrix} \delta_{HR} \\ \delta_{HL} \\ \delta_{FR} \\ \delta_{FL} \\ \delta_C \\ \delta_R \end{bmatrix}.
$$

The longitudinal and lateral aircraft state variables and controls are

$$x(t) = [v(t) \ \alpha(t) \ q(t) \ \theta(t) \ \beta(t) \ p(t) \ r(t) \ \phi(t) \ \psi(t)]^T,$$
$$u(t) = [\delta_{HR}(t) \ \delta_{HL}(t) \ \delta_{FR}(t) \ \delta_{FL}(t) \ \delta_C(t) \ \delta_R(t)]^T.$$

The matrices A and B, as well as the moments of inertia in $F(\cdot)$, are unknown. Using the longitudinal and lateral state variables and the deflections of control surfaces, the unknown aircraft parameters must be identified. To perform the identification, the reported least-squares identification algorithm is used. We define

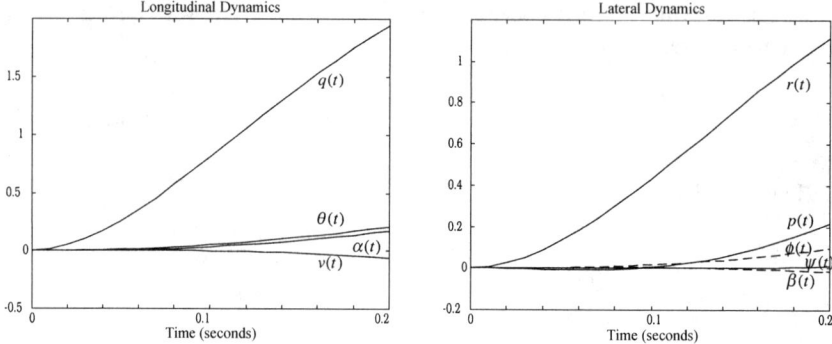

FIGURE 5.2.7.   Aircraft dynamics used for identification of the unknown parameters.

the nonlinear vector-function $f_x(x)$, which describes the aircraft nonlinearities, as

$$
f_x(x) \quad
\begin{bmatrix}
p \cos \alpha \tan \beta \\
r \sin \alpha \tan \beta \\
pr \\
p^2 \\
r^2 \\
q \cos \phi \\
r \sin \phi \\
p \sin \alpha \\
r \cos \alpha \\
qp \\
qr \\
q \tan \theta \sin \phi \\
r \tan \theta \cos \phi \\
q \cos^{-1} \theta \sin \phi \\
r \cos^{-1} \theta \cos \phi
\end{bmatrix}.
$$

This nonlinear $f_x(x)$, together with the aircraft states and controls, is used to define the augmented vector x_a for identification. Using the longitudinal and lateral dynamics, given in Figure 5.2.7, the unknown aircraft parameters (matrices A and B, as well as the aircraft's moments inertia) are identified by solving (5.2.9). We obtain

$$
A_r =
\begin{bmatrix}
-0.009 & 5.7 & -0.24 & -9.6 & -0.46 & -0.095 & -0.14 & 0 & 0 \\
-0.001 & -0.68 & 1 & 0 & 0.12 & 0.037 & 0.005 & 0 & 0 \\
0.0002 & 2.7 & -0.53 & 0 & 0.009 & 0.0062 & 0.028 & 0 & 0 \\
0 & 0 & 0 & 0 & 0 & 0 & 0 & 0 & 0 \\
-0.001 & 0.29 & 0.036 & 1.4 & -0.72 & 0.086 & -1.5 & 0 & 0 \\
0.00002 & 1.1 & 0.041 & 0.007 & -26 & -4.9 & 0.53 & 0 & 0 \\
0.00001 & -1.3 & 0.098 & 0.011 & 7.4 & -0.037 & -0.82 & 0 & 0 \\
0 & 0 & 0 & 0 & 0 & 1 & 0 & 0 & 0 \\
0 & 0 & 0 & 0 & 0 & 0 & 0 & 0 & 0
\end{bmatrix},
$$

$$
B_r = \begin{bmatrix}
0.093 & 0.093 & -0.45 & -0.45 & 0 & 0 \\
-0.24 & -0.24 & -0.41 & -0.41 & 0 & 0 \\
-12 & -12 & -4.7 & -4.7 & 0 & 0 \\
0 & 0 & 0 & 0 & 0 & 0 \\
0.057 & -0.057 & -0.003 & 0.003 & 0.85 & 0.098 \\
-4.1 & 4.1 & -5.9 & 5.9 & 2.6 & -3.7 \\
5.8 & -5.8 & 1.9 & -1.9 & 2 & -1.4 \\
0 & 0 & 0 & 0 & 0 & 0 \\
0 & 0 & 0 & 0 & 0 & 0
\end{bmatrix},
$$

$I_X = 21983$ kg-m^2, $I_Y = 154248$ kg-m^2, $I_Z = 186515$ kg-m^2, and $I_{XZ} = 2407$ kg-m^2.

5.2.2. *Time-Domain, Nonlinear, Mapping-Based Identification*

Accurate nonlinear mathematical models should be found to solve a great variety of problems in multivariable dynamics and control. Different identification methods were extensively studied, and the MATLAB identification toolbox allows one to apply different identification methods for linear systems. Nonlinear differential equations, which model multivariable systems, must be used. Many practical identification methods are based on a model-based concept. This concept uses the available nonlinear state-space differential equations, and numerous examples demonstrate the applicability, feasibility, and versatility of the model-based concept. In particular, nonlinear differential equations can be found using Kirchoff's laws, Newtonian mechanics, or the Lagrange equations of motion. In these differential equations, the unknown parameters must be identified. Identification results are typically conservative for multivariable, highly coupled nonlinear systems because nonlinear phenomena constitute a great source of problems. Parameter convergence and stability region expanding are critical issues. To design a computationally efficient identification algorithm, to enhance robustness and the stability-convergence region, as well as to ensure parameter convergence, we address the identification problem using nonlinear error mappings. This section introduces a time-domain nonlinear mapping-based identification method to solve a variety of problems encountered in nonlinear identification. The essential idea is to attain the desired convergence using nonlinear error mappings to perform identification, and the Lyapunov second method can be used to study the convergence. Robust identification algorithms must be developed to identify the unknown parameters of highly coupled nonlinear differential equations. In particular, computational efficiency, numerical stability, and noise immunity must be addressed as the central problems.

This section presents a time-domain, nonlinear, mapping-based identification method to be applied to identify the unknown parameters of nonlinear dynamic systems modeled by nonlinear differential equations. The method reported can be easily applied to identify the unknown parameters of multivariable dynamic systems, and nonlinear error mappings are used. The analysis of parameter convergence is provided, and the regions of convergence can be found using the second method of Lyapunov. Quadratic and nonquadratic Lyapunov functions are used. Analytical and numerical studies illustrate and validate the identification concept. The time-domain, nonlinear, mapping-based identification method ensures robustness and reduces major shortcomings in stability, convergence, and computational efficiency compared with many other identification algorithms. In particular, computational efficiency is guaranteed, stability and robustness are ensured, the convergence rate is improved, and the region of the convergence is expanded.

5.2.2.1.   State-Space Model and Problem Formulation

The system dynamics are modeled by the following nonlinear differential equations:

$$\dot{x}(t) = A(t) f\,[x(t), u(t)]\,, \quad t \geq 0, \quad x(t_0) = x_0,$$
$$x \in X \subset \mathbb{R}^c, \quad u \in U \subset \mathbb{R}^m, \quad A(\cdot)\colon \mathbb{R}_{\geq o} \to \mathbb{R}^{c \times n}, \qquad (5.2.11)$$

with the known smooth nonlinear map $f[x(t), u(t)]$ and the unknown parameter matrix $A(t)$.

In (5.2.11), x is the state vector that evolves in X and u is the vector of forcing functions in U.

By using a state-space model

$$\dot{x}_m(t) = A_M(t) f\,[x_M(t), u(t)]\,, \quad t \geq 0, \quad x_M(t_0) = x_{M0},$$
$$x_M \in X_M \subset \mathbb{R}^c, \quad u \in U \subset \mathbb{R}^m, \quad A_M(\cdot)\colon \mathbb{R}_{\geq o} \to \mathbb{R}^{c \times n}, \qquad (5.2.12)$$

the unknown parameters, as given by a_{Mij} (entries of the matrix A_M), must be identified. That is, the entries a_{Mij} of matrix A_M, which give the unknown parameters, must be found.

Using $\Delta x(t) = x(t) - x_M(t)$, the mismatched dynamics result, and we have the differential equation to model the evolution of the mismatched dynamics as

$$\Delta \dot{x}(t) = \quad \dot{x}(t) - \dot{x}_M(t) = \Delta A(t) f[x(t), u(t)]$$
$$+ A_M(t) \Delta f[x(t), x_M(t), u(t)], t \geq 0, \Delta x(t_0) = \Delta x_0, \quad (5.2.13)$$

where $\Delta f(\cdot)\colon \mathbb{R}^c \times \mathbb{R}^c \times \mathbb{R}^m \to \mathbb{R}^n$ is smooth,

$$\Delta f[x(t), x_M(t), u(t)] = f[x(t), u(t)] - f[x_M(t), u(t)].$$

In (5.2.13), $\Delta A(\cdot)\colon \mathbb{R}_{\geq 0} \to \mathbb{R}^{c \times n}$ is the *normalized parameter error* matrix, which is given as

$$\Delta A(t) = A(t) - A_M(t). \qquad (5.2.14)$$

Equation (5.2.13) possesses a solution $\Delta x(\cdot)$ with $x(t_0) = x_0$ and $x_M(t_0) = x_{M0}$. By making use of $\Delta \dot{x}(t)$ and $\Delta f[x(t), x_M(t), u(t)]$, one defines the error vector $e \in E \subset \mathbb{R}^c$ as

$$e(t) = \Delta \dot{x}(t) - A_M(t) \Delta f[x(t), x_M(t), u(t)] = \Delta A(t) f[x(t), u(t)]. \quad (5.2.15)$$

Taking note of (5.2.14), one concludes that the identification problem is solved if $\lim_{t \to \infty} \|\Delta A(t)\| = 0$. Let us find the expression for ΔA. Making use of the error vector e as well as the Lipschitz nonlinear map $f[x(t), u(t)]$, the evolution of the normalized parameter error matrix ΔA is given as

$$\Delta \dot{A} = -\sum_{i=0}^{\eta} e(t)^{\frac{2i+1}{2\beta+1}} f[x(t), u(t)]^T M, t \geq 0, \Delta A(t_0) = \Delta A_0. \qquad (5.2.16)$$

Here, $M \in \mathbb{R}^{n \times n}$ is the weighting matrix. Time-varying $M(t)$ can be used, and we have

$$\Delta \dot{A} = -\sum_{i=0}^{\eta} e(t)^{\frac{2i+1}{2\beta+1}} f[x(t), u(t)]^T M(t), t \geq 0, \Delta A(t_0) = \Delta A_0.$$

The elements of $M(\cdot)$: $\mathbb{R}_{\geq o} \to \mathbb{R}^{n \times n}$ are assigned to be smooth on $[t_0, \infty)$.

The nonnegative integers η and β are chosen by the designer to guarantee the convergence and to attain the desired convergence rate. It is clear that by assigning η and β, one defines the nonlinear error mappings. For example, assigning $\eta = 0$ and $\beta = 0$, one finds

$$\Delta \dot{A}(t) = -e(t) f[x(t), u(t)]^T M.$$

That is, the linear error vector $e(t)$ is used.

Assigning $\eta = 1$ and $\beta = 0$, we have

$$\Delta \dot{A}(t) = -[e(t) + e^3(t)] f[x(t), u(t)]^T M.$$

The nonlinear error mapping $e(t)^{\frac{2i+1}{2\beta+1}}$ is used to expand the stability-convergence region (expand the convergence margins), guarantee the parameter convergence, ensure robustness, as well as improve the convergence rate. The error mapping in expanded form is given as

$$e(t)^{\frac{2i+1}{2\beta+1}} = \begin{bmatrix} e_1(t)^{\frac{2i+1}{2\beta+1}} \\ e_2(t)^{\frac{2i+1}{2\beta+1}} \\ \vdots \\ e_{c-1}(t)^{\frac{2i+1}{2\beta+1}} \\ e_c(t)^{\frac{2i+1}{2\beta+1}} \end{bmatrix}.$$

Problem statement. Consider a nonlinear system, modeled by differential equations (5.2.11) in $X(X_0, U)$, and the corresponding model (5.2.12) in $X_M(X_{M0}, U)$. Solutions $x(\cdot)$, $x_M(\cdot)$, $e(\cdot)$, $A_M(\cdot)$, and $\Delta A(\cdot)$, in the given sets X, X_M, E, $\mathbb{A}_M$, and $\mathbb{A}_\Delta$, result because of nonzero initial conditions and the forcing function $u \in U$. Adjusting the matrix $A_M(t)$ of the model, we must guarantee that as time tends to infinity, the difference between the mismatch state and the error vectors Δx and e is equal to zero. From (5.2.13) and (5.2.15), $\Delta x(t) \to 0$ and $e(t) \to 0$ if $\Delta A(t) \to 0$. Furthermore, $\Delta A(t) \to 0$ if $\Delta x(t) \to 0$, $e(t) \to 0$. Hence, the identification problem is formulated as follows: Identify the unknown system parameters, as given by a_{Mij} (entries of the matrix A_M), such that for all $x \in X$, $x_M \in X_M$, $u \in U$, $A \in \mathbb{A}$, and $A_M \in \mathbb{A}_M$, the following is guaranteed:

$$\lim_{t \to \infty} \|x(t) - x_M(t)\| = 0, \quad \lim_{t \to \infty} \|e(t)\| = 0 \text{ and } \lim_{t \to \infty} \|\Delta A(t)\| = 0.$$

5.2.2.2.   Identification Method

To analyze the convergence, the second method of Lyapunov should be applied. The normalized parameter error matrix $\Delta A(t)$, as given by (5.2.14), is used to design the Lyapunov candidate. Nonquadratic C^κ ($\kappa \geq 1$), real-valued differentiable functions $V[\Delta A(t)]$ and its total derivative

$$\frac{dV[\Delta A(t)]}{dt}$$

must satisfy the Lyapunov-based criteria. That is, the sufficient conditions for convergence (stability) of the identification algorithm are formulated using the conditions imposed on the Lyapunov pair. We formulate the following theorem.

Theorem 5.2.1. *Consider a continuous-time nonlinear system (5.2.11) in $X(X_0, U)$ and its model (5.2.12) in $X_M(X_{M0}, U)$. The error vector is expressed as (5.2.15). Matrix $\Delta A(t)$ represents the difference between system and model parameters, and the evolution of $\Delta A(t)$ is given by (5.2.26). The solution of (5.2.16) for $\Delta A(\cdot)$ with $\Delta A(t_0) = \Delta A_0$ exists and is continuous in $\mathbb{A}_\Delta \subset \mathbb{R}^{c \times n}$ over $[t_0, \infty)$. Then*

- *A positive ε exists such that for a given $\Delta A_0 \in \mathbb{A}_\Delta$, the evolutions of $\Delta a_{ij}(\cdot)$ are uniformly stable in $\mathbb{A}_\Delta \subset \mathbb{R}^{c \times n}$; that is, the solution of (5.2.16) satisfies $\|\Delta A\| \leq \varepsilon$ or $\|\Delta a_{ij}\| \leq \varepsilon$, $\forall t \geq t_0$*
- *The solution $\Delta A(\cdot)$ with $\Delta A(t_0) = \Delta A_0$ defined, bounded and stable in $X \subset \mathbb{R}^c$, $X_M \subset \mathbb{R}^c$, $U \subset \mathbb{R}^m$, and $\mathbb{A}_\Delta \subset \mathbb{R}^{c \times n}$ on $[t_0, \infty)$*
- *The entries of matrix $\Delta A(t)$ converge to zero; that is, $\lim_{t \to \infty} \|\Delta a_{ij}(t)\| = 0$*

if and only if a continuous, locally Lipschitzian, real-valued function $V(\cdot)$: $\mathbb{R}^{c \times n} \to \mathbb{R}_{\geq 0}$ exists with continuous partial derivatives with respect to $\Delta A(t)$ such that for all $\Delta A \in \mathbb{A}_\Delta$ on $[t_0, \infty)$, the following criteria hold:

$$V[\Delta A(t)] \to 0, \forall \Delta A \to 0, \ V[\Delta A(t)] \to \infty, \forall \|\Delta A\| \to \infty,$$

$$V[\Delta A(t)] > 0, \ \frac{dV[\Delta A(t)]}{dt} < 0, \forall \Delta A \neq 0 \quad (5.2.17)$$

Proof. The proof is based on Lyapunov's stability theory. If criteria (5.2.17) are guaranteed by the Lipschitzian $V[\Delta A(t)]$, a C^κ-function $V[\Delta A(t)]$ is called a Lyapunov function.

The following family of real-analytic functions is used:

$$V[\Delta A(t)] = \sum_{i=1}^{\gamma} v_i [\Delta A(t)]^{\frac{2i}{2\mu+1}}, \ v_i(\cdot): \mathbb{R}^{c \times n} \to \mathbb{R}_{\geq 0},$$

$$\gamma = 1, 2, 3, \ldots, \ \mu = 0, 1, 2, \ldots \quad (5.2.18)$$

to design Lyapunov candidates. Applying (5.2.18), nonquadratic Lyapunov functions can be synthesized. For example, from (5.2.18), one obtains the following quadratic and nonquadratic functions:

$$V[\Delta A(t)] = I \Delta A(t) \Delta A^T(t) I^T, \, I = [i, \ldots, i_c] \in \mathbb{R}^{1 \times c},$$
$$V[\Delta A(t)] = I [\Delta A(t) \Delta A^T(t) + \Delta A(t) \Delta A^T(t) \Delta A(t) \Delta A^T(t)] I^T.$$

If in (5.2.16) the weighting matrix $M(\cdot)$ is time-varying, to guarantee the convergence, the time-varying Lyapunov function $V[t, \Delta A(t)]$ and its total derivative

$$\frac{dV[t, \Delta A(t)]}{dt}$$

must be bounded by the K_∞-functions ρ_1 and ρ_2, and the K-function ρ_3. Hence, the criteria (5.2.17) should be modified. In particular, the following conditions must be satisfied:

$$\rho_1 \left(\|\Delta A\| \right) \leq V[t, \Delta A(t)] \leq \rho_2 \left(\|\Delta A\| \right)$$
$$\text{and} \quad \frac{dV[t, \Delta A(t)]}{dt} \leq -\rho_3 \left(\|\Delta A\| \right). \tag{5.2.19}$$

$\square$

Theorem 5.2.2. *Consider a continuous-time nonlinear system* (5.2.11) *in* $X(X_0, U)$ *and its model* (5.2.12) *in* $X_M(X_{M0}, U)$. *Equation* (5.2.16) *possesses a unique solution for the normalized parameter error matrix for all* $\Delta A \in \mathbb{A}_\Delta$. *Nonlinear system* (5.2.11) *is identifiable in the convex bounded sets* $X \subset \mathbb{R}^c$, $X_M \subset \mathbb{R}^c$, $U \subset \mathbb{R}^m$, $\mathbb{A} \subset \mathbb{R}^{c \times n}$, *and* $\mathbb{A}_M \subset \mathbb{R}^{c \times n}$ *if a positive-definite function* (5.2.18) *exists such that* $\frac{dV[\Delta A(t)]}{dt} < 0$ *for all* $x \in X$, $x_M \in X_M$, $u \in U$, *and* $\Delta A \in \mathbb{A}_\Delta$ *on* $[t_0, \infty)$. *That is, if* $V[\Delta A(t)] > 0$ *and* $\frac{dV[\Delta A(t)]}{dt}$ *is negative-definite, the*

- *Unknown parameters of the system are identified using the nonlinear error mapping*
- *Identification problem has a unique solution in the convex bounded sets* $\mathbb{A} \subset \mathbb{R}^{c \times n}$ *and* $\mathbb{A}_M \subset \mathbb{R}^{c \times n}$, *and a matrix of the unknown parameters is found by solving the following differential equation:*

$$\dot{A}_M(t) = \sum_{i=0}^{\eta} \{\Delta \dot{x}(t) - A_M(t) \Delta f[x(t), x_M(t), u(t)]\}^{\frac{2i+1}{2\beta+1}} f[x(t), u(t)]^T M,$$

$$t \geq 0, \, A_M(t_0) = A_{M0} \tag{5.2.20}$$

- *Solution* $A_M(\cdot)$ *with* $A_M(t_0) = A_{M0}$ *is defined, bounded and stable in* $X \subset \mathbb{R}^c$, $X_M \subset \mathbb{R}^c$, $U \subset \mathbb{R}^m$ *and* $A_M \subset \mathbb{R}^{c \times n}$ *on* $[t_0, \infty)$;
- *Identification algorithm converges; that is, in the given* $\mathbb{A} \subset \mathbb{R}^{c \times n}$ *and* $\mathbb{A}_M \subset \mathbb{R}^{c \times n}$, *the unknown coefficients* a_{Mij} *of matrix* $A_M(t)$ *converge to the system parameters* a_{ij}.

Proof. From (5.2.14) and (5.2.16), one obtains

$$\dot{A}_M(t) = \dot{A}(t) + \sum_{i=0}^{\eta} e(t)^{\frac{2i+1}{2\beta+1}} f[x(t), u(t)]^T M, \, t \geq 0, \, A_M(t_0) = A_{M0},$$

and the matrix differential equation (5.2.20) results, because for time-invariant systems $\dot{A}(t) = 0, \forall A \in \mathbb{A}$.

Along with the solution of the matrix differential equation (5.2.16), one examines the convergence of the unknown parameters a_{Mij}, which are given by matrix $A_M(t)$, to the coefficients a_{ij} of $A(t)$. A family of Lyapunov candidates (5.2.18) is positive-definite; that is, $V[\Delta A(t)] > 0, \forall \Delta A \neq 0$. Furthermore, $V[\Delta A(t)] \to 0, \forall \Delta A \to 0$ and $V[\Delta A(t)] \to \infty, \forall \|\Delta A\| \to \infty$ on $[t_0, \infty)$. Hence, $\frac{dV[\Delta A(t)]}{dt}$ has to be negative-definite for all $\Delta A \in \mathbb{A}_\Delta$ with $x \in X$, $x_M \in X_M$, and $u \in U$ on $[t_0, \infty)$. Function $V[\Delta A(t)]$ is continuously differentiable, and the total derivative $\frac{dV[\Delta A(t)]}{dt}$ can be found along with the solutions of (5.2.16). If $V[\Delta A(t)]$ is a Lyapunov function, the matrix M exists such that in X, X_M, U, and $\mathbb{A}_\Delta$, we have $\frac{dV[\Delta A(t)]}{dt} < 0$. Therefore, $V[A(t)]$ satisfies the requirements of Theorem 5.2.2. Moreover, the equilibrium point of (5.2.16) is zero at $t_e \in \mathbb{R}_{\geq 0}$. Hence $\lim_{t\to\infty} \|\Delta A(t)\| = 0$, $\lim_{t\to\infty} A_M(t) = \overline{A}_M = A$, and $\lim_{t\to\infty} \|x(t) - x_M(t)\| = 0$, $\lim_{t\to\infty} \|e(t)\| = 0$. Thus, $\overline{A}_M \in \mathbb{A}_M$ is a steady-state unique solution of the nonlinear differential equation (5.2.20), and the unknown parameters are given by a_{Mij}. The above analysis implies that $\Delta x(t) \to 0$, $e(t) \to 0$, and $\Delta A(t) \to 0$. Hence, the convergence of the identification algorithm is proven. This concludes the proof.   □

Theorem 5.2.3. *Consider a continuous-time nonlinear system* (5.2.11) *in* $X(X_0, U)$ *and its model* (5.2.12) *in* $X_M(X_{M0}, U)$. *Equation*

$$\Delta\dot{A}(t) = -\sum_{i=0}^{\eta} e(t)^{\frac{2i+1}{2\beta+1}} f[x(t), u(t)]^T M(t), \, t \geq 0, \, \Delta A(t_0) = \Delta A_0$$

possesses a unique solution for the normalized parameter error matrix for all $\Delta A \in \mathbb{A}_\Delta$. *Nonlinear system* (5.2.11) *is identifiable in the convex bounded sets* $X \subset \mathbb{R}^c$, $X_M \subset \mathbb{R}^c$, $U \subset \mathbb{R}^m$, $\mathbb{A} \subset \mathbb{R}^{c \times n}$, *and* $\mathbb{A}_M \subset \mathbb{R}^{c \times n}$ *if a continuously differentiable function* $V[t, \Delta A(t)] = \sum_{i=1}^{\gamma} v_i[t, \Delta A(t)]^{\frac{2i}{2\mu+1}}$, *the* K_∞-*functions* $\rho_1(\cdot)$ *and* $\rho_2(\cdot)$, *and the* K-*function* $\rho_3(\cdot)$ *exist, such that*

$$\rho_1(\|\Delta A\|) \leq V[t, \Delta A(t)] \leq \rho_2(\|\Delta A\|) \text{ and } \frac{dV[t, \Delta A(t)]}{dt} \leq -\rho_3(\|\Delta A\|)$$

for all $x \in X$, $x_M \in X_M$, $u \in U$, *and* $\Delta A \in \mathbb{A}_\Delta$ *on* $[t_0, \infty)$. *If these criteria imposed on the Lyapunov pair are satisfied, the*

- *Unknown parameters of the system are identified using the nonlinear error mapping*
- *Identification problem has a unique solution in the convex bounded sets* $\mathbb{A} \subset \mathbb{R}^{c \times n}$ *and* $\mathbb{A}_M \subset \mathbb{R}^{c \times n}$, *and a matrix of the unknown parameters is found by*

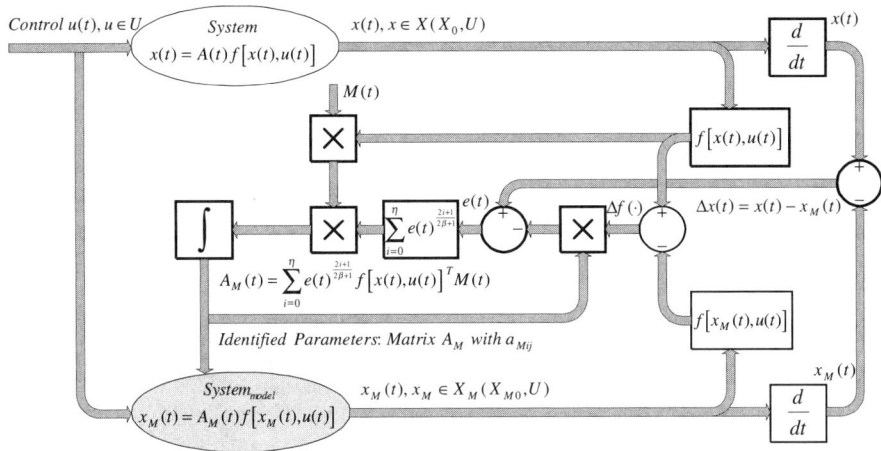

FIGURE 5.2.8.   Time-domain nonlinear mapping-based identification algorithm for multi-variable dynamic systems.

solving the following differential equation:

$$\dot{A}_M(t) = \sum_{i=0}^{\eta} \{\Delta \dot{x}(t) - A_M(t)\Delta f[x(t), x_M(t), u(t)]\}^{\frac{2i+1}{2\beta+1}}$$
$$\times f[x(t), u(t)]^T M(t), \, t \geq 0, \, A_M(t_0) = A_{M0}$$

- *Solution $A_M(\cdot)$ with $A_M(t_0) = A_{M0}$ is defined, bounded and stable in $X \subset \mathbb{R}^c$, $X_M \subset \mathbb{R}^c$, $U \subset \mathbb{R}^m$, and $\mathbb{A}_M \subset \mathbb{R}^{c \times n}$ on $[t_0, \infty)$*
- *Identification algorithm converges; that is, in the given $\mathbb{A} \subset \mathbb{R}^{c \times n}$ and $\mathbb{A}_M \subset \mathbb{R}^{c \times n}$, the unknown coefficients a_{Mij} of matrix $A_M(t)$ converge to the system parameters a_{ij}*

Figure 5.2.8 illustrates the *time-domain* nolinear *mapping-based* identification algorithm.

Convergence is the most important aspect in the identification process. Ideally, uniform asymptotic or exponential convergence is desired. However, this cannot be guaranteed. The convergence in X, X_M, U, $\mathbb{A}$, and $\mathbb{A}_M$ has been studied and proven. It is possible to use *a priori* information, which is often available in form of the "crude" parameter values, and convergence can be achieved.

5.2.2.3.   Identification of Dynamic Systems Under Reference Inputs and Disturbances

It was shown that systems can be identified using the transient behavior caused by nonzero initial conditions and control inputs. In addition, reference inputs and disturbances can be considered as the forcing function, and the systems dynamics is mapped by M_{model}. We have

$$\dot{x}(t) = F(x, r, d, u), x_0(t_0) = x_0, t \geq 0. \qquad (5.2.21)$$

Our goal is to identify the unknown system parameters by mapping nonlinear dynamics in $X \times R \times D \times U$. From the incoming measured (observed) data, a parameter set must be identified to guarantee the sufficiency criterion $M_{model} \subseteq M_{system}$ with the required accuracy. Here, M_{system} is the originally mapped (measured) information set. With the model-based framework and the mapped information sets studied, the identified dynamic model in $X \times R \times D \times U$ can easily be validated from the measured input-state $(R \times D \times U) \times X$ set using experimental data.

Using the known nonlinear quantities $f(x, r, d, u)$, which are readily found from the Lagrange equations of motion, and denoting a matrix of the unknown parameters as $A_M(\cdot)$: $\mathbb{R}_{\geq o} \to \mathbb{R}^{c \times z}$, from (5.2.21), one has

$$\dot{x}(t) = A(t) f(x, r, d, u), x(t_0) = x_0, t \geq 0. \qquad (5.2.22)$$

Here, the matrix of unknown parameters A must be identified using a state-space model

$$\dot{x}_M(t) = A_M(t) f(x_M, r, d, u), x_M(t_0) = x_{M0}, A_M(\cdot): \mathbb{R}_{\geq o} \to \mathbb{R}^{c \times z}. \qquad (5.2.23)$$

Taking note of $\Delta x = x(t) - x_M(t)$, one obtains the evolution of the mismatched dynamics as

$$\begin{aligned} \Delta \dot{x}(t) = \quad & \dot{x}(t) - \dot{x}_M(t) = \Delta A(t) f(x, r, d, u) \\ & + A_M(t) \Delta f(x, x_M, r, d, u), \Delta x(t_0) = \Delta x_0, \end{aligned} \qquad (5.2.24)$$

where $\Delta f(x, x_M, r, d, u) = f(x, r, d, u) - f(x_M, r, d, u)$.

The error vector $e \in E \subset \mathbb{R}^c$ is found to be

$$e(t) = \Delta \dot{x}(t) - A_M(t) \Delta f(x, x_M, r, d, u) = \Delta A(t) f(x, r, d, u). \qquad (5.2.25)$$

The normalized parameter error matrix ΔA evolves as

$$\Delta \dot{A}(t) = - \sum_{i=0}^{\eta} e(t)^{\frac{2i+1}{2\beta+1}} f(x, r, d, u)^T M(t), \Delta A(t_0) = \Delta A_0. \qquad (5.2.26)$$

Here, $M(\cdot)$: $\mathbb{R}_{\geq o} \to \mathbb{R}^{n \times n}$, $\gamma = 0, 1, 2, \ldots$ and $\beta = 0, 1, 2, \ldots$.
Making use of (5.2.24) and (5.2.26), one obtains

$$\dot{A}_M(t) = \dot{A}(t) + \sum_{i=0}^{\eta} e(t)^{\frac{2i+1}{2\beta+1}} f(x, r, d, u)^T M(t), A_M(t_0) = A_{M0}. \qquad (5.2.27)$$

For time-invariant systems, $\dot{A}(t) = 0$. Hence, from (5.2.27), we have

$$\dot{A}_M(t) = \sum_{i=0}^{\eta} e(t)^{\frac{2i+1}{2\beta+1}} f(x, r, d, u)^T M(t), A_M(t_0) = A_{M0}. \qquad (5.2.28)$$

That is, the unknown parameters are found by solving the nonlinear differential equation (5.2.28).

Theorem 5.2.4. *System (5.2.21) is identifiable in the convex bounded sets $X \subset \mathbb{R}^c$, $X_M \subset \mathbb{R}^c$, $U \subset \mathbb{R}^m$, $R \subset \mathbb{R}^b$, $D \subset \mathbb{R}^s$, $A \subset \mathbb{R}^{c \times n}$, and $\mathbb{A}_M \subset \mathbb{R}^{c \times n}$ if a continuously differentiable function $V[t, \Delta A(t)]$, the K_∞-functions $\rho_1(\cdot)$ and $\rho_2(\cdot)$, and the K-function $\rho_3(\cdot)$ exist such that*

$$\rho_1(\|\Delta A\|) \leq V[t, \Delta A(t)] \leq \rho_2(\|\Delta A\|) \text{ and } \frac{dV[t, \Delta A(t)]}{dt} \leq -\rho_3(\|\Delta A\|)$$

for all $x \in X$, $x_M \in X_M$, $u \in U$, and $\Delta A \in \mathbb{A}_\Delta$ on $[t_0, \infty)$. If these criteria are satisfied, a KL-function $\rho_A(\cdot)$: $\mathbb{R}_{\geq 0} \times \mathbb{R}_{\geq 0} \to \mathbb{R}_{\geq 0}$ and the K-functions $\rho_X(\cdot)$: $\mathbb{R}_{\geq 0} \to \mathbb{R}_{\geq 0}$, $\rho_U(\cdot)$: $\mathbb{R}_{\geq 0} \to \mathbb{R}_{\geq 0}$, $\rho_R(\cdot)$: $\mathbb{R}_{\geq 0} \to \mathbb{R}_{\geq 0}$, $\rho_D(\cdot)$: $\mathbb{R}_{\geq 0} \to \mathbb{R}_{\geq 0}$ exist, such that the solution $A_M(\cdot)$: $\mathbb{R}_{\geq 0} \to \mathbb{R}^{c \times n}$ exists on $[t_0, \infty)$ and satisfies

$$\begin{aligned}
\|A_M(t)\| \leq \quad & \rho_A(t, \|A_{M0}\|) + \rho_X(\|\Delta x\|) + \rho_U(\|u\|) \\
& + \rho_R(\|r\|) + \rho_D(\|d\|)
\end{aligned} \tag{5.2.29}$$

for all $x \in X$, $u \in U$, $r \in R$, $d \in D$, and $A_M \in \mathbb{A}_M$ on $[t_0, \infty)$.

If (5.2.29) is guaranteed, $\lim_{t \to \infty} \|\Delta A(t)\| = 0$, and the solution of equation (5.2.28) converges to the unknown parameters A_M; that is, $\lim_{t \to \infty} \|A_M(t)\| = A$. Furthermore, solution $A_M(\cdot)$ with $A_M(t_0) = A_{M0}$ defined, bounded and stable in $X \subset \mathbb{R}^c$, $X_M \subset \mathbb{R}^c$, $U \subset \mathbb{R}^m$, $R \subset \mathbb{R}^b$, $D \subset \mathbb{R}^s$, and $A_M \subset \mathbb{R}^{c \times n}$ on $[t_0, \infty)$.

Proof. The proof is straightforward using the results stated. The mismatched state dynamics of the system (5.2.21) and its model (5.2.22) are given by (5.2.24), and *the normalized parameter error matrix ΔA is expressed by (5.2.14), $\Delta A(t) = A(t) - A_M(t)$.* By using the error vector (5.2.25), the differential equation for ΔA was found in (5.2.26). For time-invariant dynamic systems, $\dot{A}(t) = 0$. Thus, by using (5.2.27), (5.2.28) for the matrix of unknown parameters A_M results. The stability of solution (5.2.28) guarantees the parameter convergence. The convergence is proven using the Lyapunov stability theory. □

The similarity in representations of nonlinear systems used to perform the identification via two algorithms is evident. In fact, the corresponding models are given by (5.2.8) and (5.2.23) with unknown matrices P and A_m.

Example 5.2.5. Identification of the unsteady flight dynamics

Modeling of unsteady flight is motivated by the challenging issues of flight dynamics and control to meet the specified handling and flying requirements. Extensive research has been directed toward identification of a high angle of attack flight to analyze and improve the aircraft performance. The most difficult and challenging specifications are seen in the handling and flying qualities of advanced aircraft at low speed and high angle of attack flight. To model the aircraft dynamics at high angles of attack when vortex flows, unsteady aerodynamic loads, and flexible modes occur, it is necessary to develop an accurate mathematical model of the unsteady flight. In the case of the unsteady viscous flows and flexible modes, linear longitudinal models fail to model the unsteady aerodynamics because nonlinear differential equations must be used. The following expressions are used for the lift and pitch coefficients:

$$C_L = C_L^0(v, \alpha, q, \delta, h) + \Delta C_L(v, \alpha, q, \delta, h),$$
$$C_m = C_m^0(v, \alpha, q, \delta, h) + \Delta C_m(v, \alpha, q, \delta, h),$$

where $C_L^0(v, \alpha, q, \delta, h)$ and $C_m^0(v, \alpha, q, \delta, h)$ are the steady lift and pitch coefficients that are nonlinear functions of the forward velocity (v), angle of attack (α), pitch rate (q), deflection of control surfaces (δ), and altitude (h); $\Delta C_L(v, \alpha, q, \delta, h)$ and $\Delta C_m(v, \alpha, q, \delta, h)$ are the unsteady lift and pitch coefficients caused by unsteady vortex flows and flexible modes.

The unsteady longitudinal aerodynamics is modeled by the following set of differential equations:

$$\dot{x}_1(t) = a_{11}x_1 + a_{12}x_2 + a_{13}x_3,$$
$$\dot{x}_2(t) = a_{21}x_1 + a_{22}x_2 + a_{23}x_3 + a_{24}x_4,$$
$$\dot{x}_3(t) = a_{31}x_1 + a_{32}x_2 + a_{33}x_3 + a_{34}x_5 + a_{35}x_2x_5,$$
$$\dot{x}_4(t) = a_{41}x_1 + a_{42}x_2 + a_{43}x_3 + a_{44}x_4 + a_{45}x_1^{1/3} + a_{46}x_1^3 + a_{47}x_2^{1/3} + a_{48}x_2^2$$
$$+ a_{49}x_2^3 + a_{410}x_3^{1/3} + a_{411}x_3^3 + a_{412}x_4^3,$$
$$\dot{x}_5(t) = a_{51}x_1 + a_{52}x_2 + a_{53}x_3 + a_{54}x_5 + a_{55}x_1^{1/3} + a_{56}x_1^3 + a_{57}x_2^{1/3} + a_{58}x_2^2$$
$$+ a_{59}x_2^3 + a_{510}x_3^{1/3} + a_{511}x_3^3 + a_{512}x_5^3, \qquad (5.2.30)$$

where x_1, x_2, x_3, x_4, and x_5 are the unsteady velocity Δv [m/sec], angle of attack $\Delta\alpha$ [deg], pitch rate Δq [deg/sec], lift ΔC_L, and pitch ΔC_m coefficients, respectively.

The mathematical model of the unsteady aerodynamics is

$$\dot{x}(t) = Af[x(t)] = A_1 x^{1/3} + A_2 x + x^T A_3 x + x^T A_4 x^2, \qquad (5.2.31)$$

where

$$x^{1/3} = \begin{bmatrix} x_1^{1/3} \\ x_2^{1/3} \\ x_3^{1/3} \\ x_4^{1/3} \\ x_5^{1/3} \end{bmatrix}$$

and

$$x^2 = \begin{bmatrix} x_1^2 \\ x_2^2 \\ x_3^2 \\ x_4^2 \\ x_5^2 \end{bmatrix}.$$

The unknown coefficients (entries) of matrices $A_1 \in \mathbb{R}^{5\times5}$, $A_2 \in \mathbb{R}^{5\times5}$, $A_3 \in \mathbb{R}^{5\times5}$, and $A_4 \in \mathbb{R}^{5\times5}$ represent the model structure. For illustration, we use (5.2.30) to avoid the cumbersome matrix representations. That is, differential

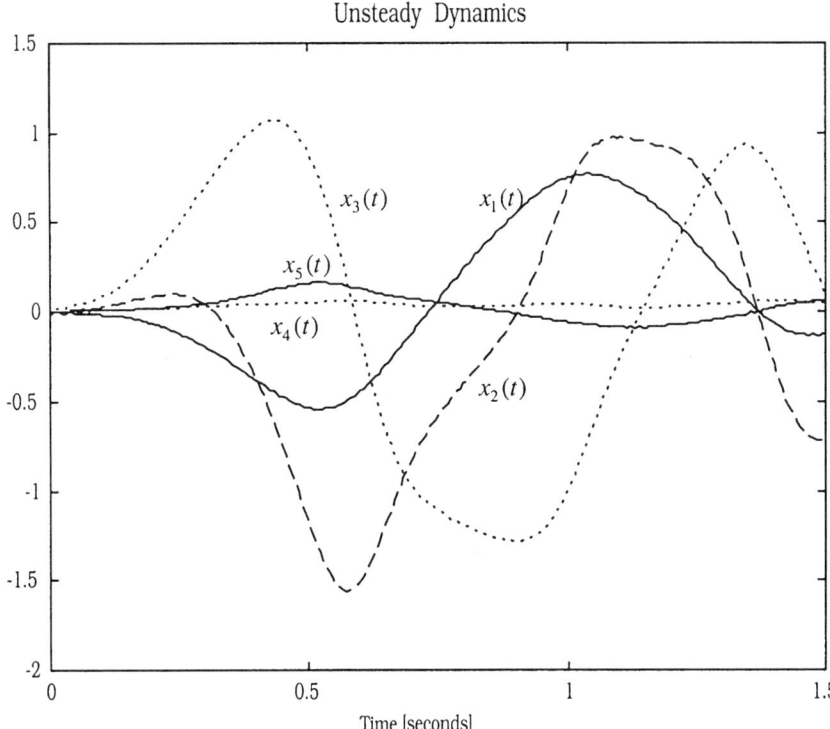

FIGURE 5.2.9. Unsteady velocity Δv, angle of attack $\Delta \alpha$, pitch rate Δq, and lift and pitch coefficients ΔC_L and ΔC_m: evolution of x_1, x_2, x_3, x_4, and x_5.

equations (5.2.30) represent the resulting state-space model. Taking note of state-space equations (5.2.21) and (5.2.30), one obtains

$$\dot{x}(t) = A f\left[x(t)\right] =$$

$$\begin{bmatrix} a_{11} & a_{12} & a_{13} & 0 & 0 & 0 & 0 & 0 & 0 & 0 & 0 & 0 & 0 & 0 \\ a_{21} & a_{22} & a_{23} & a_{24} & 0 & 0 & 0 & 0 & 0 & 0 & 0 & 0 & 0 & 0 \\ a_{31} & a_{32} & a_{33} & 0 & a_{34} & a_{35} & 0 & 0 & 0 & 0 & 0 & 0 & 0 & 0 \\ a_{41} & a_{42} & a_{43} & a_{44} & 0 & 0 & a_{45} & a_{46} & a_{47} & a_{48} & a_{49} & a_{410} & a_{111} & a_{412} & 0 \\ a_{51} & a_{52} & a_{53} & 0 & a_{54} & 0 & a_{55} & a_{56} & a_{57} & a_{58} & a_{59} & a_{510} & a_{511} & 0 & a_{512} \end{bmatrix}$$

$$\times f\left[x(t)\right], \tag{5.2.32}$$

where $f\left[x(t)\right] = \left[x_1 \ x_2 \ x_3 \ x_4 \ x_5 \ x_2 x_5 \ x_1^{1/3} \ x_1^3 \ x_2^{1/3} \ x_2^2 \ x_2^3 \ x_3^{1/3} \ x_3^3 \ x_4^3 \ x_5^3 \right]^T$.

For a scaled advanced fighter configuration, the unsteady dynamics is documented in Figure 5.2.9.

For the flight conditions ($\alpha^\circ = 40$ deg, $v^\circ = 85$ m/sec, $\rho = 1.1$ kg/m^3, and $\delta = $ const), the unknown parameters of (5.2.32) are identified. In (5.2.28) ,we select $\eta = 4$ and $\beta = 1$. Therefore, the following nonlinear matrix differential equation is solved using the nonlinear error:

$$\dot{A}_M(t) = \left[e^{1/3}(t) + e(t) + e^{5/3}(t) + e^{7/3}(t) + e^3(t)\right] f[x(t)]^T M(t),$$
$$t \geq 0, \ A_M(t_0) = A_{M0},$$

where $e(t) = \Delta \dot{x}(t) - A_M(t) \Delta f[x(t), x_M(t)]$.

The initial values of the unknown parameters are assigned to be $a_{Mij}^0 = 0$. The time-varying matrix $M(t)$ is $M(t) = \text{diag}[m_{ij}(t)] \in \mathbb{R}^{15 \times 15}$ with $m_{ij}(t) = 0.05 + e^{-10t}$. The parameters converge to steady-state values. In particular, one obtains

$a_{11} = -1.9, a_{12} = -3.1, a_{13} = -0.86, a_{21} = -4.7, a_{22} = -3, a_{23} = 18.7,$

$a_{24} = 98, a_{31} = -0.28, a_{32} = 8.1, a_{33} = -0.81, a_{34} = -14, a_{35} = -359,$

$a_{41} = -0.14, a_{42} = 0.077, a_{43} = -0.31, a_{44} = -4.2, a_{45} = -0.1,$

$a_{46} = 0.21, a_{47} = 0.35, a_{48} = 0.31, a_{49} = 0.09, a_{410} = -0.13, a_{411} = 0.2,$

$a_{412} = -0.14, a_{51} = 0.1, a_{52} = -0.039, a_{53} = -0.22, a_{54} = -4.1,$

$a_{55} = 0.064, a_{56} = 0.15, a_{57} = 0.29, a_{58} = 0.2, a_{59} = -0.051, a_{510} = 0.11,$

$a_{511} = -0.048, a_{512} = -0.094.$

The identification takes 0.28 sec.

The analysis of the convergence using a time-varying nonquadratic function

$$V[t, \Delta A(t)] = I(t)[\Delta A(t) \Delta A^T(t) + \Delta A(t) \Delta A^T(t) \Delta A(t) \Delta A^T(t)]I^T(t),$$
$$I(t) = [1 + e^{-t} \cdots 1 + e^{-t}]: \mathbb{R}_{\geq 0} \to \mathbb{R}^{1 \times 5}$$

shows that the parameter convergence is guaranteed. The initial values of the unknown parameters a_{Mij}^0 affect the convergence. Even though the convergence depends on the selection of a_{Mij}^0, analytical and numerical analysis indicates that these critical issues have been relaxed, and the convergence region is determined. The nonquadratic time-varying function $V[t, \Delta A(t)]$ is bounded by ρ_1 and ρ_2, and the total derivative of $V[t, \Delta A(t)]$ along with the trajectories of (5.2.26) is bounded by ρ_3. Hence, $\rho_1 (\|\Delta A\|) \leq V[t, \Delta A(t)] \leq \rho_2 (\|\Delta A\|)$ and $\frac{dV[t, \Delta A(t)]}{dt} \leq -\rho_3 (\|\Delta A\|)$. Thus, the convergence is guaranteed, and the unknown parameters can be identified.

5.3.   Design of Stabilizing Controllers Using the Lyapunov Theory

5.3.1.   *Lyapunov Stability Theory and Design of Control Laws*

Several methods have been developed to address and solve nonlinear design and motion control problems for multi-input/multi-output dynamic systems. The Hamilton–Jacobi theory and the Lyapunov concept are found to be the most powerful theoretical tools that allow one to design control algorithms. As was illustrated in Chapter 4, the Hamilton–Jacobi theory provides a systematic concept to solve a wide spectrum of optimization problems for dynamic systems. In particular, minimizing or maximizing the performance functional, optimal controllers are designed, and the specified performance criteria (assigned by the functionals) are achieved. However, the Hamilton–Jacobi method has several disadvantages for high-order nonlinear systems caused by difficulties associated with the solution of partial differential equations with boundary conditions. The application of Lyapunov's stability theory in nonlinear design is conceptually straightforward, and some computational drawbacks can be relaxed. However, the so-called approximation in the control space is needed because the Lyapunov theory does not support the design of controllers.

Let the system dynamics be described as

$$\dot{x}(t) = F(x, r, d) + B(x)u, \quad y = H(x), \quad u_{min} \leq u \leq u_{max},$$
$$x(t_0) = x_0, t \geq 0, \tag{5.3.1}$$

where $x \in X \subset \mathbb{R}^c$ is the state vector, $u \in U \subset \mathbb{R}^m$ is the bounded control vector, $r \in R \subset \mathbb{R}^b$ and $y \in Y \subset \mathbb{R}^b$ are the measured reference and output vectors, $d \in D \subset \mathbb{R}^s$ is the disturbance vector, $F(\cdot) \colon \mathbb{R}^c \times \mathbb{R}^b \times \mathbb{R}^s \to \mathbb{R}^c$ and $B(\cdot) \colon \mathbb{R}^c \to \mathbb{R}^{c \times m}$ are jointly continuous and Lipschitz, and $H(\cdot) \colon \mathbb{R}^c \to \mathbb{R}^b$ is the smooth map defined in the neighborhood of the origin, $H(0) = 0$.

The tracking control is the most comprehensive problem, and it was illustrated that the measured tracking error vector

$$e(t) = Nr(t) - y(t) \tag{5.3.2}$$

is used in control algorithms.

Our goal is to design controllers that not only stabilize system (5.3.1), but in addition ensure tracking (drive the error vector to a compact set) as well as guarantee disturbance attenuation. We consider a class of continuous or piecewise continuous uniformly bounded measurable reference inputs $r(\cdot) \colon [t_0, \infty) \to \mathbb{R}^b$ and disturbances $d(\cdot) \colon [t_0, \infty) \to \mathbb{R}^s$, such as

$$\|r\|_{L_p} = \left(\int_0^\infty \|r(t)\|^p \, dt \right)^{\frac{1}{p}} < \infty, \|d\|_{L_p} = \left(\int_0^\infty \|d(t)\|^p \, dt \right)^{\frac{1}{p}} < \infty, p \in [1, \infty].$$

That is, $r \in L_1[t_0, \infty) \cap L_\infty[t_0, \infty)$, $r \in R \subset \mathbb{R}^t$, and $d \in L_1[t_0, \infty) \cap L_\infty[t_0, \infty)$, $d \in D \subset \mathbb{R}^s$.

Here, $L_1[t_0, \infty) \cap L_\infty[t_0, \infty)$ is the space of the vector-valued functions that are Lebesgue measurable and uniformly bounded.

The Lyapunov theory and admissibility concept are used in the design. The constraints on control $u(\cdot) \colon [t_0, \infty) \to \mathbb{R}^m$ are imposed, and an admissible bounded controller should be designed as a continuous or piecewise continuous function within the constrained rectangular control set

$$U = \{u \in \mathbb{R}^m \colon u_{min} \leq u \leq u_{max}, u_{min} < 0, u_{max} > 0\} \subset \mathbb{R}^m.$$

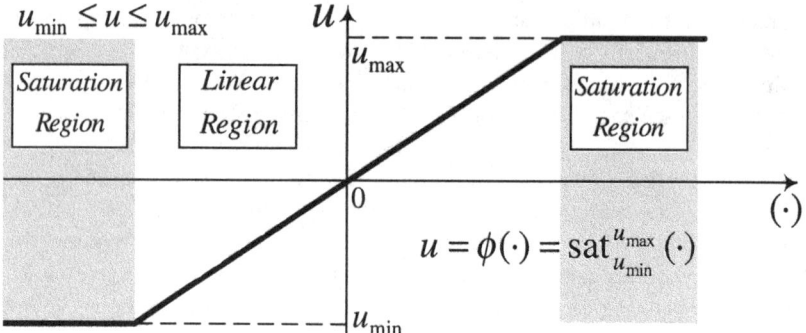

FIGURE 5.3.1.   Saturated control, $u_{\min} \le u \le u_{\max}$.

Making use of the Lyapunov candidate $V(t, x, e)$, the bounded controller is expressed as

$$
u = \phi \left(G_x(t) B(x)^T \frac{\partial V(t, x, e)}{\partial x} + G_e(t) B_e^T \frac{\partial V(t, x, e)}{\partial e} \right.
$$
$$
\left. + G_i(t) B_e^T \frac{1}{s} \frac{\partial V(t, x, e)}{\partial e} \right), \tag{5.3.3}
$$

where $\phi(\cdot)$: $\mathbb{R}_{\ge 0} \times \mathbb{R}^c \times \mathbb{R}^b \to \mathbb{R}^m$ is the continuous ($\epsilon \ge 1$) or piecewise continuous ($\epsilon = 0$) real-analytic bounded function of class C^ϵ, $\phi \in U$ for all $x \in X$ and $e \in E$ on $[t_0, \infty)$, $G_x(\cdot)$: $\mathbb{R}_{\ge 0} \to \mathbb{R}^{m \times m}$, $G_e(\cdot)$: $\mathbb{R}_{\ge 0} \to \mathbb{R}^{m \times m}$, and $G_i(\cdot)$: $\mathbb{R}_{\ge 0} \to \mathbb{R}^{m \times m}$ are the bounded symmetric matrix-functions defined on $[t_0, \infty)$, $G_x > 0$, $G_e > 0$, $G_i > 0$, and $V(\cdot)$: $\mathbb{R}_{\ge 0} \times \mathbb{R}^c \times \mathbb{R}^b \to \mathbb{R}_{\ge 0}$ is the continuously differentiable real-analytic C^κ ($\kappa \ge 1$) function with respect to $x \in X$ and $e \in E$ on $[t_0, \infty)$.

The application of the bounded function $\phi \in U$ should be illustrated. Let us assume that the control is saturated, and an admissible control $u_{\min} \le u \le u_{\max}$ is illustrated in Figure 5.3.1.

By mapping the control bounds by the piecewise continuous function sat, we have

$$
u = \mathrm{sat}_{u_{\min}}^{u_{\max}} \left(G_x(t) B(x)^T \frac{\partial V(t, x, e)}{\partial x} + G_e(t) B_e^T \frac{\partial V(t, x, e)}{\partial e} \right.
$$
$$
\left. + G_i(t) B_e^T \frac{1}{s} \frac{\partial V(t, x, e)}{\partial e} \right).
$$

For a closed-loop system (5.3.1)–(5.3.3) with $X_0 = \{x_0 \in \mathbb{R}^c\} \subseteq X \subset \mathbb{R}^c$, $u \in U \subset \mathbb{R}^m$, $r \in R \subset \mathbb{R}^b$, and $d \in D \subset \mathbb{R}^s$, it is straightforward to find an evolution set $X(X_0, U, R, D) \subset \mathbb{R}^c$.

By using the output equation, one has $X \xrightarrow{H} Y$ and the closed-loop system (5.3.1)–(5.3.3) evolves in

$$XY(X_0, U, R, D) = \{(x, y) \in X \times Y : x_0 \in X_0, u \in U, r \in R, d \in D,$$
$$t \in [t_0, \infty)\} \subset \mathbb{R}^c \times \mathbb{R}^b.$$

Approaching the design, it is important to emphasize that a reference-output map

$$\{\dot{x}(t) = F(x, r, d) + B(x)u, y = H(x), e(t) = Nr(t) - y(t),$$
$$x_0 \in X_0, e_0 \in E_0, d \in D, t \in [t_0, \infty)\} : R \mapsto Y$$

is considered. The measured error vector (tracking error)

$$e(t) = Nr(t) - y(t), e(\cdot) : [t_0, \infty) \to \mathbb{R}^b$$

represents the difference between the reference input $r(\cdot) : [t_0, \infty) \to \mathbb{R}^b$ and system output $y(\cdot) : [t_0, \infty) \to \mathbb{R}^b$. Our goal is to find the feedback gains in (5.3.3) to guarantee that the tracking error vector with $E_0 = \{e_0 \in \mathbb{R}^b\} \subseteq E \subset \mathbb{R}^b$ evolves in the specified closed set

$$S_e(\delta) = \{e \in \mathbb{R}^b : e_0 \in E_0, x \in X(X_0, U, R, D), r \in R, d \in D, y \in Y,$$
$$t \in [t_0, \infty) \mid \|e(t)\| \le \rho_e(t, \|e_0\|) + \rho_r(\|r\|) + \rho_d(\|d\|) + \rho_y(\|y\|) + \delta,$$
$$\delta \ge 0, \forall e \in E(E_0, R, D, Y), \forall t \in [t_0, \infty)\} \subset \mathbb{R}^b$$

for the considered initial conditions, control constraints, bounded reference inputs, and disturbances.

It is evident that for $r = const$ and $d = const$, the tracking error vector decays asymptotically to zero if $e \in E(E_0, R, D, Y) \subset S_e(\delta)$.

We use the following notations to define the boundedness of the error vector in $S_e(\delta)$: $\rho_e(\cdot) : \mathbb{R}_{\ge 0} \times \mathbb{R}_{\ge 0} \to \mathbb{R}_{\ge 0}$ is the KL-function; $\rho_r(\cdot) : \mathbb{R}_{\ge 0} \to \mathbb{R}_{\ge 0}, \rho_d(\cdot) : \mathbb{R}_{\ge 0} \to \mathbb{R}_{\ge 0}$, and $\rho_y(\cdot) : \mathbb{R}_{\ge 0} \to \mathbb{R}_{\ge 0}$ are the K-functions (continuous and strictly increasing).

It is important to emphasize that if the following is satisfied:

$$\|e(t)\| \le \rho_e(t, \|e_0\|) + \rho_r(\|r\|) + \rho_d(\|d\|) + \rho_y(\|y\|) + \delta,$$

one concludes that stability is guaranteed, and tracking and disturbance rejection are accomplished.

A positive-invariant domain of stability can be found for the closed-loop system (5.3.1)–(5.3.3) with $x_0 \in X_0, e_0 \in E_0, u \in U, r \in R$, and $d \in D$ as

$$S_s = \{x \in \mathbb{R}^c, e \in \mathbb{R}^b : \|x(t)\| \le \rho_x(t, \|x_0\|) + \rho_r(\|r\|) + \rho_d(\|d\|) + \delta_x,$$
$$\forall x \in X(X_0, U, R, D), \|e(t)\| \le \rho_e(t, \|e_0\|) + \rho_r(\|r\|) + \rho_d(\|d\|)$$
$$+ \rho_y(\|y\|) + \delta, \forall e \in E(E_0, R, D, Y), \forall t \in [t_0, \infty)\} \subset \mathbb{R}^c \times \mathbb{R}^b,$$

where $\rho_x(\cdot) : \mathbb{R}_{\ge 0} \times \mathbb{R}_{\ge 0} \to \mathbb{R}_{\ge 0}$ is the KL-function.

Theorem 5.3.1. *The solutions of* (5.3.1) *with* (5.3.3) *are uniformly ultimately bounded, and the equilibrium point is exponentially stable in the convex and compact set* $X(X_0, U, R, D) \subset \mathbb{R}^c$. *The tracking is ensured, and disturbance attenuation is guaranteed in* $XE(X_0, E_0, U, R, D) \subset \mathbb{R}^c \times \mathbb{R}^b$, *if a* C^κ *function* $V(t, x, e)$ *in* $S_s \subset \mathbb{R}^c \times \mathbb{R}^b$ *exists such that for all* $x \in X, e \in E, u \in U, r \in R$, *and* $d \in D$ *on* $[t_0, \infty)$

$$\rho_1 \|x\| + \rho_2 \|e\| \leq V(t, x, e) \leq \rho_3 \|x\| + \rho_4 \|e\|, \tag{5.3.4}$$

and along (5.3.1) *with* (5.3.3), *inequality*

$$\frac{dV(t, x, e)}{dt} \leq -\rho_5 \|x\| - \rho_6 \|e\| \tag{5.3.5}$$

holds. Furthermore, if (5.3.4) *and* (5.3.5) *are satisfied,* $XE \subseteq S_S$.

Here, $\rho_1(\cdot): \mathbb{R}_{\geq 0} \to \mathbb{R}_{\geq 0}, \rho_2(\cdot): \mathbb{R}_{\geq 0} \to \mathbb{R}_{\geq 0}, \rho_3(\cdot): \mathbb{R}_{\geq 0} \to \mathbb{R}_{\geq 0}$, *and* $\rho_4(\cdot): \mathbb{R}_{\geq 0} \to \mathbb{R}_{\geq 0}$ *are the* K_∞*-functions;* $\rho_5(\cdot): \mathbb{R}_{\geq 0} \to \mathbb{R}_{\geq 0}$ *and* $\rho_6(\cdot): \mathbb{R}_{\geq 0} \to \mathbb{R}_{\geq 0}$ *are the* K*-functions.*

Proof. For the closed-loop system (5.3.1)–(5.3.3), a state-error map is

$$XE(X_0, E_0, U, R, D) = \{(x, e) \in X \times E: x_0 \in X_0, e_0 \in E_0, u \in U, r \in R,$$
$$d \in D, t \in [t_0, \infty)\} \subset \mathbb{R}^c \times \mathbb{R}^b.$$

To ensure stability, an admissible domain $S_s \subset \mathbb{R}^c \times \mathbb{R}^b$ is found using the Lyapunov stability theory. In particular, the domain S_s is given by

$$S_s = \{x \in \mathbb{R}^c, e \in \mathbb{R}^b: x_0 \in X_0, e_0 \in E_0, u \in U, r \in R, d \in D, t \in [t_0, \infty)|$$
$$\rho_1 \|x\| + \rho_2 \|e\| \leq V(t, x, e) \leq \rho_3 \|x\| + \rho_4 \|e\|,$$
$$\frac{dV(t, x, e)}{dt} \leq -\rho_5 \|x\| - \rho_6 \|e\|,$$
$$\forall x \in X(X_0, U, R, D), \forall e \in E(E_0, R, D, Y),$$
$$\forall t \in [t_0, \infty)\} \subset \mathbb{R}^c \times \mathbb{R}^b.$$

The tracking and disturbance rejection are ensured if $XE(X_0, E_0, U, R, D) \subseteq S_s$ for given $X_0 \subset \mathbb{R}^c, E_0 \subset \mathbb{R}^b, U \subset \mathbb{R}^m, R \subset \mathbb{R}^b$, and $D \subset \mathbb{R}^s$. It is straightforward to show the following:

1) solutions $x(\cdot): \in [t_0, \infty) \to \mathbb{R}^c$ for system (5.3.1) with controller (5.3.3) are uniformly ultimately bounded, and the equilibrium point is exponentially stable in $S_s \subset \mathbb{R}^c \times \mathbb{R}^b$;

2) convergence of the error vector $e(\cdot): \in [t_0, \infty) \to \mathbb{R}^b$ to $S_e(\delta) \subset \mathbb{R}^b$ is ensured in the convex and compact set $XE(X_0, E_0, U, R, D) \subset \mathbb{R}^c \times \mathbb{R}^b$; that is, the tracking and disturbance attenuation are guaranteed;

3) $XE \subseteq S_s$, where XE is the state-error evolution set for the closed-loop system with bounded controller

if a C^κ Lyapunov function $V(t, x, e)$, K_∞-functions ρ_1, ρ_2, ρ_3, and ρ_4, and K-functions ρ_5 and ρ_6 exists, such that for all $x_0 \in X_0$, $e_0 \in E_0$, $u \in U$, $r \in R$, and $d \in D$ on $[t_0, \infty)$, stability criteria (5.3.4) and (5.3.5) are guaranteed.   □

Let us apply a family of nonquadratic Lyapunov candidates

$$
\begin{aligned}
V(t, x, e) = &\sum_{i=0}^{\eta} \frac{2\gamma + 1}{2(i + \gamma + 1)} \left(x^{\frac{i+\gamma+1}{2\gamma+1}} \right)^T K_{xi}(t) x^{\frac{i+\gamma+1}{2\gamma+1}} \\
&+ \sum_{i=0}^{\varsigma} \frac{2\beta + 1}{2(i + \beta + 1)} \left(e^{\frac{i+\beta+1}{2\beta+1}} \right)^T K_{ei}(t) e^{\frac{i+\beta+1}{2\beta+1}} \\
&+ \sum_{i=0}^{\sigma} \frac{2\mu + 1}{2(i + \mu + 1)} \left(e^{\frac{i+\mu+1}{2\mu+1}} \right)^T K_{si}(t) e^{\frac{i+\mu+1}{2\mu+1}}.
\end{aligned}
\tag{5.3.6}
$$

From (5.3.3) and (5.3.6), one obtains a bounded admissible controller as

$$
\begin{aligned}
u = \phi \Bigg(&G_x(t) B(x)^T \sum_{i=0}^{\eta} \text{diag}\left[x^{\frac{i-\gamma}{2\gamma+1}} \right] K_{xi}(t) x^{\frac{i+\gamma+1}{2\gamma+1}} \\
&+ G_e(t) B_e^T \sum_{i=0}^{\varsigma} \text{diag}\left[e^{\frac{i-\beta}{2\beta+1}} \right] K_{ei}(t) e^{\frac{i+\beta+1}{2\beta+1}} \\
&+ G_i(t) B_e^T \frac{1}{s} \sum_{i=0}^{\sigma} \text{diag}\left[e^{\frac{i-\mu}{2\mu+1}} \right] K_{si}(t) e^{\frac{i+\mu+1}{2\mu+1}} \Bigg).
\end{aligned}
\tag{5.3.7}
$$

Here, $K_{xi}(\cdot): \mathbb{R}_{\geq 0} \to \mathbb{R}^{c \times c}$, $K_{ei}(\cdot): \mathbb{R}_{\geq 0} \to \mathbb{R}^{b \times b}$, and $K_{si}(\cdot): \mathbb{R}_{\geq 0} \to \mathbb{R}^{b \times b}$ are the matrix-functions; $\eta = 0, 1, 2, \ldots, \gamma = 0, 1, 2, \ldots, \varsigma = 0, 1, 2, \ldots, \beta = 0, 1, 2, \ldots, \sigma = 0, 1, 2, \ldots,$ and $\mu = 0, 1, 2, \ldots$.

Plugging (5.3.7) into (5.3.1), one straightforwardly finds the total derivative of the Lyapunov candidate $V(t, x, e)$, and by solving inequality (5.3.5), the feedback gains are found. The solution of nonlinear inequalities results in convex programming.

Lemma 5.3.1. *Consider the convex set $S = \{k: f^N(k) \leq \alpha_k, N \in \mathcal{P}\}$ and differentiable convex function $f^N(\cdot): \mathbb{R}^f \to \mathbb{R}_{\geq 0}$. The finite index set of binding constraints at $k^\circ \in S$ is*

$$
\mathcal{P}(k^\circ) = \{N \in \mathcal{P}: f^N(k^\circ) = \alpha_k\}.
$$

The set of feasible direction $S_F(S, k^\circ)$ for $\{k: f^N(k) \leq f^N(k^\circ), N \in \mathcal{P}\}$, along which the convex function $f^N(\cdot): \mathbb{R}^f \to \mathbb{R}_{\geq 0}$ decreases, is found as

$$
S_F(S, k^\circ) = \Omega_{\overline{\mathcal{P}}(k^\circ)}^{\leq}(k^\circ) = \Omega_{\mathcal{P}(k^\circ)}^{<}(k^\circ) \cup \Omega_{\mathcal{P}(k^\circ)}^{=}(k^\circ).
$$

Here, N denotes the index in $\mathcal{P}$ (a set of positive integers); $\Omega_{\mathcal{P}(k^\circ)}^{<}(k^\circ)$ and $\Omega_{\mathcal{P}(k^\circ)}^{=}(k^\circ)$ are the convex cones, and $\Omega_{\mathcal{P}(k^\circ)}^{=}(k^\circ) \subset \{D_\nabla: \nabla f^N(k^\circ)^T D_\nabla = \alpha_k\}$; $\Omega_{\mathcal{P}(k^\circ)}^{<}(k^\circ)$ is the blunt convex cone, $\Omega_{\mathcal{P}(k^\circ)}^{<}(k^\circ) = \{D_\nabla: \nabla f^N(k^\circ)^T D_\nabla < \alpha_k\}$; and D_∇ is the descent direction.

Proof. The proof of this lemma is performed by using convex programming. Observe that N is the index in $\mathcal{P}$. Assume that $D_\nabla \in S_F(S, k^\circ)$. Then, positive-definite $\mu_1 > 0$ exists, and for all $\mu \in (0, \mu_1]$, we have $(k^\circ + \mu D_\nabla) \in S$.

If the directional derivative

$$\inf_{\mu > 0}\{[f^N(k^\circ + \mu D_\nabla) - f^N(k^\circ)]/\mu\} < 0,$$

then, D_∇ is the descent direction, and $f^N(k^\circ + \mu D_\nabla) \le f^N(k^\circ), \forall \mu \in (0, \mu_1]$. For $N \in \mathcal{P}$ and $\mu \in (0, \mu_1]$, we have $f^N(k^\circ + \mu D_\nabla) \le \alpha_k = f^N(k^\circ)$. Hence, one has $D_\nabla \in \Omega^{\le}_{\mathcal{P}(k^\circ)}(k^\circ)$.

Suppose that $D_\nabla \in \Omega^{\le}_{\mathcal{P}(k^\circ)}(k^\circ)$. For $N \in \mathcal{P}, \mu_2 > 0$ exists such that

$$f^N(k^\circ + \mu D_\nabla) \le f^N(k^\circ) = \alpha_k \text{ on } \mu \in (0, \mu_2].$$

Thus, it becomes evident that $D_\nabla \in S_F(S, k^\circ)$, proving Lemma 5.3.1. $\square$

Example 5.3.1. Control of a two–degree-of-freedom robot

This example introduces the application of the Lyapunov-based design method to design a controller for a two–degree-of-freedom manipulator actuated by synchronous motors. The control algorithms should be sought from basic machinery standpoints to ensure the acceptable regulating characteristics of servo-motors. Conventional design methods are limited to a certain class of servo-motors and their mathematical models. In particular, control of DC machines and AC motors (modeled in the *arbitrary* reference frame fixed with the rotor to transform the AC state and control quantities to the *quadrature-*, *direct-*, and *zero*-axis DC quantities) has been thoroughly studied in Chapter 2. However, the applied phase voltages to induction and synchronous motors are sinusoidal, and control of these servo-motors from closed-loop standpoints must be thoroughly approached. The purpose of this example is to demonstrate that the Lyapunov framework allows us to design controllers using basic electric machinery features, and common limitations can be overcome. Many industrial robots are actuated by permanent-magnet DC motors. However, steady-state and dynamic characteristics, efficiency, torque density, ruggedness, and reliability of synchronous machines exceed the corresponding characteristics of DC motors. Therefore, permanent-magnet synchronous motors have been applied in high-performance manipulators and servo-mechanisms. Control algorithms are limited by the corresponding electromagnetic and energy conversion phenomena in electric machines. Characteristics, principles, and regulating features of AC and DC motors are different. In high-performance robots, brushless synchronous motors are used (in a hazardous environment, one cannot use commutator-type motors).

As the starting point, one develops a mathematical model that describes manipulator–actuator dynamics, and the Lagrange equations of motion are used. It was shown that the application of the Lagrange concept allows us to integrate mechanical and electrical systems, as well as to enhance the clarity of coherent model developments. The objectives associated with system modeling include the development of mathematical models to analyze the robot performance and experiment with the system designed. The mathematical foundation, on which the manipulator model is based, affects the accuracy and validity of results. The Lagrange framework provides a great potential for deriving state-space models, and this concept is used to integrate mechanical and electrical systems. In particular, augmented nonlinear models can be derived to model the transient behavior of electromechanical robotic systems. This is a very important step that must be studied before approaching the design. Three major goals in motion control of robots are to guarantee stability, to ensure tracking with the desired positioning accuracy, and to accomplish disturbance attenuation.

The Lagrange equations of motion

$$\frac{d}{dt}\left(\frac{\partial \Gamma}{\partial \dot{q}_i}\right) - \frac{\partial \Gamma}{\partial q_i} + \frac{\partial D}{\partial \dot{q}_i} + \frac{\partial \Pi}{\partial q_i} = Q_i$$

are used. For one–degree-of-freedom servo-systems actuated by permanent-magnet synchronous motors, the mathematical model was developed in Chapter 2 (see Section 2.2). We recall that the independent generalized coordinates are the electric charges in the abc stator windings and the angular displacement of the rotor. The generalized forces are the applied phase voltages to the abc windings and the load torque. That is, $q_1 = \frac{i_{as}}{s}, \dot{q}_1 = i_{as}, q_2 = \frac{i_{bs}}{s}, \dot{q}_2 = i_{bs}, q_3 = \frac{i_{cs}}{s}, \dot{q}_3 = i_{cs}, q_4 = \theta_{rm}, \dot{q}_4 = \omega_{rm}, Q_1 = u_{as}, Q_2 = u_{bs}, Q_3 = u_{cs}$, and $Q_4 = -T_L$. One finds

$$\dot{x}(t) = F(x, r, d) + Bu, \quad y = Hx, \quad u_{\min} \le u \le u_{\max}, \quad x(t_0) = x_0,$$

$$x = \begin{bmatrix} i_{as} \\ i_{bs} \\ i_{cs} \\ \omega_r \\ \theta_r \end{bmatrix}, \quad u = \begin{bmatrix} u_{as} \\ u_{bs} \\ u_{cs} \end{bmatrix}, \quad d = T_L, \qquad (5.3.8)$$

where $x \in X \subset \mathbb{R}^{10}$ is the state vector, $i_{as} \in \mathbb{R}^2, i_{bs} \in \mathbb{R}^2, i_{cs} \in \mathbb{R}^2, \omega_r \in \mathbb{R}^2, \theta_r \in \mathbb{R}^2; u \in U \subset \mathbb{R}^6$ is the bounded control vector (applied phase voltages), $u_{as} \in \mathbb{R}^2, u_{bs} \in \mathbb{R}^2, u_{cs} \in \mathbb{R}^2; r \in R \subset \mathbb{R}^2$ and $y \in Y \subset \mathbb{R}^2$ are the measured reference and output vectors; $d \in D \subset \mathbb{R}^2$ is the disturbance vector (load torques); $F(\cdot): \mathbb{R}^{10} \times \mathbb{R}^2 \times \mathbb{R}^2 \to \mathbb{R}^{10}$ is a nonlinear Lipschitz map; and $B \in \mathbb{R}^{10 \times 6}$ and $H \in \mathbb{R}^{2 \times 10}$ are the matrices.

Hence, for the studied two–degree-of-freedom robot, one finds a set of 10 highly-coupled nonlinear differential equations (5.3.8), where the states and controls are

$$x = \begin{bmatrix} i_{as1} & i_{bs1} & i_{cs1} & \omega_{r1} & \theta_{r1} & i_{as2} & i_{bs2} & i_{cs2} & \omega_{r2} & \theta_{r2} \end{bmatrix}^T$$

$$\text{and } u = \begin{bmatrix} u_{as1} & u_{bs1} & u_{cs1} & u_{as2} & u_{bs2} & u_{cs2} \end{bmatrix}^T.$$

The angular rotor displacements θ_{rm1} and θ_{rm2}, and the joint manipulator variables are related as

$$\theta_1 = 0.05\theta_{rm1}, \quad \theta_2 = 0.05\theta_{rm2}.$$

Hence, for the four-pole motors used, the output equation results as

$$y = \begin{bmatrix} \theta_1 \\ \theta_2 \end{bmatrix}, \quad \theta_1 = 0.025\theta_{r1}, \quad \theta_2 = 0.025\theta_{r2}.$$

Differential equations for i_{as1}, i_{bs1}, i_{cs1}, i_{as2}, i_{bs2}, and i_{cs2} are obtained in Caushy's form and given in Example 3.2.18 with the corresponding ω_{r1}, θ_{r1}, ω_{r2}, and θ_{r2}. The following differential equations to model the torsional–mechanical dynamics are found for the first and second joints

$$\left[m_1 l_{c1}^2 + m_2 \left(l_1^2 + l_{c2}^2 + 2l_1 l_{c2} \cos \theta_2 \right) + J_1 + J_2 \right] \dot{\omega}_1$$
$$+ \left[m_2 \left(l_{c2}^2 + l_1 l_{c2} \cos \theta_2 \right) + J_2 \right] \dot{\omega}_2 - m_2 l_1 l_{c2} \sin \theta_2 \left(2\dot{\theta}_1 \dot{\theta}_2 + \dot{\theta}_2^2 \right)$$
$$+ (m_1 l_{c1} + m_2 l_1) g \cos \theta_1 + m_2 l_{c2} g \cos (\theta_1 + \theta_2)$$
$$= \tfrac{1}{2} P \psi_m \left[i_{as1} \cos \theta_{r1} + i_{bs1} \cos \left(\theta_{r1} - \tfrac{2}{3}\pi \right) + i_{cs1} \cos \left(\theta_{r1} + \tfrac{2}{3}\pi \right) \right],$$
$$\left[m_2 \left(l_{c2}^2 + l_1 l_{c2} \cos \theta_2 \right) + J_2 \right] \dot{\omega}_1 + \left(m_2 l_{c2}^2 + J_2 \right) \dot{\omega}_2 + m_2 l_1 l_{c2} \sin \theta_2 \dot{\theta}_1^2$$
$$+ m_2 l_{c2} g \cos (\theta_1 + \theta_2) = \tfrac{1}{2} P \psi_m \left[i_{as2} \cos \theta_{r2} + i_{bs2} \cos \left(\theta_{r2} - \tfrac{2}{3}\pi \right) \right.$$
$$\left. + i_{cs2} \cos \left(\theta_{r2} + \tfrac{2}{3}\pi \right) \right].$$

Hence, an augmented nonlinear model for a two–degree-of-freedom manipulator actuated by permanent-magnet synchronous motors is developed. The numerical values of parameters for the studied robot are $l_1 = 0.25$ m, $l_{c1} = 0.19$ m, $l_2 = 0.15$ m, $l_{c2} = 0.09$ m, $m_1 = 0.85$ kg, $m_2 = 0.32$ kg, $J_1 = 0.06$ kg-m^2, and $J_2 = 0.01$ kg-m^2. The rated data and parameters of a permanent-magnet synchronous motor H-232 are 135 W, 434 rad/sec, 40 V (*rms*), 0.42 N-m, 6.9 A, $r_s = 0.5\Omega$, $L_{ss} = 0.009$ H, $L_{ls} = 0.001$ H, $\psi_m = 0.069$ V-sec/rad, $B_m = 0.000013$ N-m-sec/rad, and $J = 0.000017$ kg-m^2. The applied phase voltages are bounded by $\sqrt{240}$ V.

The robot control is portioned in kinematic and motion control. The kinematic analysis is based on the relationships between the coordinates of an end-effector and the joint variables. A two-link planar manipulator is operated in the joint variable space, whereas the position of an object to be manipulated is expressed in the base frame system. The inverse kinematic problem should be solved to control the position and to realize the specified path tracking. The joint variables θ_1 and θ_2 of the planar robot, in terms of (x, y)-coordinates, are found as

$$\theta_1 = \tan^{-1} \left(\frac{y}{x} \right) - \tan^{-1} \left(\frac{l_2 \sin \theta_2}{l_1 + l_2 \cos \theta_2} \right), \quad \theta_2 = \tan^{-1} \left(\frac{\sqrt{1 - \Xi^2}}{\Xi} \right),$$

$$\Xi = \frac{x^2 + y^2 - l_1^2 - l_2^2}{2l_1 l_2}.$$

The end-effector coordinates (x, y) in the Cartesian frame (the forward kinematics) are

$$x = l_1 \cos \theta_1 + l_2 \cos(\theta_1 + \theta_2) \text{ and } y = l_1 \sin \theta_1 + l_2 \sin(\theta_1 + \theta_2).$$

Let us design a controller. We denote the reference inputs (desired angular displacement) as $r_{\theta 1}$ and $r_{\theta 2}$. Hence,

$$r = \begin{bmatrix} r_{\theta 1} \\ r_{\theta 2} \end{bmatrix}.$$

Using the reported control procedure, a bounded controller should be synthesized mapping the state variables

$$x = \begin{bmatrix} i_{as1} & i_{bs1} & i_{cs1} & \omega_{r1} & \theta_{r1} & i_{as2} & i_{bs2} & i_{cs2} & \omega_{r2} & \theta_{r2} \end{bmatrix}^T,$$

as well as the tracking errors

$$e_1(t) = r_{\theta 1}(t) - \theta_1(t) \text{ and } e_2(t) = r_{\theta 2}(t) - \theta_2(t).$$

The electromagnetic torque T_e developed by the synchronous motor is

$$T_e = \tfrac{1}{2} P \psi_m \left[i_{as} \cos \theta_r + i_{bs} \cos \left(\theta_r - \tfrac{2}{3}\pi \right) + i_{cs} \cos \left(\theta_r + \tfrac{2}{3}\pi \right) \right],$$

and it was shown that a balanced three-phase current or voltage set should be supplied.

The constraints on control $u(\cdot)$: $[t_0, \infty) \to \mathbb{R}^6$ (applied phase voltages u_{as}, u_{bs}, and u_{cs}) are imposed, and we derive an admissible bounded controller as

$$u = \begin{bmatrix} u_{as} \\ u_{bs} \\ u_{cs} \end{bmatrix} = \begin{bmatrix} \cos\theta_r & 0 & 0 \\ 0 & \cos\left(\theta_r - \tfrac{2}{3}\pi\right) & 0 \\ 0 & 0 & \cos\left(\theta_r + \tfrac{2}{3}\pi\right) \end{bmatrix}$$
$$\times \phi \left(G_x(t) B^T \frac{\partial V(t, x, e)}{\partial x} + G_e(t) B_e^T \frac{\partial V(t, x, e)}{\partial e} + G_i(t) B_e^T \frac{1}{s} \frac{\partial V(t, x, e)}{\partial e} \right),$$

(5.3.9)

where $|\phi(\cdot)| \le \sqrt{2} V_{LL}$, V_{LL} is the rated *rms* voltage, $\phi \in U$ for all $x \in X$ and $e \in E$, and $G_x(\cdot)$: $\mathbb{R}_{\ge 0} \to \mathbb{R}^{6\times6}$, $G_e(\cdot)$: $\mathbb{R}_{\ge 0} \to \mathbb{R}^{6\times6}$, and $G_i(\cdot)$: $\mathbb{R}_{\ge 0} \to \mathbb{R}^{6\times6}$ are the bounded symmetric matrix functions.

Using (5.3.6), from (5.3.9), we have

$$u = \begin{bmatrix} u_{as} \\ u_{bs} \\ u_{cs} \end{bmatrix} = \begin{bmatrix} \cos\theta_r & 0 & 0 \\ 0 & \cos\left(\theta_r - \tfrac{2}{3}\pi\right) & 0 \\ 0 & 0 & \cos\left(\theta_r + \tfrac{2}{3}\pi\right) \end{bmatrix}$$
$$\times \phi \left(G_x(t) B^T \sum_{i=0}^{\eta} \mathrm{diag}\left[x^{\frac{i-\gamma}{2\gamma+1}} \right] K_{xi}(t) x^{\frac{i+\gamma+1}{2\gamma+1}} \right.$$
$$+ G_e(t) B_e^T \sum_{i=0}^{\varsigma} \mathrm{diag}\left[e^{\frac{i-\beta}{2\beta1}} \right] K_{ei}(t) e^{\frac{i+\beta+1}{2\beta+1}}$$
$$\left. + G_i(t) B_e^T \frac{1}{s} \sum_{i=0}^{\sigma} \mathrm{diag}\left[e^{\frac{i-\mu}{2\mu+1}} \right] K_{si}(t) e^{\frac{i+\mu+1}{2\mu+1}} \right).$$

As an example, letting $\eta = \gamma = 1$, $\varsigma = \beta = 1$, and $\sigma = \mu = 1$ in (5.3.6), one finds

$$V(x, e) = \sum_{i=0}^{1} \frac{3}{2(i+2)} \left(x^{\frac{i+2}{3}} \right)^T K_{xi} x^{\frac{i+2}{3}} + \sum_{i=0}^{1} \frac{3}{2(i+2)} \left(e^{\frac{i+2}{3}} \right)^T K_{ei} e^{\frac{i+2}{3}}$$
$$+ \sum_{i=0}^{1} \frac{3}{2(i+2)} \left(e^{\frac{i+2}{3}} \right)^T K_{si} e^{\frac{i+2}{3}}.$$

We use the following Lyapunov function:

$$V(e) = \sum_{i=0}^{1} \frac{3}{2(i+2)} \left(e^{\frac{i+2}{3}}\right)^T K_{ei} e^{\frac{i+2}{3}} + \sum_{i=0}^{1} \frac{3}{2(i+2)} \left(e^{\frac{i+2}{3}}\right)^T K_{si} e^{\frac{i+2}{3}},$$

and from (5.3.9), a bounded controller is found to be

$$u = \begin{bmatrix} u_{as1} \\ u_{bs1} \\ u_{cs1} \\ u_{as2} \\ u_{bs2} \\ u_{cs2} \end{bmatrix}$$

$$= \begin{bmatrix} \cos\theta_r & 0 & 0 & 0 & 0 & 0 \\ 0 & \cos\left(\theta_{r1} - \frac{2}{3}\pi\right) & 0 & 0 & 0 & 0 \\ 0 & 0 & \cos\left(\theta_{r1} + \frac{2}{3}\pi\right) & 0 & 0 & 0 \\ 0 & 0 & 0 & \cos\theta_{r2} & 0 & 0 \\ 0 & 0 & 0 & 0 & \cos\left(\theta_{r2} - \frac{2}{3}\pi\right) & 0 \\ 0 & 0 & 0 & 0 & 0 & \cos\left(\theta_{r2} + \frac{2}{3}\pi\right) \end{bmatrix}$$

$$\times \text{sat}^{+\sqrt{240}}_{-\sqrt{240}}\left(B_e^T \sum_{i=0}^{1} \text{diag}\left[e^{\frac{i-1}{3}}\right] K_{ei}(t)e^{\frac{i+2}{3}} + \frac{1}{s} B_e^T \sum_{i=0}^{1} \text{diag}\left[e^{\frac{i-1}{3}}\right] K_{si}(t)e^{\frac{i+2}{3}}\right).$$

Solving matrix inequality (5.3.5), one finds matrices K_{e0}, K_{e1}, K_{s0}, and K_{s1}. That is, the feedback gains result. In expanded form, we have

$$u_{as1} = \cos\theta_{r1} \text{sat}^{+\sqrt{240}}_{-\sqrt{240}}\left(5.9e_1^{1/3} + 8.4e_1 + \frac{21e_1^{1/3}}{s} + \frac{72e_1}{s}\right),$$

$$u_{bs1} = \cos\left(\theta_{r1} - \frac{2}{3}\pi\right) \text{sat}^{+\sqrt{240}}_{-\sqrt{240}}\left(5.9e_1^{1/3} + 8.4e_1 + \frac{21e_1^{1/3}}{s} + \frac{72e_1}{s}\right),$$

$$u_{cs1} = \cos\left(\theta_{r1} + \frac{2}{3}\pi\right) \text{sat}^{+\sqrt{240}}_{-\sqrt{240}}\left(5.9e_1^{1/3} + 8.4e_1 + \frac{21e_1^{1/3}}{s} + \frac{72e_1}{s}\right),$$

$$u_{as2} = \cos\theta_{r2} \text{sat}^{+\sqrt{240}}_{-\sqrt{240}}\left(7.5e_2^{1/3} + 12e_2 + \frac{9e_2^{1/3}}{s} + \frac{114e_2}{s}\right), \qquad (5.3.10)$$

$$u_{bs2} = \cos\left(\theta_{r2} - \frac{2}{3}\pi\right) \text{sat}^{+\sqrt{240}}_{-\sqrt{240}}\left(7.5e_2^{1/3} + 12e_2 + \frac{9e_2^{1/3}}{s} + \frac{114e_2}{s}\right),$$

$$u_{cs2} = \cos\left(\theta_{r2} + \frac{2}{3}\pi\right) \text{sat}^{+\sqrt{240}}_{-\sqrt{240}}\left(7.5e_2^{1/3} + 12e_2 + \frac{9e_2^{1/3}}{s} + \frac{114e_2}{s}\right).$$

For a given nonquadratic function $V(e)$, sufficient criteria (5.3.4) and (5.3.5) are satisfied for all possible initial conditions, reference inputs, and disturbances in an entire operating envelope. To guide one to the manipulator dynamics, the designed algorithm (5.3.10) is verified through modeling. To illustrate the robot motions, we assign the desired orientation as

$$\text{Desired orientation} = \begin{cases} (x_{\text{initial}}, y_{\text{initial}}) = (0.4\ 0), \\ (x_{\text{desired}}, y_{\text{desired}}) = (0.2\ 0.3), \quad t \in [0\ 0.1) \text{ sec} \\ (x_{\text{initial}}, y_{\text{initial}}) = (0.2\ 0.3), \\ (x_{\text{desired}}, y_{\text{desired}}) = (0.4\ 0), \quad t \in [0.1\ 0.2] \text{ sec.} \end{cases}$$

Then, the specified angular displacements can be found. One obtains the reference inputs as

$$r_{\theta 1} = 1\ rad\ r_{\theta 2} = 0.61\ rad\ \text{ for } t \in [0\ 0.1) \text{ sec,}$$
$$\text{and } r_{\theta 1} = 0\ rad, r_{\theta 2} = 0\ rad\ \text{ for } t \in [0.1\ 0.2] \text{ sec.}$$

The transient behavior and orientation of an end-effector are illustrated in Figure 5.3.2. Stability, good dynamic performance, and precise tracking are evident. One concludes that the specified characteristics and accuracy have been achieved. The joint angles $\theta_1(t)$ and $\theta_2(t)$ follow the reference inputs $r_{\theta 1}(t)$ and $r_{\theta 2}(t)$, and the assigned orientation has been guaranteed.

The results illustrate that the controller designed guarantees stability, tracking accuracy, and disturbance attenuation.

5.3.2. Design of the Constrained Controllers for Uncertain Nonlinear Systems Using the Lyapunov Stability Theory

In this section, we study uncertain nonlinear systems when control inputs are subject to upper bounds on magnitude. The main goal is to solve the robust tracking control problem for multivariable dynamic systems. It has been shown that non-quadratic Lyapunov functions should be used to perform stability analysis, and proportional-integral tracking controllers with nonlinear error and state feedback are synthesized in this section. Robust control of uncertain nonlinear systems has received much attention, and a great number of design methods have been developed. The complexity of the maximum principle and the calculus of variations to approach the design of uncertain systems have facilitated the development of a large number of synthesis methods. The robust control theory has been extensively studied based on the Lyapunov and the Hamilton–Jacobi theories. In particular, robust performance analysis and design were researched using linear matrix inequalities and convex optimization. Viable and promising results were reported if the control is unbounded. We deal with the problem of designing bounded controllers for nonlinear systems with uncertain parameters. Although existing methods allow one to design the bounded stabilizing state-dependent controllers, robust tracking control is an important problem to be addressed and solved. For many real-world dynamic systems, one needs to study the ability of bounded partial state-feedback controllers to guarantee robust stability and tracking. For example,

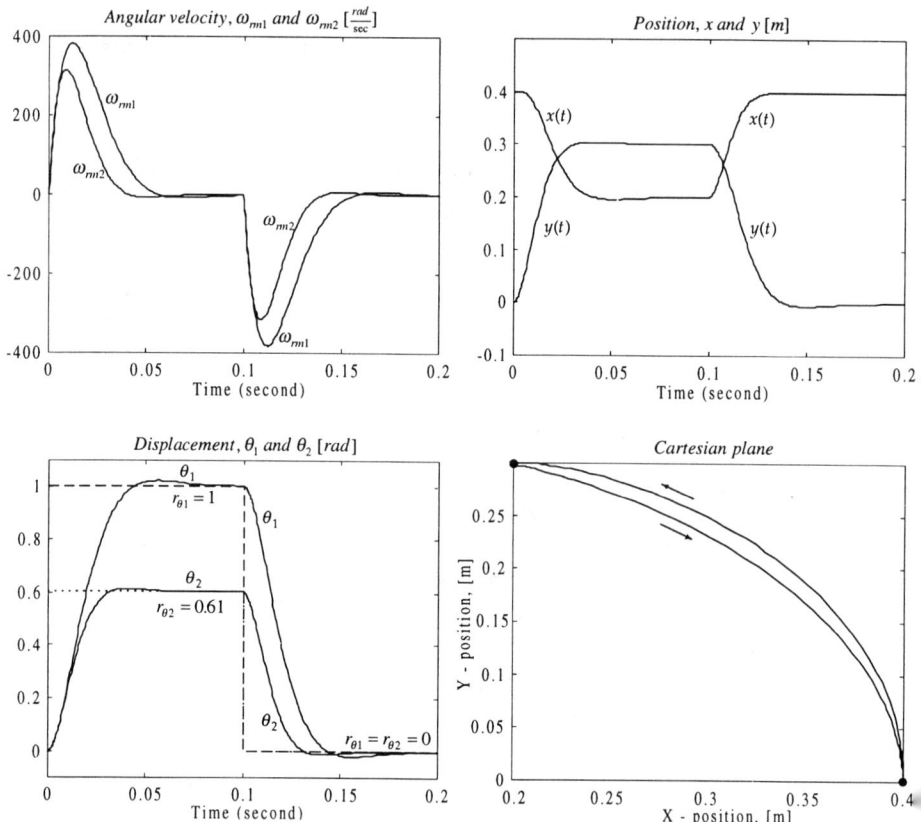

FIGURE 5.3.2.   Transient dynamics and orientation of an end-effector:
Desired orientation =

$$\begin{cases} (x_{\text{initial}}, y_{\text{initial}}) = (0.4\,0), (x_{\text{desired}}, y_{\text{desired}}) = (0.2\,0.3), t \in [0\,0.1) \text{ sec} \\ (x_{\text{initial}}, y_{\text{initial}}) = (0.2\,0.3), (x_{\text{desired}}, y_{\text{desired}}) = (0.4\,0), t \in [0.1\,0.2] \text{ sec} \end{cases}$$

partial state-feedback control laws should be desired if the measurability problem appears. This section performs the research in design of partial state-feedback control laws to attain robust stability.

In Chapter 4, it was illustrated that the Hamilton–Jacobi theory can be used to solve the constrained control problems through the application of the nonquadratic performance functionals. Under many circumstances, the dynamic programming method is a feasible and straightforward one because the ability to deal with constraints, nonlinearities, and uncertainties is in place. However, although the Hamilton–Jacobi theory is mathematically elegant and allows one analytically to design control algorithms minimizing performance functionals, a two-point boundary-value problem develops. The Lyapunov-based design methods attain reliable results and provide a simple and effective concept to study the stability, whereas the design of control algorithms is not fully supported. Much research remains to be done on strengthening the theoretical base, running the sophisticated numerical studies, and performing experimental verification. Our objectives are to study the synthesis concept using the Lyapunov theory as well as to research attendant mathematical issues in the design of robust bounded controllers.

5.3.2.1. Uncertain System Dynamics

Over the horizon $[t_0, \infty)$ we consider the dynamics of a nonlinear, n-dimensional, real-analytic, state-space system with constraints on the control inputs

$$\dot{x}(t) = F_z(t, x, r, z) + B_p(t, x, p)u, \; y = H(x), \; u_{\min} \leq u \leq u_{\max}, \; x(t_0) = x_0,$$
$$t \geq 0, \tag{5.3.11}$$

where $t \in \mathbb{R}_{\geq 0}$ is the time, $x \in X \subset \mathbb{R}^n$ is the state-space vector, $u \in U \subset \mathbb{R}^m$ is the vector of bounded control inputs, $r \in R \subset \mathbb{R}^b$ and $y \in Y \subset \mathbb{R}^b$ are the measured reference and output vectors, $z \in Z \subset \mathbb{R}^d$ and $p \in P \subset \mathbb{R}^k$ are the parameter uncertainties; functions $z(\cdot)$: $[t_0, \infty) \to Z$ and $p(\cdot)$: $[t_0, \infty) \to P$ are Lebesgue measurable and known within bounds; Z and P are the known nonempty compact sets; and $F_z(\cdot)$: $\mathbb{R}_{\geq 0} \times \mathbb{R}^n \times \mathbb{R}^b \times \mathbb{R}^d \to \mathbb{R}^n$, $B_p(\cdot)$: $\mathbb{R}_{\geq 0} \times \mathbb{R}^n \times \mathbb{R}^k \to \mathbb{R}^{n \times m}$, and $H(\cdot)$: $\mathbb{R}^n \to \mathbb{R}^b$ are the smooth mappings defined on open sets $S_F \subset \mathbb{R}_{\geq 0} \times \mathbb{R}^n \times \mathbb{R}^b \times \mathbb{R}^d$, $S_B \subset \mathbb{R}_{\geq 0} \times \mathbb{R}^n \times \mathbb{R}^k$, and $S_H \subset \mathbb{R}^n$, $F(t, 0, 0, z) = 0$, $B_u(t, 0, p) = 0$, $H(0) = 0$, $\forall t \in [t_0, \infty)$, $\forall z \in Z$, and $\forall p \in P$.

5.3.2.2. Robust Tracking

Let us formulate and solve the motion control problem by synthesizing robust controllers that guarantee stability and robust tracking. Our goal is to design control laws that robustly stabilize nonlinear systems with uncertain parameters (5.3.11) and drive the tracking error

$$e(t) = Nr(t) - y(t), \; e \in E \subset \mathbb{R}^b$$

robustly to the compact set. For systems modeled by nonlinear differential equations with parameter variations (5.3.11), the robust tracking of the measured output vector $y \in Y$ must be accomplished with respect to the measured uniformly bounded reference input vector $r \in R$. It is assumed that $Y \subset \mathbb{R}^b$ and $R \subset \mathbb{R}^b$ are convex and compact.

From (5.3.11), the nominal and uncertain dynamics are mapped by $F(\cdot)$: $\mathbb{R}_{\geq 0} \times \mathbb{R}^n \times \mathbb{R}^b \to \mathbb{R}^n$, $B(\cdot)$: $\mathbb{R}_{\geq 0} \times \mathbb{R}^n \to \mathbb{R}^{n \times m}$, and $\Xi(\cdot)$: $\mathbb{R}_{\geq 0} \times \mathbb{R}^n \times \mathbb{R}^m \times \mathbb{R}^d \times \mathbb{R}^k \to \mathbb{R}^n$. Hence, the system evolution is described as

$$\dot{x}(t) = F(t, x, r) + B(t, x)u + \Xi(t, x, u, z, p), \; y = H(x), \; u_{\min} \leq u \leq u_{\max},$$
$$x(t_0) = x_0, t \geq 0. \tag{5.3.12}$$

A norm of $\Xi(t, x, u, z, p)$ exists, and $\|\Xi(t, x, u, z, p)\| \leq \rho(t, x)$, where $\rho(\cdot)$: $\mathbb{R}_{\geq 0} \times \mathbb{R}_{\geq 0} \to \mathbb{R}_{\geq 0}$ is the continuous Lebesgue measurable function.

Our goal is to solve the motion control problem. That is, tracking controllers must be synthesized using the tracking error vector and the state variables. To guarantee robustness, to expand stability margins, to improve dynamic performance, and to meet other requirements, nonquadratic Lyapunov functions $V(t, e, x)$ will be used.

A set of admissible control $U \subset \mathbb{R}^m$ consists of the Lebesgue measurable function $u(\cdot)$: $[t_0, \infty) \to \mathbb{R}^m$. Although the Lyapunov theory does not fully support the design of control algorithms, it provides a feasible and reliable method to solve the robust regulation problem. The constraints on control inputs are imposed, and bounded continuous or discontinuous controllers should be designed within the constrained rectangular set

$$U = \{u \in \mathbb{R}^m : u_{\min} \leq u \leq u_{\max}, u_{\min} < 0, u_{\max} > 0\}.$$

Letting $u = \phi(t, e, x)$, one obtains a set of admissible controllers as given by

$$S_U = \{\phi(\cdot): \mathbb{R}_{\geq 0} \times \mathbb{R}^b \times \mathbb{R}^n \to \mathbb{R}^m | \phi(t, e, x) \in U, e \in E, x \in X,$$
$$t \in [t_0, \infty), \phi(t, 0, 0) = 0\},$$

where ϕ is the bounded, globally, Lipschitz, vector-valued, continuous ($\epsilon \geq 1$) or piecewise continuous ($\epsilon = 0$) function of class C^ϵ.

We define a family of tracking controllers as

$$u = \phi(t, e, x) = -\phi \left(G_P(t) B_E(t, x)^T \frac{\partial V(t, e, x)}{\partial e} + G_E(t) B_E(t, x)^T \right.$$
$$\left. \times \frac{1}{s} \frac{\partial V(t, e, x)}{\partial e} + G_x(t) B(t, x)^T \frac{\partial V(t, e, x)}{\partial x} \right), s = \frac{d}{dt}, \qquad (5.3.13)$$

where $G_P(\cdot)$: $\mathbb{R}_{\geq 0} \to \mathbb{R}^{m \times m}$, $G_E(\cdot)$: $\mathbb{R}_{\geq 0} \to \mathbb{R}^{m \times m}$, and $G_X(\cdot)$: $\mathbb{R}_{\geq 0} \to \mathbb{R}^{m \times m}$ are the diagonal matrix-functions defined on $[t_0, \infty)$; $B_E(\cdot)$: $\mathbb{R}_{\geq 0} \times \mathbb{R}^n \to \mathbb{R}^{b \times m}$ is the matrix function that is found from (5.3.12) with respect to the reference vector; for example, using $y = Hx$, one has $B_E(t, x) = HB(t, x)$ (matrix function B_E can be designed by the designer, and for induction motors, considered in Example 5.3.3, $B_E = \begin{bmatrix} \frac{L'_{rr}}{L_\Sigma} & \frac{L'_{rr}}{L_\Sigma} \end{bmatrix}$); and $V(\cdot)$: $\mathbb{R}_{\geq 0} \times \mathbb{R}^b \times \mathbb{R}^n \to \mathbb{R}_{\geq 0}$ is the continuous, differentiable, and real-analytic function.

The proportional and integral terms in (5.3.13)

$$G_p(t) B_E(t, x)^T \frac{1}{s} \frac{\partial V(t, e, x)}{\partial e}$$

and

$$G_E(t) B_E(t, x)^T \frac{1}{s} \frac{\partial V(t, e, x)}{\partial e}$$

accomplish the tracking of the bounded reference inputs. In addition, proportional state and error feedback results because of

$$G_X(t) B(t, x)^T \frac{\partial V(t, e, x)}{\partial x}.$$

These error and state feedback maps are applied to guarantee the robust tracking and to ensure the robust stability with respect to admissible sets.

Let us design the Lyapunov function. Nonquadratic functions $V(t, e, x)$ allow one to realize the full potential of the Lyapunov-based theory and lead one to the nonlinear feedback maps that are needed to attain stability and robustness, as well as to achieve conflicting design objectives. The following family of Lyapunov candidates is used:

$$V(t, e, x) = \sum_{i=0}^{\zeta} \frac{2\beta + 1}{2(i + \beta + 1)} \left(e^{\frac{i+\beta+1}{2\beta+1}} \right)^T K_{Ei}(t) e^{\frac{i+\beta+1}{2\beta+1}}$$

$$+ \sum_{i=0}^{\eta} \frac{2\gamma + 1}{2(i + \gamma + 1)} \left(x^{\frac{i+\gamma+1}{2\gamma+1}} \right)^T K_{Xi}(t) x^{\frac{i+\gamma+1}{2\gamma+1}}, \quad (5.3.14)$$

where $K_{Ei}(\cdot) : \mathbb{R}_{\geq 0} \to \mathbb{R}^{b \times b}$ and $K_{Xi}(\cdot) : \mathbb{R}_{\geq 0} \to \mathbb{R}^{n \times n}$ are the symmetric matrices; ζ, β, η, and γ are the nonnegative integers, $\zeta = 0, 1, 2, \ldots, \beta = 0, 1, 2, \ldots, \eta = 0, 1, 2, \ldots$, and $\gamma = 0, 1, 2, \ldots$.

One can readily apply (5.3.14) to design Lyapunov functions.

The well-known quadratic form

$$V(t, e, x) = \tfrac{1}{2} e^T K_{E0}(t) e + \tfrac{1}{2} x^T K_{X0}(t) x$$

is found by letting $\zeta = \beta = \eta = \gamma = 0$.

Using $\zeta = 1, \beta = 0, \eta = 1$, and $\gamma = 0$, one obtains

$$V(t, e, x) = \tfrac{1}{2} e^T K_{E0}(t) e + \tfrac{1}{4} e^{2^T} K_{E1}(t) e^2 + \tfrac{1}{2} x^T K_{X0}(t) x + \tfrac{1}{4} x^{2^T} K_{X1}(t) x^2.$$

Assigning $\zeta = 4, \beta = 1, \eta = 1$, and $\gamma = 0$, we have the following nonquadratic function:

$$V(t, e, x) = \tfrac{3}{4} e^{2/3^T} K_{E0}(t) e^{2/3} + \tfrac{1}{2} e^T K_{E1}(t) e + \tfrac{3}{8} e^{4/3^T} K_{E2}(t) e^{4/3}$$

$$+ \tfrac{3}{10} e^{5/3^T} K_{E3}(t) e^{5/3} + \tfrac{1}{4} e^{2^T} K_{E4}(t) e^2 + \tfrac{1}{2} x^T K_{X0}(t) x$$

$$+ \tfrac{1}{4} x^{2^T} K_{X1}(t) x^2.$$

Letting $G_P(t) = 0$, from (5.3.13) and (5.3.14), one obtains the tracking control law as

$$u = -\phi \left(G_E(t) B_E(t, x)^T \sum_{i=0}^{\zeta} \text{diag} \left[e(t)^{\frac{i-\beta}{2\beta+1}} \right] K_{Ei}(t) \frac{1}{s} e(t)^{\frac{i+\beta+1}{2\beta+1}} \right.$$

$$\left. + G_X(t) B(t, x)^T \sum_{i=0}^{\eta} \text{diag} \left[x(t)^{\frac{i-\gamma}{2\gamma+1}} \right] K_{Xi}(t) x(t)^{\frac{i+\gamma+1}{2\gamma+1}} \right), \quad (5.3.15)$$

where

$$
\mathrm{diag}\left[e(t)^{\frac{i-\beta}{2\beta+1}}\right] =
\begin{bmatrix}
e_1^{\frac{i-\beta}{2\beta+1}} & 0 & \cdots & 0 & 0 \\
0 & e_2^{\frac{i-\beta}{2\beta+1}} & \cdots & 0 & 0 \\
\vdots & \vdots & \ddots & \vdots & \vdots \\
0 & 0 & \cdots & e_{b-1}^{\frac{i-\beta}{2\beta+1}} & 0 \\
0 & 0 & \vdots & 0 & e_b^{\frac{i-\beta}{2\beta+1}}
\end{bmatrix}
$$

and

$$
\mathrm{diag}\left[x(t)^{\frac{i-\gamma}{2\gamma+1}}\right] =
\begin{bmatrix}
x_1^{\frac{i-\gamma}{2\gamma+1}} & 0 & \cdots & 0 & 0 \\
0 & x_2^{\frac{i-\gamma}{2\gamma+1}} & \cdots & 0 & 0 \\
\vdots & \vdots & \ddots & \vdots & \vdots \\
0 & 0 & \cdots & x_{n-1}^{\frac{i-\gamma}{2\gamma+1}} & 0 \\
0 & 0 & \vdots & 0 & x_n^{\frac{i-\gamma}{2\gamma+1}}
\end{bmatrix}.
$$

If matrices K_{Ei} and K_{Xi} are diagonal, we have

$$
u = -\phi\left(G_E(t)B_E(t,x)^T \sum_{i=0}^{\varsigma} K_{Ei}(t)\frac{1}{s}e(t)^{\frac{2i+1}{2\beta+1}} \right.
$$
$$
\left. + G_X(t)B(t,x)^T \sum_{i=0}^{\eta} K_{Xi}(t)x(t)^{\frac{2i+1}{2\gamma+1}} \right).
$$

It should be emphasized that controller (5.3.15) is designed using Lyapunov candidate (5.3.14). Because other Lyapunov functions are available, the control law in (5.3.13) and $V(t, e, x)$ are used to formulate the general results as given by the Lemma 5.3.2 and Theorem 5.3.2. The following lemma establishes sufficient conditions for robust tracking.

Lemma 5.3.2. *Given an uncertain system described by (5.3.11), an admissible control $u(\cdot)$: $[t_0, \infty) \to \mathbb{R}^m$, as given by (5.3.13), generates solution $x(\cdot)$: $[t_0, \infty) \to \mathbb{R}^n$, $x(t_0) = x_0$, and $x \in X(X_0, U, Z, P) \subset \mathbb{R}^n$, $0 \in IntX$. A closed-loop uncertain system is robustly stable in $X(X_0, U, Z, P) \subset \mathbb{R}^n$, and robust tracking is guaranteed in the convex and compact set $E(E_0, Y, R) \subset \mathbb{R}^b$ if for reference inputs $r \in R$ and uncertainties in Z and P, a C^κ ($\kappa \geq 1$) function $V(\cdot)$: $\mathbb{R}_{\geq 0} \times \mathbb{R}^b \times \mathbb{R}^n \to \mathbb{R}_{\geq 0}$ exists; there exist as well K_∞-functions $\rho_{X1}(\cdot)$: $\mathbb{R}_{\geq 0} \to \mathbb{R}_{\geq 0}$, $\rho_{X2}(\cdot)$: $\mathbb{R}_{\geq 0} \to \mathbb{R}_{\geq 0}$, $\rho_{E1}(\cdot)$: $\mathbb{R}_{\geq 0} \to \mathbb{R}_{\geq 0}$, $\rho_{E2}(\cdot)$: $\mathbb{R}_{\geq 0} \to \mathbb{R}_{\geq 0}$ and K-functions $\rho_{X3}(\cdot)$: $\mathbb{R}_{\geq 0} \to \mathbb{R}_{\geq 0}$, $\rho_{E3}(\cdot)$: $\mathbb{R}_{\geq 0} \to \mathbb{R}_{\geq 0}$, such that the following criteria:*

$$
\rho_{X1}(\|x\|) + \rho_{E1}(\|e\|) \leq V(t, e, x) \leq \rho_{X2}(\|x\|) + \rho_{E2}(\|e\|), \tag{5.3.16}
$$

$$
\frac{dV(t, e, x)}{dt} \leq -\rho_{X3}(\|x\|) - \rho_{E3}(\|e\|), \tag{5.3.17}
$$

are guaranteed in an invariant domain of stability $S \subset \mathbb{R}^n \times \mathbb{R}^b$ and $XE(X_0, E_0, U, R, Z, P) \subseteq S$.

Proof. The important feature should be emphasized. The bounded controller should be designed for an open-loop unstable system (5.3.11). Even for linear time-invariant systems $\dot{x}(t) = Ax + Bu$, $u \in U$, bounded controllers do not exist that globally stabilize the closed-loop systems if $\lambda(A) \subseteq \mathbb{C}^+$, where $\lambda(A)$ denote the eigenvalues of $A \in \mathbb{R}^{n \times n}$; $\mathbb{C}^+$ denotes the right-half part of the complex plane.

The closed-loop system with X_0, E_0, U, R, Z, and P evolves on $XE(X_0, E_0, U, R, Z, P)$, and the input–output map is found. An invariant domain of stability $S \subset \mathbb{R}^n \times \mathbb{R}^b$ can be obtained for the closed-loop system (5.3.11)–(5.3.13). In particular,

$$
\begin{aligned}
S = \{ x \in \mathbb{R}^n, e \in \mathbb{R}^b : \; & \|x(t)\| \leq \rho_x(\|x_0\|, t) + \rho_u(\|u\|) + \delta_x, \\
& \forall x \in X(X_0, U, Z, P), \forall t \in [t_0, \infty), \|e(t)\| \leq \rho_e(\|e_0\|, t) + \rho_y(\|y\|) \\
& + \rho_r(\|r\|) + \delta, \forall e \in E(E_0, Y, R), \forall t \in [t_0, \infty) \},
\end{aligned}
$$

where $\rho_x(\cdot) \colon \mathbb{R}_{\geq 0} \times \mathbb{R}_{\geq 0} \to \mathbb{R}_{\geq 0}$ and $\rho_e(\cdot) \colon \mathbb{R}_{\geq 0} \times \mathbb{R}_{\geq 0} \to \mathbb{R}_{\geq 0}$ are the KL-functions, and $\rho_u(\cdot) \colon \mathbb{R}_{\geq 0} \to \mathbb{R}_{\geq 0}$, $\rho_y(\cdot) \colon \mathbb{R}_{\geq 0} \to \mathbb{R}_{\geq 0}$, and $\rho_r(\cdot) \colon \mathbb{R}_{\geq 0} \to \mathbb{R}_{\geq 0}$ are the K-functions.

The state and error vectors are bounded in S, and the stability is guaranteed if $XE(X_0, E_0, U, R, Z, P) \subseteq S, \forall x_0 \in X_0, \forall e_0 \in E_0, \forall u \in U, \forall r \in R, \forall z \in Z, \forall p \in P$ on $[t_0, \infty)$. The admissible domain is found by using the Lyapunov theory, and we have

$$
\begin{aligned}
S = \{ x \in \mathbb{R}^n, e \in \mathbb{R}^b : \; & x_0 \in X_0, e_0 \in E_0, u \in U, r \in R, z \in Z, p \in P, \\
& t \in [t_0, \infty) | \rho_{X1}(\|x\|) + \rho_{E1}(\|e\|) \leq V(t, e, x) \leq \rho_{X2}(\|x\|) + \rho_{E2}(\|e\|), \\
& \frac{dV(t, e, x)}{dt} \leq -\rho_{X3}(\|x\|) - \rho_{E3}(\|e\|) \}.
\end{aligned}
$$

Robust stability and tracking in XE are studied, and exponential convergence can be proven. Under the assumption that sets X_0, E_0, U, R, Z, and P are admissible, a bounded controller exists that guarantees that the closed-loop system evolves in $XE \subseteq S$. Hence, from $XE \subseteq S$, one concludes that robust stability and tracking are guaranteed in XE. That is, the bounded real-analytic control function $u(\cdot)$ (5.3.13) guarantees robust input–output tracking performance for the resulting closed-loop system and steers the output vector to the set

$$
\begin{aligned}
S_E(\delta) = \{ e \in \mathbb{R}^b : \; & e_0 \in E_0, x \in X(X_0, U, Z, P), r \in R, y \in Y, \\
& t \in [t_0, \infty) \mid \|e(t)\| \leq \rho_e(\|e_0\|, t) + \rho_y(\|y\|) + \rho_r(\|r\|) + \delta, \\
& \delta > 0 \}. \tag{5.3.18}
\end{aligned}
$$

If $XE \subseteq S$, XE is a positive-invariant robust domain for the trajectories of system (5.3.11) with (5.3.13) on $[t_0, \infty)$. One concludes that trajectories $x(\cdot) \colon [t_0, \infty) \to \mathbb{R}^n$ and $e(\cdot) \colon [t_0, \infty) \to \mathbb{R}^b$ do not exceed an admissible domain of stability S, all solutions are bounded in XE, and convergence of the tracking error to the compact set $S_E(\delta)$ is guaranteed. Hence, controller (5.3.13), generating $x(\cdot) \colon [t_0, \infty) \to \mathbb{R}^n$ and $e(\cdot) \colon [t_0, \infty) \to \mathbb{R}^b$ of system (5.3.11), renders robust tracking with respect to *a priori* given sets if sufficient conditions (5.3.16) and (5.3.17) are guaranteed in S, and $XE \subseteq S$.   □

Definition 5.3.1. A set $S \subset \mathbb{R}^n \times \mathbb{R}^b$ is an invariant domain of robust stability for the closed-loop system (5.3.11) with (5.3.13) if $x(\cdot)$: $[t_0, \infty) \to \mathbb{R}^n$, $x(t_0) = x_0$ and $e(\cdot)$: $[t_0, \infty) \to \mathbb{R}^b$, $e(t_0) = e_0$ starting in S remain in S for all $z(\cdot)$: $[t_0, \infty) \to Z$ and $p(\cdot)$: $[t_0, \infty) \to P$.

Definition 5.3.2. The robust tracking is achieved if for admissible bounded reference inputs $r(\cdot)$: $[t_0, \infty) \to \mathbb{R}^b$, the error vector $e(\cdot)$: $[t_0, \infty) \to \mathbb{R}^b$, $e(t_0) = e_0$ is bounded, and

$$\|e(t)\| \le \rho_e(\|e_0\|, t) + \rho_y(\|y\|) + \rho_r(\|r\|) + \delta$$

for all $z(\cdot)$: $[t_0, \infty) \to Z$ and $p(\cdot)$: $[t_0, \infty) \to P$.

Definition 5.3.3. Suppose that a continuously differentiable function $V(\cdot)$: $\mathbb{R}_{\ge 0} \times \mathbb{R}^b \times \mathbb{R}^n \to \mathbb{R}_{\ge 0}$ exists such that for (5.3.11)–(5.3.13), sufficient conditions (5.3.16) and (5.3.17) are guaranteed in S. Then, the closed-loop system (5.3.11) with (5.3.13) is robustly stable in S. If $XE \subseteq S$, the equilibrium state is robustly stable, evolutions $x(\cdot)$: $[t_0, \infty) \to \mathbb{R}^n$, $x(t_0) = x_0$ and $e(\cdot)$: $[t_0, \infty) \to \mathbb{R}^b$, $e(t_0) = e_0$ are bounded, and $V(t, e, x)$ is a Lyapunov function.

Lemma 5.3.2 formulates the sufficient conditions under which the robust control problem is solvable. Computing the derivative of the $V(t, e, x)$, along with (5.3.11)–(5.3.13), the unknown coefficients of $V(t, e, x)$ can be found using (5.3.17); that is, matrices $K_{Ei}(\cdot)$: $\mathbb{R}_{\ge 0} \to \mathbb{R}^{b \times b}$ and $K_{Xi}(\cdot)$: $\mathbb{R}_{\ge 0} \to \mathbb{R}^{n \times n}$ of (5.3.14) are obtained. This problem is solved using the nonlinear inequality concept. The problem can be formulated by using the Lyapunov- and Riccati-type inequalities using quadratic Lyapunov candidates $V(t, e, x) = \frac{1}{2} e^T K_{E0}(t)e + \frac{1}{2} x^T K_{X0}(t)x$. However, it is difficult to meet the performance specifications, stability, and robustness using quadratic candidates, whereas nonquadratic Lyapunov functions offer the desired result. Hence, the nonlinear inequalities are needed to be solved. Solving these inequalities is a convex problem that consists of finding a feasible solution. Convex programming should be applied, and inequality (5.3.17), because of the affine dependence, gives $K_{Ei}(\cdot)$: $\mathbb{R}_{\ge 0} \to \mathbb{R}^{b \times b}$ and $K_{Xi}(\cdot)$: $\mathbb{R}_{\ge 0} \to \mathbb{R}^{n \times n}$.

Theorem 5.3.2. *Consider an uncertain system as modeled by* (5.3.11). *Bounded control $u(\cdot)$: $[t_0, \infty) \to \mathbb{R}^m$* (5.3.13) *and reference $r(\cdot)$: $[t_0, \infty) \to \mathbb{R}^b$ generate system solutions, and $x(\cdot)$: $[t_0, \infty) \to \mathbb{R}^n$ with $x(t_0) = x_0$ and $e(\cdot)$: $[t_0, \infty) \to \mathbb{R}^b$ with $e(t_0) = e_0$ evolve in XE. If the sufficient conditions for robust stability and tracking* (5.3.16) *and* (5.3.17) *are satisfied in S by $V(t, e, x) \in C^\kappa$ and $XE \subseteq S$, $V(\cdot)$: $\mathbb{R}_{\ge 0} \times \mathbb{R}^b \times \mathbb{R}^n \to \mathbb{R}_{\ge 0}$ is a Lyapunov function with respect to $XE(X_0, E_0, U, R, Z, P)$. The bounded C^ϵ, control law* (5.3.13) *guarantees the robust stability and tracking, and the tracking error vector $e(t)$ converges to the compact set* (5.3.18).

Proof. Nonlinearities in (5.3.11) are locally Lipschitz, and ϕ and $V(t, e, x)$ are C^ϵ and C^κ bounded functions. Then, the solution of (5.3.11) with (5.3.13) is unique on its interval of existence $[t_0, \infty)$. Sufficient conditions (5.3.16) and (5.3.17) are examined in S. Controller (5.3.13) guarantees the robust stability for system (5.3.11) and its output $y(t)$ tracks the uniformly bounded reference $r(t)$ if a real-analytic and continuous C^κ function $V(\cdot)$: $\mathbb{R}_{\geq 0} \times \mathbb{R}^b \times \mathbb{R}^n \to \mathbb{R}_{\geq 0}$ exists such that (5.3.16) and (5.3.17) are satisfied in S and $XE \subseteq S$. Lemma 5.3.2 supports the proof of this theorem, which is virtually identical to the one given by Lyapunov. It should be emphasized that if the constrained controller (5.3.13) exists, which guarantees the robust stability and tracking for (5.3.11), a Lyapunov function $V(t, e, x)$ exists. □

Lemma 5.3.3. *A smooth C^κ function $V(\cdot)$: $\mathbb{R}_{\geq 0} \times \mathbb{R}^b \times \mathbb{R}^n \to \mathbb{R}_{\geq 0}$ is a Lyapunov function for uncertain system (5.3.11)–(5.3.13) in $XE(X_0, E_0, U, R, Z, P)$ if there exist K_∞-functions $\rho_{X1}(\cdot)$: $\mathbb{R}_{\geq 0} \to \mathbb{R}_{\geq 0}$, $\rho_{X2}(\cdot)$: $\mathbb{R}_{\geq 0} \to \mathbb{R}_{\geq 0}$, $\rho_{E1}(\cdot)$: $\mathbb{R}_{\geq 0} \to \mathbb{R}_{\geq 0}$, $\rho_{E2}(\cdot)$: $\mathbb{R}_{\geq 0} \to \mathbb{R}_{\geq 0}$ and K-functions $\rho_{X3}(\cdot)$: $\mathbb{R}_{\geq 0} \to \mathbb{R}_{\geq 0}$, $\rho_{E3}(\cdot)$: $\mathbb{R}_{\geq 0} \to \mathbb{R}_{\geq 0}$ such that for all $x \in X(X_0, U, Z, P)$ and $e \in E(E_0, Y, R)$ on $[t_0, \infty)$*

$$\rho_{X1}(\|x\|) + \rho_{E1}(\|e\|) \leq V(t, e, x) \leq \rho_{X2}(\|x\|) + \rho_{E2}(\|e\|),$$

$$\frac{\partial V(t, e, x)}{\partial t} + \frac{dV(t, e, x)}{dt}^T \left[F_z(t, x, r, z) - B_P \phi \left(G_P(t) B_E(t, x)^T \frac{\partial V(t, e, x)}{\partial e} \right. \right.$$

$$\left. \left. + G_E(t) B_E(t, x)^T \frac{1}{s} \frac{\partial V(t, e, x)}{\partial e} + G_X(t) B(t, x)^T \frac{\partial V(t, e, x)}{\partial x} \right) \right]$$

$$\leq -\rho_{X3}(\|x\|) - \rho_{E3}(\|e\|)$$

Proof. Using the results of Lemma 5.3.2 and Theorem 5.3.2 the proof is straightforward. □

5.3.2.3. Partial State-Feedback Control

It is important to design partial state-feedback tracking controllers. The state vector is partitioned by using the measured and unmeasured states, and

$$x = \begin{bmatrix} x_M \\ x_U \end{bmatrix},$$

where $x_M \in \mathbb{R}^c$ and $x_U \in \mathbb{R}^q$ are the measured and unmeasured state variables.

Using the directly measured states, the tracking control law can be designed as

$$u = \phi(t, e, x_M) = -\phi \left(G_P(t) B_{EM}(t, x_M)^T \frac{\partial V(t, e, x)}{\partial e} + G_E(t) B_{EM}(t, x_M)^T \right.$$

$$\left. \times \frac{1}{s} \frac{\partial V(t, e, x)}{\partial e} + G_X(t) B_{XM}(t, x_M)^T \frac{\partial V(t, e, x)}{\partial x_M} \right),$$

where $B_{EM}(\cdot)$: $\mathbb{R}_{\geq 0} \times \mathbb{R}^c \to \mathbb{R}^{m \times b}$ and $B_{XM}(\cdot)$: $\mathbb{R}_{\geq 0} \times \mathbb{R}^c \to \mathbb{R}^{m \times c}$ are the matrix functions found using x_M.

By using B_{EM} and B_{XM}, and taking note of the Lyapunov function

$$V(t, e, x) = v_E(t, e) + v_{XM}(t, x_M) + v_{XU}(t, x_U),$$

a partial state-feedback controller results because only measured states x_M are used.

Here, $v_E(\cdot): \mathbb{R}_{\geq 0} \times \mathbb{R}^b \to \mathbb{R}_{\geq 0}$, $v_{XM}(\cdot): \mathbb{R}_{\geq 0} \times \mathbb{R}^c \to \mathbb{R}_{\geq 0}$, and $v_{XU}(\cdot): \mathbb{R}_{\geq 0} \times \mathbb{R}^q \to \mathbb{R}_{\geq 0}$ are the smooth, real-analytic functions, $v_E(t, 0) = 0$, $v_{XM}(t, 0) = 0$, and $v_{XU}(t, 0) = 0$, $\forall t \in [t_0, \infty)$.

Taking note of $x = \begin{bmatrix} x_M \\ x_U \end{bmatrix}$, the matrix $K_{Xi}(\cdot): \mathbb{R}_{\geq 0} \to \mathbb{R}^{n \times n}$ in (5.3.14) is

$$K_{Xi}(t) = \begin{bmatrix} K_{XMi}(t) & 0 \\ 0 & K_{XUi}(t) \end{bmatrix}, \quad K_{XMi}(\cdot): \mathbb{R}_{\geq 0} \to \mathbb{R}^{c \times c}, \quad K_{XUi}(\cdot): \mathbb{R}_{\geq 0} \to \mathbb{R}^{q \times q},$$

and the partial state-feedback control law results. In particular, for $G_P(t) = 0$, we have

$$u = -\phi \left(G_E(t) B_{EM}(t, x_M)^T \sum_{i=0}^{\varsigma} \text{diag}\left[e(t)^{\frac{i-\beta}{2\beta+1}} \right] K_{Ei}(t) \frac{1}{s} e(t)^{\frac{i+\beta+1}{2\beta+1}} \right.$$
$$\left. + G_X(t) B_{XM}(t, x_M)^T \sum_{i=0}^{\eta} \text{diag}\left[x_M(t)^{\frac{i-\gamma}{2\gamma+1}} \right] K_{Xi}(t) x_M(t)^{\frac{i+\gamma+1}{2\gamma+1}} \right).$$

Example 5.3.2. Control of a drive with permanent-magnet DC motor

Consider an electric drive actuated by a permanent-magnet motor JDH-2250. The parameters of this motor are $r_a(\cdot) \in [2.7_{T=20^\circ C} \quad 3.7_{T=140^\circ C}] \, \Omega$ and $L_a = 0.004$ H; back emf and torque constant is $k_a(\cdot) \in [0.11_{T=20^\circ C} \quad 0.094_{T=140^\circ C}]$ V-sec/rad (N-m/A); moment of inertia is $J = 0.0001$ kg-m^2; and viscous friction coefficient is $B_m = 0.00008$ N-m-sec/rad.

Permanent-magnet motor and step-down converter are shown in Figure 5.3.3 (the switching frequency is 50 kHz). The converter has an internal resistance $r_d = 0.05\Omega$, and a low-pass filter ($L_L = 0.0007H$ and $C_L = 0.003F$) is inserted to ensure the specified 5% voltage ripple.

We have the following nonlinear state-space model with bounded control:

$$\begin{bmatrix} \frac{du_a}{dt} \\ \frac{di_L}{dt} \\ \frac{di_a}{dt} \\ \frac{d\omega_r}{dt} \end{bmatrix} = \begin{bmatrix} 0 & \frac{1}{C_L} & -\frac{1}{C_L} & 0 \\ -\frac{1}{L_L} & 0 & 0 & 0 \\ \frac{1}{L_a} & 0 & -\frac{r_a}{L_a} & -\frac{k_a}{L_a} \\ 0 & 0 & \frac{k_a}{J} & -\frac{B_m}{J} \end{bmatrix} \begin{bmatrix} u_a \\ i_L \\ i_a \\ \omega_r \end{bmatrix} + \begin{bmatrix} 0 \\ \left(\frac{V_d}{L_L u_{t\,max}} - \frac{r_d}{L_L u_{t\,max}} i_L \right) \\ 0 \\ 0 \end{bmatrix} u_c$$

$$- \begin{bmatrix} 0 \\ 0 \\ 0 \\ \frac{1}{J} \end{bmatrix} T_L, \quad u_c \in [0 \quad 10] \, \text{V}.$$

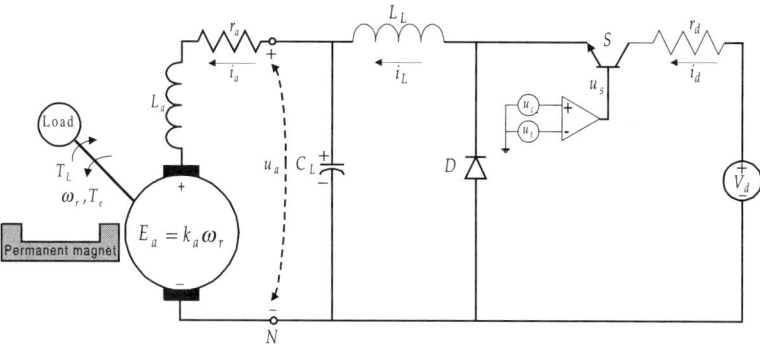

FIGURE 5.3.3.   Permanent-magnet DC motor with step-down converter.

A bounded control law should be synthesized. We apply the following Lyapunov candidate:

$$V(e,x) = \tfrac{1}{2}k_{e0}e^2 + \tfrac{1}{4}k_{e1}e^4 + \tfrac{1}{2}k_{ei0}e^2 + \tfrac{1}{4}k_{ei1}e^4 + \tfrac{1}{2}\begin{bmatrix} u_a & i_L & i_a & \omega_r \end{bmatrix}$$

$$\times K_{x0}\begin{bmatrix} u_a \\ i_L \\ i_a \\ \omega_r \end{bmatrix} + \tfrac{1}{2}\begin{bmatrix} u_a & i_L & i_a & \omega_r \end{bmatrix} K \begin{bmatrix} u_a \\ i_L \\ i_a \\ \omega_r \end{bmatrix},$$

$K_{x0} \in \mathbb{R}^{4\times4}$, and $K \in \mathbb{R}^{4\times4}$.
One obtains

$$u_c = \begin{cases} 10 & \text{for } u \geq 10, \\ u & \text{for } 0 < u < 10, \\ 0 & \text{for } u \leq 0, \end{cases}$$

$$u = 1.5e + 0.11e^3 + 6.8\int e\,dt + 0.49\int e^3\,dt - 0.95u_a - 0.38i_L - 0.62i_a$$

$$-0.0041\omega_r - (0.002u_a + 0.0006i_L + 0.001i_a + 0.00001\omega_r)i_L.$$

The feedback gains were found by solving inequality

$$\frac{dV(e,x)}{dt} \leq -\|e\|^2 - \|e\|^4 - \|x\|^2.$$

The criteria, imposed on the Lyapunov pair to guarantee robust stability, are satisfied. Hence, the robust bounded control law guarantees stability and ensures tracking and disturbance rejection. Experimental validation of stability, tracking, and disturbance attenuation is needed. The efficacy of the bounded controller is verified through comprehensive experiments. The worst dynamics are observed in case the motor temperature is high. For $T = 140°C$ (the maxim operating temperature), the stator resistance reaches the maximum value, and the torque constant k_a is minimum. Therefore, the minimum electromagnetic torque, which

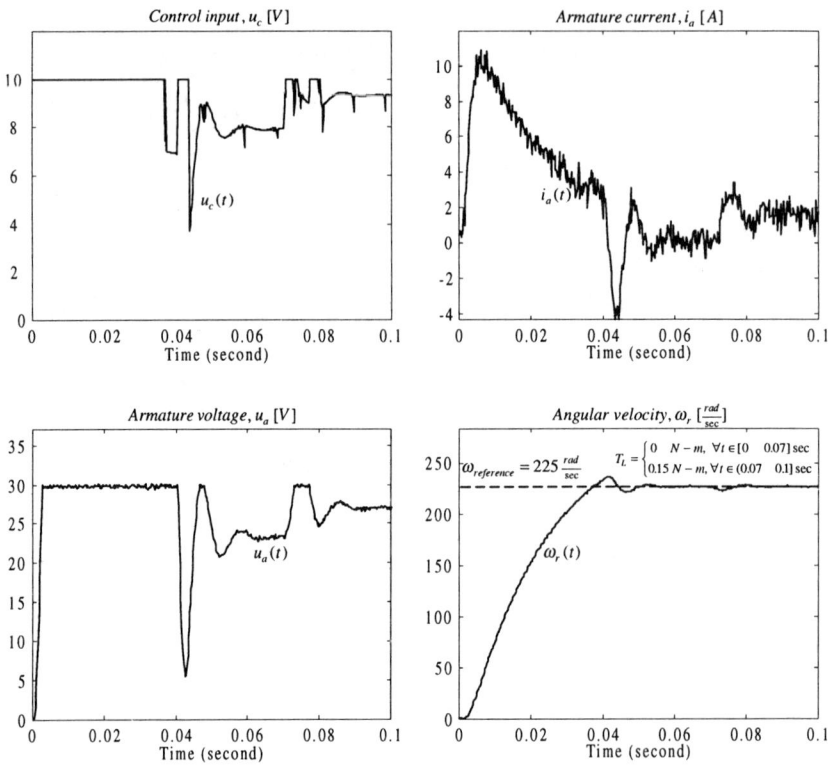

FIGURE 5.3.4.   Transient dynamics of the closed-loop servo-system.

is given as $T_e = k_a i_a$, results. One should examine the tracking error $e(t)$, which represents the differences between the actual rotor angular velocity $\omega_r(t)$ and its desired (assigned) reference value $\omega_{\text{reference}}(t)$, as well as disturbance attenuation features (it is required that the angular velocity remains equal to the reference value if the load torque T_L is applied). Figure 5.3.4 depicts the control signal-level voltage $u_c(t)$ as well as the transient dynamics for states $u_a(t)$, $i_a(t)$, $\omega_r(t)$ when $\omega_{\text{reference}} = 225\frac{\text{rad}}{\text{sec}}$ and $T = 140°\text{C}$. A motor reaches the desired (reference) angular velocity within 0.05 sec with overshoot 4.5%, and the steady-state error is zero. The analysis of the experimental results indicates that the tracking error $e(t) = \omega_{\text{reference}}(t) - \omega_r(t)$ converges to zero. The disturbance attenuation has been studied. In particular, the load torque 0.15 N-m is applied at 0.07 sec. By analyzing the angular velocity $\omega_r(t)$, one concludes that the settling time is 0.01 sec with 2.5% deflection from $\omega_{\text{reference}}(t)$, and the steady-state error is zero. From the experimental data, it follows that excellent performance has been achieved, and the angular velocity precisely follows the reference speed assigned.

Example 5.3.3. Tracking control of a servo with induction motor

With increasing performance demands being placed on servo-systems and electric drives actuated by induction motors, the solution of a robust tracking problem becomes essential. The need for high-performance servos and drives in a wide variety of industrial systems is well known. The critical features that challenge motion control are constraints, nonlinearities, measurability, observability, unmodeled dynamics, and parameter variations. These features should be incorporated to solve the nonlinear motion control problem. Control of induction motors is an extremely challenging problem, and novel algorithms are needed to be designed to guarantee robust stability, to ensure the desired dynamic performance, and to attain the specified tracking accuracy. In this example, the synthesis of a robust partial feedback controller is reported.

For a servo-system, the stator and rotor currents i_{as}, i_{bs}, i'_{ar}, and i'_{br}, as well as the electrical angular velocity and displacement ω_r and θ_r, are the state variables. The applied voltages to phases a and b of the stator and rotor u_{as}, u_{bs}, u'_{ar}, and u'_{br} are the control vector. The mathematical model is developed using the Kirchhoff voltage law. The voltage equations for two-phase induction motors are

$$
\begin{bmatrix}
\frac{di_{as}}{dt} \\
\frac{di_{bs}}{dt} \\
\frac{di'_{ar}}{dt} \\
\frac{di'_{br}}{dt} \\
\frac{d\omega_r}{dt} \\
\frac{d\theta_r}{dt}
\end{bmatrix}
=
\begin{bmatrix}
-\frac{L'_{rr}r_s}{L_\Sigma} & 0 & 0 & 0 & 0 & 0 \\
0 & -\frac{L'_{rr}r_s}{L_\Sigma} & 0 & 0 & 0 & 0 \\
0 & 0 & -\frac{L_{ss}r'_r}{L_\Sigma} & 0 & 0 & 0 \\
0 & 0 & 0 & -\frac{L_{ss}r'_r}{L_\Sigma} & 0 & 0 \\
0 & 0 & 0 & 0 & -\frac{B_m}{J} & 0 \\
0 & 0 & 0 & 0 & 1 & 0
\end{bmatrix}
\begin{bmatrix}
i_{as} \\
i_{bs} \\
i'_{ar} \\
i'_{br} \\
\omega_r \\
\theta_r
\end{bmatrix}
$$

$$
+
\begin{bmatrix}
\frac{L^2_{ms}}{L_\Sigma}i_{bs}\omega_r + \frac{L_{ms}L'_{rr}}{L_\Sigma}i'_{ar}\left(\omega_r \sin\theta_r + \frac{r'_r}{L'_{rr}}\cos\theta_r\right) \\
\quad + \frac{L_{ms}L'_{rr}}{L_\Sigma}i'_{br}\left(\omega_r \cos\theta_r - \frac{r'_r}{L'_{rr}}\sin\theta_r\right) \\[4pt]
-\frac{L^2_{ms}}{L_\Sigma}i_{as}\omega_r - \frac{L_{ms}L'_{rr}}{L_\Sigma}i'_{ar}\left(\omega_r \cos\theta_r - \frac{r'_r}{L'_{rr}}\sin\theta_r\right) \\
\quad + \frac{L_{ms}L'_{rr}}{L_\Sigma}i'_{br}\left(\omega_r \sin\theta_r + \frac{r'_r}{L'_{rr}}\cos\theta_r\right) \\[4pt]
\frac{L_{ms}L_{ss}}{L_\Sigma}i_{as}\left(\omega_r \sin\theta_r + \frac{r_s}{L_{ss}}\cos\theta_r\right) \\
\quad - \frac{L_{ms}L_{ss}}{L_\Sigma}i_{bs}\left(\omega_r \cos\theta_r - \frac{r_s}{L_{ss}}\sin\theta_r\right) - \frac{L^2_{ms}}{L_\Sigma}i'_{br}\omega_r \\[4pt]
\frac{L_{ms}L_{ss}}{L_\Sigma}i_{as}\left(\omega_r \cos\theta_r - \frac{r_s}{L_{ss}}\sin\theta_r\right) \\
\quad + \frac{L_{ms}L_{ss}}{L_\Sigma}i_{bs}\left(\omega_r \sin\theta_r + \frac{r_s}{L_{ss}}\cos\theta_r\right) + \frac{L^2_{ms}}{L_\Sigma}i'_{br}\omega_r \\[4pt]
-\frac{P^2 L_{ms}}{4J}\left[\left(i_{as}i'_{ar} + i_{bs}i'_{br}\right)\sin\theta_r + \left(i_{as}i'_{br} - i_{bs}i'_{ar}\right)\cos\theta_r\right] \\[4pt]
0
\end{bmatrix}
$$

$$+ \begin{bmatrix} \frac{L'_{rr}}{L_\Sigma} & 0 & 0 & 0 \\ 0 & \frac{L'_{rr}}{L_\Sigma} & 0 & 0 \\ 0 & 0 & \frac{L_{ss}}{L_\Sigma} & 0 \\ 0 & 0 & 0 & \frac{L_{ss}}{L_\Sigma} \\ 0 & 0 & 0 & 0 \\ 0 & 0 & 0 & 0 \end{bmatrix} \begin{bmatrix} u_{as} \\ u_{bs} \\ u'_{ar} \\ u'_{br} \end{bmatrix}$$

$$+ \begin{bmatrix} -\frac{L_{ms}}{L_\Sigma}\cos\theta_r u'_{ar} + \frac{L_{ms}}{L_\Sigma}\sin\theta_r u'_{br} \\ -\frac{L_{ms}}{L_\Sigma}\sin\theta_r u'_{ar} - \frac{L_{ms}}{L_\Sigma}\cos\theta_r u'_{br} \\ -\frac{L_{ms}}{L_\Sigma}\cos\theta_r u_{as} - \frac{L_{ms}}{L_\Sigma}\sin\theta_r u_{bs} \\ \frac{L_{ms}}{L_\Sigma}\sin\theta_r u_{as} - \frac{L_{ms}}{L_\Sigma}\cos\theta_r u_{bs} \\ 0 \\ 0 \end{bmatrix} - \begin{bmatrix} 0 \\ 0 \\ 0 \\ 0 \\ \frac{P}{2J} \\ 0 \end{bmatrix} T_L,$$

$$L_\Sigma = L_{ss}L'_{rr} - L_{ms}^2. \tag{5.3.19}$$

Using the number of poles P, the actual (mechanical) angular velocity and displacement of the rotor are found as $\omega_{rm} = \frac{2}{P}\omega_r$ and $\theta_{rm} = \frac{2}{P}\theta_r$. For squirrel-cage motors, $u'_{ar} = u'_{br} = 0$, and hence

$$B(x) = \begin{bmatrix} \frac{L'_{rr}}{L_\Sigma} & 0 \\ 0 & \frac{L'_{rr}}{L_\Sigma} \\ -\frac{L_{ms}}{L_\Sigma}\cos\theta_r & -\frac{L_{ms}}{L_\Sigma}\sin\theta_r \\ \frac{L_{ms}}{L_\Sigma}\sin\theta_r & -\frac{L_{ms}}{L_\Sigma}\cos\theta_r \\ 0 & 0 \\ 0 & 0 \end{bmatrix}.$$

Stator and rotor resistances vary because of heating. Nonlinear magnetic characteristics of the core material, magnetic saturation, flux linkage–magnetizing current relation, *fringing field*, harmonic, and other effects contribute to variations of inductances. The parameters of a four-pole induction motor are $r_s(\cdot) \in \begin{bmatrix} 0.8_{T=20°C} & 1.2_{T=140°C} \end{bmatrix} \Omega$, $r_r(\cdot) \in \begin{bmatrix} 1_{T=20°C} & 1.45_{T=140°C} \end{bmatrix} \Omega$, $L_{ms}(\cdot) \in \begin{bmatrix} 0.15 & 0.17 \end{bmatrix}$ H, $L_{mr}(\cdot) \in \begin{bmatrix} 0.14 & 0.16 \end{bmatrix}$ H, $L_{ls} = L_{lr} = 0.02$ H, and $J(\cdot) \in \begin{bmatrix} 0.0035 & 0.0092 \end{bmatrix}$ kg-m^2.

The servo-output is $y(t) = \theta(t) = 0.05\theta_r(t)$. The output $y(t)$ is compared with the reference input $r(t)$. The tracking error vector is $e(t) = r(t) - \theta(t)$. The desired tracking should be achieved synthesizing the robust partial state-feedback controller; that is, only measured states must be used to implement the control law. For squirrel-cage induction motors, one has the measured and unmeasured state variables, as well as control vector, which are defined as

$$
\text{states } x = \begin{bmatrix} i_{as} \\ i_{bs} \\ i'_{ar} \\ i'_{br} \\ \omega_r \\ \theta_r \end{bmatrix}, \qquad \text{measured states } x_M = \begin{bmatrix} i_{as} \\ i_{bs} \\ \omega_r \\ \theta_r \end{bmatrix},
$$

$$
\text{unmeasured states } x_U = \begin{bmatrix} i'_{ar} \\ i'_{br} \end{bmatrix}, \qquad \text{and control } u = \begin{bmatrix} u_{as} \\ u_{bs} \end{bmatrix}.
$$

The admissible applied voltages are

$$
-\sqrt{2}110 \le (u_{as}, u_{bs}) \le \sqrt{2}110\text{V}.
$$

The Lyapunov candidate is found from (5.3.14). Let $\zeta = 4$, $\beta = 1$, $\eta = 1$, and $\gamma = 0$. Thus,

$$
\begin{aligned}
V(e, x) = {} & \tfrac{3}{4} K_{E0} e^{4/3} + \tfrac{1}{2} K_{E1} e^2 + \tfrac{3}{8} K_{E2} e^{8/3} + \tfrac{3}{10} K_{E3} e^{10/3} + \tfrac{1}{4} K_{E4} e^4 \\
& + \tfrac{1}{2} x^T K_{X0} x + \tfrac{1}{4} x^{2^T} K_{X1} x^2,
\end{aligned}
$$

$$
K_{X0} = \begin{bmatrix} K_{XM0} & 0 \\ 0 & K_{XU0} \end{bmatrix} \in \mathbb{R}^{6\times6}, \quad K_{X1} = \begin{bmatrix} K_{XM1} & 0 \\ 0 & K_{XU1} \end{bmatrix} \in \mathbb{R}^{6\times6},
$$

$K_{XM0} \in \mathbb{R}^{4\times4}$, $K_{XM1} \in \mathbb{R}^{4\times4}$, $K_{XU0} \in \mathbb{R}^{2\times2}$ and $K_{XU1} \in \mathbb{R}^{2\times2}$.

Using this nonquadratic Lyapunov candidate, from (5.3.15), one finds

$$
G_E B_{EM}(x_M)^T \sum_{i=0}^{\zeta} \text{diag}\left[e^{\frac{i-\beta}{2\beta+1}} \right] K_{Ei} \frac{1}{s} e^{\frac{i+\beta+1}{2\beta+1}} = \begin{bmatrix} 1 & 0 \\ 0 & 1 \end{bmatrix} \begin{bmatrix} \frac{L'_{rr}}{L_\Sigma} \\ \frac{L_{rr}}{L_\Sigma} \end{bmatrix} \sum_{i=0}^{4} K_{Ei} \frac{1}{s} e^{\frac{2i+1}{3}},
$$

$$
G_X B_{XM}(x_M)^T \sum_{i=0}^{\eta} \text{diag}\left[x_M^{\frac{i-\gamma}{2\gamma+1}} \right] K_{Xi} x_M^{\frac{i+\gamma+1}{2\gamma+1}} = \begin{bmatrix} 1 & 0 \\ 0 & 1 \end{bmatrix} \begin{bmatrix} \frac{L'_{rr}}{L_\Sigma} & 0 & 0 & 0 \\ 0 & \frac{L'_{rr}}{L_\Sigma} & 0 & 0 \end{bmatrix}
$$

$$
\times \left(K_{XM0} \begin{bmatrix} i_{as} \\ i_{bs} \\ \omega_r \\ \theta_r \end{bmatrix} + \begin{bmatrix} i_{as} & 0 & 0 & 0 \\ 0 & i_{bs} & 0 & 0 \\ 0 & 0 & \omega_r & 0 \\ 0 & 0 & 0 & \theta_r \end{bmatrix} K_{XM1} \begin{bmatrix} i_{as}^2 \\ i_{bs}^2 \\ \omega_r^2 \\ \theta_r^2 \end{bmatrix} \right).
$$

Using the resulting closed-loop system, one finds $\frac{dV(e,x)}{dt}$, and the unknown coefficients K_{Ei} as well as unknown matrices K_{XM0} and K_{XM1} are found solving

(5.3.17). In the studied problem, the following inequality:

$$\frac{dV(e, x)}{dt} \leq -\sum_{i=0}^{4} \|e\|^{\frac{2i+4}{3}} - \|x\|^2 - \|x\|^4$$

was solved.

The weighting matrices are

$$G_P = \begin{bmatrix} 0 & 0 \\ 0 & 0 \end{bmatrix}, G_E = \begin{bmatrix} 1 & 0 \\ 0 & 1 \end{bmatrix}, \text{ and } G_X = \begin{bmatrix} 1 & 0 \\ 0 & 1 \end{bmatrix}.$$

Having found the unknown coefficients and matrices of the Lyapunov candidate, one obtains

$$u_{as} = \cos(\omega_f t)\text{sat}_{-\sqrt{2110}}^{+\sqrt{2110}} \left[\frac{1}{s}(0.52e^{1/3} + 0.39e + 0.14e^{5/3} + 0.061e^{7/3} + 0.019e^3) \right.$$

$$\left. -2.1i_{as} - 2.1i_{bs} - 0.018\omega_r - 0.12\theta_r - 0.27i_{as}(i_{as}^2 + i_{bs}^2 + \omega_r^2 + \theta_r^2) \right],$$

$$-\sqrt{2110} \leq u_{as} \leq \sqrt{2110},$$

$$u_{bs} = \sin(\omega_f t)\text{sat}_{-\sqrt{2110}}^{+\sqrt{2110}} \left[\frac{1}{s}(0.52e^{1/3} + 0.39e + 0.14e^{5/3} + 0.061e^{7/3} + 0.019e^3) \right.$$

$$\left. -2.1i_{as} - 2.1i_{bs} - 0.018\omega_r - 0.12\theta_r - 0.27i_{bs}(i_{as}^2 + i_{bs}^2 + \omega_r^2 + \theta_r^2) \right],$$

$$-\sqrt{2110} \leq u_{bs} \leq \sqrt{2110}. \tag{5.3.20}$$

The amplitude of applied voltages applied to the stator phases u_{as} and u_{bs} changes, and the frequency is varied to maintain the constant air-gap flux density. Controller (5.3.20) guarantees the *constant volts per hertz* operation. In (5.3.20), the directly measured states (stator currents, angular velocity, and displacement) and the tracking error are used. Furthermore, the designed control law is bounded. Sufficient conditions for robust stability and tracking are guaranteed. Hence, the $V(e, x)$ used is the Lyapunov function, controller (5.3.20) guarantees stability, and the tracking error converges to the compact set.

5.4.   Optimization of Continuous-Time Systems

5.4.1.   *Optimization of Time-Invariant Systems*

Our goal is to design unbounded and constrained control laws for time-invariant systems using the Hamilton–Jacobi theory. That is, by applying necessary and sufficient conditions for optimality, nonlinear control laws should be synthesized. We study time-invariant dynamic systems modeled as

$$\dot{x}(t) = F(x) + B(x)u, \, x(t_0) = x_0. \tag{5.4.1}$$

Consider the nonquadratic positive-definite performance functional as given by

$$J(x(\cdot), u(\cdot)) = \int_{t_0}^{t_f} (W_x(x) + \tfrac{1}{2} u^T G u) \, dt, \tag{5.4.2}$$

where $W_x(\cdot): \mathbb{R}^c \to \mathbb{R}_{\geq 0}$ is the positive-definite and continuously differentiable C^ς ($\varsigma \geq 2$) integrand function and $G \in \mathbb{R}^{m \times m}$ is the positive-definite weighting matrix.

Let us examine necessary and sufficient conditions that have to be satisfied if control is optimal. One examines the minimization problem, and the scalar Hamiltonian function is given by

$$H\left(x, u, \frac{\partial V}{\partial x}\right) = W_x(x) + \tfrac{1}{2} u^T G u + \left(\frac{\partial V}{\partial x}\right)^T (F(x) + B(x)u), \tag{5.4.3}$$

where $V(\cdot): \mathbb{R}^c \to \mathbb{R}_{\geq 0}$ is the smooth and bounded return function $V(0) = 0$.

The performance functional (5.4.2) is positive-definite because real-valued, positive-definite continuous integrands $W_x(\cdot)$ and $\tfrac{1}{2} u^T G u$ are used. Applying the necessary conditions for optimality

$$\frac{\partial H\left(x, u, \frac{\partial V}{\partial x}\right)}{\partial u} = 0 \qquad \text{(first-order necessary condition)}$$

and

$$\frac{\partial^2 H\left(x, u, \frac{\partial V}{\partial x}\right)}{\partial u \times \partial u^T} > 0 \qquad \text{(second-order necessary condition)},$$

one finds control function $u(\cdot): [t_0, t_f) \to \mathbb{R}^m$, which minimizes functional (5.4.2), and the optimality is studied.

The derivative of the Hamiltonian function $H\left(x, u, \frac{\partial V}{\partial x}\right)$ exists, and control $u(\cdot): [t_0, t_f) \to \mathbb{R}^m$ is found by using the first-order necessary condition. The Hamilton–Jacobi equation is expressed as

$$-\frac{\partial V}{\partial t} = \min_u \left[W_x(x) + \tfrac{1}{2} u^T G u + \left(\frac{\partial V}{\partial x}\right)^T (F(x) + B(x)u) \right]. \tag{5.4.4}$$

Using the first-order necessary condition, the extremum of the Hamiltonian function $H\left(x, u, \frac{\partial V}{\partial x}\right)$ is found, and the controller results.

In particular, from (5.4.4), we have

$$u^T G + \left(\frac{\partial V}{\partial x}\right)^T B(x) = 0,$$

and thus, the unbounded control law is synthesized in the following form:

$$u = -G^{-1} B^T(x) \frac{\partial V}{\partial x}. \tag{5.4.5}$$

By using the Hamiltonian function (5.4.3) and controller (5.4.5), one concludes that

$$\frac{\partial^2 H\left(x, u, \frac{\partial V}{\partial x}\right)}{\partial u \times \partial u^T} \in \mathbb{R}^{m \times m}$$

is positive-definite because

$$\frac{\partial^2 H\left(x, u, \frac{\partial V}{\partial x}\right)}{\partial u \times \partial u^T} = G > 0.$$

That is, the minimum is guaranteed. Hence, the necessary conditions for optimality are satisfied, and controller (5.4.5) is a candidate for an optimal one.

Lemma 5.4.1. *Consider state-space equations (5.4.1) that model continuous-time dynamic systems. Unbounded control law (5.4.5) guarantees a minimum for the performance functional (5.4.2). If an optimal control $u(\cdot)$: $[t_0, t_f) \to \mathbb{R}^m$ exists, it is unique and represented by (5.4.5). Furthermore, $V(\cdot)$: $\mathbb{R}^c \to \mathbb{R}_{\geq 0}$ is the solution of Hamilton–Jacobi–Bellman equation*

$$-\frac{\partial V}{\partial t} = W_x(x) + \left(\frac{\partial V}{\partial x}\right)^T F(x) - \frac{1}{2}\left(\frac{\partial V}{\partial x}\right)^T B(x) G^{-1} B^T(x) \frac{\partial V}{\partial x}$$

satisfying the boundary conditions. The positive-definite return function $V(\cdot)$: $\mathbb{R}^c \to \mathbb{R}_{\geq 0}$, $V \in C^\kappa$, $\kappa \geq 1$, is given as

$$V(x_0) = \inf_u J(x_0, u) \geq 0.$$

To verify the optimality, the sufficient conditions must be studied. We formulate the following lemma.

Lemma 5.4.2. *If a continuously differentiable positive-definite C^κ ($\kappa \geq 1$) return function $V(\cdot)$: $\mathbb{R}^c \to \mathbb{R}_{\geq 0}$, $V(x_0) = \inf_u J(x_0, u) \geq 0$ and a real-analytic control map $u(\cdot)$: $[t_0, t_f) \to \mathbb{R}^m$, as given by (5.4.5), exist such that on $[t_0, t_f)$*

$$\lim_{t \to t_f} V(x) \leq \lim_{t \to t_f^0} V(x^0) = 0,$$

$$W_x(x^0) + \left(\frac{\partial V}{\partial x^0}\right)^T F(x^0) - \frac{1}{2}\left(\frac{\partial V}{\partial x^0}\right)^T B(x^0) G B^T(x^0) \frac{\partial V}{\partial x^0} = 0,$$

$$W_x(x) + \left(\frac{\partial V}{\partial x}\right)^T F(x) - \frac{1}{2}\left(\frac{\partial V}{\partial x}\right)^T B(x) G B^T(x) \frac{\partial V}{\partial x} \leq 0,$$

then, control $u^0(\cdot)$: $[t_0, t_f) \to \mathbb{R}^m$, generating the solution $x^0(\cdot)$: $[t_0, t_f) \to \mathbb{R}^c$ with $x^0(t_0) = x_0$, is an optimal algorithm.

The nonquadratic performance functional (5.4.2) is used. From (5.4.1), (5.4.2), and (5.4.5), it follows at once that the minimum value of the functional should be expressed in the nonquadratic form. In particular, the minimum value of functional

(5.4.2) is given by smooth, continuously differentiable, real-analytic functions. Assume that (5.4.4) admits a solution. Then, $V(\cdot)$: $\mathbb{R}^c \to \mathbb{R}_{\geq 0}$ can be represented by the nonquadratic continuously differentiable and positive-definite C^K function. We have

$$V(x) = \sum_{i=0}^{\eta} \frac{2\gamma+1}{2(i+\gamma+1)} \left(x^{\frac{i+\gamma+1}{2\gamma+1}} \right)^T K_i(t) x^{\frac{i+\gamma+1}{2\gamma+1}}$$

$$+ \left(\sum_{i=0}^{\sigma} \frac{2\beta+1}{2i+4\beta+3} \left(x^{\frac{i+\beta+1}{2\beta+1}} \right)^T K_{mi}(t) x^{\frac{i+\beta+1}{2\beta+1}} \right) M|x|, \qquad (5.4.6)$$

where $K_i \in \mathbb{R}^{c\times c}$ and $K_{mi} \in \mathbb{R}^{c\times c}$ are the positive-definite matrices; $M \in \mathbb{R}^{1\times c}$; $\eta = 0, 1, 2, \ldots, \gamma = 0, 1, 2, \ldots, \sigma = 0, 1, 2, \ldots$ and $\beta = 0, 1, 2, \ldots$.

Definition 5.4.1. A real-analytic, positive-definite C^K function $V(\cdot)$: $\mathbb{R}^c \to \mathbb{R}_{\geq 0}$ is said to be a return function for the corresponding optimization problem if sufficient conditions for optimality are met.

The following theorem is formulated.

Theorem 5.4.1. *Given the time-invariant dynamic system* (5.4.1), *the nonquadratic positive-definite C^K function $V(\cdot)$: $\mathbb{R}^c \to \mathbb{R}_{\geq 0}$ (5.4.6) is the solution of the partial differential equation*

$$-\frac{\partial V}{\partial t} = W_x(x) + \left(\frac{\partial V}{\partial x} \right)^T F(x) - \left(\frac{\partial V}{\partial x} \right)^T B(x) G^{-1} B^T(x) \frac{\partial V}{\partial x}.$$

Nonlinear control $u^0(\cdot)$: $[t_0, t_f) \to \mathbb{R}^m$ (5.4.5), generating solution $x^0(\cdot)$: $[t_0, t_f) \to \mathbb{R}^c$ with $x^0(t_0) = x_0$, is optimal if sufficient conditions for optimality are satisfied.

Proof. The proof of this theorem is straightforward using the Hamilton–Jacobi and sufficiency theories. It is evident that in general, the minimum value of functional (5.4.2) must be found by using the power-series forms. Hence, nonquadratic return functions must be applied. □

Example 5.4.1.

The second-order dynamic system modeled as

$$\dot{x}_1(t) = x_2 + x_1 x_2 u,$$
$$\dot{x}_2(t) = x_2 u^2,$$

subject to the following performance functional:

$$J = \int_{t_0}^{t_f} W_x(x)\, dt.$$

The Hamilton–Jacobi equation is

$$-\frac{\partial V}{\partial t} = \min_{u} \left[W_x(x) + \left(\frac{\partial V}{\partial x}\right)^T [F(x) + B(x, u)] \right]$$

$$= \min_{u} \left[W_x(x) + \frac{\partial V}{\partial x_1}(x_2 + x_1 x_2 u) + \frac{\partial V}{\partial x_2} x_2 u^2 \right].$$

Using the first-order necessary condition, the extremum of the Hamiltonian function is found. In particular, from

$$\frac{\partial H}{\partial u} = \frac{\partial V}{\partial x_1} x_1 x_2 + 2\frac{\partial V}{\partial x_2} x_2 u = 0,$$

one has an optimal controller as expressed by

$$u = -\tfrac{1}{2} x_1 \frac{\partial V}{\partial x_1} \left(\frac{\partial V}{\partial x_2}\right)^{-1}.$$

Therefore, the following partial differential equation, needed to be solved, results:

$$-\frac{\partial V}{\partial t} = W_x(x) + x_2 \frac{\partial V}{\partial x_1} - \tfrac{1}{4} x_1^2 x_2 \left(\frac{\partial V}{\partial x_1}\right)^2 \left(\frac{\partial V}{\partial x_2}\right)^{-1}.$$

Example 5.4.2.

The major goal of this example is to illustrate that the Hamilton–Jacobi theory allows one to synthesize the bounded controllers. We study the multivariable second-order dynamic system with bounded control. In particular,

$$\begin{aligned} \dot{x}_1(t) &= x_1^3 u_1 + x_2^5, \quad -1 \le u_1 \le 1, \\ \dot{x}_2(t) &= u_2, \quad\quad\quad -1 \le u_2 \le 1. \end{aligned}$$

For the performance functional in the form

$$J = \int_{t_0}^{t_f} W_x(x)\, dt,$$

the Hamilton–Jacobi equation is found to be

$$-\frac{\partial V}{\partial t} = \min_{u \in U} \left[W_x(x) + \left(\frac{\partial V}{\partial x}\right)^T (F(x) + B(x)u) \right]$$

$$= \min_{\substack{-1 \le u_1 \le 1 \\ -1 \le u_2 \le 1}} \left[W_x(x) + \frac{\partial V}{\partial x_1}(x_1^3 u_1 + x_2^5) + \frac{\partial V}{\partial x_2} u_2 \right].$$

From the first-order necessary condition for optimality, an optimal controller is expressed by

$$u_1 = -\operatorname{sgn}\left(x_1^3 \frac{\partial V}{\partial x_1}\right) \text{ and } u_2 = -\operatorname{sgn}\left(\frac{\partial V}{\partial x_2}\right)$$

The Hamilton–Jacobi–Bellman partial differential equations is

$$-\frac{\partial V}{\partial t} = W_x(x) - \left|x_1^3 \frac{\partial V}{\partial x_1}\right| + \frac{\partial V}{\partial x_1}x_2^5 - \left|\frac{\partial V}{\partial x_2}\right|.$$

Example 5.4.3.

Let us design an optimal relay-type controller for the system

$$\dot{x}_1(t) = x_2,$$
$$\dot{x}_2(t) = u, \quad -1 \le u \le 1.$$

The control takes two values $+1$ and -1. That is, $u = 1$ or $u = -1$. If $u = 1$, from $\dot{x}_1(t) = x_2$, $\dot{x}_2(t) = 1$, we have

$$\frac{dx_2}{dx_1} = \frac{1}{x_2}.$$

The integration gives $x_2^2 = 2x_1 + c_1$.
    If $u = -1$, from $\dot{x}_1(t) = x_2$, $\dot{x}_2(t) = -1$, we have

$$\frac{dx_2}{dx_1} = -\frac{1}{x_2}.$$

The integration gives $x_2^2 = -2x_1 + c_2$.
    Because of the switching action ($u = 1$ or $u = -1$), the switching curve can be found as a function of the state variables. The comparison of $x_2^2 = 2x_1 + c_1$ and $x_2^2 = -2x_1 + c_2$ indicates that the switching curve is explicitly described by

$$-x_2^2 - 2x_1 \operatorname{sgn}(x_2) = 0,$$

which can be rewritten as
$$-x_1 - \tfrac{1}{2}x_2|x_2| = 0.$$

Because the control takes the values $+1$ or -1 at the switching curve $-x_1 - \tfrac{1}{2}x_2|x_2| = 0$ (see Figure 5.4.1), the following expression for a relay controller results:

$$u = -\operatorname{sgn}(x_1 + \tfrac{1}{2}x_2|x_2|).$$

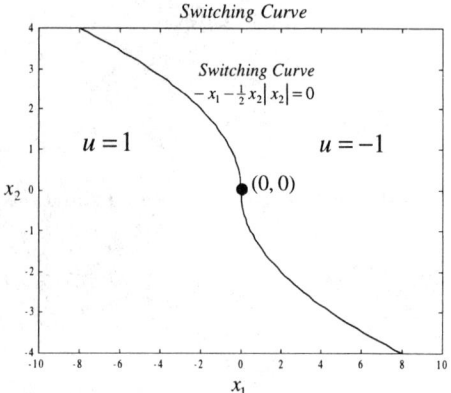

FIGURE 5.4.1.   Switching curve, $-x_1 - \frac{1}{2}x_2|x_2| = 0$.

Having found the control law using the calculus by analyzing the solutions of the differential equations with the relay controller (closed-loop system switching), we apply the Hamilton–Jacobi theory. The following functional is minimized:

$$J = \int_{t_0}^{t_f} 1 \, dt.$$

From the Hamilton–Jacobi equation

$$-\frac{\partial V}{\partial t} = \min_{-1 \le u \le 1} \left[1 + \frac{\partial V}{\partial x_1} x_2 + \frac{\partial V}{\partial x_2} u \right],$$

an optimal controller is obtained as

$$u = -\operatorname{sgn}\left(\frac{\partial V}{\partial x_2} \right).$$

The solution of the partial differential equation is found using the following return function:

$$V(x_1, x_2) = k_{11}x_1^2 + k_{12}x_1x_2 + k_{22}x_2^3|x_2|,$$

and the controller is expressed as

$$u = -\operatorname{sgn}(x_1 + \tfrac{1}{2}x_2|x_2|).$$

The SIMULINK diagram to model this closed-loop system is developed and shown in Figure 5.4.2.

The transient dynamics is analyzed. The switching curve, the phase-plane evolution of the variables, and the transient behavior for different initial conditions are shown in Figure 5.4.3.

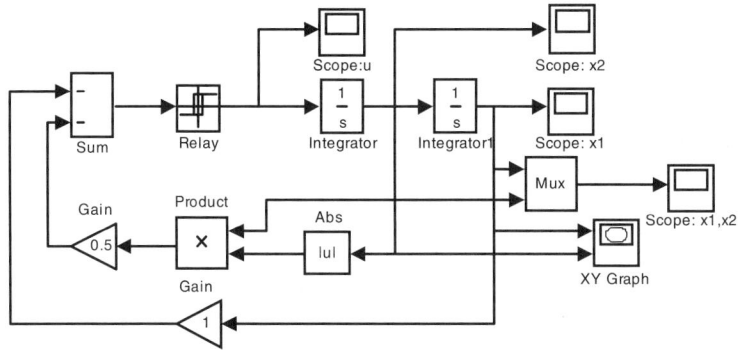

FIGURE 5.4.2.   SIMULINK diagram to model the closed-loop system.

Using different concepts, it was shown that an optimal controller is

$$u = -\operatorname{sgn}(x_1 + \tfrac{1}{2}x_2|x_2|).$$

From $\dot{x}_1(t) = x_2$, we can express the controller using proportional and derivative

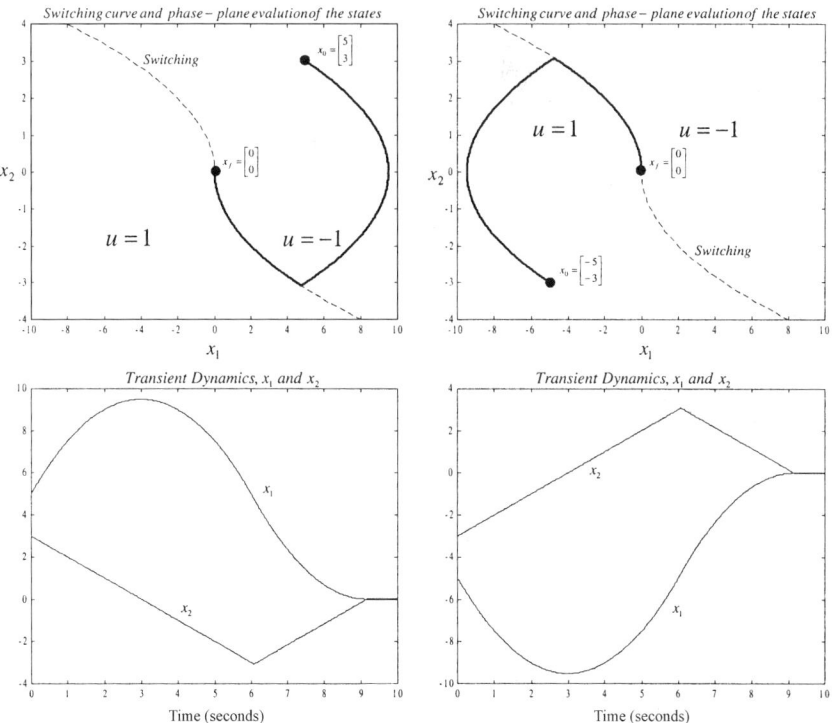

FIGURE 5.4.3.   Switching (bang-bang control) and system dynamics.

feedback. In particular,

$$u = -\operatorname{sgn}(x_1 + \tfrac{1}{2}x_2|x_2|) = -\operatorname{sgn}(x_1 + \tfrac{1}{2}\dot{x}_1|\dot{x}_1|).$$

Example 5.4.4.

Consider the third-order nonholonomic system

$$\dot{x}_1(t) = u_1,$$
$$\dot{x}_2(t) = u_2,$$
$$\dot{x}_3(t) = x_1 u_2 - x_2 u_1.$$

The positive-definite nonquadratic performance functional is represented by

$$J = \tfrac{1}{2} \int_0^\infty (x_1^2 + x_2^2 + (x_1^2 + x_2^2)x_3^4 + u_1^2 + u_2^2) \, dt.$$

The application of the first-order necessary condition leads us to

$$u_1 = -\frac{\partial V}{\partial x_1} + x_2 \frac{\partial V}{\partial x_3},$$

$$u_2 = -\frac{\partial V}{\partial x_2} - x_1 \frac{\partial V}{\partial x_3}.$$

The Hamilton–Jacobi–Bellman equation is solved by using the scalar return function

$$V(x_1, x_2, x_3) = \tfrac{1}{2}k_{11}x_1^2 + \tfrac{1}{2}k_{22}x_2^2 + \tfrac{1}{3}k_{33}x_3^2|x_3|.$$

Without difficulty, one obtains $k_{11} = k_{22} = k_{33} = 1$. This gives us a controller; in particular,

$$u_1 = -x_1 + x_2 x_3|x_3|,$$
$$u_2 = -x_2 - x_1 x_3|x_3|.$$

Clearly, upon examination of sufficient conditions, we recognize that they are guaranteed. Hence, the designed controller is an optimal one. Let us examine the transient dynamics. We assign the initial conditions as

$$\begin{bmatrix} x_{10} \\ x_{20} \\ x_{30} \end{bmatrix} = \begin{bmatrix} 10 \\ -10 \\ 10 \end{bmatrix}.$$

The closed-loop system behavior is illustrated in Figure 5.4.4. A three-dimensional plot for system states evolution is also documented.

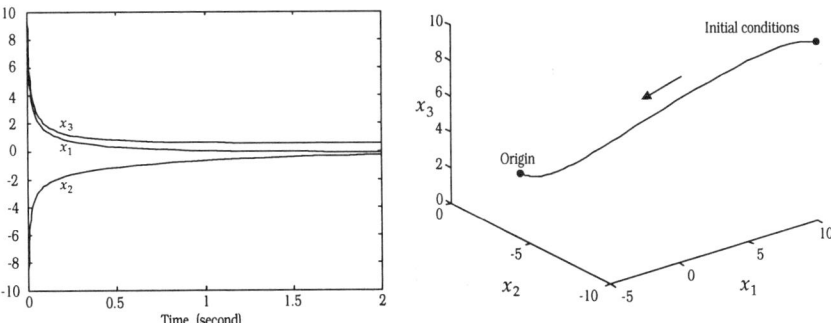

FIGURE 5.4.4.   System evolution and three-dimensional plot.

5.4.1.1. Nonlinear Control Laws Design for a Class of Nonlinear Systems

Our goal is to design a bounded control law for the dynamic systems modeled as

$$\dot{x}(t) = F(x) + B(x)u^{2w+1}, u_{\min} \leq u \leq u_{\max}, x(t_0) = x_0.$$

It is evident that if $w = 0$, we have

$$\dot{x}(t) = F(x) + B(x)u, u_{\min} \leq u \leq u_{\max}, x(t_0) = x_0.$$

The set of admissible control $U \subset \mathbb{R}^m$ consists of the Lebesgue measurable function $u(\cdot)$: $[t_0, \infty) \to \mathbb{R}^m$, and a bounded controller should be designed within the constrained compact control set

$$U = \{u \in \mathbb{R}^m | u_{i\,\min} \leq u_i \leq u_{i\,\max}, i = 1, \ldots, m\}.$$

The control bounds imposed are mapped by bounded, integrable, one-to-one globally Lipschitz, vector-valued continuous functions $\phi \in C^{\epsilon}(\epsilon \geq 1)$. The most common ϕ are the algebraic and transcendental (exponential, hyperbolic, logarithmic, trigonometric) continuously differentiable, integrable, one-to-one functions. For example, the odd one-to-one integrable function tanh with domain $(-\infty, +\infty)$ maps the control bounds. This function has the corresponding inverse function $\tanh^{-1}$ with range $(-\infty, +\infty)$.

The performance functional to be minimized is given as

$$J = \int_{t_0}^{t_f} [W_x(x) + W_u(u)] \, dt$$

$$= \int_{t_0}^{t_f} \left[W_x(x) + (2w + 1) \int (\phi^{-1}(u))^T G^{-1} u^{2w} \, du \right] dt,$$

where $G^{-1} \in \mathbb{R}^{m \times m}$ is the positive-definite weighting matrix.

Using the properties of ϕ, one concludes that inverse function ϕ^{-1} is integrable. Hence, integral $\int (\phi^{-1}(u))^T G^{-1} u^{2w} \, du$ exists. It is easy to see that $\frac{\partial W_u(u)}{\partial u} = (2w + 1)(\phi^{-1}(u))^T G^{-1} u^{2w}$.

Making use of the Hamilton–Jacobi equation

$$-\frac{\partial V}{\partial t} = \min_{u \in U} \left\{ W_x(x) + (2w+1) \int (\phi^{-1}(u))^T G^{-1} u^{2w} \, du \right.$$

$$\left. + \frac{\partial V(x)^T}{\partial x} [F(x) + B(x)u^{2w+1}] \right\},$$

the bounded controller is found as

$$u = -\phi \left(G B(x)^T \frac{\partial V(x)}{\partial x} \right), u \in U.$$

Example 5.4.5.

In this example, we study dynamic systems described by the following differential equations:

$$\frac{dx}{dt} = ax + bu, \quad u_{\min} \le u \le u_{\max}, \quad w = 0,$$

and

$$\frac{dx}{dt} = ax + bu^3, \quad u_{\min} \le u \le u_{\max}, \quad w = 1,$$

For the system

$$\frac{dx}{dt} = ax + bu, \quad u_{\min} \le u \le u_{\max}, \quad w = 0,$$

mapping the saturation by one-to-one hyperbolic tangent, we minimize

$$J = \int_{t_0}^{\infty} [W_x(x) + W_u(u)] \, dt = \int_{t_0}^{\infty} \left[W_x(x) + \int (\phi^{-1}(u))^T G^{-1} \, du \right] dt$$

$$= \int_{t_0}^{\infty} \left[W_x(x) + \int \tanh^{-1} u \, du \right] dt$$

with the positive-definite integrand

$$W_u(u) = \int \tanh^{-1} u \, du = u \tanh^{-1} u + \tfrac{1}{2} \ln(1 - u^2).$$

The bounded controller is found as

$$u = -\phi \left(b \frac{\partial V}{\partial x} \right) = -\tanh \left(b \frac{\partial V}{\partial x} \right).$$

Figure 5.4.5 illustrates the proposed integrand

$$W_u(u) = (2w+1) \int (\phi^{-1}(u))^T G^{-1} u^{2w} \, du = \int \tanh^{-1} u G^{-1} \, du$$

if $G^{-1} = 1$. For the comparison, the conventional quadratic integrand $\frac{1}{2}u^2$ is plotted in Figure 5.4.5 as well.

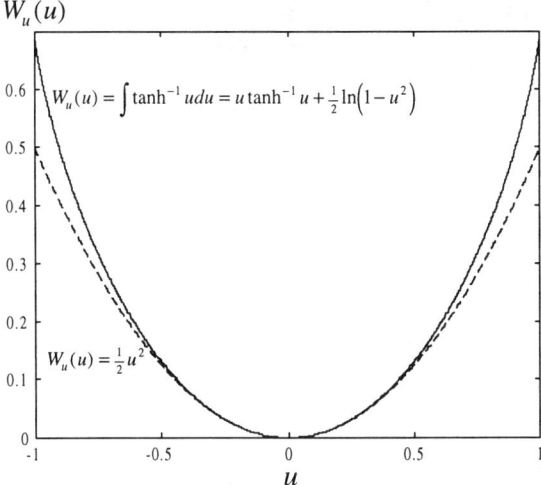

$W_u(u)$

$W_u(u) = \int \tanh^{-1} u\, du = u \tanh^{-1} u + \frac{1}{2} \ln\!\left(1 - u^2\right)$

$W_u(u) = \frac{1}{2} u^2$

u

FIGURE 5.4.5.   Plots of the nonquadratic and quadratic integrands.

In contrast, for a dynamic system

$$\frac{dx}{dt} = ax + bu^3,\ u_{\min} \le u \le u_{\max},\ w = 1,$$

the following integrand is used:

$$W_u(u) = (2w + 1) \int (\phi^{-1}(u))^T G^{-1} u^{2w}\, du = \int \tanh^{-1} u u^2\, du$$

$$= \tfrac{1}{3} u^3 \tanh^{-1} u + \tfrac{1}{6} u^2 + \tfrac{1}{6} \ln(1 - u^2),\ G^{-1} = \tfrac{1}{3}.$$

This positive-definite integrand, if $G^{-1} = \frac{1}{3}$, is plotted in Figure 5.4.6. The constrained control law is

$$u = -\tanh\left(b \frac{\partial V}{\partial x}\right).$$

In general, if the hyperbolic tangent is used to map the control bounds, for the single-input case, one has

$$\int u^{2w} \tanh^{-1} \frac{u}{k}\, du = \tfrac{1}{2w+1}\left(u^{2w+1} \tanh^{-1} \frac{u}{k} - k \int \frac{u^{2w+1}}{k^2 - u^2}\, du\right)$$

and

$$\int \frac{1}{u^{2w}} \tanh^{-1} \frac{u}{k}\, du = \tfrac{1}{2w-1}\left(-\frac{1}{u^{2w-1}} \tanh^{-1} \frac{u}{k} + k \int \frac{1}{u^{2w-1}(k^2 - u^2)}\, du\right).$$

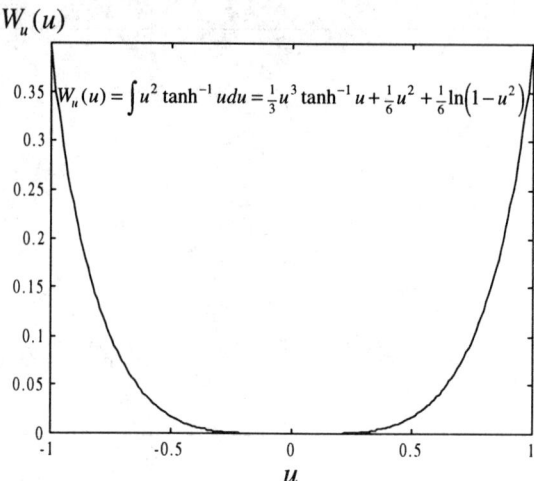

$W_u(u)$

0.35 $W_u(u) = \int u^2 \tanh^{-1} u \, du = \frac{1}{3} u^3 \tanh^{-1} u + \frac{1}{6} u^2 + \frac{1}{6} \ln\left(1 - u^2\right)$

FIGURE 5.4.6.   Plot of the nonquadratic integrand.

5.4.1.2. Tracking Control of Nonlinear Systems With Control Bounds

The Hamilton–Jacobi theory and dynamic programming have been widely used to study both weak and strong optimalities. The straightforward procedures have been developed in Chapter 4 for linear dynamic systems $\dot{x}^{system}(t) = Ax^{system} + Bu$, $y = Hx^{system}$, $x^{system}(t_0) = x_0^{system}$ with control bounds $u_{min} \leq u \leq u_{max}$. For nonlinear systems, innovative results in design of bounded control algorithms were given. In particular, the following systems were considered:

$$\dot{x}^{system}(t) = F_s(x^{system}) + B_s(x^{system})u, \quad y = H(x^{system}), \quad u_{min} \leq u \leq u_{max},$$
$$x^{system}(t_0) = x_0^{system}, \tag{5.4.7}$$

where $x^{system} \in X_s \subset \mathbb{R}^n$ is the state vector, $u \in U \subset \mathbb{R}^m$ is the vector of control inputs, $y \in Y \subset \mathbb{R}^b$ is the measured output, $F_s(\cdot): \mathbb{R}^n \to \mathbb{R}^n$, $B_s(\cdot): \mathbb{R}^n \to \mathbb{R}^{n \times m}$ and $H(\cdot): \mathbb{R}^n \to \mathbb{R}^b$ are the smooth mappings defined on open sets $S_F \subset X_s$, $S_B \subset X_s$ and $S_H \subset X_s$, $F_s(0) = 0$, $B_s(0) = 0$, and $H(0) = 0$.

The constrained optimization problem can be solved using specially structured performance functions, and the problem to find $\min_{u \in U} J(x, u)$ was solved. For a great number of dynamic systems (electromechanical systems, electric circuits, flight and underwater vehicles, internal combustion engines, robotic manipulators, etc.), the constrained optimization problem must be solved for

$$\dot{x}^{system}(t) = F_s(x^{system}) + B_s(x^{system})u^{\frac{2w+1}{2z+1}}, \quad y = H(x^{system}),$$
$$u_{min} \leq u \leq u_{max}, \quad x^{system}(t_0) = x_0^{system}, \tag{5.4.8}$$

where w and z are the nonnegative integers.

Compared with (5.4.7), the system dynamics is modeled in a more general form using the state-space differential equations (5.4.8). In fact, if $w = 0$ and $z = 1$, the model (5.4.7) results.

We study dynamic systems modeled by nonlinear differential equations (5.4.8). To design the tracking controller, we augment the system dynamics (5.4.8) with the *exogenous* dynamics

$$\dot{x}^{\text{ref}}(t) = Nr - y = Nr - H(x^{\text{system}}), \quad x^{\text{ref}} \in \mathbb{R}^b,$$

where $r \in R \subset \mathbb{R}^b$ is the measured reference vector and $N \in \mathbb{R}^{b \times b}$ is the diagonal matrix.

One obtains

$$\dot{x}(t) = F(x, r) + B(x)u^{\frac{2w+1}{2z+1}}, \quad u_{\min} \le u \le u_{\max}, \quad x(t_0) = x_0,$$
$$x = \begin{bmatrix} x^{\text{system}} \\ x^{\text{ref}} \end{bmatrix}, \tag{5.4.9}$$

where $x \in X \subset \mathbb{R}^c$ is the augmented state vector,

$$F(x, r) = \begin{bmatrix} F_s(x^{\text{system}}) \\ -H(x^{\text{system}}) \end{bmatrix} + \begin{bmatrix} 0 \\ N \end{bmatrix} r, \quad \text{and} \quad B(x) = \begin{bmatrix} B_s(x^{\text{system}}) \\ 0 \end{bmatrix}.$$

Assume that the trajectories of (5.4.9) depend continuously on the control $u \in U$, and assume $x(\cdot)$: $[t_0, \infty) \to \mathbb{R}^c$ is the unique absolutely continuous solution of (5.4.9). The set of admissible control $U \subset \mathbb{R}^m$ consists of the Lebesgue measurable, essentially bounded, vector-valued functions $u(\cdot)$: $[t_0, \infty) \to \mathbb{R}^m$ with range in a polyhedron or ellipsoid U set in $\mathbb{R}^m$. A bounded controller should be designed within the constrained compact control set $U = \{u \in \mathbb{R}^m \mid u_{\min} \le u_i \le u_{\max}, i = 1, \dots, m\}$, $U = L^1(\mathbb{R}_{\ge 0} \times \mathbb{R}^m)$. In the control set $0 \in U$, we bound the control by the measurable mapping $\phi \in U$. That is, the control constraints are mapped by a bounded, integrable, one-to-one globally Lipschitz, vector-valued continuous function $\phi \in C^\epsilon$ ($\epsilon \ge 1$). Our goal is to analytically design the bounded admissible state-feedback controller in the closed form

$$u = \phi(x), \phi(\cdot): \mathbb{R}^c \to \mathbb{R}^m.$$

The algebraic and transcendental (exponential, hyperbolic, logarithmic, trigonometric), continuously differentiable, integrable, one-to-one functions are used to map the control bounds. For example, the odd one-to-one integrable function tanh with domain $(-\infty, +\infty)$ can be used, $u = \phi(x) = \tanh(x)$. The hyperbolic tangent has the corresponding inverse function $\tanh^{-1} u$ with range $(-\infty, +\infty)$. Another example is

$$u = \phi(x) = \frac{1}{1 + e^{-cx+d}},$$

where c and d are the constants.

Let us design the performance cost to be minimized. The following functional is developed:

$$J = \int_{t_0}^{t_f} [W_x(x) + W_u(u)] \, dt$$

$$= \int_{t_0}^{t_f} \left[W_x(x) + \tfrac{2w+1}{2z+1} \int (\phi^{-1}(u))^T G^{-1} u^{\frac{2w-2z}{2z+1}} \, du \right] dt, \quad (5.4.10)$$

where $G^{-1} \in \mathbb{R}^{m \times m}$ is the positive-definite diagonal matrix.

An innovative positive-definite performance integrand

$$W_u(u) = \tfrac{2w+1}{2z+1} \int (\phi^{-1}(u))^T G^{-1} u^{\frac{2w-2z}{2z+1}} \, du$$

is used in (5.4.10).

Lemma 5.4.3. *Consider a performance functional (5.4.10) to be minimized subject to (5.4.9). The control constraints are mapped by a bounded, integrable, one-to-one, real-analytic, globally Lipschitz continuous function $\phi(\cdot)$: $\mathbb{R}^c \to \mathbb{R}^m$ of class $C^\epsilon (\epsilon \geq 1)$, $\phi \in U \subset \mathbb{R}^m$. Under the conditions imposed on $\phi(\cdot)$, an inverse integrable function $\phi^{-1}(\cdot)$ exists. Continuously differentiable integrands $W_x(\cdot)$: $\mathbb{R}^c \to \mathbb{R}_{\geq 0}$ and $W_u(\cdot)$: $\mathbb{R}^m \to \mathbb{R}_{\geq 0}$ are real valued and positive-definite. The performance functional is real valued and positive-definite for all $x \in X$ and $u \in U$ on $[t_0, t_f)$.*

Proof. To prove Lemma 5.4.3, the integrands $W_x(x)$ and $W_u(u)$ are studied. The performance integrand $W_x(x)$ is assigned to be positive-definite. For example, the following real-valued positive-definite $W_x(x)$ can be used:

$$W_x(x) = (x^{1/3})^T Q x^{1/3}, \quad W_x(t, x) = x^T Q(t) x, \quad W_x(x) = (x^2)^T Q x^2,$$

$$W_x(x) = x^T Q x + (x^2)^T Q x^2, \text{ etc.}$$

Here, $Q \in \mathbb{R}^{c \times c}$ is the semidefinite diagonal matrix $Q \geq 0$.

That is, $W_x(x)$ can be given using a great number of nonquadratic forms.

Continuously differentiable, real-analytic, bounded, one-to-one function $\phi \in U$, which intersects the origin and lies in the first and third quadrants, is used. Therefore, a continuously differentiable, one-to-one, inverse integrable function ϕ^{-1} exists. In the rectangular coordinate system, functions ϕ and ϕ^{-1} lie in the first and third quadrants of the input–output plane (this case typifies a general feature in all essential details). Using the properties of the bounded, integrable, one-to-one, globally Lipschitz, vector-valued continuous function ϕ, one concludes that the inverse function ϕ^{-1} is integrable. Hence, integral

$$\int (\phi^{-1}(u))^T G^{-1} u^{\frac{2w-2z}{2z+1}} \, du$$

exists. Furthermore, $W_u(u)$ is positive-definite.

Hence, performance integrands $W_x(\cdot)$: $\mathbb{R}^c \to \mathbb{R}_{\geq 0}$ and $W_u(\cdot)$: $\mathbb{R}^m \to \mathbb{R}_{\geq 0}$ are real-valued, positive-definite, and continuously differentiable integrand functions. Therefore, functional (5.4.10) is real valued and positive-definite for all $x \in X$ and $u \in U$ on $[t_0, t_f)$. One concludes that the continuity conditions are met by the performance functional (5.4.10).   □

To find a control law, we formulate two lemmas that state necessary and sufficient conditions. Necessary conditions for optimality are formulated using the Hamilton–Jacobi theory.

Lemma 5.4.4. *Given system (5.4.9) and positive-definite nonquadratic functional (5.4.10), necessary conditions that the control function $u(\cdot)$: $[t_0, \infty) \to \mathbb{R}^m$ guarantees a minimum to the Hamiltonian, which is given as*

$$H = W_x(x) + \tfrac{2w+1}{2z+1} \int (\phi^{-1}(u))^T G^{-1} u^{\frac{2w-2z}{2z+1}} \, du + \frac{\partial V(x)}{\partial x}^T \left[F(x, r) + B(x) u^{\frac{2w+1}{2z+1}} \right],$$

are

$$(n1) \quad \frac{\partial H}{\partial u} = 0$$

and

$$(n2) \quad \frac{\partial^2 H}{\partial u \times \partial u^T} > 0.$$

The positive-definite (minimum-cost) return function $V(\cdot)$: $\mathbb{R}^c \to \mathbb{R}_{\geq 0}$, $V \in C^\kappa, \kappa \geq 1$, is

$$V(x_0) = \inf_{u \in U} J(x_0, u) = \inf J(x_0, \phi(\cdot)) > 0.$$

An admissible optimal control is given as

$$u = -\phi \left(GB(x)^T \frac{\partial V(x)}{\partial x} \right), u \in U. \tag{5.4.11}$$

Optimal controller (5.4.11) is a local minimizer for problem (5.4.9)–(5.4.10).

Proof. The necessary conditions for optimality result from the Hamilton–Jacobi theory. The Hamilton–Jacobi equation for (5.4.9) and (5.4.10) is

$$-\frac{\partial V}{\partial t} = \min_{u \in U} \left\{ W_x(x) + \tfrac{2w+1}{2z+1} \int (\phi^{-1}(u))^T G^{-1} u^{\frac{2w-2z}{2z+1}} \, du \right.$$

$$\left. + \frac{\partial V(x)}{\partial x}^T \left[F(x, r) + B(x) u^{\frac{2w+1}{2z+1}} \right] \right\}. \tag{5.4.12}$$

The controller should be derived by finding the control function that guarantees the minimum to (5.4.12). The first-order necessary condition $(n1)$ leads us to a bounded admissible control law as given by (5.4.11). This bounded control minimizes the functional (5.4.10). The second-order necessary condition for optimality

$(n2)$ is satisfied because matrix G^{-1} is positive-definite. Hence, a unique, bounded, real-analytic, and continuous control candidate is designed, and controller (5.4.12) guarantees the optimality.

We turn our attention to the analysis of sufficiency.

Lemma 5.4.5. *Consider system (5.4.9) subject to the bounded admissible controller (5.4.11). If a proper function $V(x)$, $V(x_0) = \inf_{u \in U} J(x_0, u) = \inf J(x_0, \phi(\cdot)) > 0$ exists, which satisfies the Hamilton–Jacobi–Bellman equation (5.4.11)– (5.4.12), the resulting closed-loop system (5.4.9) with (5.4.11) is stable in the specified state $X \subset \mathbb{R}^c$ and control $U \subset \mathbb{R}^m$ sets, and tracking is ensured in the convex and compact set $XE(X_0, U, R, E_0) \subset \mathbb{R}^c \times \mathbb{R}^b$. That is, an invariant domain of stability $S \subset \mathbb{R}^c \times \mathbb{R}^b$ exists*

$$
\begin{aligned}
S = \{ x \in \mathbb{R}^c, e \in \mathbb{R}^b \colon \; &\|x(t)\| \le \varrho_x(\|x_0\|, t) + \varrho_u(\|u\|) + \delta_x, \\
&\|e(t)\| \le \varrho_e(\|e_0\|, t) + \varrho_r(\|r\|) + \varrho_y(\|y\|) + \delta, \\
&\forall x \in X(X_0, U), \forall e \in E(E_0, R, Y), \forall t \in [t_0, \infty) \} \subset \mathbb{R}^c \times \mathbb{R}^b,
\end{aligned}
$$

and control $u(\cdot) \colon [t_0, \infty) \to \mathbb{R}^m$, $u \in U$ (5.4.11) steers the system output to the set

$$
\begin{aligned}
S_E(\delta) = \{ e \in \mathbb{R}^b \colon \; &e_0 \in E_0, x \in X(X_0, U), r \in R, y \in Y, t \in [t_0, \infty) | \\
&\|e(t)\| \le \varrho_e(\|e_0\|, t) + \varrho_r(\|r\|) + \varrho_y(\|y\|) + \delta, \delta \ge 0, \\
&\forall e \in E(E_0, R, Y), \forall t \in [t_0, \infty) \} \subset \mathbb{R}^b. \quad (5.4.13)
\end{aligned}
$$

Here, ϱ_x and ϱ_e are the KL-functions and ϱ_u, ϱ_r, and ϱ_y are the K-functions.

Proof. An invariant domain of stability $S \subset \mathbb{R}^c \times \mathbb{R}^b$ can be found for dynamic systems, and the state and error vectors are bounded in the $S \subset \mathbb{R}^c \times \mathbb{R}^b$. For the closed-loop system, the states and error vectors evolve in $XE \subset \mathbb{R}^c \times \mathbb{R}^b$. It is required that $XE \subseteq S$ for all $x_0 \in X_0$, $u \in U$, $r \in R$, and $e_0 \in E_0$ on $[t_0, \infty)$. If sets $X_0 \subset \mathbb{R}^c$, $U \subset \mathbb{R}^m$, $R \subset \mathbb{R}^b$, and $E_0 \subset \mathbb{R}^b$ are admissible, the sufficiency is met, and $XE \subseteq S$. That is, the stability in X and tracking in E are guaranteed by the bounded real-analytic continuous controller (5.4.11). One concludes that the bounded control $u(\cdot) \colon [t_0, \infty) \to \mathbb{R}^m$, $u \in U$ renders the specified input–output tracking performance for the resulting closed-loop system and steers the error vector to (5.4.13).

Definition 5.4.2. Consider the system (5.4.9) with initial conditions $X_0 \subset \mathbb{R}^c$ and $E_0 \subset \mathbb{R}^b$. The convex and compact set $D \subset \mathbb{R}^c \times \mathbb{R}^b$ is a tracking domain of attraction for the evolution of states and errors if an admissible optimal control (5.4.11) exists such that for all $r \in R \subset \mathbb{R}^b$

$$
\begin{aligned}
\|x(t)\| &\le \varrho_x(\|x_0\|, t) + \varrho_u(\|u\|) + \delta_x \\
and \; \|e(t)\| &\le \varrho_e(\|e_0\|, t) + \varrho_r(\|r\|) + \varrho_y(\|y\|) + \delta.
\end{aligned}
$$

That is, if $r(t) = $ const, in D, $\lim_{t \to \infty} x(t) = 0$ and $\lim_{t \to \infty} e(t) = 0$. Furthermore, $\lim_{t \to \infty} y(t) = r(t)$.

The polyhedral and ellipsoidal domains of attraction are

$$D = \{x:\ Px \le 1\} \text{ and } D = \{x:\ x^T \Xi x \le 1\},\ \Xi > 0.$$

The positive-definite minimum-cost (return) function $V(\cdot)\colon \mathbb{R}^c \to \mathbb{R}_{\ge 0}$, $V \in C^\kappa$, $\kappa \ge 1$, is expressed as $V(x_0) = \inf_{u \in U} J(x_0, u) = \inf J(x_0, \phi(\cdot)) > 0$, and the solution of the functional equation (5.4.12) should be found using nonquadratic return functions. To obtain $V(\cdot)\colon \mathbb{R}^c \to \mathbb{R}_{\ge 0}$, the performance functional is evaluated. It was shown that the functional (5.4.10) admits a final value, and the minimum value is given by power-series forms

$$V(x) = \sum_{i=0}^{\eta} \frac{2\gamma+1}{2(i+\gamma+1)} \left(x^{\frac{i+\gamma+1}{2\gamma+1}} \right)^T K_i x^{\frac{i+\gamma+1}{2\gamma+1}},$$

$$\eta = 0, 1, 2, \ldots, \gamma = 0, 1, 2, \ldots, \tag{5.4.14}$$

where matrices $K_i \in \mathbb{R}^{c \times c}$ are found by solving the Hamilton–Jacobi–Bellman equation.

From (5.4.11) and (5.4.14), one obtains a nonlinear bounded controller $u \in U$. In particular,

$$u = -\phi \left(GB(x)^T \sum_{i=0}^{\eta} \text{diag}\left[x(t)^{\frac{i-\gamma}{2\gamma+1}} \right] K_i(t) x(t)^{\frac{i+\gamma+1}{2\gamma+1}} \right), \tag{5.4.15}$$

$$\text{diag}\left[x(t)^{\frac{i-\gamma}{2\gamma+1}} \right] = \begin{bmatrix} x_1^{\frac{i-\gamma}{2\gamma+1}} & 0 & \cdots & 0 & 0 \\ 0 & x_2^{\frac{i-\gamma}{2\gamma+1}} & \cdots & 0 & 0 \\ \vdots & \vdots & \ddots & \vdots & \vdots \\ 0 & 0 & \cdots & x_{c-1}^{\frac{i-\gamma}{2\gamma+1}} & 0 \\ 0 & 0 & \vdots & 0 & x_c^{\frac{i-\gamma}{2\gamma+1}} \end{bmatrix}.$$

5.4.1.3. Design of Performance Functionals and Optimal Control of Dynamic Systems

In this subsection, the performance functionals are synthesized, and the optimization problem is solved using the Hamilton–Jacobi theory. In particular, the synthesis of functionals is addressed with an emphasis on interpretation of innovative performance integrands to obtain a meaningful quantitative measure of the system dynamic performance in the time domain. For nonlinear continuous-time systems, minimizing innovative performance functionals, optimal controllers are designed. Illustrative examples are given to demonstrate the applicability and viability of the results reported. A tremendous amount of research has been made to extend the Hamilton–Jacobi theory and develop the basic theory to solve the optimization problems for a wide variety of dynamic systems. However, the performance functionals to be minimized have not been thoroughly studied. Performance functionals significantly influence dynamic performance, stability, and robustness of closed-loop systems, and the systematized design of performance functional is addressed to meet the specifications imposed in the time domain. The motivation for the design of performance functionals and the practical significance of this problem originate from the dependence of system performance and stability on the minimized functionals because different control algorithms are synthesized. The designer must be conscious with the significance of the performance integrands and analyze their importance.

The Hamilton–Jacobi theory and dynamic programming have been widely used to solve the optimization problems for linear and nonlinear dynamic systems

$$\dot{x}(t) = Ax + Bu, x(t_0) = x_0, \qquad (5.4.16)$$

$$\dot{x}(t) = F(x) + B(x)u, x(t_0) = x_0, \qquad (5.4.17)$$

where $x \in X \subset \mathbb{R}^n$ and $u \in U \subset \mathbb{R}^m$ are the state and control vectors, $A \in \mathbb{R}^{n \times n}$ and $B \in \mathbb{R}^{n \times m}$ are the known matrices, and $F(\cdot): \mathbb{R}^n \to \mathbb{R}^n$ and $B(\cdot): \mathbb{R}^n \to \mathbb{R}^{n \times m}$ are the smooth mappings, $F(0) = 0$ and $B(0) = 0$.

It was demonstrated that the optimization problem can be solved using specially structured performance functions. The following performance functionals are commonly used in the design of linear and constrained controllers:

$$J = \int_{t_0}^{t_f} \tfrac{1}{2}(x^T Q x + u^T G u) \, dt, \qquad (5.4.18)$$

$$J = \int_{t_0}^{t_f} \left[\tfrac{1}{2}x^T Q x + \int (\phi^{-1}(u))^T G \, du \right] dt, \qquad (5.4.19)$$

where $Q \in \mathbb{R}^{n \times n}$ and $G \in \mathbb{R}^{m \times m}$ are the diagonal weighting matrices and $\phi(\cdot): \mathbb{R}^n \to \mathbb{R}^m$ is the bounded, integrable, one-to-one, real-analytic globally Lipschitz continuous function $\phi \in U \subset \mathbb{R}^m$.

By applying the performance functionals (5.4.18) and (5.4.19), control laws can be designed. In particular, the Hamilton–Jacobi functional equations for system (5.4.17) are

$$-\frac{\partial V}{\partial t} = \tfrac{1}{2}(x^T Q x + u^T G u) + \frac{\partial V^T}{\partial x}[F(x) + B(x)u],$$

$$-\frac{\partial V}{\partial t} = \min_{u \in U} \left\{ \tfrac{1}{2}x^T Q x + \int (\phi^{-1}(u))^T G \, du + \frac{\partial V^T}{\partial x}[F(x) + B(x)u] \right\},$$

where $V(\cdot): \mathbb{R}^n \to \mathbb{R}_{\geq 0}$, is the positive-definite, continuously differentiable (minimum-cost) return function, $V \in C^\kappa, \kappa \geq 1$, $V(x_0) = \inf_{u \in U} J(x_0, u) > 0$.

For linear systems in (5.4.16) and quadratic functional (5.4.18), the solution of the functional equation is $V = \tfrac{1}{2}x^T K x$, $K \in \mathbb{R}^{n \times n}$. The controller is found to be

$$u = -G^{-1}B^T \frac{\partial V}{\partial x} = -G^{-1}B^T K x.$$

The unknown matrix $K \in \mathbb{R}^{n \times n}$ is obtained by solving the Riccati equation

$$-\dot{K} = Q + A^T K + K^T A - K^T B G^{-1} B^T K, \, K(t_f) = K_f.$$

For nonlinear system (5.4.17), minimizing (5.4.18), we have the following control law:

$$u = -G^{-1} B(x)^T \frac{\partial V(x)}{\partial x}.$$

If the bounds are imposed, by applying functional (5.4.19), for systems (5.4.16) and (5.4.17), we obtain

$$u = -\phi \left(G^{-1} B^T \frac{\partial V(x)}{\partial x} \right), \, u \in U \text{ and } u = -\phi \left(G^{-1} B(x)^T \frac{\partial V(x)}{\partial x} \right), \, u \in U.$$

5.4.1.4. Control of Linear Dynamic Systems

We study linear dynamic systems modeled by differential equations in the state-space form (5.4.16). The optimization problem will be solved using the following performance functional:

$$J = \int_{t_0}^{t_f} \tfrac{1}{2}[\omega(x)^T Q\omega(x) + \dot{\omega}(x)^T P\dot{\omega}(x)] \, dt, \qquad (5.4.20)$$

where $\omega(\cdot)$: $\mathbb{R}^n \to \mathbb{R}_{\geq 0}$ is a differentiable real-analytic continuous function and $Q \in \mathbb{R}^{n \times n}$ and $P \in \mathbb{R}^{n \times n}$ are the positive-definite diagonal weighting matrices.

The system transient performance is specified by integrands $\omega(x)^T Q\omega(x)$ and $\dot{\omega}(x)^T P\dot{\omega}(x)$, which are written in terms of the states and the rate of the state variables changes. It is evident that the performance functional depends on the transient behavior and time, control efforts, energy, and so on.

Lemma 5.4.6. *Consider a performance functional (5.4.20) to be minimized subject to the dynamic system as given by (5.4.16) or (5.4.17). A continuously differentiable function $\omega(\cdot)$: $\mathbb{R}^n \to \mathbb{R}_{\geq 0}$ is real valued, and scalar-valued integrands $\omega(x)^T Q\omega(x)$ and $\dot{\omega}(x)^T P\dot{\omega}(x)$ are real and positive-definite. The performance functional is finite, real valued, and positive-definite for all $x \in X$ on $[t_0, t_f)$.*

Proof. To prove Lemma 5.4.6, the integrands $\omega(x)^T Q\omega(x)$ and $\dot{\omega}(x)^T P\dot{\omega}(x)$ are studied. Here, $\omega(x)$ is a differentiable real-valued continuous function. For example, the following functions can be used:

$$\omega(x) = x, \, \omega(x) = x^3 \text{ or } \omega(x) = \mathrm{erf}x.$$

We have $J(x(\cdot)) > 0$ because $Q > 0$ and $P > 0$.

Therefore, functional (5.4.20) is real valued and positive-definite for all $x \in X$ on $[t_0, t_f)$, and the continuity conditions are met.

Lemma 5.4.7. *Given system* (5.4.16) *and positive-definite functional* (5.4.20), *necessary conditions that the control function* $u(\cdot)$: $[t_0, t_f) \rightarrow \mathbb{R}^m$ *guarantees a minimum to the Hamiltonian*

$$H\left(x, u, \frac{\partial V}{\partial x}\right) = \tfrac{1}{2}\omega(x)^T Q\omega(x) + \tfrac{1}{2}\dot{\omega}(x)^T P\dot{\omega}(x)$$
$$+ \frac{\partial V^T}{\partial x}(Ax + Bu),$$

are

$$\frac{\partial H}{\partial u} = 0 \text{ and } \frac{\partial^2 H}{\partial u \times \partial u^T} > 0.$$

The positive-definite return function $V(\cdot)$: $\mathbb{R}^n \rightarrow \mathbb{R}_{\geq 0}$, $V \in C^\kappa, \kappa \geq 1$, *is* $V(x_0) = \inf J(x_0) > 0$. *An optimal control is given as*

$$u = -\left(B^T \frac{\partial \omega^T}{\partial x} P \frac{\partial \omega}{\partial x} B\right)^{-1} B^T \left(\frac{\partial \omega^T}{\partial x} P \frac{\partial \omega}{\partial x} Ax + \frac{\partial V}{\partial x}\right). \tag{5.4.21}$$

This optimal controller is a minimizer for the problem (5.4.16) *and* (5.4.20).

Proof. We study the necessary conditions for optimality. The Hamilton–Jacobi–Bellman partial differential equation for (5.4.16) and (5.4.20)

$$-\frac{\partial V}{\partial t} = \tfrac{1}{2}\omega(x)^T Q\omega(x) + \tfrac{1}{2}\dot{\omega}(x)^T P\dot{\omega}(x) + \frac{\partial V^T}{\partial x}(Ax + Bu). \tag{5.4.22}$$

Taking note of

$$\dot{\omega}(x) = \frac{\partial \omega}{\partial x}\dot{x} = \frac{\partial \omega}{\partial x}(Ax + Bu),$$

one has

$$J = \int_{t_0}^{t_f} \tfrac{1}{2}\left[\omega(x)^T Q\omega(x) + (Ax + Bu)^T \frac{\partial \omega^T}{\partial x} P \frac{\partial \omega}{\partial x}(Ax + Bu)\right] dt. \tag{5.4.23}$$

From (5.4.22) and (5.4.23), we have

$$-\frac{\partial V}{\partial t} = \tfrac{1}{2}\omega(x)^T Q\omega(x) + \tfrac{1}{2}(Ax + Bu)^T \frac{\partial \omega^T}{\partial x} P \frac{\partial \omega}{\partial x}(Ax + Bu)$$
$$+ \frac{\partial V^T}{\partial x}(Ax + Bu). \tag{5.4.24}$$

The controller is derived by finding the control function that guarantees the minimum to (5.4.20). Using the first-order necessary condition for optimality $\frac{\partial H}{\partial u} = 0$, the controller in (5.4.21) is found.

The second-order necessary condition for optimality is guaranteed. In fact,

$$\frac{\partial^2 H}{\partial u \times \partial u^T} = B^T \frac{\partial \omega^T}{\partial x} P \frac{\partial \omega}{\partial x} B > 0$$

because $P > 0$. That is, a unique optimal controller, as given by (5.4.21), is designed.

The problem of design of the performance integrands should be addressed. In particular, in (5.4.20), one should choose the function $\omega(x)$.

Let $\omega(x) = x$. Then, using (5.4.20), we obtain

$$J = \int_{t_0}^{t_f} \tfrac{1}{2}[x^T Q x + (Ax + Bu)^T P (Ax + Bu)]\, dt.$$

From (5.4.21), one finds an optimal control law as

$$u = -(B^T P B)^{-1} B^T \left(P A x + \frac{\partial V}{\partial x} \right).$$

The solution of the functional equation (5.4.24) is given by the quadratic return function $V = \tfrac{1}{2} x^T K x$, where the unknown matrix $K \in \mathbb{R}^{n \times n}$ is found by solving the following nonlinear differential equation:

$$\begin{aligned}
-\dot{K} = {}& Q + A^T K + K A + A^T P A - A^T P B (B^T P B)^{-1} B^T P A \\
& - K B (B^T P B)^{-1} B^T P A - A^T P B (B^T P B)^{-1} B^T K \\
& - K B (B^T P B)^{-1} B^T K, \; K(t_f) = K_f.
\end{aligned}$$

From $B(B^T P B)^{-1} B^T P = P B (B^T P B)^{-1} B^T = I$, one obtains the following nonlinear differential equations that must to be solved to find the matrix K:

$$-\dot{K} = Q - K B (B^T P B)^{-1} B^T K, \; K(t_f) = K_f.$$

Furthermore, an optimal controller is

$$u = -(B^T P B)^{-1} B^T (P A + K) x.$$

This controller is different, compared with the linear quadratic control law, as given by

$$u = -G^{-1} B^T K x, \; -\dot{K} = Q + A^T K + K^T A - K^T B G^{-1} B^T K, \; K(t_f) = K_f.$$

This controller is found minimizing the quadratic performance functional (5.4.18). Furthermore, the equation to compute matrix K is different because the Riccati equation is

$$-\dot{K} = Q + A^T K + K^T A - K^T B G^{-1} B^T K, \; K(t_f) = K_f.$$

Example 5.4.6.

Consider the first-order system modeled as

$$\frac{dx}{dt} = ax + bu.$$

Using (5.4.23), one obtains the following performance functional:

$$J = \int_{t_0}^{\infty} \frac{1}{2} \left[Q\omega(x)^2 + P\frac{\partial^2 \omega}{\partial x^2}(a^2 x^2 + 2abxu + b^2 u^2) \right] dt.$$

Letting $\omega(x) = x$, $Q = 1$, and $P = 1$, we have

$$J = \int_{t_0}^{\infty} \frac{1}{2}(x^2 + a^2 x^2 + 2abxu + b^2 u^2)\, dt.$$

Making use of $V = \frac{1}{2}kx^2$, the following controller results:

$$u = -\frac{1}{b}(a + k)x.$$

The unknown k is found by solving the following differential equation:

$$-\dot{k} = 1 - k^2.$$

Hence, $k = 1$, and

$$u = -\frac{1}{b}(a + 1)x.$$

The closed-loop system is

$$\frac{dx}{dt} = -x.$$

Example 5.4.7.

For

$$\frac{dx}{dt} = ax + bu,$$

the functional to be minimized is given by (5.4.23).
Let $\omega(x) = \tanh x$, $Q = 1$, and $P = 1$.
Then,

$$J = \int_{t_0}^{\infty} \frac{1}{2}[\tanh^2 x + \text{sech}^4 x(a^2 x^2 + 2abxu + b^2 u^2)]\, dt.$$

For $x \ll 1$, $\tanh^2 x \approx x^2$, and $\text{sech}^4 x \approx 1$. That is, if $x \ll 1$, the functional can be viewed as the linear quadratic performance cost.
If $x \gg 1$, $\tanh^2 x \approx 1$ and $\text{sech}^4 x \approx 0$. Hence, the performance functional becomes

$$J \approx \frac{1}{2}\int_{t_0}^{\infty} dt.$$

This functional is commonly used to solve the time-optimal problem.

Taking note of the first-order necessary condition for optimality, one finds an optimal controller as

$$u = -\frac{a}{b}x - \frac{1}{b\operatorname{sech}^4 x}\frac{\partial V}{\partial x}.$$

The partial differential equation is

$$-\frac{\partial V}{\partial t} = \tfrac{1}{2}\tanh^2 x - \frac{1}{2\operatorname{sech}^4 x}\frac{\partial^2 V}{\partial x^2}.$$

Let $V = \tfrac{1}{2}kx^2$.
Then, we have

$$u = -\frac{a}{b}x - \frac{1}{b\operatorname{sech}^4 x}kx,$$

and the closed-loop system evolves as

$$\frac{dx}{dt} = -\frac{k}{\operatorname{sech}^4 x}x.$$

If $x \ll 1$, $\operatorname{sech}^4 x \approx 1$, and thus,

$$u = -\frac{a+k}{b}x.$$

Then, the system dynamics is

$$\frac{dx}{dt} = -kx.$$

For $x \gg 1$, we have the high feedback gain.

The analysis of a functional and control law illustrates that a novel concept augments different performance specifications imposed on transient dynamics. In particular, it was shown that the linear quadratic regulator and time-optimal problems were approached and solved.

5.4.1.5. Control of Nonlinear Dynamic Systems

We consider nonlinear dynamic systems as given by (5.4.17).

Lemma 5.4.8. *Given multivariable nonlinear system (5.4.17) and positive-definite performance functional (5.4.20)*

$$J = \int_{t_0}^{t_f} \tfrac{1}{2}\Big[\omega(x)^T Q\omega(x) + [F(x) + B(x)u]^T \frac{\partial \omega^T}{\partial x}$$

$$\times P\frac{\partial \omega}{\partial x}[F(x) + B(x)u]\Big]\,dt. \qquad (5.4.25)$$

The Hamiltonian is expressed as the scalar function

$$H\left(x, u, \frac{\partial V}{\partial x}\right) = \tfrac{1}{2}\omega(x)^T Q\omega(x) + \tfrac{1}{2}[F(x) + B(x)u]^T \frac{\partial \omega^T}{\partial x}$$

$$\times P \frac{\partial \omega}{\partial x}[F(x) + B(x)u] + \frac{\partial V^T}{\partial x} \times [F(x) + B(x)u]. \quad (5.4.26)$$

The positive-definite return function $V(\cdot)$: $\mathbb{R}^n \to \mathbb{R}_{\geq 0}$, $V \in C^\kappa$, $\kappa \geq 1$ *satisfies the Hamilton–Jacobi–Bellman equation*

$$-\frac{\partial V}{\partial t} = \tfrac{1}{2}\omega(x)^T Q\omega(x) + \tfrac{1}{2}[F(x) + B(x)u]^T \frac{\partial \omega^T}{\partial x}$$

$$\times P \frac{\partial \omega}{\partial x}[F(x) + B(x)u] + \frac{\partial V^T}{\partial x}[F(x) + B(x)u]. \quad (5.4.27)$$

An optimal control $u(\cdot)$: $[t_0, t_f) \to \mathbb{R}^m$ *is found as*

$$u = -\left(B(x)^T \frac{\partial \omega^T}{\partial x} P \frac{\partial \omega}{\partial x} B(x)\right)^{-1} B(x)^T \left(\frac{\partial \omega^T}{\partial x} P \frac{\partial \omega}{\partial x} F(x) + \frac{\partial V}{\partial x}\right). \quad (5.4.28)$$

The second-order necessary condition for optimality is guaranteed. Hence, control law (5.4.28) is an optimal controller for problems (5.4.17) and (5.4.25).

Proof. Making use of the Hamiltonian (5.4.26), one obtains a controller in (5.4.28).

The second-order necessary condition for optimality is satisfied because

$$\frac{\partial^2 H}{\partial u \times \partial u^T} = B(x)^T \frac{\partial \omega^T}{\partial x} P \frac{\partial \omega}{\partial x} B(x) > 0.$$

The positive-definite minimum-cost (return) function $V(\cdot)$: $\mathbb{R}^n \to \mathbb{R}_{\geq 0}$ is expressed as $V(x_0) = \inf J(x_0) > 0$, and the solution of the functional equation (5.4.27) should be found using nonquadratic return functions. To obtain $V(\cdot)$: $\mathbb{R}^n \to \mathbb{R}_{\geq 0}$, the performance functional is evaluated. The functional (5.4.25) admits a final value, and the minimum value is given by power-series forms $J_{\min} = \sum_{i=0}^\infty v(x_0)^{\frac{2(i+\gamma+1)}{2\gamma+1}}$, $\gamma = 0, 1, 2, \ldots$. The solution of the partial differential equation (5.4.27) is satisfied by a continuously differentiable, positive-definite return function

$$V(x) = \sum_{i=0}^\eta \frac{2\gamma+1}{2(i+\gamma+1)} \left(x^{\frac{i+\gamma+1}{2\gamma+1}}\right)^T K_i x^{\frac{i+\gamma+1}{2\gamma+1}}, \qquad \eta = 0, 1, 2, \ldots \quad (5.4.29)$$

Matrices $K_i \in \mathbb{R}^{n \times n}$ are found by solving the Hamilton–Jacobi–Bellman equation.

For example, letting $\eta = 1$ and $\gamma = 0$ in (5.4.29), one obtains

$$V(x) = \tfrac{1}{2}x^T K_0 x + \tfrac{1}{4}(x^2)^T K_1 x^2.$$

We have

$$V(x) = \frac{3}{4}\left(x^{\frac{2}{3}}\right)^T K_0 x^{\frac{2}{3}} + \frac{1}{2}x^T K_1 x + \frac{3}{8}\left(x^{\frac{4}{3}}\right)^T K_2 x^{\frac{4}{3}}$$
$$+ \frac{3}{10}\left(x^{\frac{5}{3}}\right)^T K_3 x^{\frac{5}{3}} + \frac{1}{4}(x^2)^T K_4 x^2 \text{ if } \eta = 4, \gamma = 1.$$

5.4.2. *Constrained Optimization of Time-Varying Systems*

This subsection provides one with systematic developments of the constrained optimization problem for nonlinear continuous-time systems. Necessary conditions for optimality, which are concerned with maximizing or minimizing the specified performance quantities and viewed applying the Hamilton–Jacobi concept, as well as sufficiency theory, constitute the basic foundation. It will be demonstrated that constrained optimization can be carried out applying the nonquadratic functionals and mapping the control bounds by smooth one-to-one functions. The results in constrained optimization are extended and unified using nonquadratic, sufficiently smooth, and real-valued positive-definite functionals that depend on the state and control variables and associated bounds. Constraints restrict a set of solutions as well as a class of control laws, from which an optimal controller, if it exists, can be found. This section illustrates how bounded smooth control laws can be designed. The Hamilton–Jacobi–Bellman equation has a series-type solution and should be solved using nonquadratic, continuously differentiable return (optimal) functions.

The calculus of variations and the optimal control theory provide a great source of workable issues. One of the major problems in optimal control is the lack of efficient methods to solve nonlinear control problems for high-order systems. The calculus of variations is not a very efficient method in this case, and it is difficult to solve the Euler–Lagrange equation with the prescribed boundary conditions. Often, the solution is obtained after many iterative computations, and it is difficult to approach the design even for simple systems that have the closed-form solutions. These have hindered the usefulness of the calculus of variations. Optimization problems can be solved using a variety of methods, and it is desirable to develop, extend, and modify different concepts. As the designer might appreciate, the Hamilton–Jacobi theory, Pontryagin's maximum principle, and dynamic programming can be used. These general and powerful concepts can be applied to a wide variety of optimization problems. Theoretical developments in optimal control are connected, and their study is remarkably remunerative. Thorough treatment of optimization problems will eventually lead us to new results. At the present time, research is proceeding from a number of standpoints.

In this section, we will apply new nonquadratic performance integrands, and control constraints are mapped by smooth one-to-one functions. In particular, control bounds will be incorporated into the design using an innovative mapping concept, and nonquadratic functionals and return functions will be employed. These extensions have advantages in obtaining the solution employing the Hamilton–Jacobi concept and sufficiency theory. An important feature is that instead of encountering difficulties, approaching constrained optimization, using the calculus of variations or Pontryagin's maximum principle, a feasible and computationally efficient optimization procedure is developed.

5.4.2.1. System Dynamics

Consider the case in which the vector state equation of nonlinear time-varying systems assumes the following form:

$$\dot{x}(t) = F(t, x) + B(t, x)u, \quad u_{min} \leq u \leq u_{max}, \quad x(t_0) = x_0, \qquad (5.4.30)$$

where $t \in \mathbb{R}_{\geq 0}$ is the time "variable," $x \in \mathbb{R}^c$ is the state vector, $u \in U \subset \mathbb{R}^m$ is the control vector, u_{min} and u_{max} are finite, and $F(\cdot)$: $\mathbb{R}_{\geq 0} \times \mathbb{R}^c \to \mathbb{R}^c$ and $B(\cdot)$: $\mathbb{R}_{\geq 0} \times \mathbb{R}^c \to \mathbb{R}^{c \times m}$ are the continuous Lipschitz maps.

A prescribed set of admissible control values is given as $U \subset \mathbb{R}^m$, and control function $u(\cdot)$: $[t_0, t_f) \to \mathbb{R}^m$ takes values in $U \subset \mathbb{R}^m$. Hence, the encumbered feature of the considered problem is that the control function is constrained to belong to a closed and bounded set $U \subset \mathbb{R}^m$. We have the state equation in (5.4.30), where control is bounded

$$U \subset \mathbb{R}^m, \quad \forall t \in [t_0, t_f), \qquad (5.4.31)$$

and $u(\cdot)$: $[t_0, t_f) \to \mathbb{R}^m$ generates solutions $x(\cdot)$: $[t_0, t_f) \to \mathbb{R}^c, x \in X \subset \mathbb{R}^c$ with initial conditions $x(t_0) = x_0 \in X_0 \subseteq X \subset \mathbb{R}^c$. Here, $X \subset \mathbb{R}^c$ denotes the set of state trajectories.

5.4.2.2. Nonlinear Mappings, Integrands, and Performance Functionals: Basic Concept and Notations

Optimization involves a great number of forms for performance integrands. The problem we consider is the minimization of the following functional:

$$J = \int_{t_0}^{t_f} (W_x(t, x) + W_u(t, u)) \, dt, \qquad (5.4.32)$$

subject to (5.4.30) and (5.4.31). In functional (5.4.32), $W_x(t, x)$ and $W_u(t, u)$ are the performance integrand functions $W_x(\cdot)$: $\mathbb{R}_{\geq 0} \times \mathbb{R}^c \to \mathbb{R}_{\geq 0}$ and $W_u(\cdot)$: $\mathbb{R}_{\geq 0} \times \mathbb{R}^m \to \mathbb{R}_{\geq 0}$.

Among an admissible control, let us find an optimal control $u^0(\cdot)$: $[t_0, t_f^0) \to \mathbb{R}^m$ and corresponding $x^0(\cdot)$: $[t_0, t_f^0) \to \mathbb{R}^c$ to guarantee a minimum to the performance functional. The key point is to develop a procedure that allows one to solve the constrained optimization problem via minimization of nonquadratic functionals. To introduce the optimization procedure, let us integrate the control bounds imposed using a nonlinear mapping concept. Our purpose is to find a critical point solution to the performance functional subject to (5.4.30) and (5.4.31). The main goal is to design a constrained control law. For a variety of reasons, we should specify and explicitly study the control bounds. The problem is how to map the constraints imposed. It is important that the basic idea and meaningful features are understood. We study the constrained optimization as a problem of finding the state feedback as a bounded $C^\epsilon(\epsilon \geq 1)$ function $\phi(\cdot)$ of the state vector and time. A control law is generated by memoryless state-feedback, and control constraints $u_{\min} \leq u \leq u_{\max}$ are mapped and smoothed by one-to-one C^ϵ, function $\phi \in U$. In particular,

$$u = \phi(f(t, x)), \quad f(\cdot): \mathbb{R}_{\geq 0} \times \mathbb{R}^c \to \mathbb{R}^m. \tag{5.4.33}$$

Although we do not attempt to provide a rigorous description of the design method, which will be performed later, it is appropriate to introduce a bounded controller at this time. The striking application of the Hamilton–Jacobi concept leads us to a controller in

$$u = \phi(f(t, x)) = -\phi\left(G(t)^{-1}B(t, x)^T \frac{\partial V}{\partial x}\right);$$

see (5.4.39).

That is,

$$f(t, x) = -G(t)^{-1}B(t, x)^T \frac{\partial V}{\partial x}.$$

In contrast, if control is not constrained, one obtains the conventional controller as

$$u = f(t, x) = -G(t)^{-1}B(t, x)^T \frac{\partial V}{\partial x}.$$

Hence, a control map, as given by (5.4.33), is bounded, and the role of a C^ϵ, real-analytic, one-to-one function $\phi(\cdot)$ is evident.

Proposition 5.4.1. *Given a smooth, continuously differentiable, real-analytic, one-to-one function $\phi(\cdot)$, which is bounded by the closed set $U \subset \mathbb{R}^m$; that is, $\phi \in U$. There exist*

$$\bar{U}_{\min} = \begin{bmatrix} u_{1\,\min} \\ \cdots \\ u_{m\,\min} \end{bmatrix} \in \mathbb{R}^m \quad and \quad \bar{U}_{\max} = \begin{bmatrix} u_{1\,\max} \\ \cdots \\ u_{m\,\max} \end{bmatrix} \in \mathbb{R}^m,$$

such that for all $x \in X \subset \mathbb{R}^c$ on $[t_0, t_f)$

$$\bar{U}_{\min} \leq \phi(f(t, x)) \leq \bar{U}_{\max}.$$

A class of memoryless bounded control laws is introduced in the general form. In particular, we have

$$u = \phi(f(t, x)), \quad \phi \in U,$$

where $\phi(\cdot): \mathbb{R}^m \to \mathbb{R}^m$ is the bounded, globally Lipschitz, continuous integrable, one-to-one function of class $C^\epsilon, \epsilon \geq 1$, and $\phi(f(t, 0)) = 0, \forall t \in [0, \infty)$, and $f(\cdot): \mathbb{R}_{\geq 0} \times \mathbb{R}^c \to \mathbb{R}^m$ is real-analytic and smooth on $\mathbb{R}^c \backslash \{0\}$.

A composition of continuous functions $\phi(\cdot)$ and $f(\cdot)$ is continuous; that is, control $u(\cdot): [t_0, t_f) \to \mathbb{R}^m, u \in U$ is continuous for all $x \in X$ on $[t_0, t_f)$.

Definition 5.4.3. Control function $u(\cdot)$: $[t_0, t_f) \to \mathbb{R}^m$ is said to be bounded if for all $x \in X$ on $[t_0, t_f)$ a KL-function $\rho_\Phi(\cdot)$: $\mathbb{R}_{\geq 0} \times \mathbb{R}_{\geq 0} \to \mathbb{R}_{\geq 0}$ exists such that

$$\|\phi(f(t, x))\| \leq \rho_\Phi(\|x\|, t).$$

Proposition 5.4.2. *The necessary and sufficient conditions that the bounded control $u(\cdot)$: $[t_0, t_f) \to \mathbb{R}^m$ is continuous are the continuity of $\phi(\cdot)$ and $f(\cdot)$ for all $x \in X$ on $[t_0, t_f)$.*

By using the map $\phi(\cdot)$, which is Lebesgue measurable for all $x \in X$ on $[t_0, t_f)$ and real-analytic on $\mathbb{R}^c \setminus \{0\}$, the set of allowable constraints is

$$\{\phi(\cdot): \mathbb{R}^m \to \mathbb{R}^m: [U_1 f(t, x) - \phi(f(t, x))]^T [U_2 f(t, x) - \phi(f(t, x))] \leq 0,$$

$$\forall x \in X, \forall t \in [t_0, t_f)\},$$

where $U_1 \in \mathbb{R}^{m \times m}$ and $U_2 \in \mathbb{R}^{m \times m}$ are the diagonal matrices.

Proposition 5.4.3. *Let $\phi(\cdot)$ be a one-to-one Lipschitz function. Then, an inverse one-to-one function $\phi^{-1}(\cdot)$ exists, and*

$$\phi[\phi^{-1}(f(t, x))] = \phi^{-1}[\phi(f(t, x))] = f(t, x), \forall x \in X, \forall t \in [t_0, t_f).$$

Proposition 5.4.4. *Let one-to-one function $\phi(\cdot)$ be continuous and integrable on the interval $\begin{bmatrix} u_{\min} & u_{\max} \end{bmatrix}$. Then, $\phi(\cdot)$ has an inverse function, and $\phi^{-1}(\cdot)$ is continuous and integrable on $\begin{bmatrix} \phi(u_{\min}) & \phi(u_{\max}) \end{bmatrix}$.*

To conform with (5.4.30), (5.4.31), and (5.4.33), one obtains

$$\dot{x}(t) = F(t, x) + B(t, x)u, u = \phi(f(t, x)), x(t_0) = x_0, x \in X \subset \mathbb{R}^c,$$

$$u \in U \subset \mathbb{R}^m. \tag{5.4.34}$$

Certain features have been examined. Our approach avoids the application of piecewise continuous functions and associated problems. An innovative mapping of control bounds using smooth bounded $\phi(\cdot)$ motivates the use of new functionals, and performance integrands should be designed, keeping in mind the constraints imposed. In constrained optimization, piecewise, continuous, and finite integrands have been widely used. In contrast, we set up our procedure in a framework in which the nonquadratic, sufficiently differentiable performance integrands are used. The purpose is to optimize the system with regard to the constraints. As a part of the design, let us specify and find the performance integrands that are correlated with the specified requirements. The functional has to be positive-definite, and we define the performance integrands as

$$W_x(t, x) = x^T Q(t)x, W_u(t, u) = \int (\phi^{-1}(u))^T G(t) \, du, \tag{5.4.35}$$

where $Q(\cdot)$: $\mathbb{R}_{\geq 0} \to \mathbb{R}^{c \times c}$ and $G(\cdot)$: $\mathbb{R}_{\geq 0} \to \mathbb{R}^{m \times m}$ are positive-definite.

It is obvious that positive-definite $W_x(t, x)$ can be given using a great number of nonquadratic forms.

Lemma 5.4.9. *Consider a system described by* (5.4.34). *Let the control constraints be mapped by an integrable, one-to-one, real-analytic, bounded function* $\phi(\cdot)$: $\mathbb{R}_{\geq 0} \times \mathbb{R}^c \to \mathbb{R}^m$ *of class* C^ϵ ($\epsilon \geq 1$), $\phi \in U \subset \mathbb{R}^m$ *on* $[t_0, t_f)$. *Under the conditions imposed on* $\phi(\cdot)$, *continuously differentiable* C^ξ ($\xi \geq 2$) *integrand*

$$W_u(t, u) = \int (\phi^{-1}(u))^T G(t) \, du$$

is positive-definite. The functional, as given by

$$J = \int_{t_0}^{t_f} \left(x^T Q(t) x + \int (\phi^{-1}(u))^T G(t) \, du \right) dt \qquad (5.4.36)$$

is real valued and positive-definite for all $x \in X$ *and* $u \in U$ *on* $[t_0, t_f)$.

Proof. To prove Lemma 5.4.9, it is sufficient to establish the asserted properties for $W_x(t, x)$ and $W_u(t, u)$. To do this, observe that $W_x(t, x)$ is real valued and positive-definite; see (5.4.35). Continuously differentiable, real-analytic, bounded, one-to-one function $\phi(\cdot)$, which intersects the origin, is used. For a given $\phi \in C^\epsilon$, a continuously differentiable, integrable, one-to-one inverse function $\phi^{-1}(\cdot)$ exists. In the rectangular coordinate system, functions $\phi(\cdot)$ and $\phi^{-1}(\cdot)$ lie in the first and third quadrants of the input–output plane, and real-valued $\int (\phi^{-1}(u))^T G(t) \, du$ is positive-definite (this case typifies a general feature in all essential details). From (5.4.35), it follows that the performance integrands $W_x(t, x)$ and $W_u(t, u)$ are real valued and positive-definite. Furthermore, $W_u \in C^\xi$. One concludes that functional (5.4.36) is real valued and positive-definite for all $x \in X$ and $u \in U$ on $[t_0, t_f)$.

The basic concept and certain features associated with the mapping of control constraints by smooth bounded functions $\phi(\cdot)$ and the application of nonquadratic integrand $W_u(t, u)$ should be demonstrated. The following examples illustrate the major perspectives.

Example 5.4.8.

Consider the so-called time-optimal and minimum-energy problems for

$$\dot{x}(t) = F(t, x) + B(t, x)u, \quad u_{\min} \leq u \leq u_{\max}, \quad U = \{|u| \leq 1\}.$$

By making use of smooth functions $\phi(\cdot)$ to map the control bounds imposed, and taking note of (5.4.35) for $G = 1$, we apply

$$W_u(u) = \int u^{21} \, du \quad \text{and} \quad W_u(u) = \int \tanh^{-1} u \, du.$$

Observe that these $W_u \in C^\xi$ are real valued and positive-definite for all $u \in U$; in particular,

$$W_u(u) = \int u^{21} \, du = \tfrac{1}{22} u^{22}$$

$$\text{and} \quad W_u(u) = \int \tanh^{-1} u \, du = u \tanh^{-1} u + \tfrac{1}{2} \log(1 - u^2).$$

The nonlinear maps and derived integrands provide an alternative avenue in the design. It is readily verified that the application of the first-order necessary condition leads one to the bounded control laws

$$u = -\left(B(t, x)^T \frac{\partial V}{\partial x}\right)^{1/21} \approx -\operatorname{sgn}\left(B(t, x)^T \frac{\partial V}{\partial x}\right)$$

and

$$u = -\tanh\left(B(t, x)^T \frac{\partial V}{\partial x}\right) \approx -\operatorname{sat}\left(B(t, x)^T \frac{\partial V}{\partial x}\right).$$

The detailed discussions of the origins and solutions of time-optimal and minimum-energy problems are well known, and controllers were designed as

$$u = -\operatorname{sgn}\left(B(t, x)^T \frac{\partial V}{\partial x}\right) \quad \text{and} \quad u = -\operatorname{sat}\left(B(t, x)^T \frac{\partial V}{\partial x}\right).$$

The designer maps the control bounds by $\phi(\cdot)$. The problem of fitting an input–output map $u = \phi(\sigma)$ to a set of data, which is given in a tabular form for a sequence of distinct values, results in different interpretations. A cubic spline approximation for a discrete set of base points, which are usually equally spaced, can be used. The smooth one-to-one functions can be found, and $\phi(\sigma)$ is represented by $\phi(\sigma) = p_i(\sigma) + R_{i+1}(\sigma)$, where $p_i(\sigma)$ is the ith-degree polynomial; $R_{i+1}(\sigma)$ is the error term.

The application of the Sturm–Liouville theorem ensures promising results.

Example 5.4.9.

Let us fit a set of input–output data

$$\text{input} = \begin{bmatrix} -2 & -1.6 & -1.2 & -0.8 & -0.4 & 0 & 0.4 & 0.8 & 1.2 & 1.6 & 2 \end{bmatrix},$$

$$\text{output} = \begin{bmatrix} -1 & -0.98 & -0.93 & -0.77 & -0.53 & 0 & 0.53 & 0.77 & 0.93 & 0.98 & 1 \end{bmatrix},$$

by trigonometric functions. A polynomial approximation gives $\phi(\sigma) \approx 1.2\sigma - 0.35\sigma^3 + 0.044\sigma^5$. It is straightforward to apply hyperbolic functions, for example, $\tanh(\sigma)$. However, to study other features, let us use the incomplete orthogonal set $a_1 \sinh^{-1} \omega_1 \sigma + a_2 \sin(\omega_2 \sigma) + a_3 \sin(\omega_3 \sigma)$. Applying the Sturm–Liouville theorem, one obtains the coefficients of the expansion; in particular, $\phi(\sigma) \approx 0.42 \sinh^{-1} 3.3\sigma + 0.075 \sin(2.1\sigma) + 0.021 \sin(5.3\sigma)$. Useful illustrations are given by Figure 5.4.7.

This example is valuable for at least two reasons. In the first place, a set of input–output data can be easily mapped. Second, it provides us with a striking example of the role of analytical and numerical methods in nonlinear mapping with the required degree of accuracy and smoothness. These aspects are viable and prominent in polynomial/spline approximations and expansion problems.

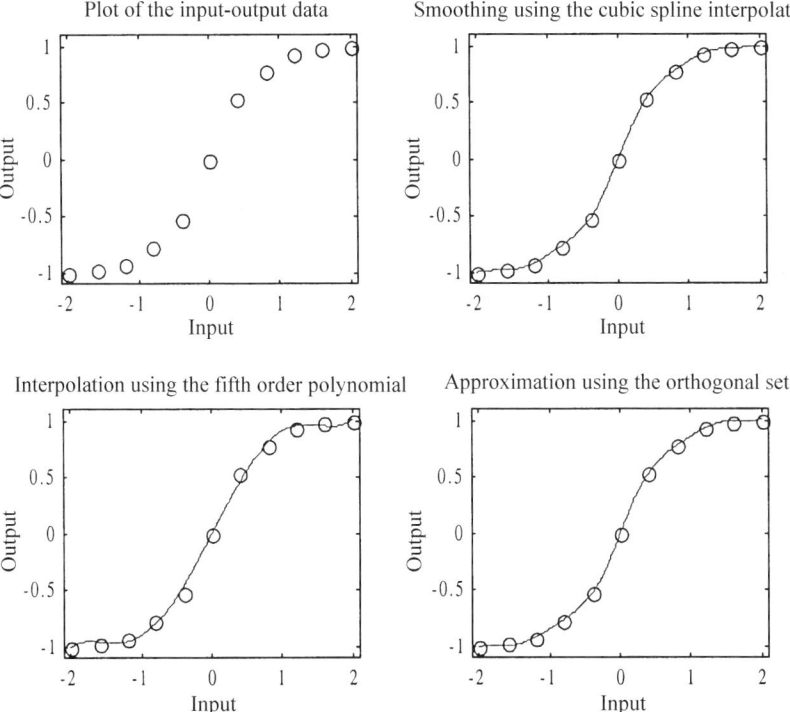

FIGURE 5.4.7.   Smoothing, interpolation, and approximation of the input–output data.

5.4.2.3. Constrained Optimization

Now, we are motivated to develop the synthesis procedure. Consider the problem of minimizing functional (5.4.36). Suppose that for (5.4.34), an optimal control $u^0(\cdot)$: $[t_0, t_f^0) \to \mathbb{R}^m$ and corresponding dynamics $x^0(\cdot)$: $[t_0, t_f^0) \to \mathbb{R}^c$ with $x^0(t_0) = x_0$ exist. The Hamilton–Jacobi–Bellman equation, corresponding to the system dynamics and performance functional, should be found in the ordinary way. In particular, for system (5.4.34) and functional (5.4.36), we have the following functional equation:

$$-\frac{\partial V}{\partial t} = \min_{u \in U} \left[x^T Q(t)x + \int (\phi^{-1}(u))^T G(t) \, du \right.$$
$$\left. + \frac{\partial V^T}{\partial x} (F(t, x) + B(t, x)u) \right]. \qquad (5.4.37)$$

This is the Hamilton–Jacobi functional equation for scalar function $V(t, x)$, $V(\cdot)$: $\mathbb{R}_{\geq 0} \times \mathbb{R}^c \to \mathbb{R}_{\geq 0}$. To see how dynamic programming relates to the maximum principle, we define the Hamiltonian scalar function $H(\cdot)$: $\mathbb{R}_{\geq 0} \times \mathbb{R}^c \times \mathbb{R}^m \times \mathbb{R}_{\geq 0} \to \mathbb{R}_{\geq 0}$ that is assumed to be of class C^γ, $\gamma \geq 2$, with respect to all of its arguments.

In particular,

$$H(t, x, u, \lambda) = x^T Q(t)x + \int (\phi^{-1}(u))^T G(t)\, du$$

$$+ \lambda^T (F(t, x) + B(t, x)u), \qquad (5.4.38)$$

where $\lambda(\cdot)\colon \mathbb{R}_{\geq 0} \times \mathbb{R}^c \to \mathbb{R}_{\geq 0}$ is the continuously differentiable multiplier vector-function, which is called the functional (or adjoint) vector.

The vector functional equation is

$$\frac{d\lambda}{dt} = -\frac{\partial H(t, x, u, \lambda)}{\partial x}$$

and

$$\frac{dx}{dt} = -\frac{\partial H(t, x, u, \lambda)}{\partial \lambda}.$$

From Hamiltonian (5.4.38) and functional equation (5.4.37), we observe a similarity. Equations (5.4.37) and (5.4.38) are identical if

$$\lambda = \frac{\partial V}{\partial x}.$$

It is important to note that the Pontryagin maximum principle can be studied using the dynamic programming concept. Using the results above, as well as necessary and sufficient conditions, let us find an optimal control law.

The Hamilton–Jacobi–Bellman equation requires continuously differentiable function $V(t, x)$. Usually, this is not satisfied in constrained optimization because piecewise continuous or finite performance integrands are used, and piecewise continuous control functions result. As a result, it was determined that dynamic programming is not well suited when hard constraints are imposed. The modification of dynamic programming is essentially appropriate and manageable. Optimization involves minimization or maximization problems. Necessary and sufficient conditions play a major role in optimal control. To introduce a method, let us formulate two lemmas that state necessary and sufficient conditions for optimality. For functional (5.4.36) subject to (5.4.34), we need to find a critical point solution. Necessary conditions for a minimum are immediate from the Hamilton–Jacobi theory.

Lemma 5.4.10. *For the dynamic system (5.4.34) and functional (5.4.36), the necessary conditions that control function $u(\cdot)\colon [t_0, t_f] \to \mathbb{R}^m$ gives a minimum to the Hamiltonian function, as given by (5.4.38), are*

(n1) $\frac{\partial H(t,x,u,\lambda)}{\partial u} = 0, \forall x \in X \subset \mathbb{R}^c, \forall u \in U \subset \mathbb{R}^m, \forall t \in [t_0, t_f),$

(n2) $\frac{\partial^2 H(t,x,u,\lambda)}{\partial u \times \partial u^T} > 0, \forall x \in X \subset \mathbb{R}^c, \forall u \in U \subset \mathbb{R}^m, \forall t \in [t_0, t_f).$

Proof. These necessary conditions have an origin in the calculus of variations.

The control law should be found as a function of the state vector and time using (n1). This criterion allows us to design a candidate for an optimal controller, and the minimum of the right-hand member of (5.4.37) corresponds to the critical point. Once a performance functional is determined in (5.4.36), the optimization problem can be solved for a critical point. This critical point, which is a relative minima, should be obtained on the basis of (5.4.37). Let us turn our attention to the following question: Under which condition is a critical point solution an optimal one? To answer this question, the sufficient conditions should be examined.

The bounded controller should be designed for open-loop unstable systems (5.4.30). Even for linear time-invariant systems $\dot{x}(t) = Ax + Bu, u \in U \subset \mathbb{R}^m$, bounded controllers do not exist that globally stabilize and optimize the closed-loop systems if $\lambda(A) \subseteq C^+$. Here, $\lambda(A)$ are the eigenvalues of $A \in \mathbb{R}^{c \times c}$ and C^+ is the right-half part of the complex plane. Therefore, the importance of sufficient conditions for optimality is magnitude by instability of system (5.4.30), which evolves in $X \subset \mathbb{R}^c$ for given $X_0 = \{x_0 \in \mathbb{R}^c\} \subset \mathbb{R}^c$ and $U \subset \mathbb{R}^m$. Sufficiency theory gives an explicit answer. Using the *field* concept, we formulate the sufficient conditions that guarantee optimality. These criteria are stated by the following lemma as a byproduct of the field theorem.

Lemma 5.4.11. *Consider the continuous-time system defined by (5.4.34). Control* $u^0(\cdot)$: $[t_0, t_f^0) \rightarrow \mathbb{R}^m, u^0 \in U \subset \mathbb{R}^m$ *and solution* $x^0(\cdot)$: $[t_0, t_f^0) \rightarrow \mathbb{R}^c$ *with* $x^0(t_0) = x_0$ *are optimal in* $U \subset \mathbb{R}^m$ *and* $X \subset \mathbb{R}^c$ *if a* $C^\kappa (\kappa \geq 1)$ *function* $V(\cdot)$: $\mathbb{R}_{\geq 0} \times \mathbb{R}^c \rightarrow \mathbb{R}_{\geq 0}, \rho_1(\|x\|) \leq V(t, x) \leq \rho_2(\|x\|), \rho_1, \rho_2 \in K_\infty$ *exists, such that for all* $x \in X$ *and* $u \in U$ *on* $[t_0, t_f)$

(s1) $\lim_{t \rightarrow t_f} V(t, x) \leq \lim_{t \rightarrow t_f^0} V(t, x^0) = 0,$

(s2) $x^{0^T} Q(t) x^0 + \int (\phi^{-1}(u^0))^T G(t) \, du^0 + \frac{\partial V^T}{\partial x^0}(F(t, x^0) + B(t, x^0) u^0) = 0,$

(s3) $x^T Q(t) x + \int (\phi^{-1}(u))^T G(t) \, du + \frac{\partial V^T}{\partial x}(F(t, x) + B(t, x) u) \leq -\rho_3(\|x\|),$
   $\rho_3 \in K.$

Proof. Except for the introduced system with *soft* control bounds, nonquadratic integrand $\int (\phi^{-1}(u))^T G(t) \, du$, and notations, all essential features involved in a general proof of this lemma are established using the Lyapunov stability theory.

Lemmas 5.4.10 and 5.4.11 state general conditions under which the constrained control problem can be solved. Although these lemmas are reminiscent of familiar results pertaining to nonlinear optimization, the formulated results obviously extend the scope of the Hamilton–Jacobi concept and sufficiency theory in constrained optimization because new control mappings are introduced, innovative performance integrands and nonquadratic return functions are offered, and a new class of control laws is synthesized.

Theorem 5.4.2. *Given the controllable continuous-time system (5.4.34) and the real-valued positive-definite functional (5.4.36), an admissible smooth control,* *which is found using the first-order necessary condition for optimality, is*

$$u = -\phi \left(G(t)^{-1} B(t, x)^T \frac{\partial V}{\partial x} \right), \, u \in U \subset \mathbb{R}^m. \tag{5.4.39}$$

The second-order necessary condition for optimality is guaranteed. The minimum value of functional (5.4.36) is given by a complete power-series truncated form. The Hamilton–Jacobi–Bellman equation (5.4.37) has a series-type solution, and the real-valued scalar functional function of class C^κ

$$V(t, x) = \sum_{i=0}^{\gamma} \frac{2\mu+1}{2(i+\mu+1)} \left(x^{\frac{i+\mu+1}{2\mu+1}} \right)^T K_i(t) x^{\frac{i+\mu+1}{2\mu+1}}, \, K_i(\cdot): \mathbb{R}_{\geq 0} \to \mathbb{R}^{c \times c}, \tag{5.4.40}$$

is the solution of equation

$$-\frac{\partial V}{\partial t} = x^T Q(t) x + \frac{\partial V^T}{\partial x} F(t, x)$$
$$- \int \left(\phi \left(G(t)^{-1} B(t, x)^T \frac{\partial V}{\partial x} \right) \right)^T d \left(B(t, x)^T \frac{\partial V}{\partial x} \right). \tag{5.4.41}$$

Control $u^0(\cdot): [t_0, t_f^0) \to \mathbb{R}^m, u^0 \in U \subset \mathbb{R}^m$, as given by (5.4.39), is optimal if the sufficient conditions are guaranteed.

Proof. Let us examine necessary and sufficient conditions that an optimal solution has to satisfy; see Lemmas 5.4.4 and 5.4.5. Application of $(n1)$ and minimization of (5.4.37) lead us to a smooth bounded controller in (5.4.39). It is evident that $u \in U$ because $\phi \in U$. One concludes that functional (5.4.36) is minimized to satisfy the control bounds imposed, and controller (5.4.39) does not exceed the constraints given by (5.4.33). Here, we begin to reap the benefits from functional (5.4.36). In fact, integrand

$$W_u(t, u) = \int (\phi^{-1}(u))^T G(t) \, du$$

is designed with the property that (5.4.33) is guaranteed by (5.4.39). Condition $(n2)$ represents further necessary criterion. If the second partial derivative of the Hamiltonian is positive-definite, the absolute minimum corresponds to a critical point. The basic features of a C^ϵ function $\phi(\cdot)$ were given, and

$$\frac{\partial^2 H(\cdot)}{\partial u \times \partial u^T}$$

is positive-definite. Hence, the first- and second-order necessary conditions for optimality are satisfied. That is, controller (5.4.39) guarantees a strong relative minimum to functional (5.4.36) and minimizes the Hamilton–Jacobi equation (5.4.37). If the Jacobi (or conjugate) condition is held, an absolute minimum is ensured. We analytically designed a bounded control as a nonlinear continuous function of the state vector and time.

Integration of a product of functions leads us to $\int g\, dh = gh - \int h\, dg$. Then, using (5.4.39), it follows that

$$
\int (\phi^{-1}(u))^T G(t)\, du
$$

$$
= \frac{\partial V^T}{\partial x} B(t,x)\phi \left(G(t)^{-1} B(t,x)^T \frac{\partial V}{\partial x} \right)
$$

$$
- \int \left(\phi \left(G(t)^{-1} B(t,x)^T \frac{\partial V}{\partial x} \right) \right)^T d\left(B(t,x)^T \frac{\partial V}{\partial x} \right). \tag{5.4.42}
$$

Using (5.4.37), (5.4.39), and (5.4.42), one finds at once (5.4.41), which must be solved. A set of sufficient conditions must be satisfied. If the hypotheses of Lemma 5.4.5 are guaranteed, control $u^0(\cdot)$: $[t_0, t_f^0) \rightarrow \mathbb{R}^m$, $u^0 \in U \subset \mathbb{R}^m$ is optimal and generates solution $x^0(\cdot)$: $[t_0, t_f^0) \rightarrow \mathbb{R}^c$, $x^0(t_0) = x_0$, $x^0 \in X \subset \mathbb{R}^c$. The desired result is obtained.

Theorem 5.4.3. *For system (5.4.34), a smooth bounded control law in (5.4.39) guarantees at least a strong relative minimum of functional (5.4.36) for all $x \in X$ and $u \in U$ on $[t_0, t_f)$. Moreover, if an optimal control exists, it is unique and given by (5.4.39).*

Proof. The proof comes from the proof of Theorem 5.4.2.

One must solve the Hamilton–Jacobi–Bellman partial differential equation (5.4.41). Let us discuss the problem of construction of return functions. To obtain the return function, the performance functional must be evaluated in $X \subset \mathbb{R}^c$. The application of quadratic return functions is an important stratagem for linear dynamic systems and minimization of quadratic functionals. It was illustrated that the solution of the Hamilton–Jacobi–Bellman equation is given by the quadratic $V(t, x)$. In nonlinear optimization, the situation is much more difficult. For a limited class of nonlinear systems and nonquadratic performance functionals applied, quadratic candidates might still be employed. However, although one may be tempted to use the quadratic return functions, this avenue is found to be conservative in practice. A great number of return functions have been developed, and a key feature is that nonquadratic return functions $V(t, x)$ should be used. Nonquadratic forms of solution can be found, and procedures are based on solving the partial differential equation using the power-series forms. Observe that the approximation in functional space has its origin in classic analysis. Even in the case of the linear quadratic regulator problem, this iterative procedure for generation of return functions is based on an initial guess of $V(x)$. To approach the constrained optimization problem for (5.4.34), the nonquadratic functional (5.4.36) was used. From the proceeding consideration, it appears that the minimum value of functional (5.4.36) should be expressed in the truncated form

$$
J_{\min} = \sum_{i=0}^{\gamma} \frac{2\mu+1}{2(i+\mu+1)} \left(x_0^{\frac{i+\mu+1}{2\mu+1}} \right)^T K_i(t) x_0^{\frac{i+\mu+1}{2\mu+1}}.
$$

Assume that the Hamilton–Jacobi–Bellman partial differential equation (5.4.41) admits a return function. Any scalar function satisfying the hypotheses of Lemma 5.4.5 can serve as a return candidate. The importance of J_{min}, as given by a complete power-series truncated form, lies in the fact that the solution of the Hamilton–Jacobi–Bellman partial differential equation can be found using a continuously differentiable function $V(t, x)$. Solution of (5.4.41) with (5.4.40) involves Γ nonlinear equations in Γ unknown coefficients of $V(t, x)$. One concludes that the constrained optimization problem has been formulated and solved using a nonlinear mappings concept minimizing nonquadratic performance functionals. A critical point solution is found, and a bounded control law is designed. These establish general results in constrained optimization and allow us to obtain an analytical solution in the closed form. The procedure is general and does not depend on particular control problems studied and control bounds imposed. The optimality is verified using the sufficiency criteria.

5.4.2.4. Discussions

Hamilton–Jacobi–Bellman partial differential equations place strong requirements on the return functions, which have to be continuously differentiable. For the constrained optimization problem, the straightforward dynamic programming is hampered if piecewise-continuous performance integrands are used because the continuity restriction on functions $V(t, x)$ and performance integrands are not satisfied, whereas Pontryagin's concept is valid. On this basis, dynamic programming, compared with the maximum principle, appears to have a lack of complete generality. This drawback has held back Bellman's concept. A modified synthesis procedure has been developed. This has been accomplished by using an innovative mapping concept, nonquadratic functionals, and return functions. New results have been researched, representing the control bounds using nonlinear smooth mappings. The extension of the control theory to account for such constraints constitutes one generalization. Another generalization is that the constrained optimization can be performed via dynamic programming using innovative performance functionals. The procedure ensures and applies nonquadratic positive-definite continuously differentiable performance integrands and scalar C^K return functions. Our avenue is one that stimulates an innovative inroad toward the further development of the Hamilton–Jacobi theory and uses the full potential of the dynamic programming approach. Many practical problems can be solved, and a variety of optimization problems can be treated through the researched modification and extension.

Example 5.4.10.

Consider a continuous-time system, which is given as

$$\dot{x}_1(t) = x_2 + 5\cos(5x_1 x_2 t),$$
$$\dot{x}_2(t) = -x_1 - (x_1^2 + x_1^2\cos(0.1x_1 t) - 1)x_2 + u, \ |u| \le u_{max}. \quad (5.4.43)$$

The control is constrained, and the saturation-type control set is

$$U = \{|u| \le 1\}. \qquad (5.4.44)$$

Let us demonstrate how the theoretical developments relate to the constrained optimization problem for system (5.4.43)–(5.4.44). It is difficult to obtain an analytical solution for the studied system using the calculus of variations or Pontryagin's principle. In contrast, the proposed technique is easy to apply. Guided by our experience in designing the functionals, let us find a performance functional using (5.4.36). The saturation is mapped by the hyperbolic tangent, and the functional to be minimized is given by

$$J = \int_{t_0}^{\infty} (x_1^2 + x_2^2 + \int \tanh^{-1} u \, du) \, dt. \qquad (5.4.45)$$

We find the Hamilton–Jacobi equation as

$$-\frac{\partial V}{\partial t} = \min_{|u| \leq 1} \left\{ x_1^2 + x_2^2 + \int \tanh^{-1} u \, du + \frac{\partial V}{\partial x_1}(x_2 + 5\cos(5x_1 x_2 t)) \right.$$
$$\left. + \frac{\partial V}{\partial x_2}[-x_1 - (x_1^2 + x_1^2 \cos(0.1x_1 t) - 1)x_2 + u] \right\}. \qquad (5.4.46)$$

An admissible control has to satisfy the necessary conditions for optimality. To obtain the feedback form of the controller, condition $(n1)$ is used. It is immediate from (5.4.46) that a feedback control law is

$$u = -\tanh \frac{\partial V}{\partial x_2},$$

and from (5.4.46), we have

$$-\frac{\partial V}{\partial t} = x_1^2 + x_2^2 + \frac{\partial V}{\partial x_1}(x_2 + 5\cos(5x_1 x_2 t))$$
$$+ \frac{\partial V}{\partial x_2}[-x_1 - (x_1^2 + x_1^2 \cos(0.1x_1 t) - 1)x_2]$$
$$- \int \tanh \frac{\partial V}{\partial x_2} \, d\frac{\partial V}{\partial x_2}, \qquad (5.4.47)$$

where $\int \tanh \frac{\partial V}{\partial x_2} \, d\frac{\partial V}{\partial x_2} = \ln\cosh \frac{\partial V}{\partial x_2}$.
Let us employ the following return function:

$$V(t, x_1, x_2) = k_{11}x_1^2 + 2k_{12}x_1 x_2 + k_{22}x_2^2 + k_{41}x_1^2 x_2^2(1 + \cos(0.1x_1 t)) \quad (5.4.48)$$

to approximate the minimum value of the functional.
Using (5.4.48) and solving (5.4.47), we find the nonlinear controller; in particular,

$$u = -\tanh(3.6x_1 + 5.2x_2 + x_1^2 x_2(1 + \cos(0.1x_1 t))), \ |u| \leq 1. \qquad (5.4.49)$$

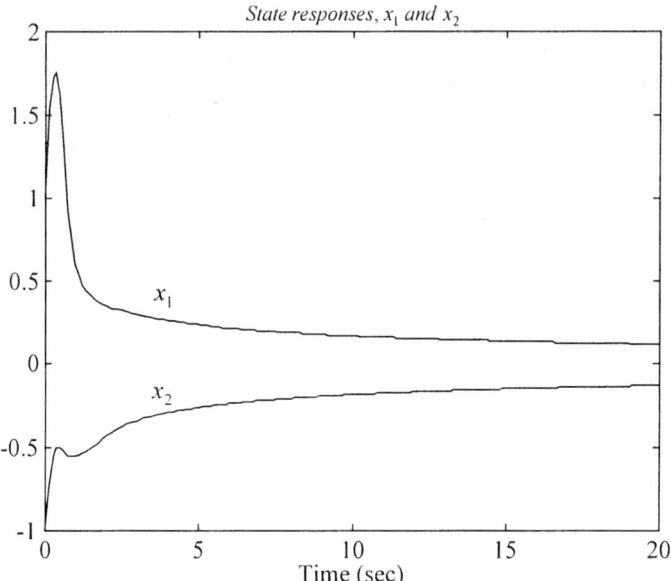

FIGURE 5.4.8.   Controlled state evolution, $\begin{bmatrix} x_{10} \\ x_{20} \end{bmatrix} = \begin{bmatrix} 1 \\ -1 \end{bmatrix}$.

The synthesized control law (5.4.49) is smooth and bounded. That is, $u \in U$. For completeness of this example, sufficient conditions must be examined and simulations should be performed. Using (5.4.43)–(5.4.49), the hypotheses of Lemma 5.4.5 are easily verified. Figure 5.4.8 shows the closed-loop system evolution caused by the initial conditions.

5.4.3. *Optimal Control of Nonlinear Continuous-Time Systems*: *Design of Bounded Controllers Via Generalized Nonquadratic Functionals*

In this subsection, by using the Hamilton–Jacobi framework and sufficiency theory, we present a solution of the constrained optimization problem for nonlinear systems with *soft* and *hard* bounds imposed on control. The application of the *generalized* nonquadratic functional and nonquadratic return functions will be emphasized. We use necessary and sufficient conditions for optimality to solve the constrained optimization problem. Specifically, necessary conditions have been used to synthesize the bounded controllers, and sufficient conditions are applied to verify the optimality.

We consider the Euclidean representation of a continuous-time system; in particular,

$$\dot{x}(t) = F(x) + B(x)u, \, u_{min} \leq u \leq u_{max}, \quad x(t_0) = x_0. \tag{5.4.50}$$

Our goal is to find a control function $u \in U \subset \mathbb{R}^m$ that minimizes a generalized nonquadratic functional

$$J = \int_{t_0}^{t_f} \left[W_x(x) + W_v \left(\frac{\partial V}{\partial x} \right) + W_u(u) \right] dt, \tag{5.4.51}$$

where the integrand functions are expressed as

$$W_x(x) = \sum_{i=0}^{\varsigma} \frac{2\beta+1}{2(i+\beta+1)} \left(x^{\frac{i+\beta+1}{2\beta+1}} \right)^T Q_i x^{\frac{i+\beta+1}{2\beta+1}}, \, Q_i \in \mathbb{R}^{c \times c},$$

$$Q_i \geq 0; \, \varsigma = 0, 1, 2, \ldots \text{ and } \beta = 0, 1, 2, \ldots, \tag{5.4.52}$$

$$W_v \left(\frac{\partial V}{\partial x} \right) = \int \left[\phi \left(G^{-1} B(x)^T \frac{\partial V}{\partial x} \right) \right]^T d \left(B(x)^T \frac{\partial V}{\partial x} \right)$$

$$- \frac{\partial V^T}{\partial x} B(x) P G^{-1} B(x)^T \frac{\partial V}{\partial x}, \quad P \in \mathbb{R}^{m \times m}, P > 0, \tag{5.4.53}$$

$$W_u(u) = \int (\phi^{-1}(u))^T G \, du, \quad G \in \mathbb{R}^{m \times m}, \quad G > 0, \tag{5.4.54}$$

and $V(\cdot): \mathbb{R}^c \to \mathbb{R}_{\geq 0}$ is a continuously differentiable, real-analytic function with respect to $x \in X$ on $[t_0, t_f)$, $V \in C^{\kappa} (\kappa \geq 1)$.

In (5.4.53) and (5.4.54), $\phi(\cdot): \mathbb{R}^c \to \mathbb{R}^m$ is the continuous integrable, one-to-one function of class $C^{\epsilon} (\epsilon \geq 1)$ with $\phi(0) = 0$. Function ϕ maps control bounds imposed; that is, $\phi \in U$ (for example, tanh, sat, sgn functions).

We should minimize (5.4.51) with respect to the state and control that are subject to the nonlinear differential equations (5.4.50). The Hamilton–Jacobi theory and sufficiency concept provide a general outline. The constraints on control function $u(\cdot): [t_0, t_f) \to \mathbb{R}^m$ are imposed. Our goal is to design a bounded control law as a continuous function within the constrained rectangular control set

$$U = \{u \in \mathbb{R}^m : u_{min} \leq u \leq u_{max}, u_{min} < 0, u_{max} > 0\} \subset \mathbb{R}^m.$$

A set of admissible controllers is given as

$$\text{Set} = \{\phi(\cdot): \mathbb{R}^c \to \mathbb{R}^m | \phi \in U, \forall x \in X, \phi(0) = 0\}.$$

The bounded controller should be designed by minimizing a generalized performance functional (5.4.44), and the basic features should be emphasized. We introduce the following weak assumption.

Assumption 5.4.1. *For nonlinear system (5.4.50) in* $X \subset \mathbb{R}^c$ *and* $U \subset \mathbb{R}^m$, *inequality*

$$W_x(x) + W_u(u) > W_v\left(\tfrac{\partial V}{\partial x}\right)$$

is guaranteed, where the integrand functions are given by (5.4.52), (5.4.53), and (5.4.54).

It should be noted that $W_x(x) > 0$ and $W_u(u) > 0$; see (5.4.52) and (5.4.54). Furthermore, the first term of the integrand $W_v\left(\tfrac{\partial V}{\partial x}\right)$ is positive-definite; that is,

$$\int \left[\phi\left(G^{-1}B(x)^T \frac{\partial V}{\partial x}\right)\right]^T d\left(B(x)^T \frac{\partial V}{\partial x}\right) > 0.$$

Hence, with the appropriate choice of the weighting matrices Q_i, G, and P, the positive-definiteness of the performance functional is guaranteed. By making use of the assumption, the following lemma results.

Lemma 5.4.12. *A generalized nonquadratic performance functional, as given by (5.4.51), is finite, real valued, and positive-definite for all* $x \in X$ *and* $u \in U$ *on* $[t_0, t_f)$.

Proof. In the functional (5.4.51), the nonquadratic integrands $W_x(x)$ and $W_u(u)$, as given by (5.4.52) and (5.4.54) are positive-definite. An innovative performance integrand $W_v\left(\tfrac{\partial V}{\partial x}\right)$ is expressed by (5.4.53) Inequality $W_x(x) + W_u(u) > W_v\left(\tfrac{\partial V}{\partial x}\right)$ must be satisfied for all $x \in X$ and $u \in U$, and one concludes that functional (5.4.51) is positive-definite.

To design the control algorithm, we apply the Hamilton–Jacobi theory. In particular, the necessary conditions for optimality are used.

Lemma 5.4.13. *Consider the nonlinear system in (5.4.50) and a generalized performance functional (5.4.51). The minimum of the right-hand side of the Hamilton–Jacobi functional equation*

$$-\frac{\partial V}{\partial t} = \min_{u \in U}\left[W_x(x) + W_v\left(\tfrac{\partial V}{\partial x}\right) + W_u(u) + \left(\frac{\partial V}{\partial x}\right)^T (F(x) + B(x)u)\right] \quad (5.4.55)$$

has to correspond to the stationary point, and the first- and second-order necessary conditions for optimality are

(n1) $\dfrac{\partial H\left(x, u, \frac{\partial V}{\partial x}\right)}{\partial u} = 0,$

(n2) $\dfrac{\partial^2 H\left(x, u, \frac{\partial V}{\partial x}\right)}{\partial u \times \partial u^T} \geq 0,$

where the Hamiltonian function is given as

$$H\left(x, u, \tfrac{\partial V}{\partial x}\right) = W_x(x) + W_v\left(\tfrac{\partial V}{\partial x}\right) + W_u(u) + \left(\frac{\partial V}{\partial x}\right)^T (F(x) + B(x)u).$$

The first-order necessary condition results in the bounded controller

$$u = -\phi\left(G^{-1}B(x)^T \frac{\partial V}{\partial x}\right), u \in U \subset \mathbb{R}^m. \tag{5.4.56}$$

The Hessian matrix is positive-definite; that is, control function $u(\cdot): [t_0, t_f) \to$ $\mathbb{R}^m$ *ensures the global minimum of* (5.4.55). *If an optimal control exists, it is unique and given by* (5.4.56).

Proof. The proof is based on necessary conditions for optimality. By applying the first-order criterion and the nonquadratic functional (5.4.51), the bounded control law (5.4.56) results. By making use of (5.4.53) and (5.4.54), one concludes that

$$\frac{\partial^2 H(\cdot)}{\partial u \times \partial u^T}$$

is positive-definite. Thus, the second-order condition is guaranteed, and the proof is completed.

By taking note of (5.4.55) and (5.4.56), one obtains the Hamilton–Jacobi–Bellman partial differential equation for $V(\cdot): \mathbb{R}^c \to \mathbb{R}_{\geq 0}$. In particular, we have

$$-\frac{\partial V}{\partial t} = \sum_{i=0}^{\varsigma} \frac{2\beta+1}{2(i+\beta+1)} \left(x^{\frac{i+\beta+1}{2\beta+1}}\right)^T Q_i x^{\frac{i+\beta+1}{2\beta+1}} + \frac{\partial V^T}{\partial x} F(x)$$

$$-\frac{\partial V^T}{\partial x} B(x) P G^{-1} B(x)^T \frac{\partial V}{\partial x}. \tag{5.4.57}$$

A generalized nonquadratic performance functional (5.4.51) is minimized, and the minimum value of (5.4.51) is expressed in power form. To solve the partial differential equation (5.4.57), the nonquadratic return function, as given by

$$V(x) = \sum_{i=0}^{\varsigma} \frac{2\beta+1}{2(i+\beta+1)} \left(x^{\frac{i+\beta+1}{2\beta+1}}\right)^T K_i x^{\frac{i+\beta+1}{2\beta+1}}, \quad K_i \in \mathbb{R}^{c \times c}, \tag{5.4.58}$$

is used.

In (5.4.58), γ, μ, and χ are the real numbers, $\gamma = 0, 1, 2, \ldots, \mu = 0, 1, 2, \ldots,$ and $\chi = 1, 2, 3, \ldots$.

As one finds $K_i \in \mathbb{R}^{c \times c}$, which must be positive-definite, feedback coefficients of the control law (5.4.56) result.

The bounded controller (5.4.56) is qualified as a candidate for an optimal one. To ensure that bounded $u(\cdot): [t_0, t_f) \to \mathbb{R}^m, u \in U$ is optimal, sufficient conditions should be examined.

Lemma 5.4.14. *If a continuously differentiable, positive-definite convex function* $V(\cdot)\colon \mathbb{R}^c \to \mathbb{R}_{\geq 0}$, $V \in C^\kappa (\kappa \geq 1)$ *exists, such that for all* $x \in X$ *on* $[t_0, t_f)$

$$\sum_{i=0}^{\varsigma} \frac{2\beta+1}{2(i+\beta+1)} \left(x^{0\frac{i+\beta+1}{2\beta+1}} \right)^T Q_i x^{0\frac{i+\beta+1}{2\beta+1}} + \frac{\partial V^T}{\partial x^0} F(x^0)$$

$$= \frac{\partial V^T}{\partial x^0} B(x^0) P G^{-1} B(x^0)^T \frac{\partial V}{\partial x^0},$$

$$\sum_{i=0}^{\varsigma} \frac{2\beta+1}{2(i+\beta+1)} \left(x^{\frac{i+\beta+1}{2\beta+1}} \right)^T Q_i x^{\frac{i+\beta+1}{2\beta+1}} + \frac{\partial V^T}{\partial x} F(x) \leq \frac{\partial V^T}{\partial x} B(x) P G^{-1} B(x)^T \frac{\partial V}{\partial x},$$

the bounded smooth control $u^0(\cdot)\colon [t_0, t_f) \to \mathbb{R}^m$, $u^0 \in U$ *as given by* (5.4.56), *is optimal. Furthermore, solution* $x^0(\cdot)\colon [t_0, t_f) \to \mathbb{R}^c$, $x^0 \in X$ *is an optimal one.*

Proof. The proof can be easily carried out by using the sufficiency theory.

Theorem 5.4.3. *For system* (5.4.50), *the bounded control law, as given by* (5.4.56), *satisfies necessary conditions for optimality and minimizes a generalized non-quadratic performance functional* (5.4.51). *Control function* $u^0(\cdot)\colon [t_0, t_f) \to \mathbb{R}^m$ *is optimal if sufficient conditions are guaranteed. The positive-definite convex function* $V(\cdot)\colon \mathbb{R}^c \to \mathbb{R}_{\geq 0}$, $V \in C^\kappa (\kappa \geq 1)$, *as given by* (5.4.58), *is a solution of the partial differential equation* (5.4.57).

Proof. The proof is straightforward by using the results of Lemmas 5.4.13 and 5.4.14.

5.4.4. Tracking Control of Continuous-Time Systems

Let us recall the major steps to solve the constrained optimization problem, and then attack and solve the tracking control problem. The following state-space equation with control bounds was used:

$$\dot{x}(t) = F(x) + B(x)u, \ u_{\min} \leq u \leq u_{\max}, \ x(t_0) = x_0. \tag{5.4.59}$$

Minimizing the performance functional

$$J = \int_{t_0}^{t_f} \left(\sum_{i=0}^{\varsigma} \frac{2\eta+1}{2(\kappa i+\eta+1)} \left(x^{\frac{\kappa i+\eta+1}{2\eta+1}} \right)^T Q_i x^{\frac{\kappa i+\eta+1}{2\eta+1}} + \int (\phi^{-1}(u))^T G \, du \right) dt,$$

the following Hamilton–Jacobi functional equation results:

$$-\frac{\partial V}{\partial t} = \min_{u \in U} \left[\sum_{i=0}^{\varsigma} \frac{2\eta+1}{2(\kappa i+\eta+1)} \left(x^{\frac{\kappa i+\eta+1}{2\eta+1}} \right)^T Q_i x^{\frac{\kappa i+\eta+1}{2\eta+1}} \right.$$

$$\left. + \int (\phi^{-1}(u))^T G \, du + \frac{\partial V^T}{\partial x} (F(x) + B(x)u) \right].$$

Using the first-order necessary condition for optimality, one obtains

$$u = -\phi\left(G^{-1}B(x)^T\frac{\partial V}{\partial x}\right), \quad u \in U,$$

and the return function is found solving the Hamilton–Jacobi–Bellman partial differential equation

$$-\frac{\partial V}{\partial t} = \sum_{i=0}^{\varsigma}\frac{2\eta+1}{2(\kappa i+\eta+1)}\left(x^{\frac{\kappa i+\eta+1}{2\eta+1}}\right)^T Q_i x^{\frac{\kappa i+\eta+1}{2\eta+1}} + \frac{\partial V^T}{\partial x}F(x)$$

$$-\int\left[\phi\left(G^{-1}B(x)^T\frac{\partial V}{\partial x}\right)\right]^T d\left(B(x)^T\frac{\partial V}{\partial x}\right),$$

$$V(x) = \sum_{i=0}^{\gamma}\frac{2\mu+1}{2(\chi i+\mu+1)}\left(x^{\frac{\chi i+\mu+1}{2\mu+1}}\right)^T K_i x^{\frac{\chi i+\mu+1}{2\mu+1}}.$$

5.4.4.1. Tracking Integral Control

To solve the motion control problem, an integral control law will be designed. The following augmented state-space equations are derived combining the system dynamics (5.4.59) with an *exogenous* system. In particular,

$$\dot{x}^{\text{system}}(t) = F(x^{\text{system}}) + B(x^{\text{system}})u, \ y = Hx^{\text{system}}, \ x_0^{\text{system}}(t_0) = x_0^{\text{system}},$$

$$\dot{x}_i^{\text{ref}}(t) = Nr - y = Nr - Hx^{\text{system}}, \quad N \in \mathbb{R}^{b\times b}.$$

Therefore, we have

$$\begin{bmatrix}\dot{x}^{\text{system}}(t)\\ \dot{x}_i^{\text{ref}}(t)\end{bmatrix} = \begin{bmatrix}F(x^{\text{system}})\\ -Hx^{\text{system}}\end{bmatrix} + \begin{bmatrix}B(x^{\text{system}})\\ 0\end{bmatrix}u + \begin{bmatrix}0\\ N\end{bmatrix}r,$$

$$y = Hx^{\text{system}}, \quad \begin{bmatrix}x^{\text{system}}(t_0)\\ x^{\text{ref}}(t_0)\end{bmatrix} = \begin{bmatrix}x_0^{\text{system}}\\ x_0^{\text{ref}}\end{bmatrix}. \qquad (5.4.60)$$

The tracking error is defined as

$$e(t) = Nr(t) - y(t),$$

and

$$x_i^{\text{ref}}(t) = \int e(t)\,dt.$$

Using a smooth bounded map

$$u = \phi\left(f\begin{bmatrix}x^{\text{system}}\\ x_i^{\text{ref}}\end{bmatrix}\right),$$

and minimizing the nonquadratic functional

$$
J = \int_{t_0}^{t_f} \left(\sum_{i=0}^{\varsigma} \frac{2\eta+1}{2(\kappa i+\eta+1)} \left[\begin{array}{c} x^{\text{system}} \frac{\kappa i+\eta+1}{2\eta+1} \\ x_i^{\text{ref}} \frac{\kappa i+\eta+1}{2\eta+1} \end{array} \right]^T Q_i \left[\begin{array}{c} x^{\text{system}} \frac{\kappa i+\eta+1}{2\eta+1} \\ x_i^{\text{ref}} \frac{\kappa i+\eta+1}{2\eta+1} \end{array} \right] \right.
$$

$$
\left. + \int (\phi^{-1}(u))^T G \, du \right) dt,
$$

$$
Q_i \in \mathbb{R}^{(c+b)\times(c+b)}, \ Q_i \geq 0, \tag{5.4.61}
$$

the bounded controller is found from the Hamilton–Jacobi functional equation. In particular,

$$
u = -\phi \left(G^{-1} \left[\begin{array}{c} B(x^{\text{system}}) \\ 0 \end{array} \right]^T \frac{\partial V \left(\left[\begin{array}{c} x^{\text{system}} \\ x_i^{\text{ref}} \end{array} \right] \right)}{\partial \left[\begin{array}{c} x^{\text{system}} \\ x_i^{\text{ref}} \end{array} \right]} \right), \ u \in U. \tag{5.4.62}
$$

The nonquadratic return function

$$
V \left(\left[\begin{array}{c} x^{\text{system}} \\ x_i^{\text{ref}} \end{array} \right] \right) = \sum_{i=0}^{\gamma} \frac{2\mu+1}{2(\chi i+\mu+1)} \left[\begin{array}{c} x^{\text{system}} \frac{\chi i+\mu+1}{2\mu+1} \\ x_i^{\text{ref}} \frac{\chi i+\mu+1}{2\mu+1} \end{array} \right]^T K_i \left[\begin{array}{c} x^{\text{system}} \frac{\chi i+\mu+1}{2\mu+1} \\ x_i^{\text{ref}} \frac{\chi i+\mu+1}{2\mu+1} \end{array} \right],
$$

$$
K_i \in \mathbb{R}^{(c+b)\times(c+b)} \tag{5.4.63}
$$

is a solution of the Hamilton–Jacobi–Bellman partial differential equation

$$
-\frac{\partial V}{\partial t} = \sum_{i=0}^{\varsigma} \frac{2\eta+1}{2(\kappa i+\eta+1)} \left[\begin{array}{c} x^{\text{system}} \frac{\kappa i+\eta+1}{2\eta+1} \\ x_i^{\text{ref}} \frac{\kappa i+\eta+1}{2\eta+1} \end{array} \right]^T Q_i \left[\begin{array}{c} x^{\text{system}} \frac{\kappa i+\eta+1}{2\eta+1} \\ x_i^{\text{ref}} \frac{\kappa i+\eta+1}{2\eta+1} \end{array} \right]
$$

$$
+ \frac{\partial V \left[\begin{array}{c} x^{\text{system}} \\ x_i^{\text{ref}} \end{array} \right]^T}{\partial \left[\begin{array}{c} x^{\text{system}} \\ x_i^{\text{ref}} \end{array} \right]} \left[\begin{array}{c} F(x^{\text{system}}) \\ -Hx^{\text{system}} \end{array} \right]
$$

$$
- \int \left[\phi \left(G^{-1} \left[\begin{array}{c} B(x^{\text{system}}) \\ 0 \end{array} \right]^T \frac{\partial V \left[\begin{array}{c} x^{\text{system}} \\ x_i^{\text{ref}} \end{array} \right]}{\partial \left[\begin{array}{c} x^{\text{system}} \\ x_i^{\text{ref}} \end{array} \right]} \right) \right]^T
$$

$$
\times d \left(\left[\begin{array}{c} B(x^{\text{system}}) \\ 0 \end{array} \right]^T \frac{\partial V \left[\begin{array}{c} x^{\text{system}} \\ x_i^{\text{ref}} \end{array} \right]}{\partial \left[\begin{array}{c} x^{\text{system}} \\ x_i^{\text{ref}} \end{array} \right]} \right). \tag{5.4.64}
$$

Example 5.4.11. Tracking control of a high-performance aircraft

Advanced, aerodynamically improved aircraft configurations are needed to improve the mission efficiency, to expand the flight envelope, to attain the required flying and handling qualities, and to achieve the desired aircraft performance. The forward-mounted canards, which share the lift load with wings, improve the pitch control capabilities. For the forward-swept wing configuration, the air flows inward, whereas for the rearward-swept wing airframe, the air flows outward. The supercritical-swept wing configuration, coupled with movable canards, significantly improves flying and handling capabilities, maneuverability, controllability, enhances the flight envelope, reduces aerodynamic drag, allowing fuel consumption, reduces engine temperature, and increases the operational range. The trailing-edge flaps, used asymmetrically, accomplish roll. The yawing moment primarily results because of rudder deflection; that is, the rudders provide yaw control. However, the roll–yaw cross-coupling effect occurs, complicating longitudinal and lateral dynamics (the Lagrange equations of motion indicate that the longitudinal and lateral directional dynamics cannot be decoupled into two independent sets). In this section, we study nonlinear identification and control for an open-loop, unstable, twin-tail, supercritical-swept wing aircraft. The following state and control vectors are used:

$$x(t) = \begin{bmatrix} v(t) & \alpha(t) & q(t) & \theta(t) & \beta(t) & p(t) & r(t) & \phi(t) & \psi(t) \end{bmatrix}^T,$$
$$u(t) = \begin{bmatrix} \delta_{HR}(t) & \delta_{HL}(t) & \delta_{FR}(t) & \delta_{FL}(t) & \delta_C(T) & \delta_R(t) \end{bmatrix}^T.$$

Hence, the aircraft state variables are

$$x^{air}(t) = \begin{bmatrix} v(t) & \alpha(t) & q(t) & \theta(t) & \beta(t) & p(t) & r(t) & \phi(t) & \psi(t) \end{bmatrix}^T \in \mathbb{R}^9,$$

and the reference and output vectors can be found using the Euler angles; that is,

$$r(t) = \begin{bmatrix} r_\theta(t) & r_\phi(t) & r_\psi(t) \end{bmatrix}^T \in \mathbb{R}^3, y(t) = \begin{bmatrix} \theta(t) & \phi(t) & \psi(t) \end{bmatrix}^T \in \mathbb{R}^3,$$
$$\text{and } N = \begin{bmatrix} 1 & 0 & 0 \\ 0 & 1 & 0 \\ 0 & 0 & 1 \end{bmatrix}.$$

Using the augmented longitudinal and lateral aircraft dynamics, given in terms of states and controls, the following differential equations result:

$$\dot{x}^{air} = \begin{bmatrix} \dot{v}(t) \\ \dot{\alpha}(t) \\ \dot{q}(t) \\ \dot{\theta}(t) \\ \dot{\beta}(t) \\ \dot{p}(t) \\ \dot{r}(t) \\ \dot{\phi}(t) \\ \dot{\psi}(t) \end{bmatrix} = Ax^{air} + F(x^{air}) + Bu$$

$$
\begin{bmatrix} v \\ \alpha \\ q \\ \theta \\ \beta \\ p \\ r \\ \phi \\ \psi \end{bmatrix} = A \begin{bmatrix} v \\ \alpha \\ q \\ \theta \\ \beta \\ p \\ r \\ \phi \\ \psi \end{bmatrix} + \begin{bmatrix} 0 \\ -p\cos\alpha\tan\beta - r\sin\alpha\tan\beta \\ \frac{1}{I_Y}[(I_Z - I_X)pr - I_{XZ}p^2 + I_{XZ}r^2] \\ q\cos\phi - r\sin\phi \\ p\sin\alpha - r\cos\alpha \\ \frac{1}{I_X I_Z - I_{XZ}^2}[I_{XZ}(I_X - I_Y + I_Z)qp + (I_Y I_Z - I_{XZ}^2 - I_Z^2)qr] \\ \frac{1}{I_X I_Z - I_{XZ}^2}[(I_X^2 - I_X I_Y + I_{XZ}^2)qp - I_{XZ}(I_X - I_Y + I_Z)qr] \\ q\tan\theta\sin\phi + r\tan\theta\cos\phi \\ q\cos^{-1}\theta\sin\phi + r\cos^{-1}\theta\cos\phi \end{bmatrix}
$$

$$
+ B \begin{bmatrix} \delta_{HR} \\ \delta_{HL} \\ \delta_{FR} \\ \delta_{FL} \\ \delta_C \\ \delta_R \end{bmatrix}. \tag{5.4.65}
$$

The matrices $A \in \mathbb{R}^{9 \times 9}$ and $B \in \mathbb{R}^{9 \times 6}$, as well as moments of inertia I_X, I_Y, I_Z, I_{XZ}, are

$$
A = \begin{bmatrix}
-0.009 & 5.7 & -0.24 & -9.6 & -0.46 & -0.095 & -0.14 & 0 & 0 \\
-0.001 & -0.68 & 1 & 0 & 0.12 & 0.037 & 0.005 & 0 & 0 \\
0.0002 & 2.7 & -0.53 & 0 & 0.009 & 0.0062 & 0.028 & 0 & 0 \\
0 & 0 & 0 & 0 & 0 & 0 & 0 & 0 & 0 \\
-0.001 & 0.29 & 0.036 & 1.4 & -0.72 & 0.086 & -1.5 & 0 & 0 \\
0.00002 & 1.1 & 0.041 & 0.007 & -26 & -4.9 & 0.53 & 0 & 0 \\
0.00001 & -1.3 & 0.098 & 0.011 & 7.4 & -0.037 & -0.82 & 0 & 0 \\
0 & 0 & 0 & 0 & 0 & 1 & 0 & 0 & 0 \\
0 & 0 & 0 & 0 & 0 & 0 & 0 & 0 & 0
\end{bmatrix},
$$

$$
B = \begin{bmatrix}
0.093 & 0.093 & -0.45 & -0.45 & -0.07 & -0.13 \\
-0.28 & -0.28 & -0.0068 & -0.0068 & 0.0049 & 0 \\
-25 & -25 & -0.59 & -0.59 & 3.5 & 0 \\
0 & 0 & 0 & 0 & 0 & 0 \\
0.015 & -0.015 & -0.36 & 0.36 & 0.085 & -0.051 \\
-0.24 & 0.24 & -9.8 & 9.8 & 0.26 & -0.37 \\
0.38 & -0.38 & 0.19 & -0.19 & 0.52 & -4.6 \\
0 & 0 & 0 & 0 & 0 & 0 \\
0 & 0 & 0 & 0 & 0 & 0
\end{bmatrix},
$$

$I_X = 21983$ kg-m^2, $I_Y = 154248$ kg-m^2, $I_Z = 186515$ kg-m^2, and $I_{XZ} = 2047$ kg-m^2.

The studied supercritical-swept wing fighter is unstable in the open-loop configuration.

To design a control law, the output equation $y = Hx^{\text{air}}$ is used. The error vector is expressed in terms of the reference inputs, as given by

$$r(t) = \begin{bmatrix} r_\theta(t) \\ r_\phi(t) \\ r_\psi(t) \end{bmatrix},$$

and output vector $y = Hx^{\text{air}}$.

Augmenting the coupled longitudinal and lateral model (5.4.65) with exogenous dynamics

$$\dot{x}_i(t) = \begin{bmatrix} 1 & 0 & 0 \\ 0 & 1 & 0 \\ 0 & 0 & 1 \end{bmatrix} \begin{bmatrix} r_\theta(t) \\ r_\phi(t) \\ r_\psi(t) \end{bmatrix} - Hx^{\text{air}}$$

$$= \begin{bmatrix} 1 & 0 & 0 \\ 0 & 1 & 0 \\ 0 & 0 & 1 \end{bmatrix} \begin{bmatrix} r_\theta(t) \\ r_\phi(t) \\ r_\psi(t) \end{bmatrix} - \begin{bmatrix} 0 & 0 & 0 & 1 & 0 & 0 & 0 & 0 & 0 \\ 0 & 0 & 0 & 0 & 0 & 0 & 0 & 1 & 0 \\ 0 & 0 & 0 & 0 & 0 & 0 & 0 & 0 & 1 \end{bmatrix} \begin{bmatrix} v(t) \\ \alpha(t) \\ q(t) \\ \theta(t) \\ \beta(t) \\ p(t) \\ r(t) \\ \phi(t) \\ \psi(t) \end{bmatrix},$$

an augmented model is found.

The following hard bounds (mechanical limits) on the deflections of control surfaces are imposed:

$$|\delta_{HR}, \delta_{HL}| \le 0.5 \text{ rad}, |\delta_{FR}, \delta_{FL}| \le 0.4 \text{ rad}, |\delta_C| \le 0.6 \text{ rad, and } |\delta_R| \le 0.5 \text{ rad}.$$

Our goal is to synthesize the bounded controller, minimizing the nonquadratic functional (5.4.61). We map the control bounds as

$$u = U_{\max} \tanh(\cdot) = \begin{bmatrix} \delta_{HR\max} & 0 & 0 & 0 & 0 & 0 \\ 0 & \delta_{HL\max} & 0 & 0 & 0 & 0 \\ 0 & 0 & \delta_{FR\max} & 0 & 0 & 0 \\ 0 & 0 & 0 & \delta_{FL\max} & 0 & 0 \\ 0 & 0 & 0 & 0 & \delta_{C\max} & 0 \\ 0 & 0 & 0 & 0 & 0 & \delta_{R\max} \end{bmatrix} \tanh(\cdot),$$

where $\delta_{HR\max} = \delta_{HL\max} = 0.5$ rad, $\delta_{FR\max} = \delta_{FL\max} = 0.4$ rad, $\delta_{C\max} = 0.6$ rad, and $\delta_{R\max} = 0.5$ rad.

Hence, control $u(\cdot)$: $[t_0, t_f] \rightarrow \mathbb{R}^6$ is bounded by the hyperbolic tangent.

The nonquadratic functional to be minimized is given by (5.4.61). Letting $\varsigma = 0$, $\eta = 0$, and $\kappa = 1$, we have

$$J = \int_{t_0}^{\infty} \left(\frac{1}{2} \begin{bmatrix} x^{air} \\ x_i^{ref} \end{bmatrix}^T Q \begin{bmatrix} x^{air} \\ x_i^{ref} \end{bmatrix} + \int (\tanh^{-1}(U_{max}^{-1} u))^T G \, du \right) dt,$$

where $Q \in \mathbb{R}^{12 \times 12}$ and $G \in \mathbb{R}^{6 \times 6}$ are positive-definite.

From (5.4.62), it is then immediate that a bounded controller results in

$$u = - \begin{bmatrix} \delta_{HR\,max} & 0 & 0 & 0 & 0 & 0 \\ 0 & \delta_{HL\,max} & 0 & 0 & 0 & 0 \\ 0 & 0 & \delta_{FR\,max} & 0 & 0 & 0 \\ 0 & 0 & 0 & \delta_{FL\,max} & 0 & 0 \\ 0 & 0 & 0 & 0 & \delta_{C\,max} & 0 \\ 0 & 0 & 0 & 0 & 0 & \delta_{R\,max} \end{bmatrix}$$

$$\times \tanh \left(G^{-1} \begin{bmatrix} B \\ 0 \end{bmatrix}^T \frac{\partial V \left(\begin{bmatrix} x^{air} \\ x_i^{ref} \end{bmatrix} \right)}{\partial \begin{bmatrix} x^{air} \\ x_i^{ref} \end{bmatrix}} \right).$$

With a positive-definite quadratic return function, find using (5.4.63) and letting $\gamma = 1$, $\mu = 4$, and $\chi = 4$,

$$V \left(\begin{bmatrix} x^{air} \\ x_i^{ref} \end{bmatrix} \right) = \frac{9}{10} \begin{bmatrix} x^{air\frac{5}{9}} \\ x_i^{ref\frac{5}{9}} \end{bmatrix}^T K_0 \begin{bmatrix} x^{air\frac{5}{9}} \\ x_i^{ref\frac{5}{9}} \end{bmatrix} + \frac{1}{2} \begin{bmatrix} x^{air} \\ x_i^{ref} \end{bmatrix}^T K_1 \begin{bmatrix} x^{air} \\ x_i^{ref} \end{bmatrix}.$$

The feedback gain matrices $K_{f0} = G^{-1} \begin{bmatrix} B \\ 0 \end{bmatrix}^T K_0 \in \mathbb{R}^{6 \times 12}$ and $K_{f1} = G^{-1} \begin{bmatrix} B \\ 0 \end{bmatrix}^T K_1 \in \mathbb{R}^{6 \times 12}$ of the bounded control law

$$u = - \begin{bmatrix} \delta_{HR\,max} & 0 & 0 & 0 & 0 & 0 \\ 0 & \delta_{HL\,max} & 0 & 0 & 0 & 0 \\ 0 & 0 & \delta_{FR\,max} & 0 & 0 & 0 \\ 0 & 0 & 0 & \delta_{FL\,max} & 0 & 0 \\ 0 & 0 & 0 & 0 & \delta_{C\,max} & 0 \\ 0 & 0 & 0 & 0 & 0 & \delta_{R\,max} \end{bmatrix}$$

$$\times \tanh \left(K_{f0} \begin{bmatrix} x^{air\frac{1}{9}} \\ x_i^{ref\frac{1}{9}} \end{bmatrix} + K_{f1} \begin{bmatrix} x^{air} \\ x_i^{ref} \end{bmatrix} \right)$$

are found solving the partial differential equation (5.4.64).

Figure 5.4.9 illustrates the aircraft's outputs when the reference inputs (given in radians) are

$$r = \begin{bmatrix} r_\theta \\ r_\phi \\ r_\psi \end{bmatrix} = \begin{cases} 0.5, t \geq 0 \\ 0.2, t \geq 0 \\ 1, t \geq 0; \end{cases}$$

that is, we study a coordinated maneuver assigning a 29°, a 11°, and a 57° turn.

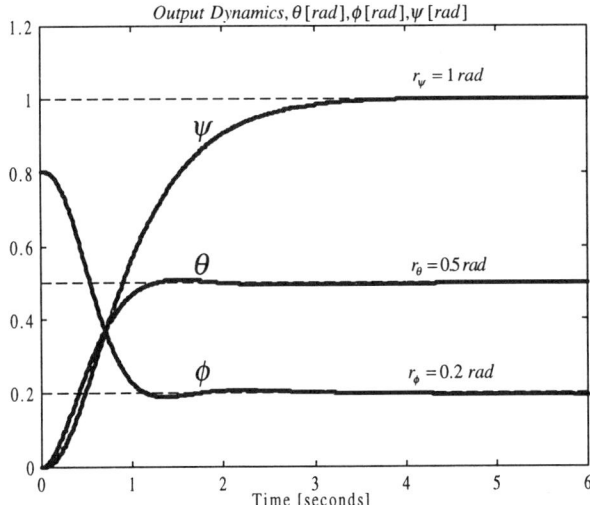

FIGURE 5.4.9.   Aircraft outputs.

5.4.5. *Robust Control of Nonlinear Continuous-Time Systems*

The major goal of this section is to design robust controllers using the Hamilton–Jacobi theory for nonlinear uncertain systems with input bounds and parameter uncertainties. The design is performed by applying necessary and sufficient conditions for optimality and stability. The synthesis is based on the constructive use of nonquadratic return functions and nonlinear performance functionals.

5.4.5.1. Uncertain Systems: Equations of Motion and Design of Bounded Controllers

Consider an uncertain system modeled by the following differential equation:

$$\dot{x}(t) = F(t, x, z) + B_u(t, x, p)u, x(t_0) = x_0, \quad t \geq 0. \tag{5.4.66}$$

The state and control vectors are subject to $x \in X$ and $u \in U$, where $X \subset \mathbb{R}^c$ is convex and closed, and $U \subset \mathbb{R}^m$ is convex and compact. It is assumed that $F(\cdot)$: $\mathbb{R}_{\geq 0} \times \mathbb{R}^c \times \mathbb{R}^e \to \mathbb{R}^c$ and $B_u(\cdot)$: $\mathbb{R}_{\geq 0} \times \mathbb{R}^c \times \mathbb{R}^d \to \mathbb{R}^{c \times m}$, together with partial derivatives, are continuous and satisfy appropriate differentiability and Lipschitz properties in $\mathbb{R}_{\geq 0} \times \mathbb{R}^c \times \mathbb{R}^e$, $\mathbb{R}_{\geq 0} \times \mathbb{R}^c \times \mathbb{R}^d$, $F(t, 0, z) = 0$, and $B_u(t, 0, p) = 0, \forall t \in [0, \infty), \forall z \in Z, \forall p \in P$. Uncertainties $z(\cdot) \in Z$ and $p(\cdot) \in P$ are assumed to be Lebesgue measurable and known within bounds, and $Z \subset \mathbb{R}^e$ and $P \subset \mathbb{R}^d$ are known, non- empty, and compact.

The control problem considered is to minimize the performance cost composed in states and controls, subject to (5.4.66) and control constraints. We express the functional as

$$J = \int_{t_0}^{t_f} \left(x^T Q(t)x + \int (\phi^{-1}(u))^T G^{-1}(t) \, du \right) dt, \qquad (5.4.67)$$

where $Q(\cdot)$: $\mathbb{R}_{\geq 0} \to \mathbb{R}^{c \times c}$, $Q(t) \geq qE, \forall t \in [0, \infty), q > 0, E \in \mathbb{R}^{c \times c}$ is the identity matrix and $G(\cdot)$: $\mathbb{R}_{\geq 0} \to \mathbb{R}^{m \times m}, G(t) \geq gI, \forall t \in [0, \infty), g > 0, I \in \mathbb{R}^{m \times m}$ is the identity matrix.

The control constraints are mapped by one-to-one, bounded integrable function $\phi(\cdot), \phi \in U$, and the real-analytic, bounded, integrable, Lipschitz inverse function $\phi^{-1}(\cdot)$ exists.

Lemma 5.4.15. *Consider the system with parameter uncertainties* (5.4.66). *The control bounds imposed are mapped by a real-analytic, integrable, bounded on $\mathbb{R}^c \backslash \{0\}$ and continuous function $\phi(\cdot), \phi \in U, \phi \in C^\epsilon (\epsilon \geq 1)$. Then, the control $u = \phi(f(t, x)), u(\cdot)$: $[t_0, t_f) \to \mathbb{R}^m$ is continuous and bounded, $u \in U$. Furthermore, the continuous integrand $\int (\phi^{-1}(u))^T G^{-1}(t) \, du \in C^\varsigma (\varsigma \geq 2)$ in cost* (5.4.67) *is real valued and positive-definite. The nonquadratic functional* (5.4.67) *is real valued and positive-definite for all $x \in X$ and $u \in U$ on $[t_0, t_f)$.*

To derive the control algorithm, we formulate two lemmas that state necessary and sufficient conditions. Necessary conditions, which allow us to find a critical point solution, follow from the Hamilton–Jacobi theory.

Lemma 5.4.16. *Given the nonlinear system in* (5.4.66) *and nonquadratic functional in* (5.4.67). *The necessary conditions that the control function $u(\cdot)$: $[t_0, t_f) \to \mathbb{R}^m$ guarantees a minimum to the Hamiltonian $H(\cdot)$, which is expressed by*

$$H\left(t, x, u, z, p, \frac{\partial V}{\partial x} \right) = x^T Q(t)x + \int (\phi^{-1}(u))^T G^{-1}(t) \, du$$

$$+ \frac{\partial V^T}{\partial x} [F(t, x, z) + B_u(t, x, p)u],$$

are

$$\frac{\partial H \left(t, x, u, z, p, \frac{\partial V}{\partial x} \right)}{\partial u} = 0, \forall x \in X, \forall u \in U, \forall t \in [t_0, t_f),$$

$$\frac{\partial^2 H \left(t, x, u, z, p, \frac{\partial V}{\partial x} \right)}{\partial u \times \partial u^T} > 0, \forall x \in X, \forall u \in U, \forall t \in [t_0, t_f).$$

Proof. These necessary conditions have their origin in the classic calculus of variations, and the proof is straightforward.

The Hamilton–Jacobi functional equation for (5.4.66) and (5.4.67) is

$$-\frac{\partial V(t, x)}{\partial t} = \min_{u \in U} \left\{ x^T Q(t)x + \int (\phi^{-1}(u))^T G^{-1}(t) \, du \right.$$

$$\left. + \frac{\partial V^T}{\partial x} [F(t, x, z) + B_u(t, x, p)u] \right\}. \qquad (5.4.68)$$

The bounded controller should be obtained by finding the control value that attains the minimum of (5.4.68). From the first-order necessary condition, it is easy to verify that (5.4.68) implies

$$u = -\phi \left(G(t) B(t, x)^T \frac{\partial V}{\partial x} \right).\tag{5.4.69}$$

That is, $u \in U$.

Matrix $G(t)$ is positive-definite. By analyzing the Hamiltonian, one concludes that the second-order necessary condition is guaranteed.

Let us turn our attention to the following question. When is the designed controller a robust one? Sufficiency theory gives an explicit answer and provides the criteria to study the robustness. These conditions are stated by the following lemma.

Lemma 5.4.17. *Consider uncertain system (5.4.66) subject to the bounded controller (5.4.69). The resulting system (5.4.66) with (5.4.69) is robustly stable in the specified sets of the states X, controls U, and uncertainties Z and P if a real-valued bounded C^κ function $V(\cdot)$: $\mathbb{R}_{\geq 0} \times \mathbb{R}^c \to \mathbb{R}_{\geq 0}$, $\rho_1(\|x\|) \leq V(t, x) \leq \rho_2(\|x\|)$ exists, such that for all $x \in X$, $z \in Z$, and $p \in P$ on $[t_0, t_f)$*

$$\frac{\partial V(t, x)}{\partial t} + \frac{\partial V^T}{\partial x} F(t, x, z)$$

$$+ \frac{\partial V^T}{\partial x} [B(t, x) - B_u(t, x, p)] \phi \left(G(t) B(t, x)^T \frac{\partial V}{\partial x} \right)$$

$$- \int \left[\phi \left(G(t) B(t, x)^T \frac{\partial V}{\partial x} \right) \right]^T d \left(B(t, x)^T \frac{\partial V}{\partial x} \right) \leq -x^T Q(t)x.\tag{5.4.70}$$

Furthermore, if $\rho_1(\|x\|) \leq V(t, x) \leq \rho_2(\|x\|)$ and (5.4.70) holds, KL-function $\rho_x(\cdot)$: $\mathbb{R}_{\geq 0} \times \mathbb{R}_{\geq 0} \to \mathbb{R}_{\geq 0}$ and K-function $\rho_u(\cdot)$: $\mathbb{R}_{\geq 0} \to \mathbb{R}_{\geq 0}$ exist such that

$$\|x(t)\| \leq \rho_x(\|x_0\|, t) + \rho_u(\|u\|) + \delta_x$$

is satisfied for all admissible solutions of (5.4.66) $x(\cdot)$: $[t_0, t_f) \to \mathbb{R}^c$, $x(t_0) = x_0$, $x \in X$, generated by control (5.4.69) $u(\cdot)$: $[t_0, t_f) \to \mathbb{R}^m$, $u \in U$, with uncertainties $z \in Z$ and $p \in P$.

Proof. The proof is based on the sufficiency theory. It can be easily verified that the criteria stated in Lemma 5.4.17 are sufficient conditions for stability. However, some mathematical deviations are needed to be shown. Recall that integration by parts gives

$$\int \Xi \, d\Sigma = \Xi\Sigma - \int \Sigma^T \, d\Xi^T.$$

Then, integrating by parts, one obtains

$$\int (\phi^{-1}(u))^T G^{-1}(t) \, du = \int \left[\phi^{-1} \left(\phi \left(G(t) B(t, x)^T \frac{\partial V}{\partial x} \right) \right) \right]^T$$

$$\times G^{-1}(t)\, d\left[\phi\left(G(t)B(t,x)^{T}\frac{\partial V}{\partial x}\right)\right]$$

$$= \frac{\partial V^{T}}{\partial x}B(t,x)\phi\left(G(t)B(t,x)^{T}\frac{\partial V}{\partial x}\right)$$

$$- \int\left[\phi\left(G(t)B(t,x)^{T}\frac{\partial V}{\partial x}\right)\right]^{T}$$

$$\times d\left(B(t,x)^{T}\frac{\partial V}{\partial x}\right). \tag{5.4.71}$$

Clearly, by using (5.4.68), (5.4.69), and (5.4.71), one obtains (5.4.70). The sufficiency criteria are defined, and the sufficiency analysis is immediate.

Lemmas 5.4.16 and 5.4.17 supply general conditions under which the robust problem can be solved. The following theorem is formulated and proven.

Theorem 5.4.4. *Given the dynamic system with parameter uncertainties* (5.4.66). *Using the real-valued, positive-definite performance measure* (5.4.67), *an admissible bounded controller, which satisfies necessary conditions and causes* $x(\cdot)$: $[t_0, t_f)$ $\rightarrow \mathbb{R}^c$, $x(t_0) = x_0$ *in* $X(X_0, U, Z, P) \subset \mathbb{R}^c$, *is found in* (5.4.69). *The minimum value of the performance cost* (5.4.67) *is given by the power-series form. Assume that a smooth solution* $V(\cdot)$: $\mathbb{R}_{\geq 0} \times \mathbb{R}^c \rightarrow \mathbb{R}_{\geq 0}$ *of the Hamilton–Jacobi–Bellman equation exists. Then, the smooth bounded control function* $u(\cdot)$: $[t_0, t_f) \rightarrow \mathbb{R}^m$, $u \in U$ (5.4.69) *attains the minimum of* (5.4.68) *and minimizes the performance cost* (5.4.67). *The function* $V(t, x)$, $V \in C^\kappa$ *in* (5.4.67) *is the solution of the partial differential equation* (5.4.70) *subject to the boundary conditions. Control* $u(\cdot)$: $[t_0, t_f) \rightarrow \mathbb{R}^m$, $u \in U$, *as given by* (5.4.69), *is robust in* $X(X_0, U, Z, P)$ *if sufficient conditions are guaranteed for all* $x \in X$, $z \in Z$, *and* $p \in P$ *on* $[t_0, t_f)$.

Proof. Let us examine necessary conditions that a robust control, if it exists, has to satisfy. From the first-order necessary condition, the state feedback is designed as a bounded real-analytic function (5.4.69). Thus, a critical point, which is a relative minimum, is obtained on the basis of the functional equation (5.4.68). Because $\frac{\partial^2 H(\cdot)}{\partial u \times \partial u^T}$ is positive-definite, one concludes that the second-order necessary condition is met. Hence, the absolute minimum corresponds to a critical point, and controller (5.4.69) gives a strong relative minimum to cost (5.4.67) and globally minimizes the Hamiltonian.

The bounded control law is a nonlinear function of the state vector and time. Substituting controller (5.4.69) into (5.4.68), and using (5.4.71), the partial differential equation (5.4.70), which should be solved, results. Observe that

$$\int\left[\phi\left(G(t)B(t,x)^{T}\frac{\partial V}{\partial x}\right)\right]^{T} d\left(B(t,x)^{T}\frac{\partial V}{\partial x}\right)$$

is smooth and continuously differentiable if $V(t, x)$ is of class C^κ functions. To obtain $V(t, x)$, the performance cost must be evaluated at the allowed values of the state variables and control. Functional (5.4.67) admits a final value, and the minimum value of the nonquadratic cost is given in the following nonquadratic power-series form:

$$J_{\min} = \sum_{i=0}^{\gamma} \frac{2\mu+1}{2(\chi i+\mu+1)} \left(x_0^{\frac{\chi i+\mu+1}{2\mu+1}} \right)^T K_i(t) x_0^{\frac{\chi i+\mu+1}{2\mu+1}}.$$

If the control problem is solvable, (5.4.70) is satisfied by a smooth function $V(t, x)$. Moreover, a C^κ solution to (5.4.70) exists, and the bounded controller (5.4.69) guarantees the robustness of (5.4.66) in $X(X_0, U, Z, P)$ if sufficient conditions are met. Assume that the Hamilton–Jacobi–Bellman equation admits a return function. Then, the solution of (5.4.70) can be found by using the continuously differentiable C^κ function $V(t, x)$. The quadratic return function candidate might be employed only if the designer is able to neglect the higher order terms in a Taylor's series expansion. Solution of (5.4.70) involves Γ nonlinear equations in Γ unknowns of $V(t, x)$. Sufficient conditions must be satisfied. If the criteria of Lemma 5.4.17 are guaranteed, the designed control function $u(\cdot)$: $[t_0, t_f) \to \mathbb{R}^m$, $u \in U$ in (5.4.69) is robust in $X(X_0, U, Z, P)$.

Theorem 5.4.5 establishes the following important lemma.

Lemma 5.4.18. *Consider the state-space equation in (5.4.66). The smooth control function $u(\cdot)$: $[t_0, t_f) \to \mathbb{R}^m$ (5.4.69), which is constrained to belong to the known and bounded set U on $[t_0, t_f)$, minimizes a real-valued, positive-definite functional (5.4.67). The C^κ function (5.4.67) is the solution of (5.4.70).*

Proof. Using the results of Theorem 5.4.5, the proof is carried out in a straightforward way.

Many promising results have been established to account for uncertainties, and several workable frameworks are available. One promising avenue to attack this problem is to represent uncertain systems using the following state equation:

$$\dot{x}(t) = A_Z(z) F_X(t, x) + B_P(p) B(t, x) u, \ x(t_0) = x_0, t \geq 0.$$

Majorants and minorants of $A_Z(\cdot)$ and $B_P(\cdot)$ can be used in the synthesis of robust controllers, and they should be found using the *modulus*. Hence, the parameter variations in (5.4.66) can be bounded by using the parameter-dependent mappings.

Other manageable results were researched, and the *matching conditions* formulation, which first appeared in the late 1970s, was largely developed. For uncertain systems, workable control algorithms and design methods in the absence of matching conditions have been established as well. A number of control techniques rely on the definition of parameter uncertainties in the worst-case domain.

Example 5.4.12. A nonlinear time-varying system

Consider an uncertain system with the following state-space model:

$$\dot{x}_1(t) = -3x_1 + (5 + \Delta a_{12}(z)) x_2^{\frac{2}{3}}, \ \Delta a_{12}(z) \in \begin{bmatrix} -4 & 4 \end{bmatrix},$$

$$\dot{x}_2(t) = -5x_2 + (15 + \Delta a_{23}(z))x_3, \ \Delta a_{23}(z) \in \left[-5 \quad 5\right],$$

$$\dot{x}_3(t) = -5x_1x_2 + (30 + \Delta a_{32}(z))x_2 - 3x_3 + 10\sin(x_1^2 x_3 t) + u,$$

$$\Delta a_{32}(z) \in \left[-5 \quad 5\right],$$

$$|x_1| \le x_{1\,\text{max}}, \ |x_2| \le x_{2\,\text{max}}, \ |x_3| \le x_{3\,\text{max}}, \ |u| \le u_{\text{max}},$$

$$X = \{x \in \mathbb{R}^3, |x_1| \le 20, |x_2| \le 20, |x_3| \le 20\}, \ U = \{u \in \mathbb{R}, |u| \le 50\}.$$

The problem is to minimize the cost

$$J = \int_{t_0}^{\infty} \left(\int \tanh^{-1}\left(\frac{x_1}{x_{1\,\text{max}}}\right) dx_1 + \int \tanh^{-1}\left(\frac{x_2}{x_{2\,\text{max}}}\right) dx_2 \right.$$

$$\left. + \int \tanh^{-1}\left(\frac{x_3}{x_{3\,\text{max}}}\right) dx_3 + \int \tanh^{-1}\left(\frac{u}{u_{\text{max}}}\right) du \right) dt,$$

subject to the nonlinear dynamics. The smooth bounded control law is found from the Hamiltonian using the first-order necessary condition. In particular, we have

$$u = -50 \tanh(0.52x_1 + 2.9x_2 + 7.6x_3 + 0.2\sin(x_1^2 x_3 t)).$$

This controller guarantees robustness because sufficient conditions are met.

5.4.6. Robust Tracking Control of Nonlinear Systems

In this section, we will solve the tracking control problem by designing integral robust control laws. By minimizing the nonquadratic performance cost and making use of necessary conditions for optimality, a bounded control law is designed. The sufficient conditions for stability are derived and examined based on the Lyapunov stability theory.

Consider dynamic systems described by

$$\dot{x}^{\text{system}}(t) = F(x^{\text{system}}, z) + B_u(x^{\text{system}}, p)u, \ y = Hx^{\text{system}}, \ u_{\text{min}} \le u \le u_{\text{max}},$$

$$x^{\text{system}}(t_0) = x_0^{\text{system}}. \tag{5.4.72}$$

Using the output equation

$$y = Hx^{\text{system}},$$

one finds that the measured error vector is expressed as

$$e(t) = Nr(t) - y(t) = Nr(t) - Hx^{\text{system}}(t),$$

where $r \in R \subset \mathbb{R}^b$ is the measured reference vector (desired displacement),

$$r(t) = \begin{bmatrix} r_1(t) \\ \vdots \\ r_b(t) \end{bmatrix};$$

$N \in \mathbb{R}^{b \times b}$ is the matrix with constant coefficients,

$$
N = \begin{bmatrix}
N_1 & 0 & \cdots & 0 & 0 \\
0 & N_2 & \cdots & 0 & 0 \\
\vdots & \vdots & \ddots & \vdots & \vdots \\
0 & 0 & \cdots & N_{b-1} & 0 \\
0 & 0 & \cdots & 0 & N_b
\end{bmatrix} \in \mathbb{R}^{b \times b}.
$$

From (5.4.72) and exogenous dynamics

$$
\dot{x}^{ref}(t) = Nr - Hx^{\text{system}},
$$

we have the following augmented state-space equations of motion:

$$
\begin{bmatrix} \dot{x}^{\text{system}}(t) \\ \dot{x}^{ref}(t) \end{bmatrix} = \begin{bmatrix} F(x^{\text{system}}, z) \\ -Hx^{\text{system}} \end{bmatrix} + \begin{bmatrix} B_u(x^{\text{system}}, p) \\ 0 \end{bmatrix} u + \begin{bmatrix} 0 \\ N \end{bmatrix} r,
$$
$$
y = Hx^{\text{system}}. \tag{5.4.73}
$$

Mapping the control bounds imposed by a one-to-one bounded function $\phi(\cdot)$, the nonquadratic cost to be minimized is given as

$$
J = \int_{t_0}^{t_f} \left(\sum_{i=0}^{\varsigma} \frac{2\eta+1}{2(\kappa i+\eta+1)} \begin{bmatrix} x^{\text{system}\ \frac{\kappa i+\eta+1}{2\eta+1}} \\ x^{ref\ \frac{\kappa i+\eta+1}{2\eta+1}} \end{bmatrix}^T Q_i \begin{bmatrix} x^{\text{system}\ \frac{\kappa i+\eta+1}{2\eta+1}} \\ x^{ref\ \frac{\kappa i+\eta+1}{2\eta+1}} \end{bmatrix} \right.
$$
$$
\left. + \int (\phi^{-1}(u))^T G\,du \right) dt, \tag{5.4.74}
$$

where $Q_i \in \mathbb{R}^{c \times c}$, $Q_i \geq 0$ and $G \in \mathbb{R}^{m \times m}$, $G > 0$ are the weighting matrices. The real nonnegative integers ς, η, and κ are assigned by the designer, and $\varsigma = 0, 1, 2, \ldots$, $\eta = 0, 1, 2, \ldots$, and $\kappa = 1, 2, 3, \ldots$. For example, letting $\varsigma = \eta = 0$ and $\kappa = 1$, we have

$$
J = \int_{t_0}^{t_f} \left(\frac{1}{2} \begin{bmatrix} x^{\text{system}} \\ x^{ref} \end{bmatrix}^T Q_0 \begin{bmatrix} x^{\text{system}} \\ x^{ref} \end{bmatrix} + \int (\phi^{-1}(u))^T G\,du \right) dt.
$$

Using the Hamilton–Jacobi equation

$$
-\frac{\partial V}{\partial t} = \min_{u \in U} \left[\sum_{i=0}^{\varsigma} \frac{2\eta+1}{2(\kappa i+\eta+1)} \begin{bmatrix} x^{\text{system}\ \frac{\kappa i+\eta+1}{2\eta+1}} \\ x^{ref\ \frac{\kappa i+\eta+1}{2\eta+1}} \end{bmatrix}^T Q_i \begin{bmatrix} x^{\text{system}\ \frac{\kappa i+\eta+1}{2\eta+1}} \\ x^{ref\ \frac{\kappa i+\eta+1}{2\eta+1}} \end{bmatrix} \right.
$$

$$
+ \int (\phi^{-1}(u))^T G\, du + \frac{\partial V\left(\begin{bmatrix} x^{\text{system}} \\ x^{\text{ref}} \end{bmatrix}\right)^T}{\partial \begin{bmatrix} x^{\text{system}} \\ x^{\text{ref}} \end{bmatrix}} \left(\begin{bmatrix} F(x^{\text{system}}, z) \\ -H x^{\text{system}} \end{bmatrix}\right.
$$

$$
+ \left.\begin{bmatrix} B_u(x^{\text{system}}, p) \\ 0 \end{bmatrix} u\right)\Bigg],
$$

one finds a bounded controller applying the first-order necessary condition for optimality. In particular,

$$
u = -\phi\left(G^{-1}\begin{bmatrix} B(x^{\text{system}}) \\ 0 \end{bmatrix}^T \frac{\partial V\left(\begin{bmatrix} x^{\text{system}} \\ x^{\text{ref}} \end{bmatrix}\right)}{\partial \begin{bmatrix} x^{\text{system}} \\ x^{\text{ref}} \end{bmatrix}}\right). \tag{5.4.75}
$$

The nonquadratic return function

$$
V\left(\begin{bmatrix} x^{\text{system}} \\ x^{\text{ref}} \end{bmatrix}\right) = \sum_{i=0}^{\gamma} \frac{2\mu+1}{2(\chi i+\mu+1)} \begin{bmatrix} x^{\text{system}\frac{\chi i+\mu+1}{2\mu+1}} \\ x^{\text{ref}\frac{\chi i+\mu+1}{2\mu+1}} \end{bmatrix}^T K_i \begin{bmatrix} x^{\text{system}\frac{\chi i+\mu+1}{2\mu+1}} \\ x^{\text{ref}\frac{\chi i+\mu+1}{2\mu+1}} \end{bmatrix} \tag{5.4.76}
$$

is a solution to the Hamilton–Jacobi–Bellman equation

$$
-\frac{\partial V}{\partial t} = \sum_{i=0}^{\varsigma} \frac{2\eta+1}{2(\kappa i+\eta+1)} \begin{bmatrix} x^{\text{system}\frac{\kappa i+\eta+1}{2\eta+1}} \\ x^{\text{ref}\frac{\kappa i+\eta+1}{2\eta+1}} \end{bmatrix}^T Q_i \begin{bmatrix} x^{\text{system}\frac{\kappa i+\eta+1}{2\eta+1}} \\ x^{\text{ref}\frac{\kappa i+\eta+1}{2\eta+1}} \end{bmatrix}
$$

$$
+ \frac{\partial V\left(\begin{bmatrix} x^{\text{system}} \\ x^{\text{ref}} \end{bmatrix}\right)^T}{\partial \begin{bmatrix} x^{\text{system}} \\ x^{\text{ref}} \end{bmatrix}} \begin{bmatrix} F(x^{\text{system}}, z) \\ -H x^{\text{system}} \end{bmatrix}
$$

$$
+ \frac{\partial V\left(\begin{bmatrix} x^{\text{system}} \\ x^{\text{ref}} \end{bmatrix}\right)^T}{\partial \begin{bmatrix} x^{\text{system}} \\ x^{\text{ref}} \end{bmatrix}} \left(\begin{bmatrix} B(x^{\text{system}}) \\ 0 \end{bmatrix} - \begin{bmatrix} B_u(x^{\text{system}}, p) \\ 0 \end{bmatrix}\right)
$$

$$
\times \phi\left(G^{-1}\begin{bmatrix} B(x^{\text{system}}) \\ 0 \end{bmatrix}^T \frac{\partial V\left(\begin{bmatrix} x^{\text{system}} \\ x^{\text{ref}} \end{bmatrix}\right)}{\partial \begin{bmatrix} x^{\text{system}} \\ x^{\text{ref}} \end{bmatrix}}\right) \tag{5.4.77}
$$

$$
- \int\left[\phi\left(G^{-1}\begin{bmatrix} B(x^{\text{system}}) \\ 0 \end{bmatrix}^T \frac{\partial V\left(\begin{bmatrix} x^{\text{system}} \\ x^{\text{ref}} \end{bmatrix}\right)}{\partial \begin{bmatrix} x^{\text{system}} \\ x^{\text{ref}} \end{bmatrix}}\right)\right]^T
$$

$$\times d \left(\begin{bmatrix} B(x^{\text{system}}) \\ 0 \end{bmatrix}^T \frac{\partial V \left(\begin{bmatrix} x^{\text{system}} \\ x^{\text{ref}} \end{bmatrix} \right)}{\partial \begin{bmatrix} x^{\text{system}} \\ x^{\text{ref}} \end{bmatrix}} \right).$$

In (5.4.76), the unknown matrices $K_i \in \mathbb{R}^{c \times c}$ are found by solving equation (5.4.77); γ, μ, and χ are the real numbers, $\gamma = 0, 1, 2, \ldots$, $\mu = 0, 1, 2, \ldots$, and $\chi = 1, 2, 3, \ldots$.

The second-order necessary condition for optimality is guaranteed because the second derivative of the Hamiltonian function $H(\cdot)$ is positive-definite; that is,

$$\frac{\partial H(\cdot)}{\partial u \times \partial u^T} > 0.$$

Hence, the absolute minimum corresponds to a critical point.

To design the return functions, the performance cost should be evaluated. The nonquadratic functional (5.4.74) admits a final value, and the minimum value of the cost $J_{\min}(\cdot)$ is found in the nonquadratic power-series form, as given by (5.4.76). For linear dynamic systems

$$\begin{bmatrix} \dot{x}^{\text{system}}(t) \\ \dot{x}^{\text{ref}}(t) \end{bmatrix} = \begin{bmatrix} A & 0 \\ -H & 0 \end{bmatrix} \begin{bmatrix} x^{\text{system}} \\ x^{\text{ref}} \end{bmatrix} + \begin{bmatrix} B \\ 0 \end{bmatrix} u + \begin{bmatrix} 0 \\ N \end{bmatrix} r, \quad y = Hx^{\text{system}},$$

the minimization of the following quadratic cost:

$$J = \int_{t_0}^{t_f} \left(\frac{1}{2} \begin{bmatrix} x^{\text{system}} \\ x^{\text{ref}} \end{bmatrix}^T Q \begin{bmatrix} x^{\text{system}} \\ x^{\text{ref}} \end{bmatrix} + \frac{1}{2} u^T Gu \right) dt$$

results in the well-known linear quadratic regulator problem. The quadratic return function

$$V \left(\begin{bmatrix} x^{\text{system}} \\ x^{\text{ref}} \end{bmatrix} \right) = \frac{1}{2} \begin{bmatrix} x^{\text{system}} \\ x^{\text{ref}} \end{bmatrix}^T K \begin{bmatrix} x^{\text{system}} \\ x^{\text{ref}} \end{bmatrix}$$

is the solution to the Hamilton–Jacobi–Bellman equation, and the controller is given by

$$u = -G^{-1} \begin{bmatrix} B \\ 0 \end{bmatrix}^T K \begin{bmatrix} x^{\text{system}} \\ x^{\text{ref}} \end{bmatrix}.$$

Letting $\varsigma = \eta = 0$ and $\kappa = 1$ in (5.4.74), the following cost results

$$J = \int_{t_0}^{t_f} \left(\frac{1}{2} \begin{bmatrix} x^{\text{system}} \\ x^{\text{ref}} \end{bmatrix}^T Q_0 \begin{bmatrix} x^{\text{system}} \\ x^{\text{ref}} \end{bmatrix} + \int (\phi^{-1}(u))^T G \, du \right) dt.$$

Assigning in (5.4.76) $\gamma = \mu = 0$ and $\chi = 1$, the quadratic return function is obtained as

$$V \left(\begin{bmatrix} x^{\text{system}} \\ x^{\text{ref}} \end{bmatrix} \right) = \frac{1}{2} \begin{bmatrix} x^{\text{system}} \\ x^{\text{ref}} \end{bmatrix}^T K_0 \begin{bmatrix} x^{\text{system}} \\ x^{\text{ref}} \end{bmatrix}.$$

Hence, the bounded controller is

$$
u = -\phi \left(G^{-1} \begin{bmatrix} B \\ 0 \end{bmatrix}^T K \begin{bmatrix} x^{\text{system}} \\ x^{\text{ref}} \end{bmatrix} \right).
$$

For nonlinear systems, quadratic return functions cannot be applied. The minimum value of (5.4.74) $J_{\min}(\cdot)$ is found using states evolution as well as control function, and (5.4.76) gives the expression for the return function. Letting $\gamma = 1$, $\mu = 4$, and $\chi = 4$, we have

$$
V \left(\begin{bmatrix} x^{\text{system}} \\ x^{\text{ref}} \end{bmatrix} \right) = \frac{9}{10} \begin{bmatrix} x^{\text{system}\frac{5}{9}} \\ x^{\text{ref}\frac{5}{9}} \end{bmatrix}^T K_0 \begin{bmatrix} x^{\text{system}\frac{5}{9}} \\ x^{\text{ref}\frac{5}{9}} \end{bmatrix} + \frac{1}{2} \begin{bmatrix} x^{\text{system}} \\ x^{\text{ref}} \end{bmatrix}^T K_1 \begin{bmatrix} x^{\text{system}} \\ x^{\text{ref}} \end{bmatrix}.
$$

As the unknown coefficients of $K_i \in \mathbb{R}^{c \times c}$ in (5.4.76) are found, the feedback gains of the bounded nonlinear control algorithm (5.4.75) result.

The sufficient conditions for robustness should be studied. By making use of (5.4.73), (5.4.75), and (5.4.77), and referring to the positive-definiteness of integrand

$$
\sum_{i=0}^{\varsigma} \frac{2\eta+1}{2(\kappa i + \eta + 1)} \begin{bmatrix} x^{\text{system}\,\frac{\kappa i + \eta + 1}{2\eta+1}} \\ x^{\text{ref}\,\frac{\kappa i + \eta + 1}{2\eta+1}} \end{bmatrix}^T Q_i \begin{bmatrix} x^{\text{system}\,\frac{\kappa i + \eta + 1}{2\eta+1}} \\ x^{\text{ref}\,\frac{\kappa i + \eta + 1}{2\eta+1}} \end{bmatrix},
$$

one concludes that if the function $V(\cdot): \mathbb{R}^c \to \mathbb{R}_{\geq 0}$ exists such that

$$
\frac{\partial V}{\partial t} + \frac{\partial V \left(\begin{bmatrix} x^{\text{system}} \\ x^{\text{ref}} \end{bmatrix} \right)^T}{\partial \begin{bmatrix} x^{\text{system}} \\ x^{\text{ref}} \end{bmatrix}} \begin{bmatrix} F(x^{\text{system}}, z) \\ -Hx^{\text{system}} \end{bmatrix} + \frac{\partial V \left(\begin{bmatrix} x^{\text{system}} \\ x^{\text{ref}} \end{bmatrix} \right)^T}{\partial \begin{bmatrix} x^{\text{system}} \\ x^{\text{ref}} \end{bmatrix}} \left(\begin{bmatrix} B(x^{\text{system}}) \\ 0 \end{bmatrix} \right.
$$

$$
\left. - \begin{bmatrix} B_u(x^{\text{system}}, p) \\ 0 \end{bmatrix} \right) \phi \left(G^{-1} \begin{bmatrix} B(x^{\text{system}}) \\ 0 \end{bmatrix}^T \frac{\partial V \left(\begin{bmatrix} x^{\text{system}} \\ x^{\text{ref}} \end{bmatrix} \right)}{\partial \begin{bmatrix} x^{\text{system}} \\ x^{\text{ref}} \end{bmatrix}} \right)
$$

$$
- \int \left[\phi \left(G^{-1} \begin{bmatrix} B(x^{\text{system}}) \\ 0 \end{bmatrix}^T \frac{\partial V \left(\begin{bmatrix} x^{\text{system}} \\ x^{\text{ref}} \end{bmatrix} \right)}{\partial \begin{bmatrix} x^{\text{system}} \\ x^{\text{ref}} \end{bmatrix}} \right) \right]^T
$$

$$
\times d \left(\begin{bmatrix} B(x^{\text{system}}) \\ 0 \end{bmatrix}^T \frac{\partial V \left(\begin{bmatrix} x^{\text{system}} \\ x^{\text{ref}} \end{bmatrix} \right)}{\partial \begin{bmatrix} x^{\text{system}} \\ x^{\text{ref}} \end{bmatrix}} \right)
$$

$$
\leq - \sum_{i=0}^{\varsigma} \frac{2\eta+1}{2(\kappa i + \eta + 1)} \begin{bmatrix} x^{\text{system}\,\frac{\kappa i + \eta + 1}{2\eta+1}} \\ x^{\text{ref}\,\frac{\kappa i + \eta + 1}{2\eta+1}} \end{bmatrix}^T Q_i \begin{bmatrix} x^{\text{system}\,\frac{\kappa i + \eta + 1}{2\eta+1}} \\ x^{\text{ref}\,\frac{\kappa i + \eta + 1}{2\eta+1}} \end{bmatrix},
$$

robustness is guaranteed with respect to X, U, Z, and P. That is, the admissibility concept is used to perform the analysis of stability needed to solve the robust motion control problem.

5.5. Sliding Mode Control

Sliding mode control is an enabling concept allowing one to design optimal and robust control algorithms approaching stabilization and tracking problems. This section gives an introduction to the sliding mode control and develops new theoretical results with design examples. Control algorithms can be designed to guarantee the evolution of system variables within the specified neighborhood of the *switching surface*. That is, the dynamic behavior is tailored by the particular choice of switching surface, and robustness to parameter uncertainties and unmodeled dynamics can be achieved. The ability to specify system performance and attain the required performance level makes sliding mode control attractive from the design perspective.

The basic idea of the sliding mode control concept is to attain the system evolution within the state-space submanifold (so-called *sliding manifold* or switching surface) assigned by the designer and expressed as

$$v(t, x) = 0. \tag{5.5.1}$$

Conventional sliding control is related to the time-optimal control because hard-switching (relay-type) control algorithms are designed to drive the state variables toward the sliding manifolds, applying discontinuous control. For example, for a time-invariant switching surface $v(x) = 0$, one has a polyhedron in the control space with 2^m vertexes. In particular,

$$u(x) = \begin{cases} u_{\max}, & \forall v(x) > 0 \\ 0, & \forall v(x) = 0, u_{\min} \le u(x) \le u_{\max}, u_{\max} > 0, u_{\min} < 0, \\ u_{\min}, & \forall v(x) < 0 \end{cases} \tag{5.5.2}$$

or $u(x) = \mathrm{sgn}[v(x)]$.

The sliding manifold and the control algorithm should be designed to attain the system evolution within the switching surface. Figure 5.5.1 illustrates the trajectories for a dynamic system mapped by $\dot{x}(t) = F(x) + B(x)u$.

Example 5.5.1.

In Example 5.4.3, we studied the optimal controller for the following dynamic system:

$$\dot{x}_1(t) = x_2,$$
$$\dot{x}_2(t) = u, -1 \le u \le 1.$$

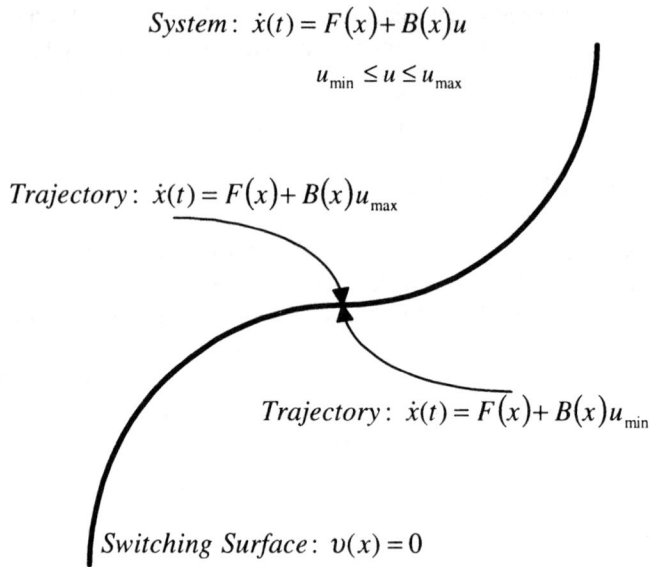

$$\text{System}: \quad \dot{x}(t) = F(x) + B(x)u$$

$$u_{min} \le u \le u_{max}$$

$$\text{Trajectory}: \quad \dot{x}(t) = F(x) + B(x)u_{max}$$

$$\text{Trajectory}: \quad \dot{x}(t) = F(x) + B(x)u_{min}$$

$$\text{Switching Surface}: \quad \upsilon(x) = 0$$

FIGURE 5.5.1.   System trajectories within time-invariant switching surface $\upsilon(x) = 0$.

Using the calculus and the Hamolton–Jacobi theory, it was illustrated that the controller is given as

$$u = -\,\text{sgn}(x_1 + \tfrac{1}{2}x_2|x_2|),$$

and the switching curve is

$$-x_1 - \tfrac{1}{2}x_2|x_2| = 0. \tag{5.5.3}$$

That is, the control takes the values +1 or −1 (hard switching) at the switching curve $-x_1 - \tfrac{1}{2}x_2|x_2| = 0$, as illustrated in Figure 5.5.2. The derived time-invariant switching surface (5.5.3) in general form was expressed by (5.5.1), and the controller was given by (5.5.2).

The simulation results, documented in Figures 5.5.3, illustrate the evolution of the system states to the origin, and the hard switching control action (+1 or −1) is shown as well.

In this section, we describe the synthesis procedures to design hard- and soft-switching continuous controllers with nonlinear switching surfaces. The objectives to be achieved through the control law design are to guarantee robustness, precise tracking, and disturbance attenuation. Dynamic systems with uncertain parameters are modeled as

$$\dot{x}(t) = F_z(t, x, r, d, z) + B_p(t, x, p)u, \; y = Hx, \, x(t_0) = x_0, \tag{5.5.4}$$

where $x \in X \subset \mathbb{R}^c$ is the state, $u \in U \subset \mathbb{R}^m$ is the control with the known closed admissible set $U = \{u \in \mathbb{R}^m | u_{min} \le u \le u_{max}, u_{max} > 0, u_{min} < 0\}, r \in R \subset \mathbb{R}^b$

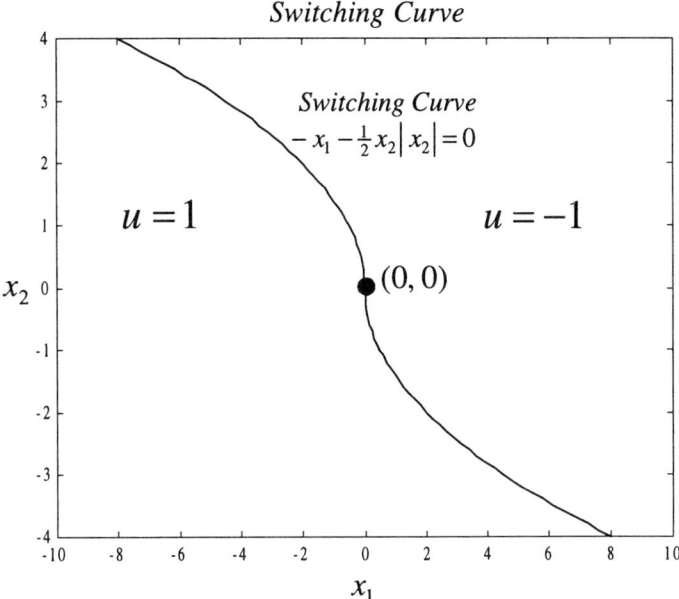

FIGURE 5.5.2.   Switching curve, $-x_1 - \frac{1}{2}x_2|x_2| = 0$.

and $y \in Y \subset \mathbb{R}^b$ are the reference and output vectors, $d \in D \subset \mathbb{R}^s$ is the disturbance vector, and $z \in Z \subset \mathbb{R}^d$ and $p \in P \subset \mathbb{R}^k$ are the time-varying bounded uncertainties.

The stabilizing and tracking admissible controllers for the system (5.5.4) are

$$u = u(t, x), u \in U \qquad \text{(stabilizing control law)} \qquad (5.5.5)$$

$$u = u(t, x, e), u \in U, \qquad \text{(tracking control law)} \qquad (5.5.6)$$

where $u(t, x)$ is the continuous (soft-switching) or discontinuous (hard-switching) control that drives the system states to the equilibrium manifold and maintains the state evolution on the equilibrium manifold, and $u(t, x, e)$ is the continuous (soft-switching) or discontinuous (hard-switching) control that drives the system states and tracking errors to the equilibrium manifold and maintains the evolution of states and tracking errors on the equilibrium manifold.

To solve the motion control problem, guarantee stability, robustness, attain tracking, and disturbance attenuation, the continuous (soft-switching) or discontinuous (hard-switching) controllers will be designed. The uncertain system (5.5.4) with (5.5.5) or (5.5.6) evolves in

$$XY(X_0, U, R, D, Z, P) = \{(x, y) \in X \times Y \colon x_0 \in X_0, u \in U, r \in R, d \in D,$$
$$z \in Z, p \in P, t \in [t_0, \infty)\} \subset \mathbb{R}^c \times \mathbb{R}^b.$$

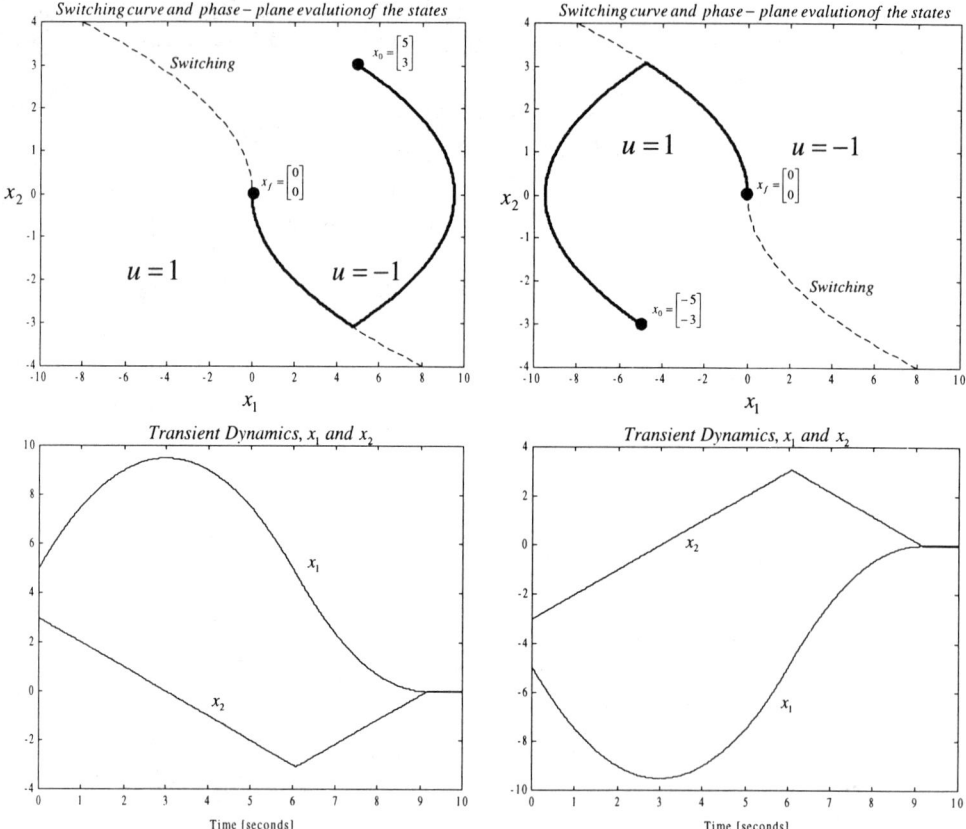

FIGURE 5.5.3.   Switching curve and system evolution.

The transient dynamics of systems (5.5.4) can be modeled as

$$\dot{x}(t) = F_x(t, x, r, d) + B_x(t, x)u + \Xi(t, x, u, z, p). \qquad (5.5.7)$$

The parameter variations are bounded, and for the time-varying bounded uncertainties $z \in Z$ and $p \in P$, a norm of $\Xi(t, x, u, z, p)$ exists such that

$$\|\Xi(t, x, u, z, p)\| \le \rho(t, x),$$

where $\rho(\cdot)$: $\mathbb{R}_{\ge 0} \times \mathbb{R}_{\ge 0} \to \mathbb{R}_{\ge 0}$ is the continuous Lebesgue measurable function.

It has been illustrated that the tracking control problem is more general problem compared with the stabilizing one. Therefore, to attain the general results, the tracking control algorithms will be synthesized.

To solve the motion control problem, the tracking error vector, as given by

$$e(t) = Nr(t) - y(t), \ N \in \mathbb{R}^{b \times b},$$

is used. Therefore, using (5.5.7), we have

$$\dot{x}(t) = F_x(t, x, r, d) + B_x(t, x)u + \Xi(t, x, u, z, p),$$
$$\dot{e}(t) = N\dot{r}(t) - \dot{y}(t) = N\dot{r}(t) - HF_x(t, x, r, d) - HB_x(t, x)u - H\Xi(t, x, u, z, p).$$

That is, in matrix form, one obtains

$$\begin{bmatrix} \dot{x}(t) \\ \dot{e}(t) \end{bmatrix} = \begin{bmatrix} F_x(t, x, r, d) \\ -HF_x(t, x, r, d) \end{bmatrix} + \begin{bmatrix} B_x(t, x) \\ -HB_x(t, x) \end{bmatrix} u$$
$$+ \begin{bmatrix} \Xi(t, x, u, z, p) \\ -H\Xi(t, x, u, z, p) \end{bmatrix} + \begin{bmatrix} 0 \\ N \end{bmatrix} \dot{r}(t). \tag{5.5.8}$$

Our goal is to synthesize a controller for system (5.5.4) such that the error vector $e(t)$ with $E_0 = \{e_0 \in \mathbb{R}^b\} \subseteq E \subset \mathbb{R}^b$ evolves in the specified closed set

$$S_e(\delta) = \{e \in \mathbb{R}^b \colon e_0 \in E_0, x \in X(X_0, U, R, D, Z, P), r \in R, d \in D, y \in Y,$$
$$t \in [t_0, \infty) \| \|e(t)\| \le \rho_e(t, \|e_0\|) + \rho_r(\|r\|) + \rho_d(\|d\|) + \rho_y(\|y\|) + \delta,$$
$$\delta \ge 0, \forall e \in E(E_0, R, D, Y), \forall t \in [t_0, \infty)\} \subset \mathbb{R}^b.$$

Here, $\rho_e(\cdot)\colon \mathbb{R}_{\ge 0} \times \mathbb{R}_{\ge 0} \to \mathbb{R}_{\ge 0}$ is the KL-function and $\rho_r(\cdot)\colon \mathbb{R}_{\ge 0} \to \mathbb{R}_{\ge 0}, \rho_d(\cdot)\colon \mathbb{R}_{\ge 0}$ $\to \mathbb{R}_{\ge 0}$ and $\rho_y(\cdot)\colon \mathbb{R}_{\ge 0} \to \mathbb{R}_{\ge 0}$ are the K-functions.

The smooth (continuous) sliding manifold is given by

$$M = \{(t, x, e) \in \mathbb{R}_{\ge 0} \times X \times E | \upsilon(t, x, e) = 0\}$$
$$= \bigcap_{j=1}^{m} \{(t, x, e) \in \mathbb{R}_{\ge 0} \times X \times E | \upsilon_j(t, x, e) = 0\}. \tag{5.5.9}$$

Here, the time-varying linear and nonlinear switching surfaces are

$$\upsilon(t, x, e) = \begin{bmatrix} K_{\upsilon x}(t) & K_{\upsilon e}(t) \end{bmatrix} \begin{bmatrix} x(t) \\ e(t) \end{bmatrix}$$
$$= K_{\upsilon x}(t)x(t) + K_{\upsilon e}(t)e(t) = 0, \tag{5.5.10}$$

$$\begin{bmatrix} \upsilon_1(t, x, e) \\ \vdots \\ \upsilon_m(t, x, e) \end{bmatrix} = \begin{bmatrix} k_{\upsilon x11}(t) & \cdots & k_{\upsilon x1c}(t) & k_{\upsilon e11}(t) & \cdots & k_{\upsilon e1b}(t) \\ \vdots & \vdots & \vdots & \vdots & \vdots & \vdots \\ k_{\upsilon xm1}(t) & \cdots & k_{\upsilon xmc}(t) & k_{\upsilon em1}(t) & \cdots & k_{\upsilon emb}(t) \end{bmatrix}$$
$$\times \begin{bmatrix} x_1(t) \\ \vdots \\ x_c(t) \\ e_1(t) \\ \vdots \\ e_b(t) \end{bmatrix} = 0,$$

and

$$v(t, x, e) = K_{vxe}(t, x, e) = 0, \qquad (5.5.11)$$

$$\begin{bmatrix} v_1(t, x, e) \\ \vdots \\ v_m(t, x, e) \end{bmatrix} = \begin{bmatrix} K_{vxe1}(t, x, e) \\ \vdots \\ K_{vxem}(t, x, e) \end{bmatrix} = 0.$$

The total derivative of the time-varying switching surface is expressed as

$$\frac{dv(t, x, e)}{dt} = \frac{\partial v(t, x, e)}{\partial t} + \frac{\partial v(t, x, e)^T}{\partial x} \dot{x}(t) + \frac{\partial v(t, x, e)^T}{\partial e} \dot{e}(t). \qquad (5.5.12)$$

From (5.5.4) and (5.5.12), one finds

$$\begin{aligned}
\frac{dv(t, x, e)}{dt} &= \frac{\partial v}{\partial t} + \frac{\partial v(t, x, e)^T}{\partial x} [F_z(t, x, r, d, z) + B_p(t, x, p)u] \\
&\quad + \frac{\partial v(t, x, e)^T}{\partial e} [N\dot{r}(t) - HF_z(t, x, r, d, z) \\
&\quad - HB_p(t, x, p)u],
\end{aligned} \qquad (5.5.13)$$

and the so-called *equivalent* control can be derived from

$$\frac{dv(t, x, e)}{dt} = 0.$$

The sufficient conditions for robust stability are given by the Lyapunov stability theory. Let a C^κ function $V(t, x, e)$ exist, as well as K_∞-functions ρ_1, ρ_2, ρ_3, and ρ_4 and K-functions ρ_5 and ρ_6, such that for all $x_0 \in X_0, e_0 \in E_0, u \in U, r \in R, d \in D, z \in Z$, and $p \in P$ on $[t_0, \infty)$, the following criteria are satisfied in $S_S \subset \mathbb{R}^c \times \mathbb{R}^b$:

$$\rho_1 \|x\| + \rho_2 \|e\| \le V(t, x, e) \le \rho_3 \|x\| + \rho_4 \|e\|, \qquad (5.5.14)$$

$$\frac{dV(t, x, e)}{dt} \le -\rho_5 \|x\| - \rho_6 \|e\|. \qquad (5.5.15)$$

Then, the solution of (5.5.4) with (5.5.6) is uniformly ultimately bounded in the convex and compact set $X(X_0, U, R, D, Z, P) \subset \mathbb{R}^c$, and the robust tracking is guaranteed in $XE(X_0, E_0, U, R, D, Z, P) \subset \mathbb{R}^c \times \mathbb{R}^b$ if $XE \subseteq S_S$.

The total derivative of the $V(t, x, e)$ is found to be

$$\begin{aligned}
\frac{dV(t, x, e)}{dt} &= \frac{\partial V(t, x, e)}{\partial t} + \frac{\partial V(t, x, e)^T}{\partial x} [F_z(t, x, r, d, z) + B_p(t, x, p)u] \\
&\quad + \frac{\partial V(t, x, e)^T}{\partial e} [N\dot{r}(t) - HF_z(t, x, r, d, z) - HB_p(t, x, p)u].
\end{aligned}$$

If the system evolves in XE, and $XE \subseteq S_S$, one concludes that the sufficient conditions for robust stability are guaranteed in XE. That is, the closed-loop system with tracking controller (5.5.6) is robustly stable and the convergence of the tracking error $e(t)$ is guaranteed.

The Lyapunov functions must be designed.

For example, the following nonquadratic functions can be applied:

$$V(t, x, e) = \tfrac{1}{2} \upsilon^T (t, x, e) K(t, x, e) \upsilon(t, x, e), \qquad (5.5.16)$$

$$V(t, x, e) = \tfrac{1}{2} \upsilon^T (t, x, e) K(t, x, e) \, \mathrm{sgn}(\upsilon(t, x, e)),$$

and

$$V(t, x, e) = \tfrac{1}{2} \upsilon^{T^2} (t, x, e) K(t, x, e) \upsilon^2(t, x, e),$$

where $K(\cdot)$ is the continuous symmetric matrix function.

Using (5.5.16), one obtains

$$\frac{dV(t, x, e)}{dt} = \upsilon^T (t, x, e) K(t, x, e) \frac{d\upsilon(t, x, e)}{dt}$$
$$+ \tfrac{1}{2} \upsilon^T (t, x, e) \frac{dK(t, x, e)}{dt} \upsilon(t, x, e),$$

where

$$\frac{d\upsilon(t, x, e)}{dt} = \frac{\partial \upsilon(t, x, e)}{\partial t} + \frac{\partial \upsilon(t, x, e)^T}{\partial x} \dot{x}(t) + \frac{\partial \upsilon(t, x, e)^T}{\partial e} \dot{e}(t).$$

By using the matrix calculus, one finds

$$\frac{dK(t, x, e)}{dt}.$$

The controller should be designed to guarantee

$$\frac{dV(t, x, e)}{dt} \le -\rho_5 \|x\| - \rho_6 \|e\|.$$

Thus, the following inequality must be solved:

$$\upsilon^T (t, x, e) K(t, x, e) \frac{d\upsilon(t, x, e)}{dt} + \tfrac{1}{2} \upsilon^T (t, x, e) \frac{dK(t, x, e)}{dt} \upsilon(t, x, e)$$
$$\le -\rho_5 \|x\| - \rho_6 \|e\|.$$

The discontinuous (hard-switching) tracking controllers with constant and varying gains are designed as

$$u(t, x, e) = -G \, \mathrm{sgn}(\upsilon), \; G > 0, \; G \in \mathbb{R}^{m \times m}, \qquad (5.5.17)$$

$$u(t, x, e) = -G(t, x, e) \, \mathrm{sgn}(\upsilon), \; G(\cdot): \mathbb{R}_{\ge 0} \times \mathbb{R}^c \times \mathbb{R}^b \to \mathbb{R}^{m \times m}. \qquad (5.5.18)$$

The simplest hard-switching tracking controller is given as (see Example 5.5.1)

$$
u(t, x, e) = \begin{cases} u_{\max}, & \forall \upsilon(t, x, e) > 0 \\ 0, & \forall \upsilon(t, x, e) = 0, u_{\min} \le u(t, x, e) \le u_{\max}, u_{\max} > 0, u_{\min} < 0, \\ u_{\min}, & \forall \upsilon(t, x, e) < 0 \end{cases}
$$

$$(5.5.19)$$

and a polyhedron in the control space results.

Constant and state-dependent gain discontinuous controllers, as well as state-feedback algorithms with switching gains are synthesized; see (5.5.17) and (5.5.18). Singularity, chattering, sensitivity, and degraded efficiency of the hard-switching sliding controllers have challenged the application of sliding mode algorithms. The innovative tracking controllers with soft switching are given by

$$
u(t, x, e) = -G\phi(\upsilon), \tag{5.5.20}
$$

$$
u(t, x, e) = -G(t, x, e)\phi(\upsilon), \tag{5.5.21}
$$

where $\phi(\cdot)$ is the continuous real-analytic function of class C^{ϵ} ($\epsilon \ge 1$).

Using the hyperbolic tangent and `erf`, one finds the following control laws:

$$
u(t, x, e) = u_{\max} \tanh^{\frac{1}{q}} \upsilon \text{ or } u(t, x, e) = u_{\max} \tanh^{q} \upsilon,
$$
$$
-u_{\max} \le u(t, x, e) \le u_{\max}, q = 3, 5, 7, \ldots,
$$

$$
u(t, x, e) = u_{\max}\text{erf}^{\frac{1}{q}} \upsilon \text{ or } u(t, x, e) = u_{\max}\text{erf}^{q} \upsilon, -u_{\max} \le u(t, x, e) \le u_{\max},
$$

whereas applying the sigmoid function, we have

$$
u(t, x, e) = u_{\max} \frac{1}{1 + e^{-a(\upsilon - b)}}, 0 \le u(t, x, e) \le u_{\max}.
$$

That is, the controller is bounded, and $u \in U$ for all $x \in X$ and $e \in E$ on $[t_0, \infty)$.

The proposed soft-switching controllers are different compared with discontinuous (hard-switching) control. Compared with hard-switching discontinuous controllers, the advantages of the soft-switching continuous control are that the singularity and sensitivity problems are avoided, robustness and stability are improved, chattering (high-frequency switching) is eliminated, and so forth. These allow the designer to significantly expand operating envelopes, enhance robustness, and improve efficiency of controlled dynamic systems.

Let us study the stability, robustness, tracking, and disturbance attenuation for closed-loop systems with soft-switching controllers that used nonlinear switching surfaces given in terms of time, state variables, and tracking error vector.

In $XE(X_0, E_0, U, R, D, Z, P) \subset \mathbb{R}^c \times \mathbb{R}^b$, one considers the evolution of system (5.5.4) with (5.5.20) or (5.5.21). The robust tracking, stability, and disturbance rejection in XE are guaranteed if $XE \subseteq S_s$ for given $X_0 \subset \mathbb{R}^c$, $E_0 \subset \mathbb{R}^b$, $U \subset \mathbb{R}^m$, $R \subset \mathbb{R}^b$, $D \subset \mathbb{R}^s$, $Z \subset \mathbb{R}^d$, and $P \subset \mathbb{R}^k$. The admissible set is found by using the criteria (5.5.14) and (5.5.15). That is,

$$
\begin{aligned}
S_s = \{ & x \in \mathbb{R}^c, e \in \mathbb{R}^b \colon x_0 \in X_0, e_0 \in E_0, u \in U, r \in R, d \in D, z \in Z, p \in P, \\
& t \in [t_0, \infty) | \rho_1 \|x\| + \rho_2 \|e\| \leq V(t, x, e) \leq \rho_3 \|x\| + \rho_4 \|e\|, \\
& \frac{dV(t, x, e)}{dt} \leq -\rho_5 \|x\| - \rho_6 \|e\|, \forall x \in X(X_0, U, R, D, Z, P), \\
& \forall e \in E(E_0, R, D, Y), \forall t \in [t_0, \infty) \} \subset \mathbb{R}^c \times \mathbb{R}^b.
\end{aligned}
$$

For admissible references, disturbances, and uncertainties, we have $XE \subseteq S_s$. That is, all solutions of the closed-loop system $x(\cdot) \colon [t_0, \infty) \to \mathbb{R}^c$ are robustly bounded, and convergence of the tracking error vector $e(\cdot) \colon [t_0, \infty) \to \mathbb{R}^b$ to the $S_e(\delta)$ is guaranteed.

Example 5.5.2. Soft switching control of a permanent-magnet synchronous motor

In the rotor reference frame, the following model of permanent-magnet synchronous motors is found (see Section 5.7 for details):

$$
\frac{di_{qs}^r}{dt} = -\frac{r_s}{L_{ss}} i_{qs}^r - \frac{\psi_m}{L_{ss}} \omega_r - i_{ds}^r \omega_r + \frac{1}{L_{ss}} u_{qs}^r,
$$

$$
\frac{di_{ds}^r}{dt} = -\frac{r_s}{L_{ss}} i_{ds}^r + i_{qs}^r \omega_r + \frac{1}{L_{ss}} u_{ds}^r,
$$

$$
\frac{di_{os}^r}{dt} = -\frac{r_s}{L_{1s}} i_{os}^r + \frac{1}{L_{1s}} u_{os}^r,
$$

$$
\frac{d\omega_r}{dt} = \frac{3 P^2 \psi_m}{8 J} i_{qs}^r - \frac{B_m}{J} \omega_r - \frac{P}{2J} T_L,
$$

where $i_{qs}^r, i_{ds}^r, i_{os}^r$ and $u_{qs}^r, u_{ds}^r, u_{os}^r$ are the *quadrature-*, *direct-*, and *zero*-axis current and voltage components.

That is, the state and control variables are the quadrature, direct, and zero currents and voltages, as well as the angular velocity. In particular,

$$
x = \begin{bmatrix} i_{qs}^r \\ i_{ds}^r \\ i_{os}^r \\ \omega_r \end{bmatrix} \quad \text{and} \quad u = \begin{bmatrix} u_{qs}^r \\ u_{ds}^r \\ u_{os}^r \end{bmatrix}.
$$

The rated data and parameters of a permanent-magnet synchronous motor are 135 W, 434 rad/sec, 40 V, 0.42 N-m, 6.9 A, $r_s(\cdot) \in [0.5_{T=20^\circ C} \quad 0.75_{T=130^\circ C}]\Omega$, $L_{ss}(\cdot) \in [0.009 \quad 0.01]$H, $L_{1s} = 0.001$H, $\psi_m(\cdot) \in [0.069_{T=20^\circ C} \quad 0.055_{T=130^\circ C}]$ V-sec/rad or $\psi_m(\cdot) \in [0.069_{T=20^\circ C} \quad 0.055_{T=130^\circ C}]$ N-m/A, and $B_m = 0.000013$ N-m-sec/rad, and the equivalent moment of inertia of the motor and attached load is $J(\cdot) \in [0.0001 \quad 0.0003]$ kg-m^2.

The quadrature-, direct-, and zero-axis voltage and current components have a DC form. Let us study the stability. For the uncontrolled motor, we have

$$u^r_{qs} = u^r_{ds} = u^r_{os} = 0.$$

The total derivative of the positive-definite quadratic function

$$V(i^r_{qs}, i^r_{ds}, i^r_{os}, \omega_r) = \tfrac{1}{2}i^{r2}_{qs} + \tfrac{1}{2}i^{r2}_{ds} + \tfrac{1}{2}i^{r2}_{os} + \tfrac{1}{2}\omega^2_r$$

is

$$
\begin{aligned}
\frac{dV(i^r_{qs}, i^r_{ds}, i^r_{os}, \omega_r)}{dt} &= -\frac{r_s}{L_{ss}}i^{r2}_{qs} - \frac{r_s}{L_{ss}}i^{r2}_{ds} - \frac{r_s}{L_{1s}}i^{r2}_{os} \\
&\quad - \frac{\psi_m(8J - 3P^2 L_{ss})}{8J L_{ss}}i^r_{qs}\omega_r - \frac{B_m}{J}\omega^2_r.
\end{aligned}
$$

The motor parameters are time-varying, and $r_s(\cdot) \in [r_{s\,min}\quad r_{s\,max}]$, $L_{ss}(\cdot) \in [L_{ss\,min}\quad L_{ss\,max}]$, $\psi_m(\cdot) \in [\psi_{m\,min}\quad \psi_{m\,max}]$, and $J(\cdot) \in [J_{min}\quad J_{max}]$. However, $r_{s\,min} > 0$, $L_{ss\,min} > 0$, $\psi_{m\,min} > 0$, and $J_{min} > 0$. Hence, the open-loop system is uniformly robustly asymptotically stable in the large because the total derivative of a positive-definite function $V(i^r_{qs}, i^r_{ds}, i^r_{os}, \omega_r)$ is negative. Thus, it is not necessary to apply the linearizing feedback to transform a nonlinear motor model into a linear one by canceling the beneficial internal nonlinearities $-i^r_{ds}\omega_r$ and $i^r_{qs}\omega_r$.

In general, because of the bounds imposed on the voltages, as well as variations of L_{ss}, one cannot attain the feedback linearization. More important is that from the electric machinery standpoints, to attain the balanced voltage set applied to the stator windings, one controls only u^r_{qs}, whereas $u^r_{ds} = 0$. In fact, to guarantee the balanced operation, $u^r_{ds} = 0$. That is, feedback linearization cannot be achieved.

Our goal is to design a soft-switching controller. The applied phase voltages are bounded by 40 V. Using (5.5.10) and (5.5.11), time-invariant linear and nonlinear switching surfaces of stabilizing controllers are obtained as functions of the state variables i^r_{qs}, i^r_{ds}, and ω_r. In particular,

$$
\begin{aligned}
\upsilon(i^r_{qs}, i^r_{ds}, \omega_r) &= -0.00049i^r_{qs} - 0.00049i^r_{ds} - 0.0014\omega_r = 0, \\
\upsilon(i^r_{qs}, i^r_{ds}, \omega_r) &= -0.00049i^r_{qs} - 0.00049i^r_{ds} - 0.000017\omega_r \\
&\quad -0.000025\omega_r|\omega_r| = 0.
\end{aligned}
$$

Only quadrature voltage is regulated, and we denote

$$u = u^r_{qs}.$$

A discontinuous, hard-switching stabilizing controller is found to be

$$
u = \mathrm{sgn}^{+40}_{-40}\,\upsilon(i^r_{qs}, i^r_{ds}, \omega_r) =
\begin{cases}
+40, & \upsilon(i^r_{qs}, i^r_{ds}, \omega_r) > 0 \\
0, & \upsilon(i^r_{qs}, i^r_{ds}, \omega_r) = 0 \\
-40, & \upsilon(i^r_{qs}, i^r_{ds}, \omega_r) < 0.
\end{cases}
$$

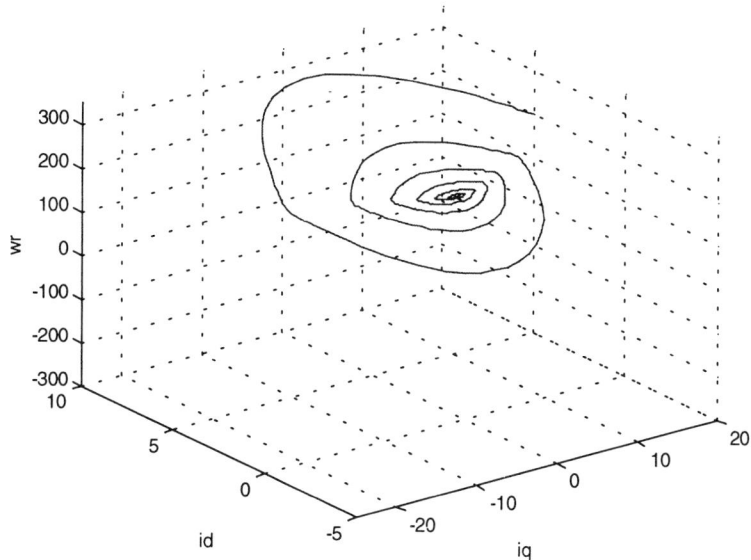

FIGURE 5.5.4.   Three-dimensional state evolution caused by initial conditions.

To avoid the singularity, this discontinuous algorithm is *regularized* as

$$u = 40 \frac{\upsilon(i_{qs}^r, i_{ds}^r, \omega_r)}{|\upsilon(i_{qs}^r, i_{ds}^r, \omega_r)| + \varepsilon}, \quad \varepsilon = 0.001.$$

A soft-switching stabilizing controller is designed as

$$u = 40 \tanh^{1/9} \upsilon(i_{qs}^r, i_{ds}^r, \omega_r).$$

Nonlinear simulation is performed to study the state evolution caused by initial conditions. We assign

$$\begin{bmatrix} i_{qs}^r(0) \\ i_{ds}^r(0) \\ \omega_r(0) \end{bmatrix} = \begin{bmatrix} 20 \\ 5 \\ 200 \end{bmatrix},$$

and the three-dimensional plot for i_{qs}^r, i_{ds}^r, and ω_r is plotted in Figure 5.5.4.

The tracking controller is synthesized using the states i_{qs}^r, i_{ds}^r, ω_r, and the tracking error $e = \omega_{\text{reference}} - \omega_{rm}$.

The nonlinear time-invariant switching surface is

$$\upsilon(i_{qs}^r, i_{ds}^r, \omega_r, e) = -0.00049 i_{qs}^r - 0.00049 i_{ds}^r - 0.000027 \omega_r + 0.0015 e$$
$$+ 0.00009 e^3 = 0,$$

and a soft-switching tracking controller is given by

$$u = 40 \tanh^{1/9} \upsilon(i_{qs}^r, i_{ds}^r, \omega_r, e).$$

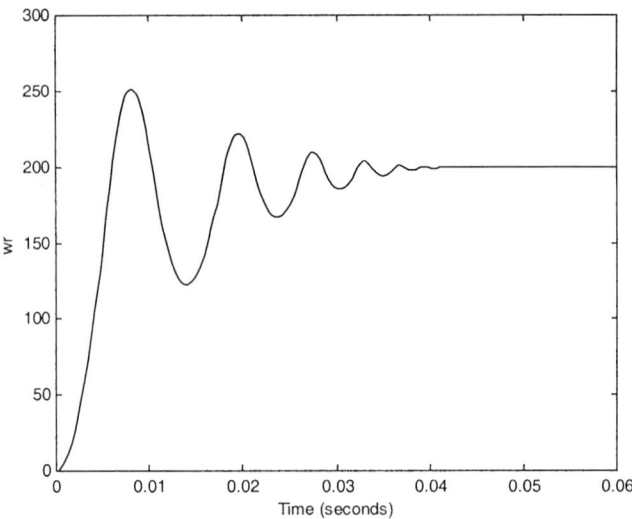

FIGURE 5.5.5.   Transient dynamic of the motor angular velocity if $\omega_{\text{reference}} = 200\frac{\text{rad}}{\text{sec}}$.

Figure 5.5.5 illustrates the dynamics of a closed-loop system with the designed tracking controller if the reference angular velocity is $\omega_{\text{reference}} = 200\frac{\text{rad}}{\text{sec}}$.

The settling time is 0.04 sec. Excellent dynamic performance is evident, and the tracking error is zero. It must be emphasized that because of soft switching, the singularity and sensitivity problems are avoided, robustness and stability are improved, and the chattering effect (high-frequency switching) is eliminated.

5.6.   Control of Discrete-Time Systems

A great interest in digital control of dynamic systems emerges because controllers are usually implemented using DSPs and microprocessors. There is a growing interest to solve the motion control problem, and integral controllers can be designed to ensure tracking, guarantee stability, and accomplish other desired criteria. This section shows that optimization and tracking can be solved by minimizing non-quadratic performance indexes, and a complete solution of the motion control problem is found for a general nonlinear model of discrete-time systems. Necessary conditions allow us to design a nonlinear controller and verify the optimality. In particular, the Hamilton–Jacobi theory is used to design a nonlinear algorithm that is optimal to a specified performance index, corresponds to the desired transient dynamics, stabilizes the systems, and ensures the tracking. Following these, this section is devoted to design unbounded and constrained controllers for nonlinear discrete-time systems using the Hamilton–Jacobi theory. Motion control and tracking are important problems, and these issues are emphasized and studied.

The optimal state-feedback nonlinear tracking problem is formulated and solved using necessary conditions for optimality, and a nonquadratic performance index is minimized.

5.6.1. *Tracking Control*

We study a class of discrete-time systems that are described by nonlinear difference equations. In particular, consider

$$x_s(k+1) = F_s(x_s(k)) + B_s(x_s(k))u(k), \qquad (5.6.1)$$

where $x_s(k) \in X \subset \mathbb{R}^n$ is the state-space vector of the system, $u(k) \subset \mathbb{R}^m$ is the control vector, and $F(\cdot): \mathbb{R}^n \to \mathbb{R}^n$ and $B_s(\cdot): \mathbb{R}^n \to \mathbb{R}^{n \times m}$ are the smooth maps.

For system (5.6.1), the output equation is represented as

$$y(k) = Hx_s(k), \qquad (5.6.2)$$

where $y(k) \in Y \subset \mathbb{R}^b$ is the system output and $H \in \mathbb{R}^{b \times n}$ is the matrix with constant coefficients.

The reference vector $r(k) \in R \subset \mathbb{R}^b$ should be considered to approach the motion control problem because the tracking error vector is given by

$$e(k) = r(k) - y(k), \, e(k) \in E \subset \mathbb{R}^b.$$

To solve the motion control problem, which the primary objective is to guarantee stability and to ensure tracking, an integral control law will be designed. The *exosystem* state equation is

$$x_i(k) = x_i(k-1) + r(k) - y(k);$$

that is,

$$\begin{aligned} x_i(k+1) &= x_i(k) + r(k+1) - y(k+1) \\ &= -Hx_s(k+1) + x_i(k) + r(k+1). \end{aligned}$$

The following augmented state-space difference equation is found using (5.6.1) and (5.6.2):

$$x(k+1) = \begin{bmatrix} x_s(k+1) \\ x_i(k+1) \end{bmatrix} = F(x(k)) + B(x(k))u(k) + B_r r(k+1),$$

$$F(x(k)) = \begin{bmatrix} F_s(x_s(k)) \\ -HF_s(x_s(k)) + Ix_i(k) \end{bmatrix}, \, B(x(k)) = \begin{bmatrix} B_s(x_s(k)) \\ -HB_s(x_s(k)) \end{bmatrix},$$

$$B_r = \begin{bmatrix} 0 \\ I \end{bmatrix}, \qquad (5.6.3)$$

where $x(k) \in X \subset \mathbb{R}^c$ is the vector of the system variables, $c = n + b$, $x(k) = \begin{bmatrix} x_s(k) \\ x_i(k) \end{bmatrix}$, and $I \in \mathbb{R}^{b \times b}$ is the identity matrix.

The derived augmented nonlinear system (5.6.3) is applied to design a control law minimizing the following nonquadratic performance index:

$$J = \sum_{k=0}^{N-1} \left[\sum_{i=0}^{\varsigma} \left(x(k)^{\frac{i+\beta+1}{2\beta+1}} \right)^{T} Q_i x(k)^{\frac{i+\beta+1}{2\beta+1}} + u(k)^{T} Gu(k) \right], \tag{5.6.4}$$

where $Q_i \in \mathbb{R}^{c \times c}$ and $G \in \mathbb{R}^{m \times m}$ are the Hermitian matrices and ς and β are the real integers assigned by the designer, $\varsigma = 0, 1, 2, \ldots, \beta = 0, 1, 2, \ldots$.

Using the dynamic programming principle, the Hamilton–Jacobi–Bellman recursive equation is given as

$$V(x(k)) = \min_{u(k)} \left[\sum_{i=0}^{\varsigma} \left(x(k)^{\frac{i+\beta+1}{2\beta+1}} \right)^{T} Q_i x(k)^{\frac{i+\beta+1}{2\beta+1}} + u(k)^{T} Gu(k) \right. $$
$$\left. + V(x(k+1)) \right], \tag{5.6.5}$$

where the positive-definite return function, which represents the minimum value of the performance index, is assigned in the quadratic form. That is,

$$V(x(k)) = x(k)^{T} P(k) x(k), \tag{5.6.6}$$

where $P(k) \in \mathbb{R}^{c \times c}$ is the unknown matrix.

Taking note of (5.6.5) and (5.6.6), and using the first-order necessary condition for optimality, one obtains a nonlinear control law in

$$u(k) = -(B(x(k))^{T} P(k+1) B(x(k)) + G)^{-1} B(x(k))^{T} P(k+1) F(x(k)). \tag{5.6.7}$$

It is evident that the second-order necessary condition for optimality is guaranteed because matrix G is assigned to be positive-definite. In fact,

$$\frac{\partial^2 (u(k)^{T} Gu(k))}{\partial u(k) \times \partial u(k)^{T}} = G > 0.$$

Substituting the nonlinear controller (5.6.7) in (5.6.5), one obtains an equation that must be solved to find the unknown matrix $P(k)$. In particular, we have

$$x(k)^{T} P(k) x(k) = \sum_{i=0}^{\varsigma} \left(x(k)^{\frac{i+\beta+1}{2\beta+1}} \right)^{T} Q_i x(k)^{\frac{i+\beta+1}{2\beta+1}} + F(x(k))^{T} P(k+1) F(x(k))$$
$$- F(x(k))^{T} P(k+1) B(x(k)) (B(x(k))^{T} P(k+1)$$
$$\times B(x(k)) + G)^{-1} B(x(k))^{T} P(k+1) F(x(k)). \tag{5.6.8}$$

Solving (5.6.8), one derives the feedback gains in (5.6.7).

5.6.2. *Constrained Optimization*

In this subsection, the constrained optimization problem is researched and solved for nonlinear discrete-time systems. The Hamilton–Jacobi theory is applied to design a new class of bounded controllers, and an innovative nonquadratic performance index is minimized. For open-loop unstable systems, the constrained optimization problem is solvable if the sufficient conditions are satisfied. This leads to the application of the admissibility framework, and the maximal, positively invariant admissible set of stability S is found by applying the Lyapunov stability criteria. A great number of methods have been developed to design bounded controllers for dynamic systems. In particular, necessary and sufficient conditions for optimality and stability were derived using the Hamilton–Jacobi framework and Lyapunov's stability theory. In this subsection, we apply nonquadratic performance indexes to analytically design a new class of bounded controllers in the closed form. It is shown that the developed design method provides an efficient avenue of synthesizing bounded controllers for multi-input/multi-output systems.

We study nonlinear discrete-time systems that are described as

$$x_{n+1} = F(x_n) + B(x_n)u_n, \quad u_{n\,\min} \le u_n \le u_{n\,\max}, \tag{5.6.9}$$

where $x_n \in X \subset \mathbb{R}^c$ is the state vector with the *auxiliary* conditions $x_{n0} \in X_0$, $u_n \in U \subset \mathbb{R}^m$ is the control vector, and $F(\cdot): \mathbb{R}^c \to \mathbb{R}^c$ and $B(\cdot): \mathbb{R}^c \to \mathbb{R}^{c \times m}$ are the smooth and Lipschitz maps.

To design a nonlinear admissible controller $u_n \in U$, let us map the imposed control bounds by a continuous integrable, one-to-one, bounded function $\phi \in U$. The following nonquadratic performance index:

$$J = \sum_{n=0}^{N-1} \left[x_n^T Q_n x_n - u_n^T B(x_n)^T K_{n+1} B(x_n) u_n \right.$$
$$\left. + 2 \int (\phi^{-1}(u_n))^T G_n \, du_n \right] \tag{5.6.10}$$

will be minimized. Making use of the Hamilton–Jacobi theory, the first- and second-order necessary conditions for optimality should be examined.

In the performance index (5.6.10), $Q_n \in \mathbb{R}^{c \times c}$ and $G_n \in \mathbb{R}^{m \times m}$ are the positive-semidefinite and positive-definite matrices, respectively. It should be emphasized that the positive-definiteness of index (5.6.10) is guaranteed if for all $x_n \in X$ and $u_n \in U$

$$\left(x_n^T Q_n x_n + 2 \int (\phi^{-1}(u_n))^T G_n \, du_n \right) > u_n^T B(x_n)^T K_{n+1} B(x_n) u_n,$$

where integrand $2 \int (\phi^{-1}(u_n))^T G_n \, du_n$ is positive-definite because the integrable, one-to-one function ϕ lies in the first and third quadrants, integrable function ϕ^{-1} exists, and matrix G_n is assigned to be positive-definite. The positive-definiteness of the performance index can be studied as positive-definite matrix $K_{n+1} \in \mathbb{R}^{c \times c}$

is found. The inequality, which guarantees the positive definiteness of the index (5.6.10) in X and U, can be easily guaranteed assigning the weighting matrices $Q_n \in \mathbb{R}^{c \times c}$ and $G_n \in \mathbb{R}^{m \times m}$.

Using the quadratic return function

$$V(x_n) = x_n^T K_n x_n,$$

we have

$$V(x_{n+1}) = x_{n+1}^T K_{n+1} x_{n+1} = (F(x_n) + B(x_n)u_n)^T K_{n+1}(F(x_n) + B(x_n)u_n),$$

and the Hamilton–Jacobi–Bellman recursive equation for system (5.6.9) and index (5.6.10) is given as

$$x_n^T K_n x_n = \min_{u_n \in U} \left[x_n^T Q_n x_n - u_n^T B(x_n)^T K_{n+1} B(x_n)u_n + 2 \int (\phi^{-1}(u_n))^T G_n \, du_n \right.$$

$$\left. + (F(x_n) + B(x_n)u_n)^T K_{n+1}(F(x_n) + B(x_n)u_n) \right]. \qquad (5.6.11)$$

Minimizing (5.6.11), using the first-order necessary condition for optimality, a bounded control law results. In particular, using (5.6.11), the striking application of the first-order necessary condition for optimality leads one to the following bounded controller:

$$u_n = -\phi(G_n^{-1} B(x_n)^T K_{n+1} F(x_n)), \, u_n \in U. \qquad (5.6.12)$$

It is evident that

$$\frac{\partial^2 (2 \int (\phi^{-1}(u_n))^T G_n \, du_n)}{\partial u_n \times \partial u_n^T}$$

is positive-definite because ϕ and ϕ^{-1} lie in the first and third quadrants, and G_n is positive-definite. Hence, one concludes that the second-order necessary condition for optimality is satisfied.

Substituting (5.6.12) in (5.6.11), we have

$$x_n^T K_n x_n = x_n^T Q_n x_n + 2 \int F(x_n)^T K_{n+1} B(x_n) \, d(\phi(G_n^{-1} B(x_n)^T K_{n+1} F(x_n)))$$

$$+ F(x_n)^T K_{n+1} F(x_n)$$

$$- 2F(x_n)^T K_{n+1} B(x_n)\phi(G_n^{-1} B(x_n)^T K_{n+1} F(x_n)). \qquad (5.6.13)$$

Using (5.6.13), and integrating by parts

$$2 \int F(x_n)^T K_{n+1} B(x_n) \, d(\phi(G_n^{-1} B(x_n)^T K_{n+1} F(x_n)))$$

$$= 2F(x_n)^T K_{n+1} B(x_n)\phi(G_n^{-1} B(x_n)^T K_{n+1} F(x_n))$$

$$- 2 \int (\phi(G_n^{-1} B(x_n)^T K_{n+1} F(x_n)))^T \, d(B(x_n)^T K_{n+1} F(x_n)),$$

one obtains an equation to find the unknown matrix $K_{n+1} \in \mathbb{R}^{c \times c}$. In particular, the following equation results:

$$x_n^T K_n x_n = x_n^T Q_n x_n + F(x_n)^T K_{n+1} F(x_n)$$
$$-2 \int (\phi(G_n^{-1} B(x_n)^T K_{n+1} F(x_n)))^T d(B(x_n)^T K_{n+1} F(x_n)). \quad (5.6.14)$$

Solving (5.6.14), matrix $K_{n+1} \in \mathbb{R}^{c \times c}$ can be found.

Thus, the feedback gains of a bounded control law (5.6.12) are derived.

By mapping the control bounds imposed by the continuous integrable, one-to-one, bounded functions $\phi \in U$, one finds

$$2 \int (\phi(G_n^{-1} B(x_n)^T K_{n+1} F(x_n)))^T d(B(x_n)^T K_{n+1} F(x_n)).$$

For example, using the hyperbolic tangent to map the saturation-type constraints, we have

$$\int \tanh z \, dz = \log \cosh z \text{ and } \int \tanh^g z \, dz = -\frac{\tanh^{g-1} z}{g-1} + \int \tanh^{g-2} z \, dz, g \neq 1.$$

If open-loop system (5.6.9) is unstable, the admissibility concept must be applied to verify the stability of the resulting closed-loop system. The closed-loop system (5.6.9) with (5.6.12) evolves in $X \subset \mathbb{R}^c$, which is given as

$$\{x_{n+1} = F(x_n) - B(x_n)\phi(G_n^{-1} B(x_n)^T K_{n+1} F(x_n)), x_{n0} \in X_0\} \in X(X_0, U).$$

Using the Lyapunov stability theory, the domain of stability $S \subset \mathbb{R}^c$ can be found applying the sufficient conditions under which the discrete-time system is stable. The positive-definite quadratic function $V(x_n)$ is used. To guarantee the stability, the first difference $\Delta V(x_n)$, which is found to be

$$\begin{aligned}\Delta V(x_n) = {} & V(x_{n+1}) - V(x_n) = F(x_n)^T K_{n+1} F(x_n) \\ & -2F(x_n)^T K_{n+1} B(x_n)\phi(G_n^{-1} B(x_n)^T K_{n+1} F(x_n)) \\ & +\phi(G_n^{-1} B(x_n)^T K_{n+1} F(x_n))^T B(x_n)^T K_{n+1} B(x_n) \\ & \times \phi(G_n^{-1} B(x_n)^T K_{n+1} F(x_n)) - x_n^T K_n x_n\end{aligned}$$

must be negative-definite for all $x_n \in X$. That is, we have

$$\begin{aligned}S = {} & \{x_n \in R^c: x_{n0} \in X_0, u \in U | V(0) = 0, V(x_n) > 0, \Delta V(x_n) < 0, \\ & \forall x \in X(X_0, U)\}.\end{aligned}$$

The sufficiency analysis must be performed by studying $S \subset \mathbb{R}^c$ and $X(X_0, U) \subset \mathbb{R}^c$.

The constrained optimization problem is solved via the bounded admissible control law (5.6.12), and for open-loop unstable system, stability is guaranteed if $X \subseteq S$.

The theorem is formulated.

Theorem 5.6.1. *Given a nonlinear, open-loop, unstable system in (5.6.9), a non-quadratic performance index is given as (5.6.10) by mapping the control bounds imposed by the continuous integrable, one-to-one, bounded function $\phi \in U$. The first-order necessary condition for optimality gives a bounded admissible control law (5.6.12). The constrained optimization problem is solved via the bounded control law, and open-loop system (5.6.9) with (5.6.12), which evolves in $X(X_0, U)$, is stable if $X \subseteq S$. The maximal admissible domain of stability S is found by applying the sufficient conditions for stability. In particular, the first difference $\Delta V(x_n)$ of the positive-definite function $V(x_n)$ must be negative-definite for all $x_n \in X$ and $u_n \in U$.*

Proof. As was illustrated, the proof is straightforward using the Hamilton–Jacobi and Lyapunov theories.

Example 5.6.1. Nonlinear multivariable tracking control of an aircraft

Innovative design methods and algorithms are needed for advanced aircraft in response to requirements toward substantial performance improvements. The following state-space aircraft model in matrix form is used:

$$\dot{x}(t) = \begin{bmatrix} \dot{v}(t) \\ \dot{\alpha}(t) \\ \dot{q}(t) \\ \dot{\theta}(t) \\ \dot{\beta}(t) \\ \dot{p}(t) \\ \dot{r}(t) \\ \dot{\phi}(t) \\ \dot{\psi}(t) \end{bmatrix} = A \begin{bmatrix} v \\ \alpha \\ q \\ \theta \\ \beta \\ p \\ r \\ \phi \\ \psi \end{bmatrix}$$

$$+ \begin{bmatrix} 0 \\ -p \cos\alpha \tan\beta - r \sin\alpha \tan\beta \\ \frac{1}{I_Y}[(I_Z - I_X)pr - I_{XZ}p^2 + I_{XZ}r^2] \\ q \cos\phi - r \sin\phi \\ p \sin\alpha - r \cos\alpha \\ \frac{1}{I_X I_Z - I_{XZ}^2}[I_{XZ}(I_X - I_Y + I_Z)qp + (I_Y I_Z - I_{XZ}^2 - I_Z^2)qr] \\ \frac{1}{I_X I_Z - I_{XZ}^2}[(I_X^2 - I_X I_Y + I_{XZ}^2)qp - I_{XZ}(I_X - I_Y + I_Z)qr] \\ q \tan\theta \sin\phi + r \tan\theta \cos\phi \\ q \cos^{-1}\theta \sin\phi + r \cos^{-1}\theta \cos\phi \end{bmatrix}$$

$$+ \begin{bmatrix} F_v \\ F_\alpha \\ \frac{M_M}{I_Y} \\ 0 \\ F_\beta \\ \frac{I_{XZ}M_N + I_Z M_L}{I_X I_Z - I_{XZ}^2} \\ \frac{I_{XZ}M_L + I_X M_N}{I_X I_Z - I_{XZ}^2} \\ 0 \\ 0 \end{bmatrix},$$

where $A \in \mathbb{R}^{9 \times 9}$ is the matrix of coefficients and F_v, F_α, and F_β are the applied forces.

This state-space model in the form of differential equations can be discretized. In particular, for continuous-time nonlinear differential equations

$$\dot{x}(t) = F_c(x) + B_c(x)u,$$

the Euler's formula

$$\frac{x_{n+1} - x_n}{T_s} - (F_c(x) + B_c(x)u) = 0$$

is applied. Equally spaced sampling instants are used; that is, the sampling time T_s is fixed, and $t = nT_s$. Then, one finds

$$x_{n+1}^{air} = F(x_n^{air}) + B(x_n^{air})u_n.$$

The aircraft state variables are denoted as

$$x_n^{air} = \begin{bmatrix} v & \alpha & q & \theta & \beta & p & r & \phi & \psi \end{bmatrix}^T.$$

For the studied airframe configuration, six control inputs are

$$u_n = \begin{bmatrix} \delta_{HR} & \delta_{HL} & \delta_{FR} & \delta_{FL} & \delta_C & \delta_R \end{bmatrix}^T.$$

The following hard bounds (mechanical limits) on the deflections of control surfaces are imposed: $|\delta_{HR}, \delta_{HL}| \leq 0.44$ rad, $|\delta_{FR}, \delta_{FL}| \leq 0.35$ rad, $|\delta_C| \leq 0.47$ rad, and $|\delta_R| \leq 0.52$ rad.

The aircraft dynamics is modeled by the following difference equation in the matrix form:

$$x_{n+1}^{air} = F(x_n^{air}) + B_n u_n, \quad F(x_n^{air}) = A_n x_n^{air} + F_n(x_n^{air}),$$

where A_n, B_n, and $F(x_n^{\text{air}})$ for the sampling period 0.03 sec are

$$A_n = \begin{bmatrix} 0.9995 & 0.2457 & -0.0273 & -0.2885 & -0.0391 & -0.0075 & -0.002 & 0 & 0 \\ -0.0001 & 0.9663 & 0.0291 & 0 & 0.0011 & 0.0017 & 0.0003 & 0 & 0 \\ 0 & 0.1135 & 0.9765 & 0 & 0.0007 & 0.0002 & 0.0012 & 0 & 0 \\ 0 & 0.0017 & 0.0296 & 1 & 0 & 0 & 0 & 0 & 0 \\ -0.0001 & 0.0048 & 0.0011 & 0.0288 & 0.977 & 0.0037 & -0.0268 & 0 & 0 \\ 0.0001 & 0.0165 & 0.0005 & -0.0198 & -1.3515 & 0.8977 & 0.025 & 0 & 0 \\ 0 & -0.0268 & 0.0015 & 0.0041 & 0.2716 & -0.0003 & 0.9811 & 0 & 0 \\ 0 & 0.0003 & 0 & -0.0002 & -0.0207 & 0.0285 & 0.0003 & 1 & 0 \\ 0 & -0.0004 & 0 & 0 & 0.0041 & 0 & 0.0297 & 0 & 1 \end{bmatrix},$$

$$B_n = \begin{bmatrix} 0.071 & 0.0067 & -0.011 & -0.0117 & -0.0004 & -0.0001 \\ -0.009 & -0.0088 & -0.0091 & -0.009 & 0 & 0 \\ -0.2818 & -0.2819 & -0.0746 & -0.0746 & 0 & 0 \\ -0.0042 & -0.0042 & -0.0011 & -0.0011 & 0 & 0 \\ -0.001 & 0.0007 & -0.0006 & 0.0005 & 0.0123 & 0.0018 \\ -0.0822 & 0.0819 & -0.0881 & 0.0879 & 0.0122 & 0.025 \\ 0.0921 & -0.0924 & 0.0233 & -0.0232 & 0.02 & -0.0132 \\ -0.0013 & 0.013 & -0.013 & 0.013 & 0.0002 & 0.0004 \\ 0.0014 & -0.0014 & 0.0003 & -0.0003 & 0.0003 & -0.0002 \end{bmatrix},$$

$$F_n(x_n^{\text{air}}) = \begin{bmatrix} 0 \\ -0.03p\cos\alpha\tan\beta - 0.03r\sin\alpha\tan\beta \\ 0.028pr - 0.00018p^2 + 0.00018r^2 \\ 0.03q\cos\phi - 0.03r\sin\phi \\ 0.03p\sin\alpha - 0.03r\cos\alpha \\ 0.00026qp - 0.017qr \\ -0.025qp - 0.00026qr \\ 0.03q\tan\theta\sin\phi + 0.03r\tan\theta\cos\phi \\ 0.03q\cos^{-1}\theta\sin\phi + 0.03r\cos^{-1}\theta\cos\phi \end{bmatrix}.$$

The tracking problem has to be solved. The aircraft outputs are the Euler angles θ, ϕ, and ψ, and the output equation is

$$y_n = H_n x_n^{\text{air}} = \begin{bmatrix} 0 & 0 & 0 & 1 & 0 & 0 & 0 & 0 & 0 \\ 0 & 0 & 0 & 0 & 0 & 0 & 0 & 1 & 0 \\ 0 & 0 & 0 & 0 & 0 & 0 & 0 & 0 & 1 \end{bmatrix} \begin{bmatrix} v \\ \alpha \\ q \\ \theta \\ \beta \\ p \\ r \\ \phi \\ \psi \end{bmatrix}.$$

Using the reference inputs (the desired Euler's angles r_θ, r_ϕ, r_ψ are assigned by the pilot or mission computer)

$$r_n = \begin{bmatrix} r_\theta \\ r_\phi \\ r_\psi \end{bmatrix},$$

and the aircraft output vector

$$
y_n = \begin{bmatrix} \theta \\ \phi \\ \psi \end{bmatrix},
$$

an augmented aircraft model is found as

$$
x_{n+1} = \begin{bmatrix} x_{n+1}^{air} \\ x_{n+1}^{ref} \end{bmatrix} = \begin{bmatrix} F(x_n^{air}) \\ -H_n F(x_n^{air}) + I x_n^{ref} \end{bmatrix} + \begin{bmatrix} B_n \\ -H_n B_n \end{bmatrix} u_n
$$
$$
+ \begin{bmatrix} 0 \\ I \end{bmatrix} r_{n+1},
$$

where $I \in \mathbb{R}^{3 \times 3}$ is the identity matrix.

The bounded controller should be designed by minimizing the nonquadratic performance index. The control bounds imposed (the mechanical limits on the deflections of control surfaces) are mapped by the hyperbolic tangent, and the index to be minimized is

$$
J = \sum_{n=0}^{\infty} \left[x_n^T Q_n x_n + 2 \int (\tanh^{-1}(U_{\max}^{-1} u_n))^T G_n \, du_n \right.
$$
$$
\left. -u_n^T \begin{bmatrix} B_n \\ -H_n B_n \end{bmatrix}^T K_{n+1} \begin{bmatrix} B_n \\ -H_n B_n \end{bmatrix} u_n \right],
$$

where

$$
U_{\max} = \begin{bmatrix} \delta_{HR\,\max} & 0 & 0 & 0 & 0 & 0 \\ 0 & \delta_{HL\,\max} & 0 & 0 & 0 & 0 \\ 0 & 0 & \delta_{FR\,\max} & 0 & 0 & 0 \\ 0 & 0 & 0 & \delta_{FL\,\max} & 0 & 0 \\ 0 & 0 & 0 & 0 & \delta_{C\,\max} & 0 \\ 0 & 0 & 0 & 0 & 0 & \delta_{R\,\max} \end{bmatrix},
$$

$\delta_{HR\,\max} = \delta_{HL\,\max} = 0.44$ rad, $\delta_{FR\,\max} = \delta_{FL\,\max} = 0.35$ rad, $\delta_{C\,\max} = 0.47$ rad, and $\delta_{R\,\max} = 0.52$ rad.

Minimizing the Hamilton–Jacobi equation, one finds a bounded control law as given by

$$
u_n = -U_{\max} \tanh \left(G_n^{-1} \begin{bmatrix} B_n \\ -H_n B_n \end{bmatrix}^T K_n \begin{bmatrix} F(x_n^{air}) \\ -H_n F(x_n^{air}) + I x_n^{ref} \end{bmatrix} \right), \quad u_n \in U.
$$

Hence, a bounded controller was found by using the first-order necessary condition for optimality. Using identity matrices $I_{q1} \in \mathbb{R}^{9 \times 9}$, $I_{q2} \in \mathbb{R}^{3 \times 3}$, and $I_g \in \mathbb{R}^{6 \times 6}$, and assigning the weighting matrices to be $Q_n = \begin{bmatrix} I_{q1} & 0 \\ 0 & 10 I_{q2} \end{bmatrix}$

and $G_n = 20I_g$, equation

$$
x_n^T K_n x_n = x_n^T Q_n x_n + \left[\begin{array}{c} F(x_n^{\text{air}}) \\ -H_n F(x_n^{\text{air}}) + I x_n^{\text{ref}} \end{array} \right]^T K_n \left[\begin{array}{c} F(x_n^{\text{air}}) \\ -H_n F(x_n^{\text{air}}) + I x_n^{\text{ref}} \end{array} \right]
$$

$$
-2 \int \left(\tanh \left(G_n^{-1} \left[\begin{array}{c} B_n \\ -H_n B_n \end{array} \right]^T K_n \left[\begin{array}{c} F(x_n^{\text{air}}) \\ -H_n F(x_n^{\text{air}}) + I x_n^{\text{ref}} \end{array} \right] \right) \right)
$$

$$
\times d \left(\left[\begin{array}{c} B_n \\ -H_n B_n \end{array} \right]^T K_n \left[\begin{array}{c} F(x_n^{\text{air}}) \\ -H_n F(x_n^{\text{air}}) + I x_n^{\text{ref}} \end{array} \right] \right)
$$

was solved to find $K_n \in \mathbb{R}^{12 \times 12}$.

The nonquadratic index is positive-definite because

$$
x_n^T Q_n x_n + 2 \int (\tanh^{-1}(U_{\max}^{-1} u_n))^T G_n \, du_n > u_n^T \left[\begin{array}{c} B_n \\ -H_n B_n \end{array} \right]^T K_n \left[\begin{array}{c} B_n \\ -H_n B_n \end{array} \right] u_n.
$$

An invariant domain of stability S is found, and the sufficient criteria for stability are guaranteed in the full-fighter operating envelope. That is, $X \subseteq S$ for all $x_n \in X$.

The aircraft dynamics is studied, and a bounded controller is verified via nonlinear simulations. Figure 5.6.1 illustrates the fighter output dynamics when the reference inputs are assigned to be

$$
r_n = \left[\begin{array}{c} r_\theta \\ r_\phi \\ r_\psi \end{array} \right] = \left\{ \begin{array}{l} 0.5, n \geq 0 \\ 1, n \geq 0 \\ 0, n \geq 0. \end{array} \right.
$$

The initial conditions for the Euler angles are

$$
\left[\begin{array}{c} \theta_0 \\ \phi_0 \\ \psi_0 \end{array} \right] = \left\{ \begin{array}{l} 0 \text{ rad} \\ 0 \text{ rad} \\ 1.5 \text{ rad}. \end{array} \right.
$$

5.7. Nonlinear Control of Permanent-Magnet Synchronous Motors

A broad spectrum of electric machines is widely used in electromechanical systems. In addition to the required steady-state and dynamic characteristics, frequently, particular features, such as power, torque densities, efficiency, controllability, versatility, flexibility, absence of commutators, brushes, simplicity, ruggedness, reliability, cost, weight-to-torque ratio, and starting capabilities, are essential and commonly considered. The final decision involves a compromise between the variety of available options, and permanent-magnet synchronous motors are used in a wide range of electromechanical systems. Permanent-magnet synchronous

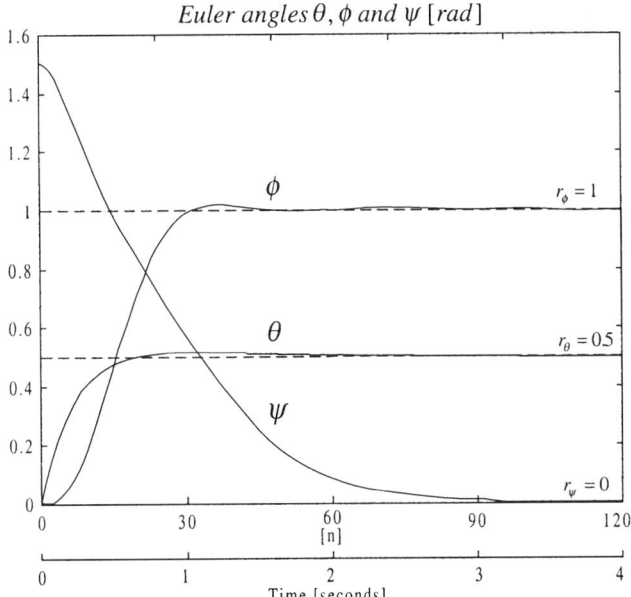

FIGURE 5.6.1.   Aircraft output dynamics for the reference inputs $r_\theta = 0.5, r_\phi = 1$ and $r_\psi = 0$.

motors are simple electromechanical devices, and they can be easily controlled. The steady-state torque-speed characteristics fulfill the *controllability* criteria over an entire envelope of operation. In contrast, induction and variable reluctance motors require sophisticated control algorithms to achieve suitable steady-state characteristics. Permanent-magnet synchronous motors are used when severe starting conditions and variable loads (disturbances) occur, and these motors have superior advantages from control capabilities in variable speed drives and in high-performance servo-systems. Commercially available permanent-magnet synchronous motors exceed the functional characteristics of brush-type DC and hydraulic motors, and therefore, these machines are used in various applications previously exclusively associated with DC and hydraulic motors (sparking and leaks reinforce the competitive position and restrict the specific areas of applicability). This strong competitive position of permanent-magnet synchronous motors demands further comprehensive studies from nonlinear analysis, modeling, and control perspectives. Innovative analytical, numerical, and experimental results are under development, and new methods are extremely urgent to improve the existing and design advanced electromechanical systems. Three primary issues are studied in this section. In particular, we perform nonlinear modeling and analysis, controllers design, and validate the theoretical results.

5.7.1. Nonlinear Motor Dynamics

A mathematical model of three-phase, two-pole permanent-magnet synchronous motors should be developed. Three-phase, two-pole permanent-magnet synchronous motor is illustrated in Figure 5.7.1.

For the magnetically coupled abc stator windings, we apply the Kirchhoff voltage law to find a set of the following differential equations:

$$u_{as} = r_s i_{as} + \frac{d\psi_{as}}{dt}, \quad u_{bs} = r_s i_{bs} + \frac{d\psi_{bs}}{dt}, \quad u_{cs} = r_s i_{cs} + \frac{d\psi_{cs}}{dt},$$

where u_{as}, u_{bs}, and u_{cs} are the applied phase voltages and i_{as}, i_{bs}, i_{cs} and ψ_{as}, ψ_{bs}, ψ_{cs} are the currents and flux linkages in the abc windings.

That is, we have

$$\frac{d\psi_{as}}{dt} = -r_s i_{as} + u_{as}, \quad \frac{d\psi_{bs}}{dt} = -r_s i_{bs} + u_{bs}, \quad \frac{d\psi_{cs}}{dt} = -r_s i_{cs} + u_{cs},$$

where the flux linkages are

$$\begin{bmatrix} \psi_{as} \\ \psi_{bs} \\ \psi_{cs} \end{bmatrix} = \begin{bmatrix} L_{1s} + \bar{L}_m & -\frac{1}{2}\bar{L}_m & -\frac{1}{2}\bar{L}_m \\ -\frac{1}{2}\bar{L}_m & L_{1s} + \bar{L}_m & -\frac{1}{2}\bar{L}_m \\ -\frac{1}{2}\bar{L}_m & -\frac{1}{2}\bar{L}_m & L_{1s} + \bar{L}_m \end{bmatrix} \begin{bmatrix} i_{as} \\ i_{bs} \\ i_{cs} \end{bmatrix} + \psi_m \begin{bmatrix} \sin\theta_r \\ \sin(\theta_r - \frac{2}{3}\pi) \\ \sin(\theta_r + \frac{2}{3}\pi) \end{bmatrix},$$

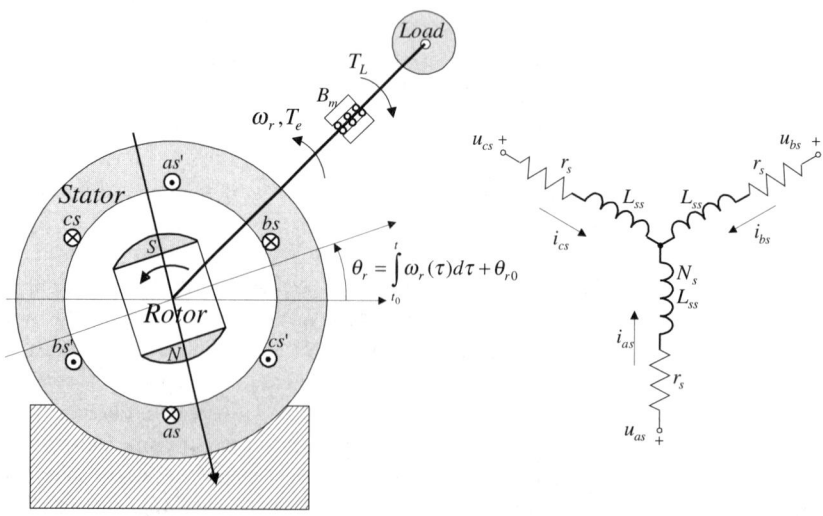

FIGURE 5.7.1.　Two-pole permanent-magnet synchronous motor.

r_s is the stator resistance, L_{1s} and $\bar{L}_m$ are the leakage and magnetizing inductances $L_{ss} = L_{1s} + \frac{3}{2}\bar{L}_m$, and ψ_m is the amplitude of the flux linkages established by the permanent magnet.

The torsional–mechanical dynamics is modeled using the following differential equations:

$$
\begin{aligned}
\frac{d\omega_r}{dt} &= \frac{P}{2J}(T_e - T_f - T_L) \\
&= \frac{P}{2J}\left[\frac{P\psi_m}{2}\left(i_{as}\cos\theta_r + i_{bs}\cos\left(\theta_r - \tfrac{2}{3}\pi\right)\right. \right. \\
&\qquad \left.\left. + i_{cs}\cos\left(\theta_r + \tfrac{2}{3}\pi\right)\right) - \frac{2B_m}{P}\omega_r - T_L\right],
\end{aligned}
$$

$$
\frac{d\theta_r}{dt} = \omega_r,
$$

where ω_r and θ_r are the electrical angular velocity and displacement; T_e, T_f, and T_L are the electromagnetic, friction, and load torques,

$$
T_e = \frac{P\psi_m}{2}\left[i_{as}\cos\theta_r + i_{bs}\cos\left(\theta_r - \tfrac{2}{3}\pi\right) + i_{cs}\cos\left(\theta_r + \tfrac{2}{3}\pi\right)\right]
$$

and $T_f = \frac{2}{P}B_m\omega_r$; B_m is the viscous friction coefficient; J is the equivalent moment of inertia; P is the number of poles; and the rotor angular velocity and displacement are $\omega_{rm} = \frac{2}{P}\omega_r$ and $\theta_{rm} = \frac{2}{P}\theta_r$.

Using differential equations for magnetically coupled stator circuitry and torsional–mechanical dynamics, one obtains the following nonlinear mathematical model of permanent-magnet synchronous motors as given by five first-order differential equations in Cauchy's form:

$$
\begin{bmatrix} \frac{di_{as}}{dt} \\ \frac{di_{bs}}{dt} \\ \frac{di_{cs}}{dt} \\ \frac{d\omega_r}{dt} \\ \frac{d\theta_r}{dt} \end{bmatrix} =
\begin{bmatrix}
-\frac{r_s(2L_{ss}-\bar{L}_m)}{2L_{ss}^2-L_{ss}\bar{L}_m-\bar{L}_m^2} & -\frac{r_s\bar{L}_m}{2L_{ss}^2-L_{ss}\bar{L}_m-\bar{L}_m^2} & -\frac{r_s\bar{L}_m}{2L_{ss}^2-L_{ss}\bar{L}_m-\bar{L}_m^2} & 0 & 0 \\
-\frac{r_s\bar{L}_m}{2L_{ss}^2-L_{ss}\bar{L}_m-\bar{L}_m^2} & -\frac{r_s(2L_{ss}-\bar{L}_m)}{2L_{ss}^2-L_{ss}\bar{L}_m-\bar{L}_m^2} & -\frac{r_s\bar{L}_m}{2L_{ss}^2-L_{ss}\bar{L}_m-\bar{L}_m^2} & 0 & 0 \\
-\frac{r_s\bar{L}_m}{2L_{ss}^2-L_{ss}\bar{L}_m-\bar{L}_m^2} & -\frac{r_s\bar{L}_m}{2L_{ss}^2-L_{ss}\bar{L}_m-\bar{L}_m^2} & -\frac{r_s(2L_{ss}-\bar{L}_m)}{2L_{ss}^2-L_{ss}\bar{L}_m-\bar{L}_m^2} & 0 & 0 \\
0 & 0 & 0 & -\frac{B_m}{J} & 0 \\
0 & 0 & 0 & 1 & 0
\end{bmatrix}
$$

$$
\times \begin{bmatrix} i_{as} \\ i_{bs} \\ i_{cs} \\ \omega_r \\ \theta_r \end{bmatrix}
$$

$$
+ \begin{bmatrix}
-\frac{\psi_m(2L_{ss}-\bar{L}_m)}{2L_{ss}^2-L_{ss}\bar{L}_m-\bar{L}_m^2}\omega_r & -\frac{\psi_m\bar{L}_m}{2L_{ss}^2-L_{ss}\bar{L}_m-\bar{L}_m^2}\omega_r & -\frac{\psi_m\bar{L}_m}{2L_{ss}^2-L_{ss}\bar{L}_m-\bar{L}_m^2}\omega_r \\
-\frac{\psi_m\bar{L}_m}{2L_{ss}^2-L_{ss}\bar{L}_m-\bar{L}_m^2}\omega_r & -\frac{\psi_m(2L_{ss}-\bar{L}_m)}{2L_{ss}^2-L_{ss}\bar{L}_m-\bar{L}_m^2}\omega_r & -\frac{\psi_m\bar{L}_m}{2L_{ss}^2-L_{ss}\bar{L}_m-\bar{L}_m^2}\omega_r \\
-\frac{\psi_m\bar{L}_m}{2L_{ss}^2-L_{ss}\bar{L}_m-\bar{L}_m^2}\omega_r & -\frac{\psi_m\bar{L}_m}{2L_{ss}^2-L_{ss}\bar{L}_m-\bar{L}_m^2}\omega_r & -\frac{\psi_m(2L_{ss}-\bar{L}_m)}{2L_{ss}^2-L_{ss}\bar{L}_m-\bar{L}_m^2}\omega_r \\
\frac{P^2\psi_m}{4J}i_{as} & \frac{P^2\psi_m}{4J}i_{bs} & \frac{P^2\psi_m}{4J}i_{cs} \\
0 & 0 & 0
\end{bmatrix}
$$

$$
\times \begin{bmatrix} \cos\theta_r \\ \cos(\theta_r - \frac{2}{3}\pi) \\ \cos(\theta_r + \frac{2}{3}\pi) \end{bmatrix}
$$

$$
+ \begin{bmatrix} \frac{2L_{ss}-\bar{L}_m}{2L_{ss}^2-L_{ss}\bar{L}_m-\bar{L}_m^2} & \frac{\bar{L}_m}{2L_{ss}^2-L_{ss}\bar{L}_m-\bar{L}_m^2} & \frac{\bar{L}_m}{2L_{ss}^2-L_{ss}\bar{L}_m-\bar{L}_m^2} \\ \frac{\bar{L}_m}{2L_{ss}^2-L_{ss}\bar{L}_m-\bar{L}_m^2} & \frac{2L_{ss}-\bar{L}_m}{2L_{ss}^2-L_{ss}\bar{L}_m-\bar{L}_m^2} & \frac{\bar{L}_m}{2L_{ss}^2-L_{ss}\bar{L}_m-\bar{L}_m^2} \\ \frac{\bar{L}_m}{2L_{ss}^2-L_{ss}\bar{L}_m-\bar{L}_m^2} & \frac{\bar{L}_m}{2L_{ss}^2-L_{ss}\bar{L}_m-\bar{L}_m^2} & \frac{2L_{ss}-\bar{L}_m}{2L_{ss}^2-L_{ss}\bar{L}_m-\bar{L}_m^2} \\ 0 & 0 & 0 \\ 0 & 0 & 0 \end{bmatrix}
$$

$$
\times \begin{bmatrix} u_{as} \\ u_{bs} \\ u_{cs} \end{bmatrix} - \begin{bmatrix} 0 \\ 0 \\ 0 \\ \frac{P}{2J} \\ 0 \end{bmatrix} T_L. \tag{5.7.1}
$$

The Park transformation serves as a viable tool in analysis of synchronous machines. In particular, the arbitrary reference frame is applied to derive a mathematical model using the quadrature-, direct-, and zero-quantities. Fixing the reference frame with the rotor, the direct Park transformation is

$$
\begin{bmatrix} i_{qs}^r \\ i_{ds}^r \\ i_{os}^r \end{bmatrix} = \frac{2}{3} \begin{bmatrix} \cos\theta_r & \cos(\theta_r - \frac{2}{3}\pi) & \cos(\theta_r + \frac{2}{3}\pi) \\ \sin\theta_r & \sin(\theta_r - \frac{2}{3}\pi) & \sin(\theta_r + \frac{2}{3}\pi) \\ \frac{1}{2} & \frac{1}{2} & \frac{1}{2} \end{bmatrix} \begin{bmatrix} i_{as} \\ i_{bs} \\ i_{cs} \end{bmatrix},
$$

$$
\begin{bmatrix} u_{qs}^r \\ u_{ds}^r \\ u_{os}^r \end{bmatrix} = \frac{2}{3} \begin{bmatrix} \cos\theta_r & \cos(\theta_r - \frac{2}{3}\pi) & \cos(\theta_r + \frac{2}{3}\pi) \\ \sin\theta_r & \sin(\theta_r - \frac{2}{3}\pi) & \sin(\theta_r + \frac{2}{3}\pi) \\ \frac{1}{2} & \frac{1}{2} & \frac{1}{2} \end{bmatrix} \begin{bmatrix} u_{as} \\ u_{bs} \\ u_{cs} \end{bmatrix}.
$$

Applying the Park transformation, we have the following expression for the electromagnetic torque:

$$
\begin{aligned}
T_e &= \frac{P\psi_m}{2} \left[i_{as}\cos\theta_r + i_{bs}\cos\left(\theta_r - \frac{2}{3}\pi\right) + i_{cs}\cos\left(\theta_r + \frac{2}{3}\pi\right) \right] \\
&= \frac{3P\psi_m}{4} i_{qs}^r.
\end{aligned}
$$

Using (5.7.1) and the Park transformation, one obtains the following differential equations to model permanent-magnet synchronous motors in the rotor reference frame:

$$
\begin{aligned}
\frac{di_{qs}^r}{dt} &= -\frac{r_s}{L_{1s} + \frac{3}{2}\bar{L}_m} i_{qs}^r - \frac{\psi_m}{L_{1s} + \frac{3}{2}\bar{L}_m}\omega_r - i_{ds}^r\omega_r \\
&\quad + \frac{1}{L_{1s} + \frac{3}{2}\bar{L}_m} u_{qs}^r,
\end{aligned}
$$

$$
\frac{di_{ds}^r}{dt} = -\frac{r_s}{L_{1s} + \frac{3}{2}\bar{L}_m} i_{ds}^r + i_{qs}^r\omega_r + \frac{1}{L_{1s} + \frac{3}{2}\bar{L}_m} u_{ds}^r,
$$

$$\frac{di_{0s}^r}{dt} = -\frac{r_s}{L_{1s}}i_{0s}^r + \frac{1}{L_{1s}}u_{0s}^r,$$

$$\frac{d\omega_r}{dt} = \frac{3P^2\psi_m}{8J}i_{qs}^r - \frac{B_m}{J}\omega_r - \frac{P}{2J}T_L,$$

where i_{qs}^r, i_{ds}^r, i_{0s}^r and u_{qs}^r, u_{ds}^r, u_{0s}^r are the quadrature-, direct-, and zero-axis current and voltage components.

The analysis of permanent-magnet synchronous motors in the arbitrary reference frame using the quadrature-, direct-, and zero-quantities is simple. The electromagnetic torque is a function of the quadrature current i_{qs}^r, and the differential equation for the zero current i_{0s}^r can be omitted from the analysis. We have

$$\frac{di_{qs}^r}{dt} = -\frac{r_s}{L_{1s} + \frac{3}{2}\bar{L}_m}i_{qs}^r - \frac{\psi_m}{L_{1s} + \frac{3}{2}\bar{L}_m}\omega_r - i_{ds}^r\omega_r$$

$$+ \frac{1}{L_{1s} + \frac{3}{2}\bar{L}_m}u_{qs}^r,$$

$$\frac{di_{ds}^r}{dt} = -\frac{r_s}{L_{1s} + \frac{3}{2}\bar{L}_m}i_{ds}^r + i_{qs}^r\omega_r + \frac{1}{L_{1s} + \frac{3}{2}\bar{L}_m}u_{ds}^r,$$

$$\frac{d\omega_r}{dt} = \frac{3P^2\psi_m}{8J}i_{qs}^r - \frac{B_m}{J}\omega_r - \frac{P}{2J}T_L. \qquad (5.7.2)$$

Let us study the stability of the open-loop system (5.7.2). For $u_{qs}^r = u_{ds}^r = 0$ and $T_L = 0$, using a positive-definite quadratic function

$$V(i_{qs}^r, i_{ds}^r, \omega_r) = \tfrac{1}{2}(i_{qs}^{r2} + i_{ds}^{r2} + \omega_r^2),$$

one obtains

$$\frac{dV(i_{qs}^r, i_{ds}^r, \omega_r)}{dt} = -\frac{r_s}{L_{ss}}\left(i_{qs}^{r2} + i_{ds}^{r2}\right) - \frac{B_m}{J}\omega_r^2$$

$$- \frac{\psi_m(8J - 3P^2L_{ss})}{8JL_{ss}}i_{qs}^r\omega_r.$$

That is, the total derivative of a positive-definite quadratic function $V(i_{qs}^r, i_{ds}^r, \omega_r)$ is negative. Hence, an open-loop system is uniformly asymptotically stable. It should be emphasized that it is possible to prove that the equilibrium state is exponentially stable.

5.7.2. Feedback Linearization and Control

As a first step toward the design, we mathematically set up the design problem. It is easy to verify that the *linearizability* condition is guaranteed.

Let

$$u_{qs}^r = u_{qs}^{r\,\text{lin}} + u_{qs}^{r\,\text{cont}} \text{ and } u_{ds}^r = u_{ds}^{r\,\text{lin}} + u_{ds}^{r\,\text{cont}}. \qquad (5.7.3)$$

Using the following linearizing feedback:

$$u_{qs}^{r\text{lin}} = \left(L_{1s} + \tfrac{3}{2}\bar{L}_m\right) i_{ds}^r \omega_r \text{ and } u_{ds}^{r\text{lin}} = -\left(L_{1s} + \tfrac{3}{2}\bar{L}_m\right) i_{qs}^r \omega_r, \qquad (5.7.4)$$

model (5.7.2) is linearized to

$$\frac{di_{qs}^r}{dt} = -\frac{r_s}{L_{1s} + \tfrac{3}{2}\bar{L}_m} i_{qs}^r - \frac{\psi_m}{L_{1s} + \tfrac{3}{2}\bar{L}_m} \omega_r$$
$$+ \frac{1}{L_{1s} + \tfrac{3}{2}\bar{L}_m} u_{qs}^r,$$

$$\frac{di_{ds}^r}{dt} = -\frac{r_s}{L_{1s} + \tfrac{3}{2}\bar{L}_m} i_{ds}^r + \frac{1}{L_{1s} + \tfrac{3}{2}\bar{L}_m} u_{ds}^r,$$

$$\frac{d\omega_r}{dt} = \frac{3P^2\psi_m}{8J} i_{qs}^r - \frac{B_m}{J}\omega_r - \frac{P}{2J}T_L. \qquad (5.7.5)$$

For (5.7.5), if $u_{qs}^{r\text{cont}} = 0$ and $u_{ds}^{r\text{cont}} = 0$, the eigenvalues are

$$\lambda_1 = -\frac{\left(2B_m L_{ss} + 2r_s J - \sqrt{4B_m^2 L_{ss}^2 - 8B_m L_{ss} r_s J + 4r_s^2 J^2 - 6L_{ss} J^3 \psi_m^2 P^2}\right)}{4L_{ss} J},$$

$$\lambda_2 = -\frac{\left(2B_m L_{ss} + 2r_s J + \sqrt{4B_m^2 L_{ss}^2 - 8B_m L_{ss} r_s J + 4r_s^2 J^2 - 6L_{ss} J^3 \psi_m^2 P^2}\right)}{4L_{ss} J},$$

and

$$\lambda_3 = -\frac{r_s}{L_{ss}}.$$

The real parts of these eigenvalues are negative. Thus, the stability is guaranteed. For the resulting linear system (5.7.5), one can design a nonlinear controller as

$$u_{qs}^{r\text{cont}} = -\begin{bmatrix} k_{iq} & k_{id} & k_\omega \end{bmatrix} \begin{bmatrix} i_{qs}^r \\ i_{ds}^r \\ \omega_r \end{bmatrix} \text{ and } u_{ds}^{r\text{cont}} = 0. \qquad (5.7.6)$$

The following eigenvalues for the closed-loop system (5.7.5)–(5.7.6), or (5.7.2)–(5.7.4)–(5.7.6), are found

$$\lambda_1 = -\frac{2k_{iq} J + 2B_m L_{ss} + 2r_s J - \sqrt{\begin{array}{c}4k_{iq}^2 J^2 - 8k_{iq} J B_m L_{ss} + 8k_{iq} J^2 r_s + 4B_m^2 L_{ss}^2 \\ -8B_m L_{ss} r_s J + 4r_s^2 J^2 - 6L_{ss} J^3 \psi_m P^2(\psi_m + k_\omega)\end{array}}}{4L_{ss} J},$$

$$\lambda_2 = -\frac{2k_{iq} J + 2B_m L_{ss} + 2r_s J + \sqrt{\begin{array}{c}4k_{iq}^2 J^2 - 8k_{iq} J B_m L_{ss} + 8k_{iq} J^2 r_s + 4B_m^2 L_{ss}^2 \\ -8B_m L_{ss} r_s J + 4r_s^2 J^2 - 6L_{ss} J^3 \psi_m P^2(\psi_m + k_\omega)\end{array}}}{4L_{ss} J},$$

$$\lambda_3 = -\frac{r_s}{L_{ss}}.$$

Hence, applying the linearizing nonlinear feedback and controller (5.7.3) with (5.7.4) and (5.7.6), one can tentatively conclude that stability, specified transient quantities and dynamic performance are attained.

Remark. In pole-placement design, the specification of optimum (desired) transient responses in terms of system models and feedback coefficients is equivalent to the specification imposed on desired transfer functions of closed-loop systems. Clearly, the desired eigenvalues can be specified by the designer, and these eigenvalues are used to find the corresponding feedback gains. However, the pole-placement concept, while guaranteeing the desired location of the characteristic eigenvalues can lead to positive feedback coefficients and control constraints. Hence, the stability, robustness to parameter variations, and system performance are significantly degraded.

Mathematically, feedback linearization reduces the complexity of the corresponding analysis and design. However, even from mathematical standpoints, the simplification and "*optimum*" performance would be achieved in expense of large control efforts required because of linearizing feedback (5.7.4). This leads to saturation. It must be emphasized that the need to linearize (5.7.2) does not exist because the open-loop system is uniformly asymptotically stable.

The most critical problem is that the linearizing feedback

$$u_{ds}^{r\,lin} = -\left(L_{1s} + \tfrac{3}{2}\bar{L}_m\right) i_{qs}^r \omega_r$$

cannot be implemented.

Recall that to guarantee the balanced operation, one must supply

$$u_{ds}^r = 0 \text{ and } u_{0s}^r = 0.$$

That is, nonlinear linearizing feedback $u_{ds}^{lin} = -\left(L_{1s} + \tfrac{3}{2}\bar{L}_m\right) i_{qs}^r \omega_r$ is not realizable.

Hence, the feedback linearizing controllers cannot be implemented to control synchronous machines. It is desirable, therefore, to develop other methods to solve the motion control problem, methods that do not entail the applied voltages to the saturation limits to cancel beneficial nonlinearities $-i_{ds}^r \omega_r$ and $i_{qs}^r \omega_r$, and methods that do not lead to unbalanced motor operation.

5.7.3. *The Lyapunov-Based Approach*

In this section, the design is approached using a nonlinear model. Using (5.7.2), we have the following matrix form:

$$\dot{x}(t) = \begin{bmatrix} -\dfrac{r_s}{L_{1s}+\frac{3}{2}\bar{L}_m} & 0 & -\dfrac{\psi_m}{L_{1s}+\frac{3}{2}\bar{L}_m} \\[2ex] 0 & -\dfrac{r_s}{L_{1s}+\frac{3}{2}\bar{L}_m} & 0 \\[2ex] \dfrac{3P^2\psi_m}{8J} & 0 & -\dfrac{B_m}{J} \end{bmatrix} \begin{bmatrix} i_{qs}^r \\[1ex] i_{ds}^r \\[1ex] \omega_r \end{bmatrix}$$

$$+ \begin{bmatrix} -i_{ds}^r \omega_r \\ i_{qs}^r \omega_r \\ 0 \end{bmatrix} + \begin{bmatrix} \dfrac{1}{L_{1s}+\frac{3}{2}\bar{L}_m} & 0 \\ 0 & \dfrac{1}{L_{1s}+\frac{3}{2}\bar{L}_m} \\ 0 & 0 \end{bmatrix} \begin{bmatrix} u_{qs}^r \\ u_{ds}^r \end{bmatrix}$$

$$- \begin{bmatrix} 0 \\ 0 \\ \dfrac{P}{2J} \end{bmatrix} T_L$$

To attain the balanced operation, one feeds

$$u_{qs}^r \neq 0, \ u_{ds}^r = 0 \ \text{and} \ u_{0s}^r = 0.$$

We choose the control law candidate as

$$u_{qs}^r = -k_\omega \omega_r, \ u_{ds}^r = 0 \ \text{and} \ u_{0s}^r = 0. \tag{5.7.7}$$

With this control algorithm, the closed-loop dynamics is

$$\dot{x}(t) = \begin{bmatrix} -\dfrac{r_s}{L_{1s}+\frac{3}{2}\bar{L}_m} & 0 & -\dfrac{\psi_m+k_\omega}{L_{1s}+\frac{3}{2}\bar{L}_m} \\ 0 & -\dfrac{r_s}{L_{1s}+\frac{3}{2}\bar{L}_m} & 0 \\ \dfrac{3P^2\psi_m}{8J} & 0 & -\dfrac{B_m}{J} \end{bmatrix} \begin{bmatrix} i_{qs}^r \\ i_{ds}^r \\ \omega_r \end{bmatrix}$$

$$+ \begin{bmatrix} -i_{ds}^r \omega_r \\ i_{qs}^r \omega_r \\ 0 \end{bmatrix} - \begin{bmatrix} 0 \\ 0 \\ \dfrac{P}{2J} \end{bmatrix} T_L.$$

Choosing the quadratic positive-definite Lyapunov function

$$V(i_{qs}^r, i_{ds}^r, \omega_r) = \tfrac{1}{2}(i_{qs}^{r2} + i_{ds}^{r2} + \omega_r^2),$$

we have

$$\frac{dV(i_{qs}^r, i_{ds}^r, \omega_r)}{dt} = -\frac{r_s}{L_{ss}}\left(i_{qs}^{r2}+i_{ds}^{r2}\right) - \frac{B_m}{J}\omega_r^2 - \frac{8J(\psi_m+k_\omega)-3P^2L_{ss}\psi_m}{8JL_{ss}} i_{qs}^r \omega_r.$$

Hence,

$$\frac{dV(i_{qs}^r, i_{ds}^r, \omega_r)}{dt} < 0.$$

One concludes that the control law (5.7.7) guarantees uniform asymptotic stability.

As an extension of the introduced stabilization problem, let us approach and solve the tracking problem. Our goal is to design a controller to ensure stability and tracking. For electric drives, the measured error vector is

$$e(t) = \omega_{\text{reference}}(t) - \omega_{rm}(t).$$

Our goal is to design controllers that stabilize electromechanical systems and drive the output error $e(t)$ to zero. The following proportional-integral and PID bounded control algorithms are introduced:

$$u_{qs}^r = \text{sat}_{u_{\min}}^{u_{\max}} \left(k_p e + k_i \int e \, dt \right), \quad u_{ds}^r = 0, u_{0s}^r = 0,$$

and

$$u_{qs}^r = \text{sat}_{u_{\min}}^{u_{\max}} \left(k_p e + k_i \int e \, dt + k_d \frac{de}{dt} \right), \quad u_{ds}^r = 0, u_{0s}^r = 0.$$

The feedback coefficients k_p, k_i, and k_d can be found by solving nonlinear matrix inequalities. Applying the Lyapunov stability theory and generalizing the results above, the stability of the resulting closed-loop system can be examined studying the criteria imposed on the Lyapunov function. For the bounded reference signal, using the positive-definite quadratic function

$$V(i_{qs}^r, i_{ds}^r, \omega_r, e) = \tfrac{1}{2} \left(i_{qs}^{r2} + i_{ds}^{r2} + \omega_r^2 + e^2 \right) > 0,$$

we have

$$\frac{dV(i_{qs}^r, i_{ds}^r, \omega_r, e)}{dt} < 0.$$

The given tracking controllers extend the applicability of the stabilizing algorithms, and allows one to solve the motion control problem for electromechanical systems driven by permanent-magnet synchronous motors. Using the *inverse* Park transformation, one derives the control laws in the *machine* (*abc*) variables. In particular, the bounded PID controller is given as

$$u_{as} = \text{sat}_{u_{\min}}^{u_{\max}} \left(k_p e + k_i \int e \, dt + k_d \frac{de}{dt} \right) \cos \theta_r,$$

$$u_{bs} = \text{sat}_{u_{\min}}^{u_{\max}} \left(k_p e + k_i \int e \, dt + k_d \frac{de}{dt} \right) \cos(\theta_r - \tfrac{2}{3}\pi),$$

$$u_{cs} = \text{sat}_{u_{\min}}^{u_{\max}} \left(k_p e + k_i \int e \, dt + k_d \frac{de}{dt} \right) \cos(\theta_r + \tfrac{2}{3}\pi). \qquad (5.7.8)$$

5.7.4. *Analytical and Numerical Results*

In this subsection, we design a tracking controller for a electromechanical system. We use a Kollmorgen four-pole permanent-magnet synchronous motor H-232 with the following rated data and parameters: 135 W, 434 rad/sec, 40 V, 0.42 N-m, 6.9 A, $r_s = 0.5\Omega$, $L_{ss} = 0.001$ H, $L_{1s} = 0.0001$ H, $\bar{L}_m = 0.0006$ H, $\psi_m = 0.069$ V-sec/rad or $\psi_m = 0.069$ N-m/A, $B_m = 0.0000115$ N-m-sec/rad, and $J = 0.000017$ kg-m^2.

The output equation is the mechanical angular velocity $\omega_{rm}(t)$, and the error is expressed as

$$e(t) = \omega_{\text{reference}}(t) - \omega_{rm}(t).$$

The maximum rated voltage is constrained as $\sqrt{240}$ V. This bound imposed can significantly deteriorate the stability, tracking accuracy, dynamic performance, and disturbance attenuation. Therefore, the control constraint must be incorporated in the analysis and design. The state-space permanent-magnet motor model is described by (5.7.2). Solving nonlinear inequality

$$\frac{dV(i_{qs}^r, i_{ds}^r, \omega_r, e)}{dt} < e^2,$$

one finds the feedback coefficient of a constrained control law. We have

$$u_{as} = \text{sat}_{-\sqrt{240}}^{+\sqrt{240}} \left(9e + 165 \int e\, dt + 0.0035 \frac{de}{dt} \right) \cos \theta_r,$$

$$u_{bs} = \text{sat}_{-\sqrt{240}}^{+\sqrt{240}} \left(9e + 165 \int e\, dt + 0.0035 \frac{de}{dt} \right) \cos \left(\theta_r - \tfrac{2}{3}\pi \right),$$

$$u_{cs} = \text{sat}_{-\sqrt{240}}^{+\sqrt{240}} \left(9e + 165 \int e\, dt + 0.0035 \frac{de}{dt} \right) \cos \left(\theta_r + \tfrac{2}{3}\pi \right).$$

This controller is bounded. The sufficient criteria for stability are satisfied. To study the transient behavior, a controller is verified through comprehensive simulations. Different reference velocity, loads, and initial conditions were studied to analyze the tracking performance of the resulting system. Figures 5.7.2 and 5.7.3 illustrate the dynamics of the closed-loop drive for the following reference speed and load torque:

$$\omega_{\text{reference}}(t) = 300 \tfrac{\text{rad}}{\text{sec}}, \quad t \in [0 \quad 0.02] \text{ sec}$$

and

$$T_L = \begin{cases} 0 \text{ N-m}, t \in [0 \ 0.01] \text{ sec} \\ 0.5 \text{ N-m}, t \in (0.01 \ 0.02] \text{ sec}. \end{cases}$$

The applied phase voltages and the resulting phase currents in the as, bs, and cs windings are illustrated in Figure 5.7.2. Figure 5.7.3 documents the motor mechanical angular velocity. The settling time for the motor angular velocity as motor starts from stall is 0.0025 sec. The disturbance attenuation features are evident. In particular, the assigned angular velocity with zero steady-state error has been guaranteed when the rated load torque was applied.

5.8.   Case Study in Nonlinear Control of Multivariable Systems: Motion Control of Induction Motors

Mathematical models of dynamic systems are used to design controllers. Mathematical models of induction motors are developed in this section using the abc (*machine*) and qd0 (quadrature, direct, and zero) variables. That is, conventional and arbitrary reference frames are used, and differential equations in Cauchy's

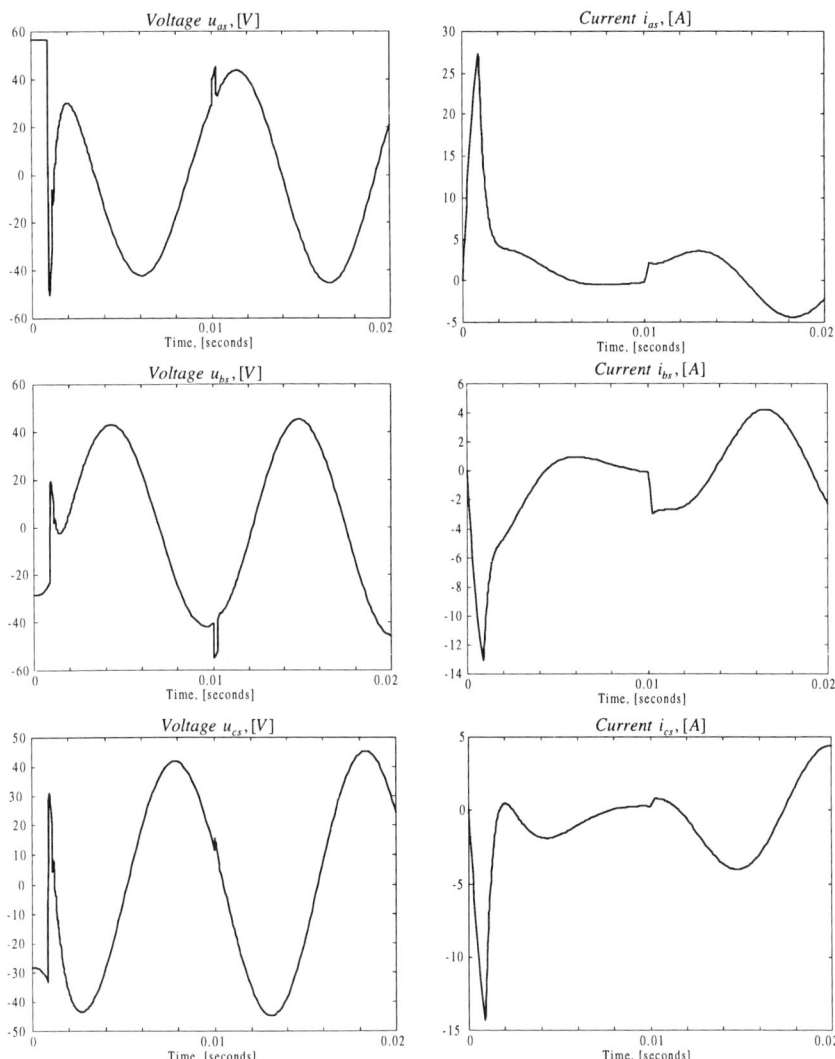

FIGURE 5.7.2.   Phase voltages and currents in the *as*, *bs*, and *cs* stator windings.

form are found to model the transient dynamics of induction motors. To control
induction motors, one regulates the magnitude and frequency of the phase voltages
applied or currents fed to the motor windings. It should be emphasized that if the
field orientation (*vector* control) is used, one finds the quadrature- and direct-axis
components of the voltages to be supplied or currents to be fed, and the *qd* variables (voltages or currents) must be transformed to the *abc* variables using the
Park transformations in real time. Hence, field-oriented control of induction motors is achieved by changing the magnitude and frequency of the *abc* voltages or

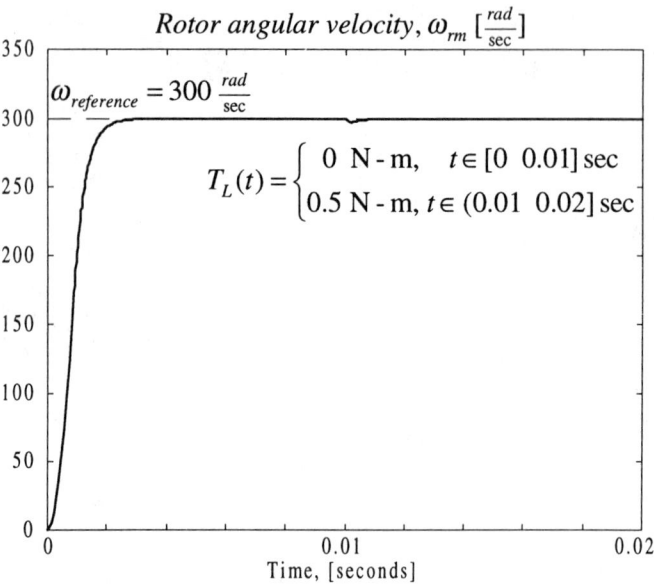

FIGURE 5.7.3.   Transient behavior of the motor angular velocity.

currents. The maximum air-gap flux density and maximum electromagnetic torque can be maintained if the specified quadrature- and direct-axis components of the stator currents or the voltage-frequency ratio are ensured, and the torque-speed characteristic curves are shaped. Induction motors can be operated at the rated value of the stator flux linkages by using the *constant volts per hertz* control, and optimum performance is attained because the maximum electromagnetic torque is developed with minimum current and winding losses. Hence, the high-efficiency results. The *vector* control can be used, and voltage- and current-fed inverters are applied. This section approaches and solves the motion control problem through the use of voltage-frequency and vector control operational principles. Control laws are synthesized for a three-phase squirrel-cage induction motor, and experimental results are documented.

5.8.1. Mathematical Models of Induction Motors

5.8.1.1. Equations in the Machine Variables

The stator and rotor currents in the *abc* windings, electrical angular velocity ω_r, and displacement θ_r are used as the state variables to develop the mathematical model. For three-phase induction motors, shown in Figure 5.8.1, using Kirchhoff's second law, one obtains the coupled stator-rotor circuitry model in Cauchy's form as

$$
\begin{bmatrix} \dfrac{di_{as}}{dt} \\[4pt] \dfrac{di_{bs}}{dt} \\[4pt] \dfrac{di_{cs}}{dt} \\[4pt] \dfrac{di_{ar}}{dt} \\[4pt] \dfrac{di_{br}}{dt} \\[4pt] \dfrac{di_{cr}}{dt} \end{bmatrix}
= \frac{1}{L_{\Sigma L}}
\begin{bmatrix}
-r_s L_{\Sigma m} & -\tfrac{1}{2} r_s L_{ms} & -\tfrac{1}{2} r_s L_{ms} & 0 & 0 & 0 \\
-\tfrac{1}{2} r_s L_{ms} & -r_s L_{\Sigma m} & -\tfrac{1}{2} r_s L_{ms} & 0 & 0 & 0 \\
-\tfrac{1}{2} r_s L_{ms} & -\tfrac{1}{2} r_s L_{ms} & -r_s L_{\Sigma m} & 0 & 0 & 0 \\
0 & 0 & 0 & -r_r L_{\Sigma m} & -\tfrac{1}{2} r_r L_{ms} & -\tfrac{1}{2} r_r L_{ms} \\
0 & 0 & 0 & -\tfrac{1}{2} r_r L_{ms} & -r_r L_{\Sigma m} & -\tfrac{1}{2} r_r L_{ms} \\
0 & 0 & 0 & -\tfrac{1}{2} r_r L_{ms} & -\tfrac{1}{2} r_r L_{ms} & -r_r L_{\Sigma m}
\end{bmatrix}
$$

$$
\times
\begin{bmatrix} i_{as} \\ i_{bs} \\ i_{cs} \\ i_{ar} \\ i_{br} \\ i_{cr} \end{bmatrix}
+ \frac{1}{L_{\Sigma L}}
$$

$$
\times
\begin{bmatrix}
0 & 0 & 0 & r_r L_{ms}\cos\theta_r & r_r L_{ms}\cos(\theta_r + \tfrac{2}{3}\pi) & r_r L_{ms}\cos(\theta_r - \tfrac{2}{3}\pi) \\
0 & 0 & 0 & r_r L_{ms}\cos(\theta_r - \tfrac{2}{3}\pi) & r_r L_{ms}\cos\theta_r & r_r L_{ms}\cos(\theta_r + \tfrac{2}{3}\pi) \\
0 & 0 & 0 & r_r L_{ms}\cos(\theta_r + \tfrac{2}{3}\pi) & r_r L_{ms}\cos(\theta_r - \tfrac{2}{3}\pi) & r_r L_{ms}\cos\theta_r \\
r_s L_{ms}\cos\theta_r & r_r L_{ms}\cos(\theta_r - \tfrac{2}{3}\pi) & r_r L_{ms}\cos(\theta_r + \tfrac{2}{3}\pi) & 0 & 0 & 0 \\
r_r L_{ms}\cos(\theta_r + \tfrac{2}{3}\pi) & r_s L_{ms}\cos\theta_r & r_r L_{ms}\cos(\theta_r - \tfrac{2}{3}\pi) & 0 & 0 & 0 \\
r_r L_{ms}\cos(\theta_r - \tfrac{2}{3}\pi) & r_r L_{ms}\cos(\theta_r + \tfrac{2}{3}\pi) & r_s L_{ms}\cos\theta_r & 0 & 0 & 0
\end{bmatrix}
$$

$$
\times
\begin{bmatrix} i_{as} \\ i_{bs} \\ i_{cs} \\ i_{ar} \\ i_{br} \\ i_{cr} \end{bmatrix}
+ \frac{1}{L_{\Sigma L}}
$$

$$
\times
\begin{bmatrix}
0 & 1.3 L_{ms}^{2}\omega_r & -1.3 L_{ms}^{2}\omega_r & L_{\Sigma ms}\omega_r \sin\theta_r & L_{\Sigma ms}\omega_r \sin(\theta_r + \tfrac{2}{3}\pi) & L_{\Sigma ms}\omega_r \sin(\theta_r - \tfrac{2}{3}\pi) \\
-1.3 L_{ms}^{2}\omega_r & 0 & 1.3 L_{ms}^{2}\omega_r & L_{\Sigma ms}\omega_r \sin(\theta_r - \tfrac{2}{3}\pi) & L_{\Sigma ms}\omega_r \sin\theta_r & L_{\Sigma ms}\omega_r \sin(\theta_r + \tfrac{2}{3}\pi) \\
1.3 L_{ms}^{2}\omega_r & -1.3 L_{ms}^{2}\omega_r & 0 & L_{\Sigma ms}\omega_r \sin(\theta_r + \tfrac{2}{3}\pi) & L_{\Sigma ms}\omega_r \sin(\theta_r - \tfrac{2}{3}\pi) & L_{\Sigma ms}\omega_r \sin\theta_r \\
L_{\Sigma ms}\omega_r \sin\theta_r & L_{\Sigma ms}\omega_r \sin(\theta_r - \tfrac{2}{3}\pi) & L_{\Sigma ms}\omega_r \sin(\theta_r + \tfrac{2}{3}\pi) & 0 & -1.3 L_{ms}^{2}\omega_r & 1.3 L_{ms}^{2}\omega_r \\
L_{\Sigma ms}\omega_r \sin(\theta_r + \tfrac{2}{3}\pi) & L_{\Sigma ms}\omega_r \sin\theta_r & L_{\Sigma ms}\omega_r \sin(\theta_r - \tfrac{2}{3}\pi) & 1.3 L_{ms}^{2}\omega_r & 0 & -1.3 L_{ms}^{2}\omega_r \\
L_{\Sigma ms}\omega_r \sin(\theta_r - \tfrac{2}{3}\pi) & L_{\Sigma ms}\omega_r \sin(\theta_r + \tfrac{2}{3}\pi) & L_{\Sigma ms}\omega_r \sin\theta_r & -1.3 L_{ms}^{2}\omega_r & 1.3 L_{ms}^{2}\omega_r & 0
\end{bmatrix}
$$

$$
\times
\begin{bmatrix} i_{as} \\ i_{bs} \\ i_{cs} \\ i_{ar} \\ i_{br} \\ i_{cr} \end{bmatrix}
+ \frac{1}{L_{\Sigma L}}
$$

$$
\times
\begin{bmatrix}
2L_{ms} + L_{lr} & \tfrac{1}{2} L_{ms} & \tfrac{1}{2} L_{ms} & -L_{ms}\cos\theta_r & -L_{ms}\cos(\theta_r + \tfrac{2}{3}\pi) & -L_{ms}\cos(\theta_r - \tfrac{2}{3}\pi) \\
\tfrac{1}{2} L_{ms} & 2L_{ms} + L_{lr} & \tfrac{1}{2} L_{ms} & -L_{ms}\cos(\theta_r - \tfrac{2}{3}\pi) & -L_{ms}\cos\theta_r & -L_{ms}\cos(\theta_r + \tfrac{2}{3}\pi) \\
\tfrac{1}{2} L_{ms} & \tfrac{1}{2} L_{ms} & 2L_{ms} + L_{lr} & -L_{ms}\cos(\theta_r + \tfrac{2}{3}\pi) & -L_{ms}\cos(\theta_r - \tfrac{2}{3}\pi) & -L_{ms}\cos\theta_r \\
-L_{ms}\cos\theta_r & -L_{ms}\cos(\theta_r - \tfrac{2}{3}\pi) & -L_{ms}\cos(\theta_r + \tfrac{2}{3}\pi) & 2L_{ms} + L_{lr} & \tfrac{1}{2} L_{ms} & \tfrac{1}{2} L_{ms} \\
-L_{ms}\cos(\theta_r + \tfrac{2}{3}\pi) & -L_{ms}\cos\theta_r & -L_{ms}\cos(\theta_r - \tfrac{2}{3}\pi) & \tfrac{1}{2} L_{ms} & 2L_{ms} + L_{lr} & \tfrac{1}{2} L_{ms} \\
-L_{ms}\cos(\theta_r - \tfrac{2}{3}\pi) & -L_{ms}\cos(\theta_r + \tfrac{2}{3}\pi) & -L_{ms}\cos\theta_r & \tfrac{1}{2} L_{ms} & \tfrac{1}{2} L_{ms} & 2L_{ms} + L_{lr}
\end{bmatrix}
$$

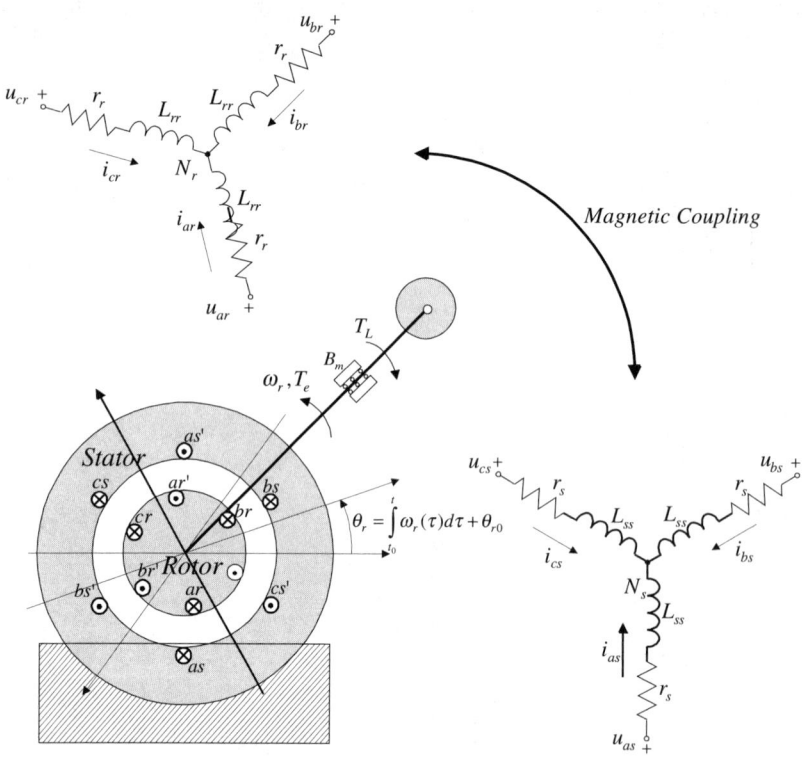

FIGURE 5.8.1.   Three-phase wye-connected symmetrical induction motor with load.

$$\times \begin{bmatrix} u_{as} \\ u_{bs} \\ u_{cs} \\ u_{ar} \\ u_{br} \\ u_{cr} \end{bmatrix} \qquad\qquad (5.8.1)$$

Here, r_s and r_r are the varying (caused by heating) stator and rotor resistances, $r_s(\cdot) \in \begin{bmatrix} r_{s\min} & r_{s\max} \end{bmatrix}, r_r(\cdot) \in \begin{bmatrix} r_{r\min} & r_{r\max} \end{bmatrix}$; L_{ls}, L_{lr}, L_{ms}, and L_{mr} are the stator and rotor leakage and magnetizing inductances, caused by nonlinear magnetizing characteristic $L_{ms}(\cdot) \in \left[L_{ms\,min} \quad L_{ms\,max} \right]$ and $L_{mr}(\cdot) \in \left[L_{mr\,min} \quad L_{mr\,max} \right]$; L_{sr} is the magnitude of the mutual inductance between stator and rotor windings. In (5.8.1), the following notations are used:

$$L_{\Sigma L} = (3L_{ms} + L_{lr})L_{lr}, \; L_{\Sigma m} = 2L_{ms} + L_{lr} \text{ and } L_{\Sigma ms} = \tfrac{3}{2}L_{ms}^2 + L_{ms}L_{lr}.$$

Using Newton's second law of motion, the tortional–mechanical dynamics are

$$
\begin{aligned}
\frac{d\omega_r}{dt} &= -\frac{P^2 L_{ms}}{4J} \Big\{ \big[i_{as}\left(i_{ar} - \tfrac{1}{2}i_{br} - \tfrac{1}{2}i_{cr}\right) - i_{bs}\left(\tfrac{1}{2}i_{ar} - i_{br} + \tfrac{1}{2}i_{cr}\right) \\
&\quad - i_{cs}\left(\tfrac{1}{2}i_{ar} + \tfrac{1}{2}i_{br} - i_{cr}\right)\big] \sin\theta_r + \tfrac{\sqrt{3}}{2}[i_{as}(i_{br} - i_{cr}) - i_{bs}(i_{ar} - i_{cr}) \\
&\quad + i_{cs}(i_{ar} - i_{br})]\cos\theta_r \Big\} - \frac{B_m}{J}\omega_r - \frac{P}{2J}T_L,
\end{aligned}
$$

$$\frac{d\theta_r}{dt} = \omega_r, \tag{5.8.2}$$

where B_m is the viscous friction coefficient, J is the equivalent moment of inertia $J(\cdot) \in \left[J_{min} \quad J_{max} \right]$, P is the number of poles, and T_L is the load torque.
The mechanical angular velocity and displacement are found as

$$\omega_{rm} = \frac{2}{P}\omega_r, \; \theta_{rm} = \frac{2}{P}\theta_r.$$

Augmenting (5.8.1) and (5.8.2), one finds a nonlinear model for three-phase induction motors.

5.8.1.2. Mathematical Model in the Arbitrary Reference Frame

Using the direct and inverse Park transformations, the *abc* variables (stator and rotor currents and voltages) can be represented using the quadrature-, direct-, and zero-axis components of the corresponding variables, and vise versa. The direct transformation for the stator variables is

$$\mathbf{u}_{qdos} = \mathbf{K}_s \mathbf{u}_{abcs}, \; \mathbf{i}_{qdos} = \mathbf{K}_s \mathbf{i}_{abcs},$$

whereas to change the rotor variables, the following transformation is applied:

$$\mathbf{u}_{qdor} = \mathbf{K}_r \mathbf{u}_{abcr}, \; \mathbf{i}_{qdor} = \mathbf{K}_r \mathbf{i}_{abcr}.$$

The transformation matrices are

$$
\mathbf{K}_s = \tfrac{2}{3}
\begin{bmatrix}
\cos\theta & \cos\left(\theta - \tfrac{2}{3}\pi\right) & \cos\left(\theta + \tfrac{2}{3}\pi\right) \\
\sin\theta & \sin\left(\theta - \tfrac{2}{3}\pi\right) & \sin\left(\theta + \tfrac{2}{3}\pi\right) \\
\tfrac{1}{2} & \tfrac{1}{2} & \tfrac{1}{2}
\end{bmatrix},
$$

$$
\mathbf{K}_r = \tfrac{2}{3}
\begin{bmatrix}
\cos(\theta - \theta_r) & \cos\left(\theta - \theta_r - \tfrac{2}{3}\pi\right) & \cos\left(\theta - \theta_r + \tfrac{2}{3}\pi\right) \\
\sin(\theta - \theta_r) & \sin\left(\theta - \theta_r - \tfrac{2}{3}\pi\right) & \sin\left(\theta - \theta_r + \tfrac{2}{3}\pi\right) \\
\tfrac{1}{2} & \tfrac{1}{2} & \tfrac{1}{2}
\end{bmatrix},
$$

where θ is the angular displacement of the reference frame with respect to the machine axis.

Using differential equations (5.8.1) and (5.8.2), the motor dynamics in the arbitrary reference frame is found by applying the direct Park transformations for the stator and rotor variables.

We have a model in Cauchy's form as

$$\frac{di_{qs}}{dt} = \frac{1}{L_{SM}L_{RM} - M^2}[-L_{RM}r_s i_{qs} - (L_{SM}L_{RM} - M^2)\omega i_{ds}$$
$$+ Mr_r i_{qr} - M(Mi_{ds} + L_{RM}i_{dr})\omega_r + L_{RM}u_{qs} - Mu_{qr}],$$

$$\frac{di_{ds}}{dt} = \frac{1}{L_{SM}L_{RM} - M^2}[(L_{SM}L_{RM} - M^2)\omega i_{qs} - L_{RM}r_s i_{ds}$$
$$+ Mr_r i_{dr} + M(Mi_{qs} + L_{RM}i_{qr})\omega_r + L_{RM}u_{ds} - Mu_{dr}],$$

$$\frac{di_{os}}{dt} = \frac{1}{L_{ls}}(-r_s i_{os} + u_{os}),$$

$$\frac{di_{qr}}{dt} = \frac{1}{L_{SM}L_{RM} - M^2}[Mr_s i_{qs} - L_{SM}r_r i_{qr} - (L_{SM}L_{RM} - M^2)\omega i_{dr}$$
$$+ L_{SM}(Mi_{ds} + L_{RM}i_{dr})\omega_r - Mu_{qs} + L_{SM}u_{qr}],$$

$$\frac{di_{dr}}{dt} = \frac{1}{L_{SM}L_{RM} - M^2}[Mr_s i_{ds} + (L_{SM}L_{RM} - M^2)\omega i_{qr} - L_{SM}r_r i_{dr}$$
$$- L_{SM}(Mi_{qs} + L_{RM}i_{qr})\omega_r - Mu_{ds} + L_{SM}u_{dr}],$$

$$\frac{di_{or}}{dt} = \frac{1}{L_{lr}}(-r_r i_{or} + u_{or}),$$

$$\frac{d\omega_r}{dt} = \frac{3P^2}{8J}M(i_{qs}i_{dr} - i_{ds}i_{qr}) - \frac{B_m}{J}\omega_r - \frac{P}{2J}T_L, \qquad (5.8.3)$$

where ω is the angular velocity of the reference frame, $M = \frac{3}{2}L_{ms}$, $L_{SM} = L_{ls} + M$, and $L_{RM} = L_{lr} + M$.

The expression for the electromagnetic torque developed by induction motors is

$$T_e = \frac{3P}{4}M(i_{qs}i_{dr} - i_{ds}i_{qr}).$$

To find mathematical models of induction motors in the rotor, stationary, and synchronous reference frames, one substitutes

$$\omega = \omega_r, \omega = 0 \text{ and } \omega = \omega_e.$$

In field-oriented control, induction motors are usually modeled in the synchronous reference frame. Then, using $\omega = \omega_e$, from (5.8.3), we have the following set of highly coupled nonlinear differential equations to model the motor dynamics in the synchronous reference frame:

$$\frac{di_{qs}^e}{dt} = \frac{1}{L_{SM}L_{RM} - M^2}[-L_{RM}r_s i_{qs}^e - (L_{SM}L_{RM} - M^2)\omega_e i_{ds}^e$$
$$+ Mr_r i_{qr}^e - M(Mi_{ds}^e + L_{RM}i_{dr}^e)\omega_r + L_{RM}u_{qs}^e - Mu_{qr}^e],$$

$$\frac{di_{ds}^e}{dt} = \frac{1}{L_{SM}L_{RM} - M^2}[(L_{SM}L_{RM} - M^2)\omega_e i_{qs}^e - L_{RM}r_s i_{ds}^e$$
$$+ Mr_r i_{dr}^e + M(Mi_{qs}^e + L_{RM}i_{qr}^e)\omega_r + L_{RM}u_{ds}^e - Mu_{dr}^e],$$

$$\frac{di_{os}^e}{dt} = \frac{1}{L_{ls}}(-r_s i_{os}^e + u_{os}^e),$$

$$\frac{di_{qr}^e}{dt} = \frac{1}{L_{SM}L_{RM} - M^2}[Mr_s i_{qs}^e - L_{SM}r_r i_{qr}^e - (L_{SM}L_{RM} - M^2)\omega_e i_{dr}^e$$
$$+ L_{SM}(Mi_{ds}^e + L_{RM}i_{dr}^e)\omega_r - Mu_{qs}^e + L_{SM}u_{qr}^e],$$

$$\frac{di_{dr}^e}{dt} = \frac{1}{L_{SM}L_{RM} - M^2}[Mr_s i_{ds}^e + (L_{SM}L_{RM} - M^2)\omega_e i_{qr}^e - L_{SM}r_r i_{dr}^e$$
$$- L_{SM}(Mi_{qs}^e + L_{RM}i_{qr}^e)\omega_r - Mu_{ds}^e + L_{SM}u_{dr}^e],$$

$$\frac{di_{or}^e}{dt} = \frac{1}{L_{lr}}(-r_r i_{or}^e + u_{or}^e),$$

$$\frac{d\omega_r}{dt} = \frac{3P^2}{8J}M(i_{qs}^e i_{dr}^e - i_{ds}^e i_{qr}^e) - \frac{B_m}{J}\omega_r - \frac{P}{2J}T_L.$$

5.8.1.3. Dynamics of Induction Motors

Variations of moment of inertia, stator and rotor resistances and inductances, and mutual inductance result in changes of the parameter-dependent coefficients of differential equations. These variations lead to the application of models with uncertain parameters because it is impossible to precisely predict or describe these variations that develop because of a great variety of diverse phenomena (nonlinear magnetization, disturbances, heating, kinematic changes, etc.). One concludes that the system dynamics is described by nonlinear differential equations with uncertain parameters. Using the models developed in (5.8.1)–(5.8.2) or (5.8.3), and taking note of the output equation $y = \omega_{rm}$, for three-phase induction motors, we have

$$\dot{x}(t) = F_x(x, \omega_{ref}, z) + B_u(p)u, u_{min} \le u \le u_{max}, x(t_0) = x_0,$$
$$y = \omega_{rm}. \qquad (5.8.4)$$

Here, $x \in X \subset \mathbb{R}^n$ is the state-space vector, and in the abc variables, one obtains $x = \begin{bmatrix} i_{as} & i_{bs} & i_{cs} & i_{ar} & i_{br} & i_{cr} & \omega_r & \theta_r \end{bmatrix}^T$, whereas in the arbitrary reference frame, $x = \begin{bmatrix} i_{qs} & i_{ds} & i_{os} & i_{qr} & i_{dr} & i_{or} & \omega_r \end{bmatrix}^T$; $u \in U \subset \mathbb{R}^m$ is the vector of the voltages applied, and in the *abc* and *qd*0 variables for squirrel-cage induction machines,

$$u = \mathbf{u}_{abcs} = \begin{bmatrix} u_{as} \\ u_{bs} \\ u_{cs} \end{bmatrix}$$

and

$$u = \mathbf{u}_{qdos} = \begin{bmatrix} u_{qs} \\ u_{ds} \\ u_{os} \end{bmatrix},$$

whereas for wound-rotor induction motors

$$u = \begin{bmatrix} \mathbf{u}_{abcs} \\ \mathbf{u}_{abcr} \end{bmatrix} = \begin{bmatrix} u_{as} & u_{bs} & u_{cs} & u_{ar} & u_{br} & u_{cr} \end{bmatrix}^T \text{ and}$$

$$u = \begin{bmatrix} \mathbf{u}_{qdos} \\ \mathbf{u}_{qdor} \end{bmatrix} = \begin{bmatrix} u_{qs} & u_{ds} & u_{os} & u_{qr} & u_{dr} & u_{or} \end{bmatrix}^T ;$$

$\omega_{ref} \in \Omega_{ref} \subset \mathbb{R}^1$ and $\omega_{rm} \in \Omega_{rm} \subset \mathbb{R}^1$ are the reference (assigned) and actual (mechanical) angular velocities; $z \in Z \subset \mathbb{R}^d$ and $p \in \mathcal{P} \subset \mathbb{R}^k$ are the bounded parameter variations; functions $z(\cdot)$: $[t_0, \infty) \rightarrow Z \subset \mathbb{R}^d$ and $p(\cdot)$: $[t_0, \infty) \rightarrow \mathcal{P} \subset \mathbb{R}^k$ are Lebesgue measurable and known within bounds; $Z \subset \mathbb{R}^d$ and $\mathcal{P} \subset \mathbb{R}^k$ are known nonempty compact sets; $F_x(\cdot)$: $\mathbb{R}^n \times \mathbb{R}^1 \times \mathbb{R}^d \rightarrow \mathbb{R}^n$ and $B_u(\cdot)$: $\mathbb{R}^k \rightarrow \mathbb{R}^{n \times m}$ are smooth.

5.8.2. *Control of Induction Motors*

5.8.2.1. Motor Control in the Machine Variables

The expression for the electromagnetic torque developed by three-phase induction motors is

$$T_e = \frac{P}{2} \mathbf{i}_{abcs}^T \frac{\partial \mathbf{L}_{sr}(\theta_r)}{\partial \theta_r} \mathbf{i}_{abcr} = -\frac{P L_{ms}}{2} \begin{bmatrix} i_{as} & i_{bs} & i_{cs} \end{bmatrix}$$

$$\times \begin{bmatrix} \sin\theta_r & \sin\left(\theta_r + \frac{2}{3}\pi\right) & \sin\left(\theta_r - \frac{2}{3}\pi\right) \\ \sin\left(\theta_r - \frac{2}{3}\pi\right) & \sin\theta_r & \sin\left(\theta_r + \frac{2}{3}\pi\right) \\ \sin\left(\theta_r + \frac{2}{3}\pi\right) & \sin\left(\theta_r - \frac{2}{3}\pi\right) & \sin\theta_r \end{bmatrix} \begin{bmatrix} i_{ar} \\ i_{br} \\ i_{cr} \end{bmatrix}.$$

Using Newton's second law of motion

$$\frac{d\omega_r}{dt} = \frac{P}{2J} T_e - \frac{B_m}{J} \omega_r - \frac{P}{2J} T_L,$$

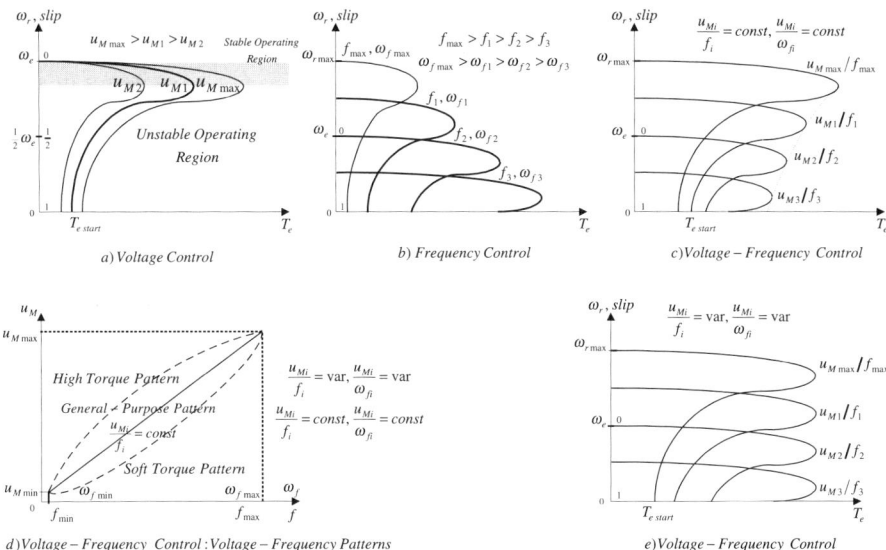

FIGURE 5.8.2.   Torque-speed characteristic curves of induction motors: a) voltage control, b) frequency control, c) voltage-frequency control, constant volts per hertz control; d) voltage-frequency patterns, and e) variable voltage-frequency control.

it is evident that the angular velocity of induction motors is regulated by changing the magnitude u_M and the frequency f of the voltages supplied to the $as, bs,$ and cs windings.

A balanced three-phase set (a set of equal magnitude sinusoidal voltages displaced by $120°$) is given as

$$
\begin{aligned}
u_{as}(t) &= \sqrt{2}u_M \cos(\omega_f t),\\
u_{bs}(t) &= \sqrt{2}u_M \cos\left(\omega_f t - \tfrac{2}{3}\pi\right),\\
u_{cs}(t) &= \sqrt{2}u_M \cos\left(\omega_f t + \tfrac{2}{3}\pi\right),
\end{aligned}
$$

where the angular frequency is $\omega_f = 2\pi f$.

The following principles are used to control the angular velocity of induction motors.

Voltage control. By changing the magnitude u_M of the applied phase voltages to the stator windings, the angular velocity can be regulated in the stable operating region; see Figure 5.8.2.a.

Frequency control. The magnitude of the applied phase voltages is assigned to be constant u_M^{constant}, and the angular velocity is regulated above and below the synchronous angular velocity ω_e, $\omega_e = \frac{4\pi f}{P}$ by changing the frequency of the supplied voltages. The speed of the general purpose induction motors can be regulated from $0.05\omega_e$ to $2\omega_e$ (the mechanical limit is imposed on the maximum speed). The torque-speed characteristic curves for different values of the frequency are shown in Figure 5.8.2.b.

Voltage-frequency control. The angular frequency ω_f is proportional to the frequency of the supplied voltages, and $\omega_f = 2\pi f$. The torque-speed characteristic curves are shown in Figure 5.8.2.b if one changes the frequency. To minimize losses, the magnitude u_M of the applied voltages should be regulated if the frequency is changed. In particular, the magnitude of phase voltages applied to the stator windings can be decreased linearly with decreasing the frequency. That is, one maintains the following relationship:

$$\frac{u_{Mi}}{f_i} = \text{const or } \frac{u_{Mi}}{\omega_{fi}} = \text{const}$$

to attain the so-called constant volts per Hertz control. The corresponding torque-speed characteristics are plotted in Figure 5.8.2.c. It must be emphasized that by regulating the voltage-frequency patterns, one shapes the torque-speed curves. For example, the following relation:

$$\sqrt{\frac{u_{Mi}}{f_i}} = \text{const}$$

is commonly used to adjust the magnitude u_M and frequency f. The standard (general purpose) voltage-frequency pattern is illustrated in Figure 5.8.2.d (the solid line). For electric drives and electromechanical systems, to guarantee the acceleration and settling time specified, overshoot and rise time needed, and stability and robustness in the operating envelope, the general purpose-, soft-, and high-starting torque patterns are implemented based on the requirements and criteria imposed (see the soft- and high-torque patterns as shown in Figure 5.8.2.d by the dashed lines). In particular, assigning $\omega_f = \varphi(u_M)$ with domain $u_{M\min} < u_M < u_{M\max}$ and range $\omega_{f\min} < \omega_f < \omega_{f\max}$, one maintains $\frac{u_{Mi}}{f_i} = \text{var or } \frac{u_{Mi}}{\omega_{fi}} = \text{var}$. For example, the specified torque-speed characteristic curves for general purpose electric drives, as plotted in Figure 5.8.2.e, can be easily achieved.

Our goal is to design the robust control algorithms, and the control bounds imposed on the magnitude u_M of the phase voltages must be incorporated; that is, $u_{M\min} \leq u_M \leq u_{M\max}$. Robust tracking, needed to be achieved, is guaranteed if the tracking error $e(t) = \omega_{\text{ref}}(t) - \omega_{rm}(t)$, which represents the difference between the reference speed ω_{ref} and the actual angular velocity ω_{rm}, is bounded, and $e \in E \subset \mathbb{R}^1$. The initial condition set is given as $X_0 = \{x_0 \in \mathbb{R}^n\} \subseteq X \subset \mathbb{R}^n$. For the given ω_{rm0} and assigned ω_{ref}, the corresponding $E_0 = \{e_0 \in \mathbb{R}^1\} \subseteq E \subset \mathbb{R}^1$ results.

Let $t \in \mathbb{R}_{\geq 0}$, and then

$$\left\{(t, \mathbf{u}_{abcs}): \mathbf{u}_{abcs} = \begin{bmatrix} u_{as} \\ u_{bs} \\ u_{cs} \end{bmatrix} = \begin{bmatrix} \cos(\omega_f t) \\ \cos\left(\omega_f t - \frac{2}{3}\pi\right) \\ \cos\left(\omega_f t + \frac{2}{3}\pi\right) \end{bmatrix} \sqrt{2} u_M \right\},$$

$$u_M \neq 0, \omega_f \neq 0 \qquad (5.8.5)$$

is a three-phase, balanced sinusoidal set with amplitude $|\sqrt{2} u_M|$ and angular frequency ω_f.

In expanded form, we have

$$u_{as} = \sqrt{2} u_M \cos(\omega_f t), \ u_{bs} = \sqrt{2} u_M \cos\left(\omega_f t - \frac{2}{3}\pi\right),$$

$$u_{cs} = \sqrt{2} u_M \cos\left(\omega_f t + \frac{2}{3}\pi\right).$$

In general, induction motors are controlled by changing u_M and ω_f. Hence, one obtains

$$\mathbf{u}_{abcs} = \Xi(u_M, \omega_f).$$

The phase voltages, applied to the *abc* stator windings of induction motors are bounded as $\mathbf{u}_{abcs\,min} \leq \mathbf{u}_{abcs} \leq \mathbf{u}_{abcs\,max}$.

It is immediate that $u_{M\,min} \leq u_M \leq u_{M\,max}$, where $u_{M\,max} = V_{LL}$.

Here, V_{LL} is the rated *rms* voltage.

One defines the admissible control set as

$$U = \{\mathbf{u}_{abcs}: \mathbf{u}_{abcs\,min} \leq \mathbf{u}_{abcs} \leq \mathbf{u}_{abcs\,max}\} \subset \mathbb{R}^3,$$

and

$$\mathbf{u}_{abcs} \in U.$$

To attain the desired voltage-frequency pattern, the magnitude u_M and the angular frequency ω_f are related. In particular, the angular frequency is a nonlinear function of u_M, and

$$\omega_f = \varphi(u_M).$$

Here, a real-analytic continuous or piecewise continuous bounded function $\varphi(\cdot)$ should be found to guarantee the specified voltage-frequency patterns. Because $\omega_{f\,min} \leq \omega_f \leq \omega_{f\,max}$, one obtains

$$\omega_f \in \Omega_f,$$

where the admissible angular frequency set is $\Omega_f = \{\omega_f: \omega_{f\,min} \leq \omega_f \leq \omega_{f\,max}\}$.

Hence, $\varphi \in \Omega_f$.

As an illustration, assume that the constant volts per hertz pattern is desired. One finds ω_f using the following relationship: $\omega_f = \omega_e \frac{u_M}{V_{LL}}$. It is evident that the angular frequency ω_f changes linearly with u_M, and ω_f does not exceed ω_e because $u_M \leq V_{LL}$. In general, $\omega_{f\,min} \leq \omega_f \leq \omega_{f\,max}$, and for general-purpose induction machines, $\omega_f \in [0.05\omega_e \quad 2\omega_e]$; that is, $0.05\omega_e \leq \omega_f \leq 2\omega_e$. Hence, to attain the constant volts per hertz pattern, we have $\omega_f = k_f \omega_e \frac{u_M}{V_{LL}}$, $\omega_{f\,min} \leq \omega_f \leq k_f \omega_e$. Here, k_f is the admissible frequency coefficient, $k_f = \frac{\omega_{f\,max}}{\omega_e}$, and for general-purpose induction motors, usually, $k_f = 2$.

Variable voltage-frequency control is needed to ensure high- and soft-starting torque patterns, and we have

$$\omega_f = \varphi\left(\omega_e \frac{u_M}{V_{LL}}\right), \ \omega_{f\,min} \le \omega_f \le \omega_{f\,max}, \ \varphi \in \Omega_f.$$

Using (5.8.5), a controller can be found. In particular, for squirrel-cage induction motors, the bounded control law $\mathbf{u}_{abcs} \in U$, as given by

$$\mathbf{u}_{abcs} = \begin{bmatrix} u_{as} \\ u_{bs} \\ u_{cs} \end{bmatrix} = \begin{bmatrix} \cos(\omega_f t) \\ \cos\left(\omega_f t - \frac{2}{3}\pi\right) \\ \cos\left(\omega_f t + \frac{2}{3}\pi\right) \end{bmatrix} \sqrt{2} u_M,$$

$$u_M = -\phi\left(G_e L_{ms} \frac{\partial V(e, x)}{\partial e} + G_i L_{ms} \frac{1}{s} \frac{\partial V(e, x)}{\partial e}\right),$$

$$s = \frac{d}{dt}, \tag{5.8.6}$$

guarantees

- Voltage control if $\omega_f = $ const
- Frequency control if $u_M = u_M^{constant}$, and

$$\mathbf{u}_{abcs} = \begin{bmatrix} u_{as} \\ u_{bs} \\ u_{cs} \end{bmatrix} = \begin{bmatrix} \cos(\omega_f t) \\ \cos\left(\omega_f t - \frac{2}{3}\pi\right) \\ \cos\left(\omega_f t + \frac{2}{3}\pi\right) \end{bmatrix} \sqrt{2} u_M^{constant},$$

$$\omega_f = \varphi\left(\omega_e \frac{u_M}{V_{LL}}\right)$$

- Constant volts per hertz control if

$$\mathbf{u}_{abcs} = \begin{bmatrix} u_{as} \\ u_{bs} \\ u_{cs} \end{bmatrix} = \begin{bmatrix} \cos(\omega_f t) \\ \cos\left(\omega_f t - \frac{2}{3}\pi\right) \\ \cos\left(\omega_f t + \frac{2}{3}\pi\right) \end{bmatrix} \sqrt{2} u_M$$

and

$$\omega_f = k_f \omega_e \frac{u_M}{V_{LL}}$$

- Variable voltage-frequency control if

$$\mathbf{u}_{abcs} = \begin{bmatrix} u_{as} \\ u_{bs} \\ u_{cs} \end{bmatrix} = \begin{bmatrix} \cos(\omega_f t) \\ \cos\left(\omega_f t - \frac{2}{3}\pi\right) \\ \cos\left(\omega_f t + \frac{2}{3}\pi\right) \end{bmatrix} \sqrt{2} u_M$$

and

$$\omega_f = \varphi\left(\omega_e \frac{u_M}{V_{LL}}\right)$$

In (5.8.6), $\phi(\cdot)$ is the real-analytic, continuous or piecewise continuous, bounded function that lies in the first quadrant, and $\phi \in U_\phi$ to guarantee $u_{M\,\min} \leq U_M \leq u_{M\,\max}$; U_ϕ is the admissible set $U_\phi = \{u_M: u_{M\,\min} \leq u_M \leq u_{M\,\max}\}$; G_e and G_i are the weighting coefficients; and $V(e,x)$ is the continuously differentiable, positive-definite function $V(\cdot)\colon \mathbb{R}^1 \times \mathbb{R}^n \to \mathbb{R}_{\geq 0}$.

Voltage control. If in (5.8.6) $\omega_f = \text{const}$, one regulates the magnitude u_M of the phase voltages, which is given as

$$u_M = -\phi\left(G_e L_{ms} \frac{\partial V(e,x)}{\partial e} + G_i L_{ms} \frac{1}{s} \frac{\partial V(e,x)}{\partial e}\right).$$

That is, the voltage control is in place.

Frequency control. Assigning the magnitude of the applied phase voltages to be constant and equal to u_M^{constant}, we have

$$\mathbf{u}_{abcs} = \begin{bmatrix} u_{as} \\ u_{bs} \\ u_{cs} \end{bmatrix} = \begin{bmatrix} \cos(\omega_f t) \\ \cos\left(\omega_f t - \frac{2}{3}\pi\right) \\ \cos\left(\omega_f t + \frac{2}{3}\pi\right) \end{bmatrix} \sqrt{2} u_M^{\text{constant}}.$$

Because $\omega_f = \varphi\left(\omega_e \frac{u_M}{V_{LL}}\right)$, the angular frequency is changed as a function of u_M. That is, the ω_f is controlled, and the frequency control is maintained.

Voltage-frequency control. The magnitude of the applied phase voltages is found as

$$u_M = -\phi\left(G_e L_{ms} \frac{\partial V(e,x)}{\partial e} + G_i L_{ms} \frac{1}{s} \frac{\partial V(e,x)}{\partial e}\right).$$

The angular frequency ω_f, which is a function of u_M, is expressed by $\omega_f = \varphi\left(\omega_e \frac{u_M}{V_{LL}}\right)$. Then, the constant volts per hertz operation is guaranteed if $\omega_f = k_f \omega_e \frac{u_M}{V_{LL}}$. The phase voltages u_{as}, u_{bs}, and u_{cs} are bounded because the real-analytic, continuous or piecewise continuous function $\phi \in U_\phi$ is used. For example, making use of the `saturation` function, the bounds imposed on the phase voltages are incorporated, and one obtains $u_{M\,\min} \leq u_M \leq u_{M\,\max} = V_{LL}$. Hence, $-\sqrt{2}V_{LL} \leq u_{as} \leq \sqrt{2}V_{LL}$, $-\sqrt{2}V_{LL} \leq u_{bs} \leq \sqrt{2}V_{LL}$, and $-\sqrt{2}V_{LL} \leq u_{cs} \leq \sqrt{2}V_{LL}$. If the angular frequency is regulated to guarantee the constant volts per hertz operation, $\omega_f = k_f \omega_e \frac{u_M}{V_{LL}}$. It is evident that making use of $k_f = 1$, because $u_M \leq V_{LL}$, the ω_f does not exceed the synchronous angular velocity $\omega_e = \frac{4\pi f}{P}$.

In high-performance drives, soft- and high-torque patterns are used to attain the desired specifications, requirements, and objectives. The variable voltage-frequency control is guaranteed by controller (5.8.6). Making use the bounded function $\varphi \in \Omega_f$, it is immediate that the soft- and hard-torque patterns are attained, and $\frac{u_{Mi}}{\omega_{fi}} = \text{var}$. In particular, the angular frequency ω_f, which is bounded as $\omega_{f\,\min} \leq \omega_f \leq \omega_{f\,\max}$, is calculated to guarantee the voltage-frequency pattern desired; see Figure 5.8.2.d. That is, the patterns specified are assigned by using $\varphi(\cdot)$.

The evolution set XE can be easily found. Controller (5.8.6) robustly stabilizes the electric drive (5.8.1)–(5.8.2) in $XE \subset \mathbb{R}^n \times \mathbb{R}^1$ and steers the error vector to a compact set

$$
\begin{aligned}
S_e(\delta) \; = \; & \{e \in \mathbb{R}^1 \colon \; e_0 \in E_0, x \in X(X_0, U, Z, \mathcal{P}), \omega_{ref} \in \Omega_{ref}, \\
& \omega_{rm} \in \Omega_{rm}, t \in [t_0, \infty) \\
& \|e(t)\| \leq \rho_e(t, \|e_0\|) + \rho_r(\|\omega_{ref}\|) + \rho_y(\|\omega_{rm}\|) + \delta, \delta \geq 0, \\
& \forall e \in E(E_0, \Omega_{ref}, \Omega_{rm}), \forall t \in [t_0, \infty)\} \subset \mathbb{R},
\end{aligned}
$$

if a continuously differentiable, positive-definite function $V(e, x)$ exists, such that for all $x \in X$, $u \in U$, $\omega_{ref} \in \Omega_{ref}$, $z \in Z$, and $p \in \mathcal{P}$ in XE

$$
\frac{\partial V(e, x)}{\partial t} + \left(\frac{\partial V(e, x)}{\partial x} \right)^T (F_x(x, \omega_{ref}, z) + B_u(p)\mathbf{u}_{abcs}) \leq 0. \qquad (5.8.7)
$$

Furthermore, if for a positive-definite function $V(e, x)$, inequality (5.8.7) is guaranteed in XE, one concludes that the set XE is an invariant domain of stability and robust tracking. The stability of the closed-loop system can be proven by using the second method of Lyapunov and the admissibility concept. We use the following nonquadratic, positive-definite Lyapunov candidate:

$$
V(e, x) = \sum_{j=0}^{\varsigma} \frac{2\beta+1}{2(j+\beta+1)} K_{ej} e^{\frac{2(j+\beta+1)}{2\beta+1}} + \sum_{j=0}^{\sigma} \frac{2\mu+1}{2(j+\mu+1)} K_{ij} e^{\frac{2(j+\mu+1)}{2\mu+1}} + eK_x x + \frac{1}{2} x^T K x,
$$

$$(5.8.8)$$

where K_{ej} and K_{ij} are the unknown coefficients; $K_x \in \mathbb{R}^{1 \times n}$ and $K \in \mathbb{R}^{n \times n}$ are the unknown matrices; and ς, β, σ, and μ are the nonnegative integers assigned by the designer, $\varsigma = 0, 1, 2, \ldots$, $\beta = 0, 1, 2, \ldots$, $\sigma = 0, 1, 2, \ldots$, and $\mu = 0, 1, 2, \ldots$ It must be emphasized that K_{ei}, K_{eii}, K_x, and K are found by solving the nonlinear inequality (5.8.7).

Assigning $\varsigma = 0$, $\beta = 0$, $\sigma = 0$, and $\mu = 0$, one finds commonly used the quadratic Lyapunov function

$$
V(e, x) = \frac{1}{2} K_{e0} e^2 + \frac{1}{2} K_{i0} e^2 + eK_x x + \frac{1}{2} x^T K x.
$$

Nonlinear controllers improve performance, guarantee robustness and disturbance attenuation, expand operating envelopes, and so on. For example, for $\varsigma = 1$, $\beta = 0$, $\sigma = 1$, and $\mu = 0$, one finds the following Lyapunov candidate:

$$
V(e, x) = \frac{1}{2} K_{e0} e^2 + \frac{1}{4} K_{e1} e^4 + \frac{1}{2} K_{i0} e^2 + \frac{1}{4} K_{i1} e^4 + eK_x x + \frac{1}{2} x^T K x,
$$

and a nonlinear control algorithm results.

Making use of (5.8.6) and (5.8.8), a bounded controller is found to be

$$
\mathbf{u}_{abcs} = \begin{bmatrix} u_{as} \\ u_{bs} \\ u_{cs} \end{bmatrix} = \begin{bmatrix} \cos(\omega_f t) \\ \cos\left(\omega_f t - \frac{2}{3}\pi\right) \\ \cos\left(\omega_f t + \frac{2}{3}\pi\right) \end{bmatrix} \sqrt{2} u_M,
$$

$$u_M = -\phi \left[G_e L_{ms} \left(\sum_{j=0}^{\varsigma} K_{ej} e^{\frac{2j+1}{2\beta+1}} + K_x x \right) \right.$$

$$\left. + \frac{1}{s} G_i L_{ms} \sum_{j=0}^{\sigma} K_{ij} e^{\frac{2j+1}{2\mu+1}} \right],$$

$$\phi \in U_\phi, \omega_f = \varphi \left(\omega_e \frac{u_M}{V_{LL}} \right), \varphi \in \Omega_f. \qquad (5.8.9)$$

Control algorithm (5.8.9) guarantees the basic principles of operation for induction motors, and as shown, constant and variable voltage-frequency controls are attained. Furthermore, assigning $\omega_f = $ const and $u_M = u_M^{\text{constant}}$, the voltage and frequency operating principles are in place, respectively.

5.8.2.2. Control in the Arbitrary Reference Frame

The speed and so-called torque control of induction motors have found a wide application, and field orientation can be achieved using the voltages and currents as the controlled variables. To regulate the angular velocity, the electromagnetic torque and flux linkages are controlled, and proportional-integral controllers are applied. Our main goal is to depart from conventional linear, proportional-integral controllers, and use a nonlinear, robust, proportional-integral family of control algorithms with nonlinear error and state feedback mappings to improve motor performance and tracking accuracy, expand the operating envelope and stability margins, and so on. It should be emphasized that the quadrature- and direct-axis components of voltages are bounded. Therefore, constrained control algorithms must be designed. To implement the vector control concept, the following family of bounded controllers is introduced:

$$\mathbf{u}_{qds} = \begin{bmatrix} u_{qs} \\ u_{ds} \end{bmatrix} = -\phi \left(G_{qde} B_{qds} \frac{\partial V(e, x)}{\partial e} + G_{qdi} B_{qds} \frac{1}{s} \frac{\partial V(e, x)}{\partial e} \right),$$

$$\phi \in U_\phi, \qquad (5.8.10)$$

where $G_{qde} \in \mathbb{R}^{2 \times 2}$ and $G_{qdi} \in \mathbb{R}^{2 \times 2}$ are the weighting matrices,

$$G_{qde} = \begin{bmatrix} g_{qe} & 0 \\ 0 & g_{de} \end{bmatrix}$$

and

$$G_{qdi} = \begin{bmatrix} g_{qi} & 0 \\ 0 & g_{di} \end{bmatrix};$$

$B_{qds} \in \mathbb{R}^{2 \times 1}$ is the matrix $B_{qds} = \begin{bmatrix} L_{RM} \\ L_{RM} \end{bmatrix}$.

This control algorithm is related to conventional linear, proportional-integral controllers. However, nonlinear error maps and state feedback are used to guarantee the superior dynamic performance, to expand the operating envelope, to meet the tracking specifications, and to ensure the disturbance attenuation. In (5.8.10),

$$G_{qde}B_{qds}\frac{\partial V(e,x)}{\partial e}$$

and

$$G_{qdi}B_{qds}\frac{1}{s}\frac{\partial V(e,x)}{\partial e}$$

accomplish the tracking of the reference speed using the tracking error e and the measured state variables. The integral control action together with nonlinear proportional error mapping and state feedback guarantee tracking and ensure robust stability. Computing the total derivative of $V(e,x)$, along solutions (5.8.3) with (5.8.10), the unknown coefficients of $V(e,x)$, as given by (5.8.8), are found by solving the matrix inequality. Taking note of (5.8.10) and (5.8.8), the bounded controller results. In particular,

$$\mathbf{u}_{qds} = \begin{bmatrix} u_{qs} \\ u_{ds} \end{bmatrix} = -\phi \left[G_{qde}B_{qds} \left(\sum_{j=0}^{\varsigma} K_{ej} e^{\frac{2j+1}{2\beta+1}} + K_x x \right) \right.$$

$$\left. + \frac{1}{s} G_{qdi}B_{qds} \sum_{j=0}^{\sigma} K_{ij} e^{\frac{2j+1}{2\mu+1}} \right], \phi \in U_\phi.$$

The stability of the closed-loop system (5.8.3) with (5.8.10) can be easily proven using the Lyapunov theory.

5.8.2.3. Vector Control of Induction Motors

Using the rotor flux linkages as the motor variables, the equation for the electromagnetic torque is found to be

$$T_e = \frac{3PM}{4L_{RM}}(i_{qs}\psi_{dr} - i_{ds}\psi_{qr}).$$

In the synchronous reference frame, we have $T_e = \frac{3PM}{4L_{RM}}(i_{qs}^e \psi_{dr}^e - i_{ds}^e \psi_{qr}^e)$.
Assume that the rotor flux linkages are within the direct axis. Thus, the quadrature-component of the rotor flux is zero, $\psi_{qr}^e = Mi_{qs}^e + (M+L_{lr})i_{qr}^e = 0$.
Therefore, we have $T_e = \frac{3PM}{4L_{RM}}i_{qs}^e \psi_{dr}^e$.
Taking note of the slip relationship

$$\omega_e - \omega_r = \frac{r_r}{L_{RM}}\frac{i_{qs}^e}{i_{ds}^e} = -r_r \frac{i_{qr}^e}{\psi_{dr}^e}$$

and using the circuitry equations, one finds that the direct-axis component of the rotor flux linkages ψ_{dr}^e can be controlled by changing the i_{ds}^e. In particular,

$$\frac{d\psi_{dr}^e}{dt} = -\frac{r_r}{L_{RM}}\psi_{dr}^e + \frac{r_r M}{L_{RM}}i_{ds}^e.$$

Direct and indirect field orientation concepts are used to control induction motors. As the desired angular velocity is assigned, the voltages needed to be supplied to the stator phase windings are found using the inverse Park transformation. The most challenging problem is the parameter variations. In fact, the rotor resistance and inductance are time-varying, and these parameters must be estimated.

Control algorithms in the arbitrary reference frame (synchronous, rotor, and stationary frames of reference) are related to the controllers found in the machine variables. For example, in the synchronous reference frame, one obtains the quadrate-, direct-, and zero-axis components of voltages u_{qs}^e, u_{ds}^e, and u_{os}^e to attain the balanced conditions. Using the Park transformation and taking note that in the synchronous reference frame $\theta = \theta_e$ we have

$$\mathbf{u}_{qdos}^e = \mathbf{K}_s^e \mathbf{u}_{abcs},$$

where

$$\mathbf{K}_s^e = \frac{2}{3} \begin{bmatrix} \cos\theta_e & \cos\left(\theta_e - \frac{2}{3}\pi\right) & \cos\left(\theta_e + \frac{2}{3}\pi\right) \\ \sin\theta_e & \sin\left(\theta_e - \frac{2}{3}\pi\right) & \sin\left(\theta_e + \frac{2}{3}\pi\right) \\ \frac{1}{2} & \frac{1}{2} & \frac{1}{2} \end{bmatrix}.$$

Therefore,

$$\begin{bmatrix} u_{qs}^e \\ u_{ds}^e \\ u_{os}^e \end{bmatrix} = \frac{2}{3} \begin{bmatrix} \cos\theta_e & \cos\left(\theta_e - \frac{2}{3}\pi\right) & \cos\left(\theta_e + \frac{2}{3}\pi\right) \\ \sin\theta_e & \sin\left(\theta_e - \frac{2}{3}\pi\right) & \sin\left(\theta_e + \frac{2}{3}\pi\right) \\ \frac{1}{2} & \frac{1}{2} & \frac{1}{2} \end{bmatrix} \begin{bmatrix} u_{as} \\ u_{bs} \\ u_{cs} \end{bmatrix}.$$

To guarantee the balanced operation for the *counterclockwise* direction of rotation, one supplies the phase voltages as given by (5.8.5). From $\theta_e = \int_{t_0}^t \omega_e(\tau) \, d\tau + \theta_{e0}$ and assuming that $\theta_{e0} = 0$, we have $\theta_e = \omega_f t$. Hence,

$$u_{qs}^e(t) = \sqrt{2} u_M, \, u_{ds}^e(t) = 0, \, u_{os}^e(t) = 0.$$

One concludes that only $u_{qs}^e(t)$ is controlled. Control algorithms are designed by taking note of

$$u_{qs}^e = -\phi \left(g_{qe} L_{RM} \frac{\partial V(e, x)}{\partial e} + g_{qi} L_{RM} \frac{1}{s} \frac{\partial V(e, x)}{\partial e} \right), \quad \phi \in U_\phi.$$

Admissible controllers are found, and the Lyapunov theory is used to prove the stability of closed-loop systems.

5.8.3. *Control of a Three-Phase Squirrel-Cage Induction Motor: Analytical and Experimental Results*

To validate the analytical results in analysis, modeling, and control, consider an electric drive with a high-performance, four-pole, wye-connected, 60-Hz, 200-V, 0.75-kW squirrel-cage induction motor. Our goal is to design control laws to guarantee the precise tracking with *optimum* performance, stability, and robustness. The studied motor has the following parameters: $r_s(\cdot) \in \begin{bmatrix} 1.4 & 2 \end{bmatrix} \Omega$, $r_r(\cdot) \in \begin{bmatrix} 2.1 & 2.9 \end{bmatrix} \Omega$, $L_{ls} = L_{lr} = 0.007$ H, and $L_{ms}(\cdot) \in \begin{bmatrix} 0.27 & 0.38 \end{bmatrix}$H, and the equivalent moment of inertia is $J = 0.017$ kg-m^2. The main objective is to maintain the desired speed tracking by synthesizing the robust controllers using the measured states in spite of model uncertainties and control bounds. The rotor angular velocity $\omega_{rm} = \frac{1}{2}\omega_r$, as measured by the optical encoder, is compared with the reference input ω_{ref} to find the tracking error, and $e = \omega_{ref} - \omega_m$. The measured states are $i_{as}, i_{bs}, i_{cs}, \omega_{rm}, \theta_{rm}$, and the unmeasured state variables are the rotor currents i_{ar}, i_{br}, and i_{cr}. An admissible set is defined by using the motor data. The following bounds on states and applied voltages are imposed:

$$\mathbf{i}_{abcs} \in \begin{bmatrix} -6.5 & 6.5 \end{bmatrix} \text{A} , \mathbf{i}_{abcr} \in \begin{bmatrix} -4.7 & 4.7 \end{bmatrix} \text{A} ,$$

$$\omega_{rm} \in \begin{bmatrix} -377 & 377 \end{bmatrix} \text{rad/sec, and } \mathbf{u}_{abcs} \in \begin{bmatrix} -\sqrt{2200} & \sqrt{2200} \end{bmatrix} \text{V}.$$

The mathematical model of an electric drive with a three-phase squirrel-cage induction motor in the *abc* variables is described by differential equations (5.8.1) and (5.8.2). The control law, as given by

$$\mathbf{u}_{abcs} = \begin{bmatrix} u_{as} \\ u_{bs} \\ u_{cs} \end{bmatrix} = \begin{bmatrix} \cos(\omega_f t) \\ \cos\left(\omega_f t - \frac{2}{3}\pi\right) \\ \cos\left(\omega_f t + \frac{2}{3}\pi\right) \end{bmatrix} \sqrt{2}u_M,$$

$$u_M = -\text{sat}_2^{200}\left[G_e L_{ms}\left(\sum_{j=0}^{\varsigma} K_{ej} e^{\frac{2j+1}{2\beta+1}} + K_x x \right) \right.$$

$$\left. + G_i L_{ms} \sum_{j=0}^{\sigma} K_{ij} \frac{1}{s} e^{\frac{2j+1}{2\mu+1}} \right], \quad 2 \leq u_M \leq 200 \text{ V}, \quad (5.8.11)$$

guarantees the constant volts per hertz operation if

$$\omega_f = k_f \times 377 \frac{u_M}{200} = 754 \frac{u_M}{200}, 7.5 \leq \omega_f \leq 754 \text{ rad/sec},$$

$$\omega_f \in \Omega_f. \quad (5.8.12)$$

The variable voltage-frequency control with a high-torque pattern is maintained making use of

$$\omega_f = 754 \frac{0.005 u_M^2}{200}, 7.5 \leq \omega_f \leq 754 \text{ rad/sec}, \omega_f \in \Omega_f. \quad (5.8.13)$$

From (5.8.8), letting $\varsigma = 1, \beta = 0, \sigma = 1$, and $\mu = 0$, the nonquadratic Lyapunov function is

$$V(e, x) = \frac{1}{2}K_{e0}e^2 + \frac{1}{4}K_{e1}e^4 + \frac{1}{2}K_{ei0}e^2 + \frac{1}{4}K_{ei1}e^4 + k_{x1}ei_{as}$$

$$+ k_{x2}ei_{bs} + k_{x3}ei_{cs} + k_{x7}e\omega_r + k_{x8}e\theta_r$$

$$+ \frac{1}{2}\begin{bmatrix} i_{as} & i_{bs} & i_{cs} & i_{ar} & i_{br} & i_{cr} & \omega_r & \theta_r \end{bmatrix}$$

$$\times K \begin{bmatrix} i_{as} & i_{bs} & i_{cs} & i_{ar} & i_{br} & i_{cr} & \omega_r & \theta_r \end{bmatrix}^T.$$

Solving nonlinear matrix inequality (5.8.7), one obtains the feedback gains, and controller (5.8.11) in expanded form is

$$u_{as} = \cos(\omega_f t)\mathbf{sat}_2^{200}\left(3.4e + 1.7e^3 + 0.95\frac{1}{s}e + 0.42\frac{1}{s}e^3 - 0.17i_{as}\right.$$

$$\left. - 0.17i_{bs} - 0.17i_{cs} - 0.5\omega_{rm} - 0.28\theta_{rm}\right),$$

$$e = \omega_{ref} - \omega_{rm},$$

$$u_{bs} = \cos\left(\omega_f t - \tfrac{2}{3}\pi\right)\mathbf{sat}_2^{200}\left(3.4e + 1.7e^3 + 0.95\frac{1}{s}e + 0.42\frac{1}{s}e^3\right.$$

$$-0.17i_{as} - 0.17i_{bs} - 0.17i_{cs} - 0.5\omega_{rm}$$

$$\left. - 0.28\theta_{rm}\right),$$

$$u_{cs} = \cos\left(\omega_f t + \tfrac{2}{3}\pi\right)\mathbf{sat}_2^{200}\left(3.4e + 1.7e^3 + 0.95\frac{1}{s}e + 0.42\frac{1}{s}e^3\right.$$

$$-0.17i_{as} - 0.17i_{bs} - 0.17i_{cs} - 0.5\omega_{rm}$$

$$\left. - 0.28\theta_{rm}\right). \tag{5.8.14}$$

The amplitude of the applied phase voltages to the abc windings and the angular frequency ω_f vary to maintain the constant air-gap flux density (synchronous angular velocity is $\omega_e = 377$ rad/sec) if the ω_f is expressed by (5.8.12). To guarantee high-performance motor capabilities, the high-torque patterns should be implemented, and variable voltage-frequency control is in place as ω_f is found from (5.8.13). The efficacy of the bounded algorithm (5.8.14), which does not need rotor flux or currents measurements or estimations, is examined. Figure 5.8.3 illustrates the dynamics of the closed-loop electric drive if the constant volts per hertz controller (5.8.14) with (5.8.12) is implemented, and the reference angular velocity is assigned to be 150 rad/sec. Good dynamic performance, precise speed tracking with zero steady-state error, and disturbance rejection are achieved. The motor starts from stall and reaches the reference speed within 0.37 sec. The load torque 1 N-m is applied at 0.6 sec. In particular,

$$T_L = \begin{cases} 0 \text{ N-m}, & t \in [0\ 0.6) \text{ sec} \\ 1 \text{ N-m}, & t \in [0.6\ 0.8] \text{ sec}. \end{cases}$$

The maximum deflection of the angular velocity is 4.1 rad/sec with the following disturbance attenuation (zero steady-state error), and the settling time is 0.12 sec; see Figure 5.8.3.

The variable voltage-frequency controller, as given by (5.8.14) with (5.8.13), is implemented and tested. Unloaded motor starts from stall and reaches the assigned angular velocity faster compared with the constant voltage per Hertz control because a high-torque pattern is implemented; see Figure 5.8.4. The settling time is 0.32 sec. The load torque is applied at 0.6 sec, and the maximum

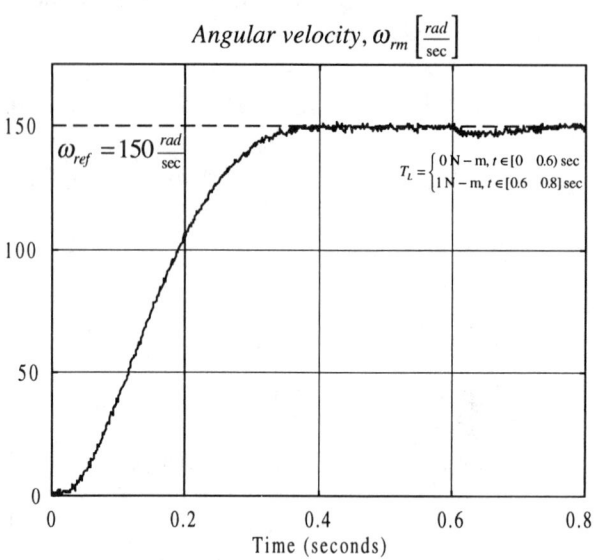

FIGURE 5.8.3.   Constant volts per Hertz control: electric drive dynamics, ω_{ref} = 150 rad/sec.

angular velocity deflection is 1.9 rad/sec. These excellent starting and disturbance rejection capabilities are achieved because of a high-torque pattern implemented using the variable voltage-frequency control.

To implement the vector control algorithms, advanced DSPs with state-of-the-art software are used because the direct and inverse Park transformations must be performed in real time. The nonlinear control law was designed, and non-quadratic Lyapunov functions lead to nonlinear error feedback; see (5.8.10). For the three-phase induction motor that has been studied, the mathematical model in the synchronous reference frame was developed. From (5.8.10), the closed-loop form of a bounded control algorithm is found. Using the nonquadratic Lyapunov function

$$V(e, x) = \tfrac{1}{2}K_{e0}e^2 + \tfrac{1}{4}K_{e1}e^4 + \tfrac{1}{2}K_{ei0}e^2 + \tfrac{1}{4}K_{ei1}e^4$$
$$+ k_{x1}ei_{qs}^e + k_{x2}ei_{ds}^e + k_{x7}e\omega_r$$
$$+ \tfrac{1}{2}\begin{bmatrix} i_{qs}^e & i_{ds}^e & i_{os}^e & i_{qr}^e & i_{dr}^e & i_{or}^e & \omega_r \end{bmatrix}$$
$$\times K \begin{bmatrix} i_{qs}^e & i_{ds}^e & i_{os}^e & i_{qr}^e & i_{dr}^e & i_{or}^e & \omega_r \end{bmatrix}^T$$

and solving nonlinear matrix inequality (5.8.7), the feedback coefficients result.

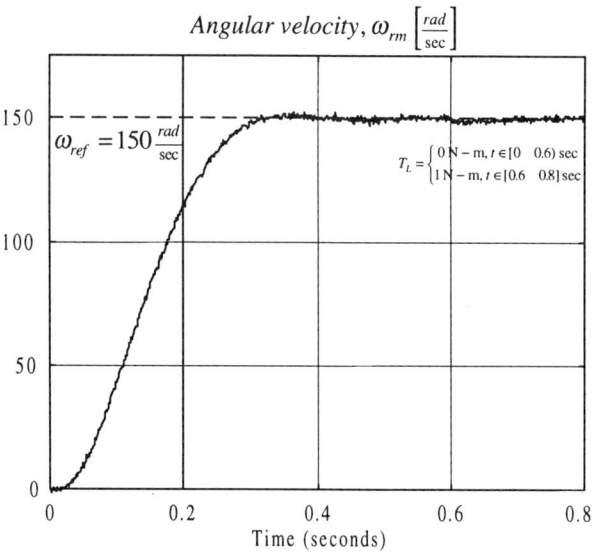

FIGURE 5.8.4. Variable voltage-frequency control with a high-torque pattern: electric drive dynamics, $\omega_{ref} = 150$ rad/sec.

The controller is found to be

$$u_{qs}^e = \mathbf{sat}_2^{200}\left(49e + 7.8e^3 + 21\frac{1}{s}e + 4.9\frac{1}{s}e^3 - 0.81i_{qs}^e - 0.32i_{ds}^e - 2.9\omega_{rm}\right),$$

$$u_{ds}^e = 0, \ u_{os}^e = 0. \tag{5.8.15}$$

The as, bs, and cs phase voltages, which must be applied to the stator windings of induction motors, are obtained using the quadrature- and direct-axis components of voltages. Hence, the inverse Park transformation

$$\mathbf{u}_{abcs} = \mathbf{K}_s^{e^{-1}}\mathbf{u}_{qdos}^e,$$

$$\begin{bmatrix} u_{as} \\ u_{bs} \\ u_{cs} \end{bmatrix} = \begin{bmatrix} \cos\theta_e & \sin\theta_e & 1 \\ \cos\left(\theta_e - \frac{2}{3}\pi\right) & \sin\left(\theta_e - \frac{2}{3}\pi\right) & 1 \\ \cos\left(\theta_e + \frac{2}{3}\pi\right) & \sin\left(\theta_e + \frac{2}{3}\pi\right) & 1 \end{bmatrix} \begin{bmatrix} u_{qs}^e \\ u_{ds}^e \\ u_{os}^e \end{bmatrix}$$

is used.

In addition, one measures the as, bs, and cs currents i_{as}, i_{bs}, and i_{cs}, and the quadrature- and direct-axis components of currents i_{qs}^e and i_{ds}^e are used in controller (5.8.15).

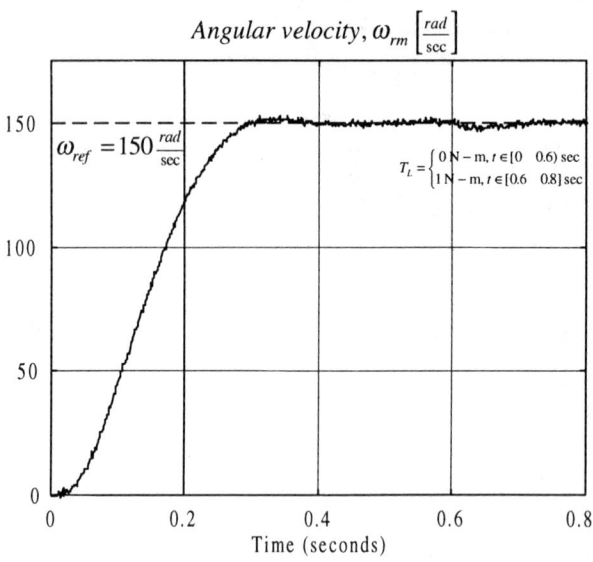

FIGURE 5.8.5.   Vector control: electric drive dynamics, $\omega_{\text{ref}} = 150$ rad/sec.

We have

$$\mathbf{i}^e_{qdos} = \mathbf{K}^e_s \mathbf{i}_{abcs},$$

$$\begin{bmatrix} i^e_{qs} \\ i^e_{ds} \\ i^e_{os} \end{bmatrix} = \tfrac{2}{3} \begin{bmatrix} \cos\theta_e & \cos\left(\theta_e - \tfrac{2}{3}\pi\right) & \cos\left(\theta_e + \tfrac{2}{3}\pi\right) \\ \sin\theta_e & \sin\left(\theta_e - \tfrac{2}{3}\pi\right) & \sin\left(\theta_e + \tfrac{2}{3}\pi\right) \\ \tfrac{1}{2} & \tfrac{1}{2} & \tfrac{1}{2} \end{bmatrix} \begin{bmatrix} i_{as} \\ i_{bs} \\ i_{cs} \end{bmatrix}.$$

That is, the direct and inverse Park transformations are embedded. Controller (5.8.15) was implemented, and the transient dynamics for the motor angular velocity is documented in Figure 5.8.5.

The transient behavior for the motor angular velocity, shown in Figure 5.8.5, illustrates that the settling time is 0.31 sec. The angular velocity follows the reference speed, and the steady-state error is zero. The maximum deflection of ω_{rm} from ω_{ref}, as the load torque is applied, is 2.2 rad/sec. The comparison of the experimental data illustrates that the tracking performance and the angular velocity waveforms are similar if one applies the variable voltage-frequency (with high-torque patterns) and vector control concepts. As high-torque patterns can be implemented by using the innovative control law developed, a spectrum of performance capabilities of induction motors are significantly improved, and robust stability and tracking are guaranteed.

5.9. Control of Servo-Systems With Permanent-Magnet DC Motors

Servo-systems actuated by permanent-magnet DC motors are modeled by three differential equations. It was illustrated that the power converter dynamics is described by differential equations, and the armature voltage applied to the motor winding is controlled by using the duty ratio of the amplifier d_D. Four-quadrant power converters are manufactured as monolithic integrated circuits, and the output dynamics is very fast. Therefore, the first-order differential equation is used to model the power amplifier. The following set of differential equations are used to model the transient behavior of servos with permanent-magnet DC motors:

$$\frac{du_a}{dt} = -\frac{1}{T_{IC}}u_a + \frac{k_{IC}}{T_{IC}}d_D, \quad \frac{di_a}{dt} = \frac{1}{L_a}u_a - \frac{r_a}{L_a}i_a - \frac{k_a}{L_a}\omega_r,$$

$$\frac{d\omega_r}{dt} = \frac{k_a}{J}i_a - \frac{B_m}{J}\omega_r - \frac{1}{J}T_L, \quad \frac{d\theta_r}{dt} = \omega_r, \qquad (5.9.1)$$

where u_a and i_a are the armature voltage and current, ω_r and θ_r are the rotor angular velocity and rotor displacement, d_D is the duty ratio (control variable), T_L is the load torque, k_{IC} and T_{IC} are the coefficient and time constant of the power amplifier, r_a and L_a are the armature resistance and inductance, J is the equivalent moment of inertia, k_a is the back emf (torque) constant, and B_m is the viscous friction coefficient.

From (5.9.1), in matrix form, we have

$$\dot{x}(t) = Ax + Bu,$$

$$\begin{bmatrix} \frac{du_a}{dt} \\ \frac{di_a}{dt} \\ \frac{d\omega_r}{dt} \\ \frac{d\theta_r}{dt} \end{bmatrix} = \begin{bmatrix} -\frac{1}{T_{IC}} & 0 & 0 & 0 \\ \frac{1}{L_a} & -\frac{r_a}{L_a} & -\frac{k_a}{L_a} & 0 \\ 0 & \frac{k_a}{J} & -\frac{B_m}{J} & 0 \\ 0 & 0 & 1 & 0 \end{bmatrix}\begin{bmatrix} u_a \\ i_a \\ \omega_r \\ \theta_r \end{bmatrix} + \begin{bmatrix} \frac{k_{IC}}{T_{IC}} \\ 0 \\ 0 \\ 0 \end{bmatrix}d_D + \begin{bmatrix} 0 \\ 0 \\ -\frac{1}{J} \\ 0 \end{bmatrix}T_L. \quad (5.9.2)$$

The output equation of the system is $y(t) = k_{gear}\theta_r(t)$. Our goal is to design and verify different control algorithms, and four different control laws are synthesized. Specifically, we study a servo with the permanent-magnet motor JDH-2250 and integrated circuit monolithic driver. The parameters are $k_{IC} = 30$, $T_{IC} = 0.0001$ sec, $r_a(\cdot) \in \left[2.7_{T=20°C} \quad 3.7_{T=140°C}\right]\Omega$, $L_a = 0.004$ H, $k_a(\cdot) \in \left[0.11_{T=20°C} \quad 0.094_{T=140°C}\right]$ V-sec/rad (N-m/A); $J = 0.0001$ kg-m^2, $B_m = 0.00008$ N-m-sec/rad, and $k_{gear} = 0.1$. The duty ratio is bounded as ± 1, $-1 \leq d_D \leq +1$.

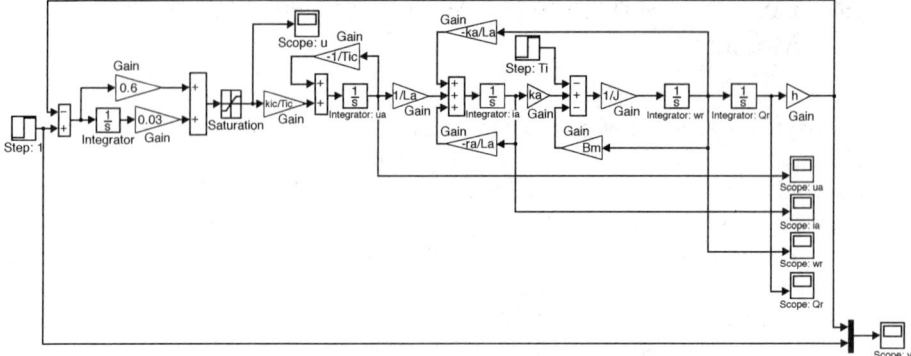

FIGURE 5.9.1.   SIMULINK diagram to model the servo with proportional-integral controller.

5.9.1. *Proportional-Integral Control*

Using the results reported in Section 4.2, the bounded proportional-integral controller is

$$ u = \mathrm{sat}_{-1}^{+1} \left(0.6e + 0.03 \int e\, dt \right), \quad -1 \le u \le 1. $$

Here, the tracking error is $e(t) = r(t) - y(t) = \theta_{\mathrm{reference}}(t) - k_{\mathrm{gear}}\theta_r(t)$.

The control signal (duty ratio) is constrained, and $-1 \le u \le 1$. The analysis of the transient behavior is performed. The SIMULINK diagram is documented in Figure 5.9.1 (in this diagram, we denote $h = k_{\mathrm{gear}}$). Figure 5.9.2 depicts the evolution of the state variables and the output $y(t)$ if the reference input is $r(t) = 2$. The settling time is 0.3 sec, and for the worst-case performance (if the motor temperature is 140° C), the overshoot for $y(t)$ is 15%.

5.9.2. *Tracking Integral Control*

The tracking integral control algorithms can be designed by applying the concept given in Section 4.4. In particular, augmenting the system dynamics (5.9.2) $\dot{x}(t) = Ax + Bu$ with that of the exogeneous system $\dot{x}^{\mathrm{ref}}(t) = \theta_{\mathrm{reference}}(t) - y(t) = \theta_{\mathrm{reference}}(t) - k_{\mathrm{gear}}\theta_r(t)$, one obtains

$$ \dot{x}_\Sigma(t) = A_\Sigma x_\Sigma + B_\Sigma u + N_\Sigma r, \ y = hx, \ x_\Sigma(t_0) = x_{\Sigma 0}, $$

where

$$ x_\Sigma = \begin{bmatrix} x \\ x^{\mathrm{ref}} \end{bmatrix} \in \mathbb{R}^5 $$

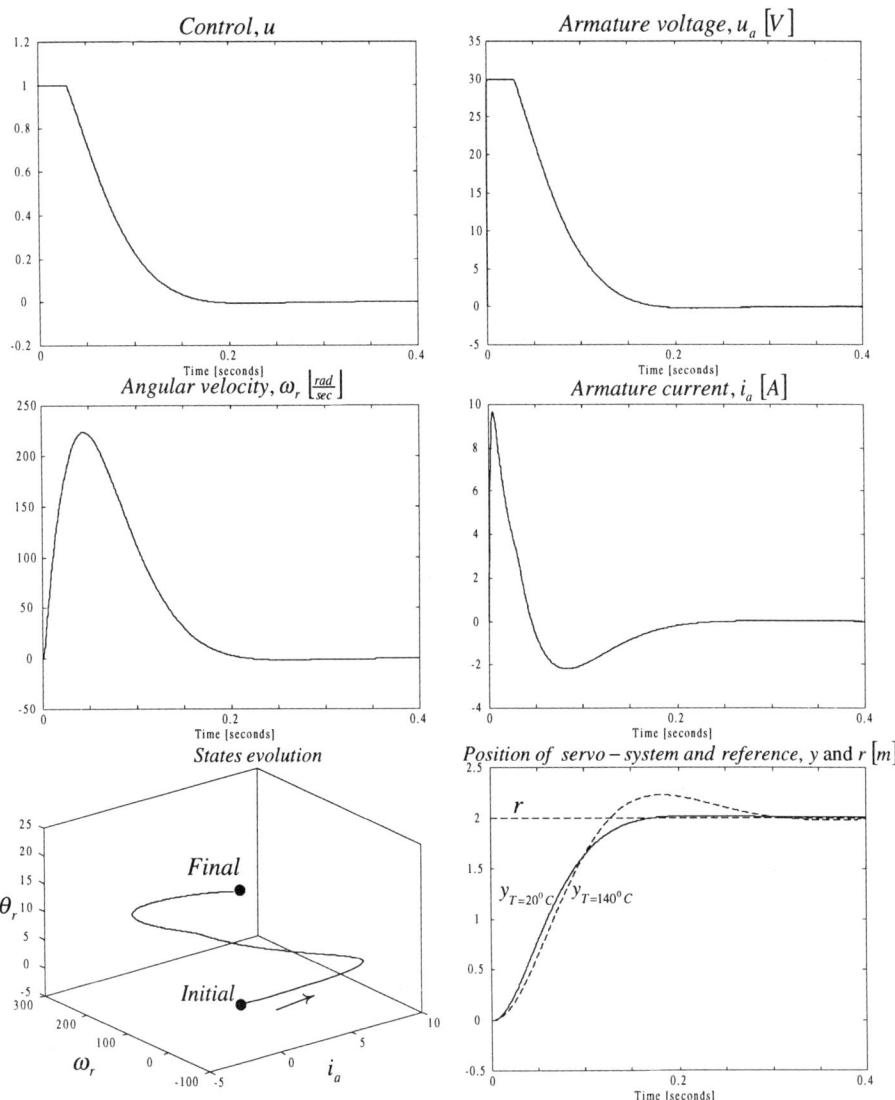

FIGURE 5.9.2.   System dynamics with proportional-integral control law.

is the augmented state vector,

$$A_\Sigma = \begin{bmatrix} A & 0 \\ -h & 0 \end{bmatrix} \in \mathbb{R}^{5\times 5}, \quad B_\Sigma = \begin{bmatrix} B \\ 0 \end{bmatrix} \in \mathbb{R}^{5\times 1}, \quad \text{and}$$

$$N_\Sigma = \begin{bmatrix} \mathbf{0} \\ N \end{bmatrix} \in \mathbb{R}^{5\times 1}$$

are the time-invariant matrices of coefficients.

We minimize the quadratic performance functional

$$J(x_\Sigma(\cdot), u(\cdot)) = \tfrac{1}{2} \int_{t_0}^{t_f} (x_\Sigma^T Q x_\Sigma + u^T G u)\, dt,$$

where $Q \in \mathbb{R}^{5 \times 5}$ is the positive-semidefinite constant-coefficient matrix and $G \in \mathbb{R}^1$ is the positive-definite weighting coefficient.

Then, one finds the control law using the first-order necessary condition for optimality as

$$u = -G^{-1} B_\Sigma^T \frac{\partial V}{\partial x_\Sigma} = -G^{-1} \begin{bmatrix} B \\ 0 \end{bmatrix}^T \frac{\partial V}{\partial x_\Sigma}.$$

The solution of the Hamilton–Jacobi–Bellman equation

$$-\frac{\partial V}{\partial t} = \tfrac{1}{2} x_\Sigma^T Q x_\Sigma + \left(\frac{\partial V}{\partial x_\Sigma}\right)^T A x_\Sigma - \tfrac{1}{2}\left(\frac{\partial V}{\partial x_\Sigma}\right)^T B_\Sigma G^{-1} B_\Sigma^T \frac{\partial V}{\partial x_\Sigma}$$

is satisfied by the quadratic return function $V = \tfrac{1}{2} x_\Sigma^T K x_\Sigma$, where $K \in \mathbb{R}^{5 \times 5}$ is the symmetric matrix found by solving the following nonlinear differential equation:

$$-\dot{K} = Q + A_\Sigma^T K + K^T A_\Sigma - K^T B_\Sigma G^{-1} B_\Sigma^T K,\ K(t_f) = K_f.$$

The controller is found as

$$u = -G^{-1} B_\Sigma^T K x_\Sigma = -G^{-1} \begin{bmatrix} B \\ 0 \end{bmatrix}^T K x_\Sigma.$$

From $\dot{x}^{\text{ref}}(t) = e(t)$, one has $x^{\text{ref}}(t) = \int e(t)\, dt$. Therefore, we obtain the integral control

$$u(t) = -G^{-1} \begin{bmatrix} B \\ 0 \end{bmatrix}^T K \begin{bmatrix} x(t) \\ \int e(t)\, dt \end{bmatrix}.$$

In this control algorithm, the error vector is used in addition to the state feedback.

It was illustrated that the duty ratio is bounded, and $-1 \le u \le 1$. Let us approximate the saturation characteristic by the hyperbolic tangent. To analytically design a constrained controller, one minimizes the nonquadratic functional

$$J(x_\Sigma(\cdot), u(\cdot)) = \int_{t_0}^{t_f} \left(x_\Sigma^T Q x_\Sigma + G \int \tan^{-1} u\, du\right) dt,$$

and the constrained controller is

$$u(t) = -\tanh\left(G^{-1} \begin{bmatrix} B \\ 0 \end{bmatrix}^T K \begin{bmatrix} x(t) \\ \int e(t)\, dt \end{bmatrix}\right)$$

$$\approx -\text{sat}_{-1}^{+1}\left(G^{-1} \begin{bmatrix} B \\ 0 \end{bmatrix}^T K \begin{bmatrix} x(t) \\ \int e(t)\, dt \end{bmatrix}\right), \quad -1 \le u \le 1.$$

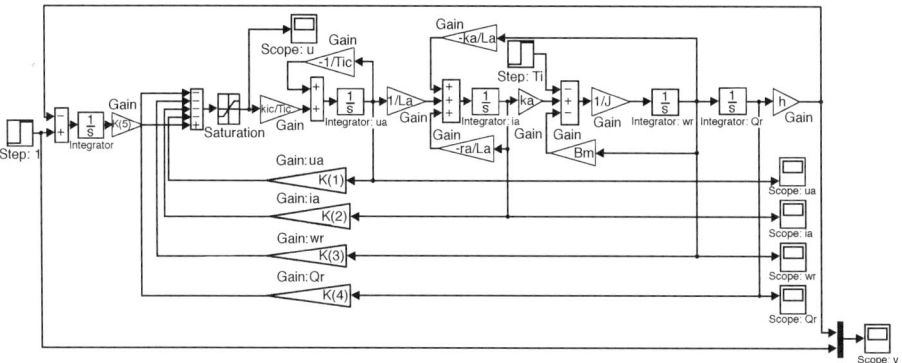

FIGURE 5.9.3. SIMULINK diagram to model the servo with a tracking integral controller.

Assigning $q_{11} = 0.0001$, $q_{22} = 0.007$, $q_{33} = 0.00001$, $q_{44} = 0.01$, $q_{55} = 4 \times 10^8$, and $G = 0.0001$, the following controller is designed:

$$u = \mathrm{sat}_{-1}^{+1}\left(-3.1u_a - 19.9i_a - 1.3\omega_r - 31.6\theta_r + 2818.8\int e\, dt\right),$$

$$-1 \le u \le 1.$$

Figure 5.9.3 illustrates the SIMULINK diagram used in simulations, and the transient dynamics are plotted in Figure 5.9.4.

5.9.3. *Relay (Time-Optimal) Control*

The third control algorithm synthesized is the relay controller with dead zone. We minimize the following performance functional:

$$J(x_\Sigma(\cdot)) = \tfrac{1}{2}\int_{t_0}^{t_f}(x_\Sigma^T Q x_\Sigma)\, dt.$$

The Hamilton–Jacobi equation is

$$-\frac{\partial V}{\partial t} = \min_{-1 \le u \le 1}\left[\tfrac{1}{2}x_\Sigma^T Q x_\Sigma + \left(\frac{\partial V}{\partial x_\Sigma}\right)^T (A x_\Sigma + B_\Sigma u)\right].$$

Hence, one finds the control law as

$$u = -\,\mathrm{sgn}\left(B_\Sigma^T \frac{\partial V(x_\Sigma)}{\partial x_\Sigma}\right), \quad -1 \le u \le 1.$$

This relay-type (minimum-time) control algorithm cannot be implemented because of the chattering phenomenon. The relay-type control laws with dead zone are used, and the following controller is found:

$$u = \mathrm{sgn}_{-1}^{+1}(-0.001u_a - 0.0001i_a - 0.00085\omega_r - 0.0002\theta_r + 1.5e)|_{\mathrm{Dead\, Zone}\pm 0.035}.$$

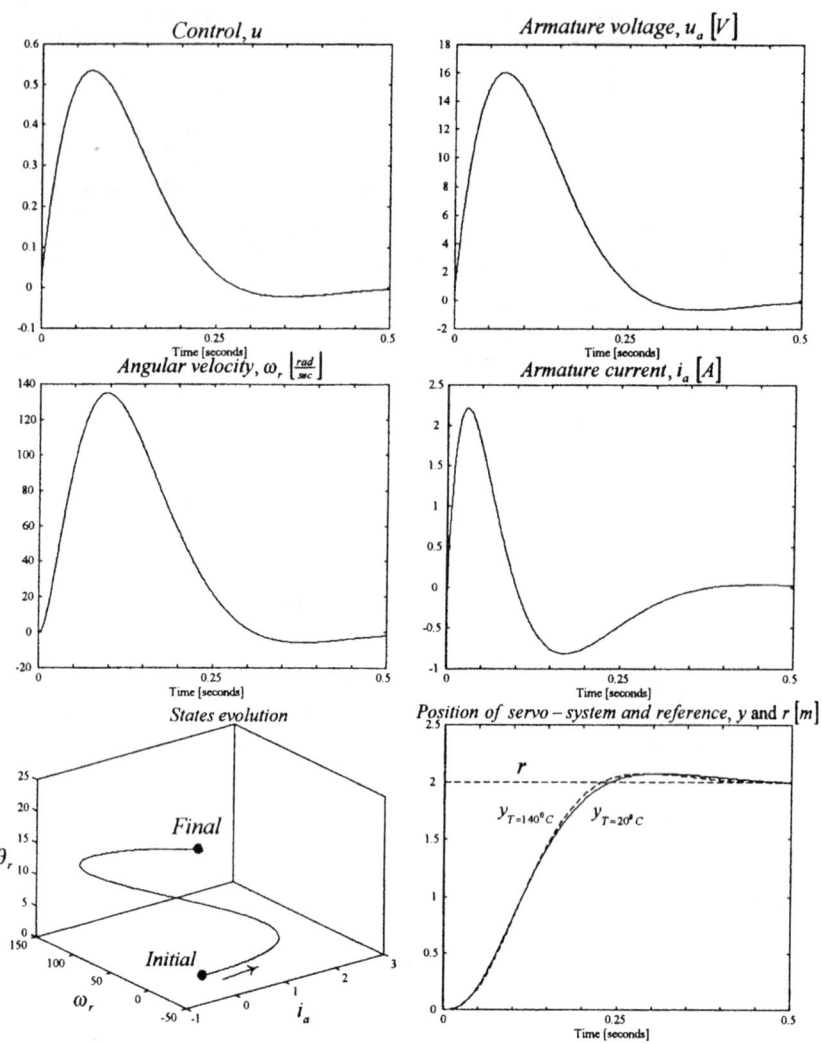

FIGURE 5.9.4.   System dynamics with tracking integral control law.

The modeling of the closed-loop system with this relay-type controller was performed, and the SIMULINK diagram is demonstrated in Figure 5.9.5. The servo-system transients are documented in Figure 5.9.6. For $r(t) = 2$, the settling time is 0.12 sec and the overshoot for $y(t)$ is 6% for the rated motor parameters (the motor temperature is 20 °C). Using the motor parameters which correspond to the worst-case performance (the motor temperature is 140 °C), the overshoot for $y(t)$ is 13%, and the settling time is 0.15 sec.

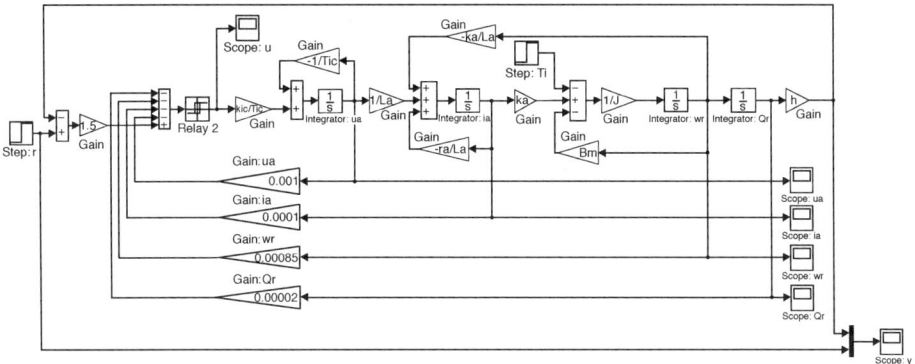

FIGURE 5.9.5.   SIMULINK diagram to model the servo-system with relay-type controller.

5.9.4. *Sliding Mode Control With Soft Switching*

Using the concept to design the variable structure systems with soft switching, as given in Section 5.5, the design of a sliding mode controller is performed.

The nonlinear switching surface is

$$k_1 u_a + k_2 i_a + k_3 \omega_r + k_4 \theta_r + k_5 e + k_6 e^{\frac{1}{3}} = 0.$$

Using the Lyapunov stability theory, the feedback gains are found, and a control law with soft switching is found to be

$$u = \tanh^{\frac{1}{3}}(-0.001 u_a - 0.01 i_a - 0.17 \omega_r - 0.0002\theta_r + 23e + 600e^{\frac{1}{3}}),$$
$$-1 \le u \le 1.$$

The transient dynamics is modeled in the SIMULINK environment; see Figure 5.9.7. Figure 5.9.8 illustrates the dynamics for the closed-loop system with the sliding controller. The settling time is 0.1 sec with no overshoot. For the worst-case performance (the motor temperature is 140 °C), the settling time and overshoot for $y(t)$ are 0.14 sec and 11%.

As a final analysis, we study the disturbance attenuation capabilities. Let the initial displacement be 1 m, and the load torque be

$$T_L = \begin{cases} 0 \text{ N-m}, & \forall t \in [0 \ 0.08) \\ 0.75 \text{ N-m}, & \forall t \in [0.08 \ 0.12) \\ -0.75 \text{ N-m}, & \forall t \in [0.12 \ 0.16] \end{cases} .$$

Figure 5.9.9 illustrates the system responses, and the disturbance attenuation features are evident by analyzing the output dynamics for $y(t)$.

The transient dynamics and three-dimensional state evolutions are documented for different control algorithms. It is illustrated that sliding-mode soft-switching algorithms provide good performance, and the chattering effect is eliminated. It

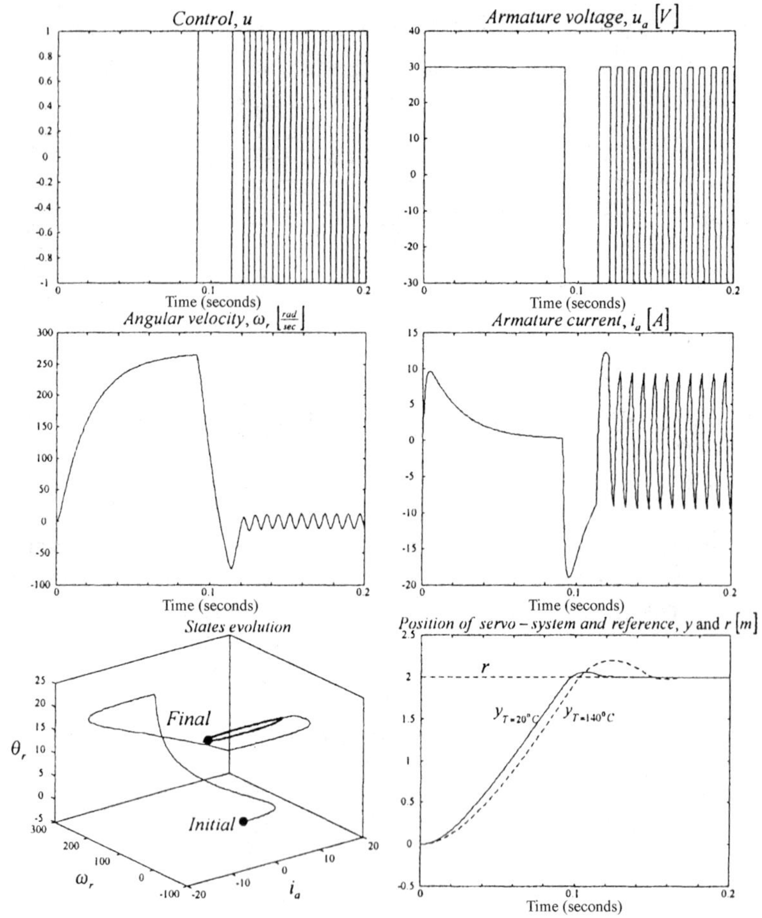

FIGURE 5.9.6.   System dynamics with relay-type control law.

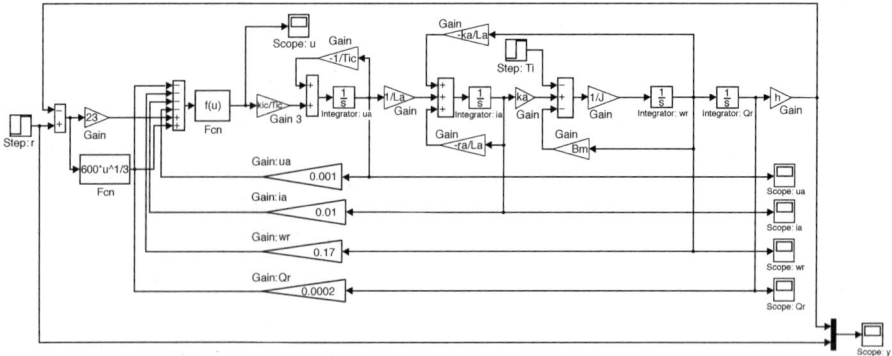

FIGURE 5.9.7.   SIMULINK diagram to model the servo-system with sliding mode controller.

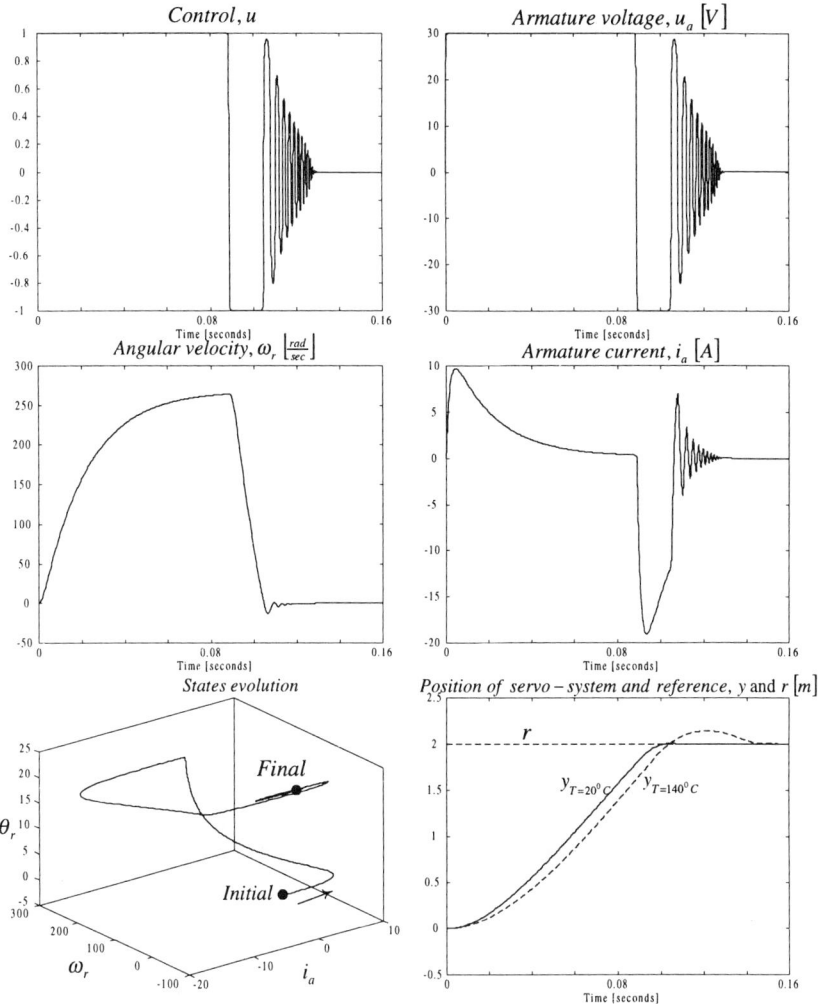

FIGURE 5.9.8.   System dynamics with sliding mode control law.

should be emphasized that proportional-integral control laws (which do not lead to the chattering) can be easily implemented, and industrial power converters integrate the PID-type controllers (that is, there is no need for additional hardware and software to implement control laws). Therefore, PID controllers are widely used in practice, as was illustrated in this section, using proportional-integral controllers, the designer may satisfy the requirements imposed on the closed-loop system performance.

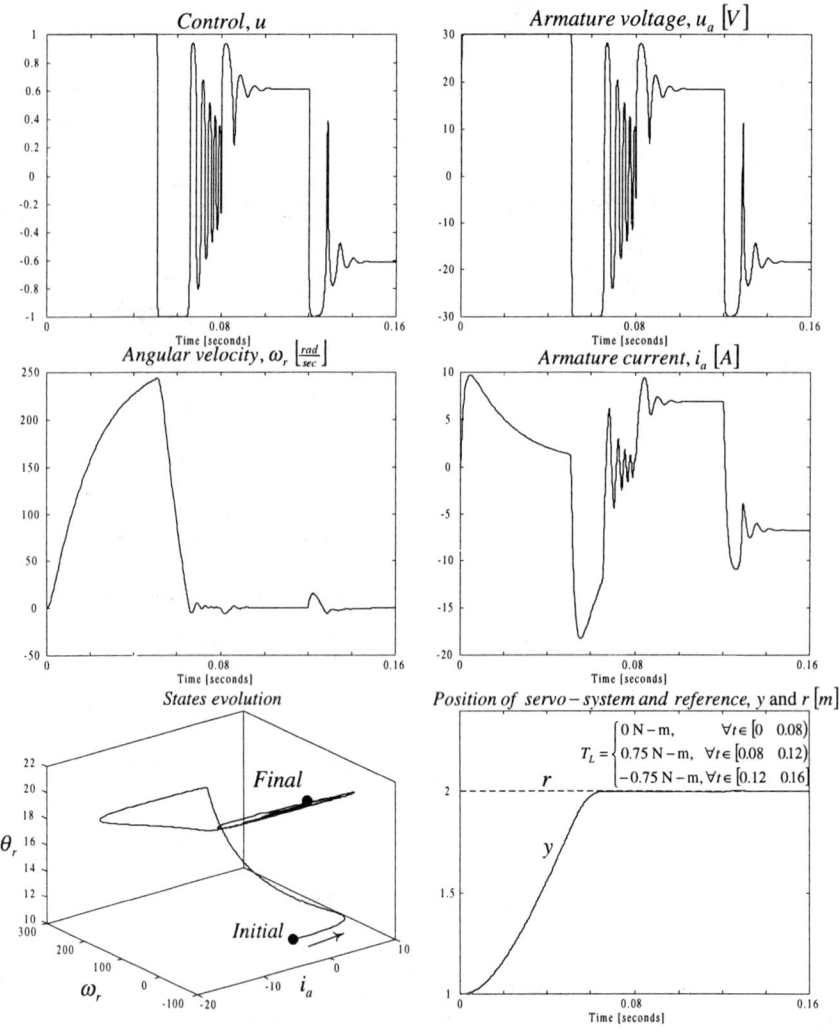

FIGURE 5.9.9.   System dynamics with sliding mode control law: disturbance attenuation analysis.

References

Applied Mathematics

[1]   E. Kreyszig, *Advanced Engineering Mathematics*, John Wiley and Sons, New York, 1999.

[2]   C. R. Wylie and L. C. Barrett, *Advanced Engineering Mathematics*, McGraw-Hill, New York, 1995.

Books in Control

[3]   M. Athans and P.L. Falb, *Optimal Control: An Introduction to the Theory and its Applications.* McGraw-Hill Book Company, New York, 1966.

[4]   R. E. Bellman, *Dynamic Programming.* Princeton University Press, Princeton, NJ, 1957.

[5]   J. J. D'Azzo and C. H. Hoipis, *Linear Control System Analysis and Design: Conventional and Modern.* McGraw-Hill, Inc., New York, 1995.

[6]   R. C. Dorf and R. H. Bishop, *Modern Control Systems.* Addison-Wesley Publishing Company, Reading, MA, 1995.

[7]   J. F. Franklin, J. D. Powell, and A. Emami-Naeini, *Feedback Control of Dynamic Systems.* Addison-Wesley Publishing Company, Reading, MA, 1994.

[8]   B. Friedland, *Advanced Control System Design.* Prentice-Hall, Upper Saddle River, NJ, 1996.

[9]   M. R. Hestenes, *Calculus of Variations and Optimal Control Theory.* John Wiley and Sons, New York, 1966.

[10] J. C. Hsu and A. U. Meyer, *Modern Control Principles and Applications.* McGraw-Hill Book Company, New York, 1968.

[11] P. A. Ioannou and J. Sun, *Robust Adaptive Control.* Prentice-Hall, Upper Saddle River, NJ, 1996.

[12] H. K. Khalil, *Nonlinear Systems.* Prentice-Hall, Upper Saddle River, NJ, 1996.

[13] N. N. Krasovski, *On the Theory of Motion Control.* Nauka, Moscow, 1968.

[14] B. C. Kuo, *Automatic Control Systems.* Prentice-Hall, Englewood Cliffs, New Jersey, 1987.

[15] G. Leitmann, *The Calculus of Variations and Optimal Control.* Plenum Press, NY, 1981.

[16] A. M. Letov, *Flight Dynamics and Control*. Nauka, Moscow, 1969.

[17] W. S. Levine, *Control Handbook*. CRC Press, FL, 1996.

[18] L. Ljung, *System Identification: Theory for the User*. Prentice-Hall, Englewood Cliffs, New Jersey, 1987.

[19] B. S. Mordukhovich, *Approximation Methods in Problems of Optimization and Control*. Nauka, Moscow, 1988.

[20] K. Ogata, *Discrete-Time Control Systems*. Prentice-Hall, Upper Saddle River, NJ, 1995.

[21] K. Ogata, *Modern Control Engineering*. Prentice-Hall, Upper Saddle River, NJ, 1997.

[22] C. L. Phillips and R. D. Harbor, *Feedback Control Systems*. Prentice-Hall, Englewood Cliffs, NJ, 1996.

[23] L. S. Pontryagin, V. G. Boltyanskii, R. V. Gamrelidze, and E. F. Mishchenko, *The Mathematical Theory of Optimal Processes*. Interscience Publishers, Inc., New York, 1962.

[24] E. P. Ryan, *Optimal Relay and Saturating Control System Synthesis*. Peregrinus, London, 1982.

[25] A. P. Sage and C. C. White, *Optimum Systems Control*. Prentice-Hall, Inc., Englewood Cliffs, NJ, 1977.

[26] M. Vidyasagar, *Nonlinear Systems Analysis*. Prentice-Hall, Englewood Cliffs, NJ, 1993.

Papers in Control

[27] D. S. Bernstein, Optimal nonlinear, but continuous, feedback control of systems with saturating actuators, *Int. Journal Control.*, vol. 62, no. 5, pp. 1209–1216, 1995.

[28] J. W. Brewer, Kronecker product and matrix calculus in system theory, *IEEE Trans. Circuit Systems*, vol. 25, no. 9, pp. 772–783, 1978.

[29] M. J. Corless and G. Leitmann, Bounded controllers for robust exponential convergence, *Journal Optimization Theory Applications*, vol. 76, no. 1, pp. 1–12, 1993.

[30] M. J. Corless and G. Leitmann, Continuous state feedback guaranteeing uniform ultimate boundedness for uncertain dynamic systems, *IEEE Trans. Automatic Control*, vol. 26, no. 5, pp. 1139–1144, 1981.

[31] R. A. DeCarlo, S. H. Zak, and G. P. Matthews, Variable structure control of nonlinear multivariable systems: A tutorial, *Proceedings of the IEEE*, vol. 76, no. 3, pp. 212–232, 1988.

[32] S. V. Drakunov and V. I. Utkin, Sliding mode control in dynamic systems, *Int. Journal of Control*, vol. 55, no. 4, pp. 1029–1037, 1992.

[33] R. F. Hartl, S. P. Sethi and R. G. Vickson, A survey of the maximum principles for optimal control problems with state constraints, *SIAM Review*, vol. 37, no. 2, pp. 181–218, 1995.

[34]   E. G. Gilbert and K. T. Tan, Linear systems with state and control constraints: the theory and applications of maximal output admissible sets, *IEEE Trans. Automatic Control*, vol. 36, no. 9, pp. 1008–1020, 1991.

[35]   V. F. Kudin and S. E. Lyshevski, Extending the solution of a class of problem of analytical design of nonlinear controllers, *Automation and Remote Control*, vol. 51, pp. 880–888, 1990.

[36]   W. E. Larimore, The optimality of canonical variate identification by example, *Proc. of the* 10*th IFAC Symposium in System Identification*, Copenhagen, Denmark, vol. 2, pp. 151–156, 1994.

[37]   S. E. Lyshevski, Design of the constrained controllers for uncertain nonlinear systems using the Lyapunov stability theory, *Journal of the Franklin Institute*, vol. 336, issue 7, pp. 1075–1092, 1999.

[38]   S. E. Lyshevski, Robust control of nonlinear continuous-time systems with parameter uncertainties and input bounds. *Intern. Journal of Systems Science*, vol. 39, no. 3, pp. 247–259, 1999.

[39]   S. E. Lyshevski, Control of linear dynamic systems with constraints: optimization issues and applications of nonquadratic functionals, *Proc. Conference on Decision and Control*, Kobe, Japan, vol. 3, pp. 3206–3211, 1996.

[40]   S. E. Lyshevski, Constrained optimization and control of nonlinear systems: new results in optimal control, *Proc. Conference on Decision and Control*, Kobe, Japan, vol. 1, pp. 541–546, 1996.

[41]   S. E. Lyshevski, Identification of nonlinear flight dynamics: theory and practice, *IEEE Trans. Aerospace and Electronic Systems*, vol. 36, no. 2, pp. 383–392, 2000.

[42]   S. E. Lyshevski, Nonlinear discrete-time systems: constrained optimization and application of nonquadratic costs, *Proc. American Control Conference*, Philadelphia, PA, vol. 6, pp. 3699–3703, 1998.

[43]   S. E. Lyshevski, Optimal control of nonlinear continuous-time systems: design of bounded controllers via generalized nonquadratic functionals, *Proc. American Control Conference*, Philadelphia, PA, vol. 1, pp. 205–209, 1998.

[44]   S. E. Lyshevski, Optimization of nonlinear time-varying systems. *Proc. Conf. Decision and Control*, Tampa, FL, vol. 2, pp. 1798–1803, 1998.

[45]   S. E. Lyshevski, Sliding modes and soft switching control in dynamic systems, *Proc. American Control Conference*, Chicago, IL, pp. 646–650, 2000.

[46]   S. E. Lyshevski, State-space model identification of deterministic nonlinear systems: nonlinear mapping technology and application of the Lyapunov theory, *Automatica*, vol. 34, no. 5, pp. 659–664, 1998.

[47]   D. Q. Mayne and W. R. Schroeder, Nonlinear control of constrained linear systems, *Int. J. Control*, vol. 60, no. 5, pp. 1035–1043, 1994.

[48]   V. Pappano, S. E. Lyshevski and B. Friedland, Nonlinear identification of induction motor parameters, *Proc. American Control Conf.*, San Diego, CA, pp. 3569–3573, 1999.

[49] Z. Pan and T. Başar, Parameter identification for uncertain linear systems with partial state measurements under an H^∞ criterion, *Proc. Conf. Decision and Control*, New Orleans, LA, vol. 1, pp. 709–714, 1995.

[50] S. Rangan, G. Wolodkin and K. Poolla, New results for Hammerstein system identification, *Proc. Conf. Decision and Control*, New Orleans, LA, vol. 1, pp. 697–702, 1995.

[51] A. Saberi, Z. Lin, and A. R. Teel, Control of linear systems with saturating actuators, *IEEE Trans. Automatic Control*, vol. 41, no. 3, pp. 368–378, 1996.

[52] E. D. Sontag, Control of systems without drift via generic loops, *IEEE Trans. Automatic Control*, vol. 40, no. 7, pp. 1210–1219, 1995.

[53] E. Sontag and H. J. Sussmann, Nonsmooth control—Lyapunov functions, *Proc. Conf. Decision and Control*, New Orleans, LA, vol. 3, pp. 2799–2805, 1995.

[54] H. J. Sussmann, E. D. Sontag, and Y. Yang, A general result on the stabilization of linear systems using bounded controls, *IEEE Trans. Automatic Control*, vol. 39, pp. 2411–2425, 1994.

[55] C. J. Wan and D. S. Bernstein, Nonlinear feedback control with global stabilization, *Dynamics and Control*, vol. 5, no. 4, pp. 321–346, 1995.

MATLAB

[56] J. B. Dabney and T. L. Harman, *Mastering* SIMULINK 2. Prentice Hall, Upper Saddle River, NJ, 1998.

[57] D. Hanselman and B. Littlefield, *The Student Edition of* MATLAB. Prentice Hall, Upper Saddle River, NJ, 1997.

[58] D. Hanselman and B. Littlefield, *Mastering* MATLAB 5. Prentice Hall, Upper Saddle River, NJ, 1998.

Electromechanical Systems

[59] P. C. Krause and O. Wasynczuk, *Electromechanical Motion Devices*. Mc-Graw-Hill, New York, 1989.

[60] S. E. Lyshevski, *Electromechanical Systems, Electric Machines, and Applied Mechatronics*. CRC Press, FL, 1999.

[61] S. E. Lyshevski, *Nano- and Microelectrical Systems*: *Fundamentals of Nano- and Microengineering*. CRC Press, FL, 2000.

[62] S. E. Lyshevski, Nonlinear control of servo-systems actuated by permanent-magnet synchronous motors, *Automatica*, vol. 34, no. 10, pp. 1231–1238, 1998.

[63] S. E. Lyshevski, Nonlinear control of mechatronic systems with permanent-magnet DC motors, *Mechatronics*, vol. 9, pp. 539–552, 1999.

[64] White D. C. and Woodson H. H., *Electromechanical Energy Conversion*. Wiley, New York, 1959.

Index